AF689837

Student Solutions Manual to Accompany General Chemistry

FOURTH EDITION

Student Solutions Manual to Accompany General Chemistry

Fourth Edition

CAROLE H. McQUARRIE

University Science Books
www.uscibooks.com

University Science Books
20 Edgehill Road
Mill Valley, CA 94941
www.uscibooks.com

About the Cover: Dalton's symbols for chemical elements. Some of these "elements" are now known to be compounds, not elements (e.g., lime, soda, and potash). Some of Dalton's atomic masses were in error because of incorrect assumptions regarding the relative numbers of atoms in compounds. Reproduced with permission from Photo Researchers, Inc., all rights reserved.

Production Manager: Jennifer Uhlich at Wilsted & Taylor
Book and Cover Design: Yvonne Tsang at Wilsted & Taylor
Composition: Carole H. McQuarrie / MPS Limited
Illustrator: Laurel Muller, Cohographics
Editorial Development: Mervin Hanson
Printer & Binder: Transcontinental

This book is printed on acid-free paper.

Copyright © 2011 by University Science Books

Reproduction or translation of any part of this work beyond that permitted by Section 107 or 108 of the 1976 United States Copyright Act without the permission of the copyright owner is unlawful. Requests for permission or further information should be addressed to the Permissions Department, University Science Books.

ISBN 978-1-891389-73-3

Library of Congress Control Number 2010928414

Printed in Canada
10 9 8 7 6 5 4 3 2

Contents

Student Solutions Manual to Accompany General Chemistry

FOURTH EDITION

CHAPTER 1. Chemistry and the Scientific Method

THE SCIENTIFIC METHOD

1-2. First the scientists must verify the results of the original experiment to be sure that no errors were made. Second, if the results are confirmed, then the theory must be either revised to accommodate the new data or discarded and replaced by a new theory.

1-4. What is meant by the statement is that no two snowflakes have yet been observed to be the same.

SCIENTIFIC NOTATION

1-6. We have in decimal form

(a) 299 800 000

(b) 0.000 548 580

(c) 0.000 000 000 05292

(d) 155 000 000 000 0000

1-8. 556.0×10^3

UNITS

1-10. (a) 10^{-12}; (b) 10^9; (c) 10^{-9}; (d) 10^3; (e) 10^{-18}; (f) 10^{-15}

1-12. The easiest way to order these quantities is to convert all the volumes to units of liters. We use the conversion factors according to the prefix as indicated in Table 1.2.

(a) 10 L

(b) $100 \times 10^{-3}\ \text{L} = 0.10\ \text{L}$

(c) $0.10 \times 10^6\ \text{L} = 1.0 \times 10^5\ \text{L}$

(d) $1.0 \times 10^3 \times 10^{-6}\ \text{L} = 1.0 \times 10^{-3}\ \text{L}$

(e) $20 \times 10^{-2}\ \text{L} = 0.20\ \text{L}$

(f) $10 \times 10^4 \times 10^{-9}\ \text{L} = 1.0 \times 10^{-5}\ \text{L}$

Thus, the order is f < d < b < e < a < c.

1-14. The volume of the cube is given by $V = l^3$. If we substitute in the given value of l, then we have that

$$V = l^3 = (200.0\ \text{pm})^3 = (200.0 \times 10^{-12}\ \text{m})^3 = 8.000 \times 10^{-30}\ \text{m}^3$$

1-16. The density is given by

$$d = \frac{m}{V}$$

Thus, the volume is given by

$$V = \frac{m}{d} = \frac{31.1\ \text{g}}{19.3\ \text{g}\cdot\text{cm}^{-3}} = 1.61\ \text{cm}^3$$

1-18. We use Equations 1.1 and 1.2

$$T(\text{in K}) = t(\text{in °C}) + 273.15 \quad \text{and} \quad t(\text{in °C}) = \frac{5}{9}[\,t(\text{in °F}) - 32.0]$$

$$t(\text{in °C}) = \frac{5}{9}(451 - 32) = 233$$

$$T(\text{in K}) = 233 + 273.15 = 506$$

The title would be *Celsius 233* or *Kelvin 506*.

CONSERVATION OF ENERGY

1-20. The energy (all potential energy) when the piton is dropped is

$$E_{\text{p}} = mgh = (56\ \text{g})\left(\frac{1\ \text{kg}}{1000\ \text{g}}\right)(9.81\ \text{m}\cdot\text{s}^{-2})(375\ \text{m}) = 206\ \text{kg}\cdot\text{m}^2\cdot\text{s}^{-2} = 206\ \text{J}$$

When the piton strikes the ground, all its energy is kinetic energy, $E_{\text{k}} = mv^2/2 = 206$ J. Solving for v gives

$$v = \left(\frac{2E_{\text{k}}}{m}\right)^{1/2} = \left(\frac{412\ \text{J}}{0.056\ \text{kg}}\right)^{1/2} = \left(\frac{412\ \text{kg}\cdot\text{m}^2\cdot\text{s}^{-2}}{0.056\ \text{kg}}\right)^{1/2} = 86\ \text{m}\cdot\text{s}^{-1}$$

1-22. We equate the kinetic energy of the car to its potential energy at the top of the hill

$$\frac{1}{2}mv^2 = mgh$$

Thus, the maximum height is given by

$$h = \frac{v^2}{2g} = \frac{\left[(89 \times 10^3\ \text{m}\cdot\text{h}^{-1})\left(\frac{1\ \text{h}}{3600\ \text{s}}\right)\right]^2}{(2)(9.81\ \text{m}\cdot\text{s}^{-2})} = 31\ \text{m}$$

1-24. The energy used in operating the lamp five hours per day for a year is given by

$$\begin{aligned} E &= (300\ \text{W})(1\ \text{J}\cdot\text{s}^{-1}\cdot\text{W}^{-1})(3600\ \text{s}\cdot\text{h}^{-1})(5\ \text{h}\cdot\text{day}^{-1})(365\ \text{day}) \\ &= 1.97 \times 10^9\ \text{J} = 1.97 \times 10^6\ \text{kJ} \\ &= (1.97 \times 10^6\ \text{kJ})\left(\frac{1\ \text{kW}}{1\ \text{kJ}\cdot\text{s}^{-1}}\right)\left(\frac{1\ \text{h}}{3600\ \text{s}}\right) = 550\ \text{kW}\cdot\text{h} \end{aligned}$$

which cost \$55 at \$0.10 per kilowatt hour. Fluorescent lamps use less energy than halogen lamps and so are less expensive to operate.

PERCENTAGE ERROR

1-26. We use the equation

$$\text{percentage error} = \frac{\text{observed value} - \text{true value}}{\text{true value}} \times 100$$

$$= \frac{352\ \text{m}\cdot\text{s}^{-1} - 344\ \text{m}\cdot\text{s}^{-1}}{344\ \text{m}\cdot\text{s}^{-1}} \times 100 = \frac{8\ \text{m}\cdot\text{s}^{-1}}{344\ \text{m}\cdot\text{s}^{-1}} \times 100 = 2\%$$

Note that we have applied the rules for subtraction and division to obtain the result to the proper number of significant figures.

1-28. The average is given by

$$\text{average} = \frac{(13.56 + 13.62 + 13.59)\ \text{g}\cdot\text{cm}^{-3}}{3} = 13.59\ \text{g}\cdot\text{cm}^{-3}$$

$$\text{percentage error} = \frac{\text{average value} - \text{accepted value}}{\text{accepted value}} \times 100$$

$$= \frac{13.59\ \text{g}\cdot\text{cm}^{-3} - 13.509\ \text{g}\cdot\text{cm}^{-3}}{13.509\ \text{g}\cdot\text{cm}^{-3}} \times 100 = 0.6\%$$

where the value 13.509 $\text{g}\cdot\text{cm}^{-3}$ comes from the *CRC Handbook*. Other sources may give slightly different values.

SIGNIFICANT FIGURES

1-30. (a) $\underline{578}$ has three significant figures.

(b) 0.000 $\underline{578}$ has three significant figures. The zeros are not significant figures.

(c) The integers 1000 and 1 are exact numbers.

(d) π = $\underline{3.141\ 592\ 65}$ has nine significant figures as stated here.

(e) $\underline{93}$ 000 000 has two significant figures. The zeros are not significant figures.

1-32. (a) $z = 33209$ (The result cannot be more accurate than zero digits past the decimal point.)

(b) $z = 254$ (The result cannot be more accurate than zero digits past the decimal point.)

(c) $z = 0.0143\ 877$ (The result cannot be expressed to more than six significant figures.)

(d) We must convert the three numbers in the first set of parentheses to the same power of ten; in this case, 10^{-27} before the addition and subtraction steps. Thus,

$$z = (0.0009109390 \times 10^{-27} + 1.67262 \times 10^{-27} - 1.67493 \times 10^{-27})(2.997\ 925 \times 10^{8})^2$$

$$= (-0.00140 \times 10^{-27})(2.997\ 925 \times 10^{8})^2 = -1.26 \times 10^{-13}$$

Notice that the result of the first step cannot be expressed to more than five figures past the decimal point. (The final result cannot be expressed to more than three significant figures.)

1-34. $V = l^3 = (133\ \text{pm})^3 = 2.35 \times 10^6\ \text{pm}^3$

UNIT CONVERSIONS AND DIMENSIONAL ANALYSIS

1-36. See Table 1.2 for the meaning of the prefixes used.

(a) $(1.259 \times 10^3\ \text{J})\left(\frac{1\ \text{kJ}}{10^3\ \text{J}}\right) = 1.259\ \text{kJ}$

(b) $(2.18 \times 10^{-18}\ \text{J})\left(\frac{1\ \text{aJ}}{10^{-18}\ \text{aJ}}\right) = 2.18\ \text{aJ}$

(c) $(5.5\ \text{MJ})\left(\frac{10^6\ \text{J}}{1\ \text{MJ}}\right)\left(\frac{1\ \text{kJ}}{10^3\ \text{J}}\right) = 5.5 \times 10^3\ \text{kJ}$

(d) $(7.5 \times 10^{-3}\ \text{ns})\left(\frac{10^{-9}\ \text{s}}{1\ \text{ns}}\right)\left(\frac{1\ \text{fs}}{10^{-15}\ \text{s}}\right) = 7.5 \times 10^3\ \text{fs}$

(e) $(2.0\ \text{m}^3)\left(\frac{100\ \text{cm}}{1\ \text{m}}\right)^3\left(\frac{1\ \text{mL}}{1\ \text{cm}^3}\right) = 2.0 \times 10^6\ \text{mL}$

1-38. (a) $(325\ \text{ft})\left(\frac{1\ \text{yd}}{3\ \text{ft}}\right)\left(\frac{1\ \text{m}}{1.0936\ \text{yd}}\right) = 99.1\ \text{m}$

(b) $(1.54\ \text{Å})\left(\frac{10^{-10}\ \text{m}}{1\ \text{Å}}\right)\left(\frac{1\ \text{pm}}{10^{-12}\ \text{m}}\right) = 154\ \text{pm}$

and $(1.54\ \text{Å})\left(\frac{10^{-10}\ \text{m}}{1\ \text{Å}}\right)\left(\frac{1\ \text{nm}}{10^{-9}\ \text{m}}\right) = 0.154\ \text{nm}$

(c) $(175\ \text{lb})\left(\frac{0.45359\ \text{kg}}{1\ \text{lb}}\right) = 79.4\ \text{kg}$

1-40. The volume in cubic centimeters is

$$\text{volume} = (454\ \text{in}^3)\left(\frac{2.540\ \text{cm}}{1\ \text{in}}\right)^3 = 7.44 \times 10^3\ \text{cm}^3$$

The volume in liters is

$$\text{volume} = (7.44 \times 10^3\ \text{cm}^3)\left(\frac{1\ \text{mL}}{1\ \text{cm}^3}\right)\left(\frac{1\ \text{L}}{10^3\ \text{mL}}\right) = 7.44\ \text{L}$$

1-42. $\text{time} = (12.8\ \text{Mbyte})\left(\frac{10^6\ \text{byte}}{1\ \text{Mbyte}}\right)\left(\frac{1\ \text{kbyte}}{10^3\ \text{byte}}\right)\left(\frac{1\ \text{s}}{75\ \text{kbyte}}\right)\left(\frac{1\ \text{min}}{60\ \text{s}}\right) = 2.8\ \text{min}$

1-44. (a) The number of milliliters in one cubic meter is

$$(1\ \text{m})^3\left(\frac{100\ \text{cm}}{1\ \text{m}}\right)^3\left(\frac{1\ \text{mL}}{1\ \text{cm}^3}\right) = 10^6\ \text{mL}$$

(b) The number of liters in one cubic meter is

$$(10^6\ \text{mL})\left(\frac{1\ \text{L}}{10^3\ \text{mL}}\right) = 10^3\ \text{L}$$

1-46. The mass of liquid is 297.1 g − 215.8 g = 81.3 g. Thus,

$$\text{density} = \frac{m}{V} = \frac{81.3\ \text{g}}{125\ \text{mL}} = 0.650\ \text{g}\cdot\text{mL}^{-1}$$

GUGGENHEIM NOTATION

1-48. We shall use Time/s and Distance/ft as table headings. Thus, the table is

Time/s	Distance/ft
0	0
2.0	51
4.0	204
6.0	459

1-50. To convert a height in feet to a height in centimeters, we use the following conversion factors (see inside back cover):

$$(\text{height in ft})\left(\frac{12\ \text{in}}{1\ \text{ft}}\right)\left(\frac{2.54\ \text{cm}}{1\ \text{in}}\right) = \text{height in cm}$$

Thus, the table is

Height/cm	Time/s
27	6.09
42.7	11.65
60.7	18.11
129	30.41

ADDITIONAL PROBLEMS

1-52. We would classify each as follows:

i. quantitative ii. qualitative iii. qualitative

iv. quantitative v. qualitative vi. quantitative

1-54. The number of significant figures indicates the precision of a measurement.

1-56. Yes. Counted values, such as the number of people in a car, have 100% precision.

1-58. The age of the rock is precise to only one significant figure; namely one, which is easy to see in scientific notation, 1×10^6 years. So, of course, you cannot add six years to one million and get anything but one million.

1-60. (a) The average for the class is

$$(\text{average})_{\text{class}} = \frac{\text{sum of the nine measurements}}{9} = \frac{101.8}{9} = 11.31\ \text{g}\cdot\text{cm}^{-3}$$

$$(\%\ \text{error})_{\text{class}} = \frac{(11.31 - 11.3)\ \text{g}\cdot\text{cm}^{-3}}{11.3\ \text{g}\cdot\text{cm}^{-3}} \times 100 = 0.00\%$$

(b) The overall class's data is more accurate because the percentage error is zero. Using more data often improves the accuracy of the average value.

1-62. The temperature is 88.2°C.

1-64. From Equation 1.3, we have

$$V = \frac{m}{d} = \frac{55\text{ g}}{0.879\text{ g}\cdot\text{cm}^{-3}} = 63\text{ cm}^3 = 63\text{ mL}$$

1-66. The energy required is equivalent to the potential energy of the 180-pound man at a height of 1776 feet, or

$$E_\text{p} = mgh = (180\text{ lb})\left(\frac{1\text{ kg}}{2.20\text{ lb}}\right)(9.81\text{ m}\cdot\text{s}^{-1})(1776\text{ ft})\left(\frac{1\text{ yd}}{3\text{ ft}}\right)\left(\frac{1\text{ m}}{1.0936\text{ yd}}\right)$$
$$= 4.3\times10^5\text{ kg}\cdot\text{m}^2\cdot\text{s}^{-2} = 4.3\times10^5\text{ J} = 430\text{ kJ}$$

1-68. (a) A typical home's daily use energy requirement is

$$\text{energy use per day} = (40\text{ kW}\cdot\text{h})(3600\text{ s}\cdot\text{h}^{-1}) = 1.44\times10^5\text{ kW}\cdot\text{s} = 1.44\times10^5\text{ kJ}$$

Let A be the area of the solar panels. Then

$$\text{daily energy requirement} = 1.44\times10^5\text{ kJ} = (0.18)(0.5\text{ kW}\cdot\text{m}^{-2})(8\text{ h})(3600\text{ s}\cdot\text{h}^{-1})\,A$$

Solving for A gives

$$A = \frac{1.44\times10^5\text{ kJ}}{(0.18)(0.5\text{ kW}\cdot\text{m}^{-2})(8\text{ h})(3600\text{ s}\cdot\text{h}^{-1})} = 56\text{ m}^2$$

Thus, the panel would easily fit on the roof of a typical single-family home.

(b) \$0.10 per kilowatt hour corresponds to $\$2.78\times10^{-5}$ per kilojoule (see Problem 1-67). The cost for a typical home per day is

$$\text{cost per day} = (\$2.78\times10^{-5}\text{ kJ}^{-1})(1.44\times10^5\text{ kJ}) = \$4.00$$

The solar panels would pay for themselves in 3750 days, or about 10 years.

1-70. We have that

$$\Delta E = c^2\Delta m = (3.00\times10^8\text{ m}\cdot\text{s}^{-1})^2(1.0\text{ g})\left(\frac{1\text{ kg}}{10^3\text{ g}}\right)$$
$$= 9.0\times10^{13}\text{ J} = 9.0\times10^{10}\text{ kJ}$$

1-72. The mass of air in the lungs is given by

$$\text{mass of air} = dV = (1.20\text{ g}\cdot\text{L}^{-1})(6.0\text{ L}) = 7.2\text{ g}$$

The number of grams of oxygen in the lungs is

$$\text{mass of oxygen} = \left(\frac{20\%}{100\%}\right)(7.2\text{ g}) = 1.4\text{ g}$$

1-74. A temperature of 1200°C corresponds to the following Kelvin and Fahrenheit temperatures:

$$T(\text{in K}) = t(\text{in }/^\circ\text{C}) + 273.15 = 1200 + 273 = 1500$$

and

$$t(\text{in }/^\circ\text{F}) = \frac{9}{5}t(\text{in }/^\circ\text{C}) + 32 = \frac{9}{5}(1200) + 32 = 2200$$

1-76. We have

$$\text{volume container} = \frac{m}{d} = \frac{6780\text{ g}}{13.6\text{ g}\cdot\text{mL}^{-1}} = 498.5\text{ mL}$$

The density of carbon tetrachloride is given by

$$d = \frac{m}{V} = \frac{797\text{ g}}{498.5\text{ mL}} = 1.60\text{ g}\cdot\text{mL}^{-1}$$

1-78. The number of acres in one hectare is

$$1\text{ hectare} = (100\text{ m})^2\left(\frac{1\text{ acre}}{4046\text{ m}^2}\right) = 2.472\text{ acre}$$

1-80. Let x be the thickness of the foil. Then the volume of the sheet is

$$\text{volume} = (20.5\text{ cm})(15.2\text{ cm})x$$

and

$$\text{mass} = 1.683\text{ g} = Vd = (20.5\text{ cm})(15.2\text{ cm})x(2.70\text{ g}\cdot\text{cm}^{-3})$$

Solving for x gives

$$x = \frac{1.683\text{ g}}{(20.5\text{ cm})(15.2\text{ cm})(2.70\text{ g}\cdot\text{cm}^{-3})} = 2.00\times10^{-3}\text{ cm} = 0.0200\text{ mm}$$

1-82. To determine your total body volume, immerse yourself completely in a cylinder of water and measure the volume of water displaced. Your volume is equal to the increase in the height of the water in the cylinder times the cross-sectional area of the cylinder ($V = hA$). Next weigh yourself. Your density is then calculated using Equation 1.3.

1-84. The area covered in square centimeters is

$$\text{area covered} = (350\text{ ft}^2)\left(\frac{12\text{ in}}{1\text{ ft}}\right)^2\left(\frac{2.54\text{ cm}}{1\text{ in}}\right)^2 = 3.25\times10^5\text{ cm}^2$$

The volume of paint in one gallon is

$$\text{volume of paint} = (1.0\text{ gal})\left(\frac{3.7854\text{ L}}{1\text{ gal}}\right)\left(\frac{10^3\text{ mL}}{1\text{ L}}\right)\left(\frac{1\text{ cm}^3}{1\text{ mL}}\right) = 3.79\times10^3\text{ cm}^3$$

Thus,

$$\text{thickness} = \frac{V}{A} = \frac{3.79\times10^3\text{ cm}^3}{3.25\times10^5\text{ cm}^2} = 0.012\text{ cm} = 0.12\text{ mm}$$

1-86. The volume of a cylinder is $V = \pi r^2 h$. The modern penny is a cylinder with a diameter of 19.05 mm and a height (thickness in this case) of 1.224 mm. Thus, the volume of a modern penny in cubic centimeters is

$$V = \pi \left(\frac{19.05 \text{ mm}}{2}\right)^2 (1.224 \text{ mm}) \left(\frac{1 \text{ cm}}{10 \text{ mm}}\right)^3 = 0.3489 \text{ cm}^3$$

The volume of a penny is equal to the volume of copper plus the volume of zinc in the penny. We can find the volume of each from the relationship $V = d/m$. Thus, we have

$$V_{\text{penny}} = V_{\text{Cu}} + V_{\text{Zn}} = \frac{m_{\text{Cu}}}{8.96 \text{ g}\cdot\text{cm}^{-3}} + \frac{m_{\text{Zn}}}{7.13 \text{ g}\cdot\text{cm}^{-3}} = 0.3489 \text{ cm}^3 \qquad (1)$$

We also are given that

$$\text{mass of penny} = m_{\text{Cu}} + m_{\text{Zn}} = 2.500 \text{ g}$$

Let $m_{\text{Zn}} = 2.500 \text{ g} - m_{\text{Cu}}$ and substitute this result into equation 1 to obtain

$$\frac{m_{\text{Cu}}}{8.96 \text{ g}\cdot\text{cm}^{-3}} + \frac{2.500 \text{ g} - m_{\text{Cu}}}{7.13 \text{ g}\cdot\text{cm}^{-3}} = 0.3489 \text{ cm}^3$$

or

$$(7.13 \text{ g}\cdot\text{cm}^{-3})\, m_{\text{Cu}} + (2.500 \text{ g} - m_{\text{Cu}})(8.96 \text{ g}\cdot\text{cm}^{-3}) = (0.3489 \text{ cm}^3)(7.13 \text{ g}\cdot\text{cm}^{-3})(8.96 \text{ g}\cdot\text{cm}^{-3})$$

Multiplying these terms and eliminating common units gives

$$7.13 \text{ g}\, m_{\text{Cu}} + 22.40 \text{ g}^2 - 8.96 \text{ g}\, m_{\text{Cu}} = 22.29 \text{ g}^2$$

Collecting like terms gives

$$-1.83\, m_{\text{Cu}} = -0.11 \text{ g}$$

or $m_{\text{Cu}} = 0.060$ g. Thus, the percentage by mass of copper in a modern penny is

$$\text{mass } \% = \frac{\text{mass of copper}}{\text{mass of a penny}} \times 100 = \frac{0.060 \text{ g}}{2.500 \text{ g}} \times 100 = 2.0\%$$

CHAPTER 2. Atoms and Molecules

CHEMICAL SYMBOLS

2-2. The chemical symbols are as follows:

(a) Sn (b) Au (c) Zr (d) Bi (e) Cu
(f) Rb (g) Br (h) Ne (i) Sb (j) As

See the alphabetical list of the elements on the inside front cover.

2-4. The names of the elements are as follows:

(a) platinum (b) tellurium (c) lead (d) tantalum (e) barium
(f) titanium (g) rhenium (h) lanthanum (i) europium (j) praseodymium

See the alphabetical list of the elements on the inside front cover.

SEPARATIONS

2-6. Dissolve the table salt in water and filter off the sand. Sand is insoluble in water.

2-8. A liquid that readily vaporizes is said to be volatile.

2-10. In distillation, a solution of two liquids is separated into its components by vaporizing and condensing the solution at the boiling point of the more volatile component.

2-12. A mixture is said to be heterogeneous if it has different properties in different regions of the mixture, and can be separated by mechanical means into two or more components.

MASS PERCENTAGE IN COMPOUNDS

2-14. The mass percentage of lanthanum in lanthanum oxide is given by

$$\text{mass percentage of La} = \frac{\text{mass of La}}{\text{mass of lanthanum oxide}} \times 100$$

$$= \frac{7.08\text{ g}}{8.29\text{ g}} \times 100 = 85.4\%$$

The mass percentage of oxygen in lanthanum oxide is given by

$$\text{mass percentage of O} = \frac{\text{mass of O}}{\text{mass of lanthanum oxide}} \times 100$$
$$= \frac{1.21\text{ g}}{8.29\text{ g}} \times 100 = 14.6\%$$

2-16. The mass percentage of tin in stannous fluoride is given by

$$\text{mass percentage of Sn} = \frac{\text{mass of Sn}}{\text{mass of stannous fluoride}} \times 100$$
$$= \frac{1.358\text{ g}}{1.793\text{ g}} \times 100 = 75.74\%$$

The mass percentage of fluorine in stannous fluoride is given by

$$\text{mass percentage of F} = 100\% - \text{mass percentage of Sn}$$
$$= 100\% - 75.74\% = 24.26\%$$

Note that 100% is an exact number.

2-18. The respective mass percentages are

$$\text{mass percentage of C} = \frac{\text{mass of C}}{\text{mass of ethanol}} \times 100$$
$$= \frac{1.93\text{ g}}{3.70\text{ g}} \times 100 = 52.2\%$$
$$\text{mass percentage of H} = \frac{\text{mass of H}}{\text{mass of ethanol}} \times 100$$
$$= \frac{0.49\text{ g}}{3.70\text{ g}} \times 100 = 13\%$$
$$\text{mass percentage of O} = \frac{\text{mass of O}}{\text{mass of ethanol}} \times 100$$
$$= \frac{(3.70 - 1.93 - 0.49)\text{ g}}{3.70\text{ g}} \times 100 = \frac{1.28\text{ g}}{3.70\text{ g}} \times 100 = 34.6\%$$

Due to the use of significant figures, the total percentages may not add up to 100%.

NOMENCLATURE

2-20. Using Table 2.6 and the list of the elements on the inside front cover, we have

(a) barium fluoride (b) magnesium nitride
(c) cesium chloride (d) calcium sufide

2-22. Using Table 2.6 and the list of the elements on the inside front cover, we have

(a) magnesium fluoride (b) aluminum nitride
(c) magnesium selenide (d) lithium phosphide

2-24. (a) antimony trichloride and antimony pentachloride
(b) iodine trichloride and iodine pentachloride

(c) selenium dioxide and selenium trioxide
(d) carbon monosulfide and carbon disulfide. Note that we have named CS analogous to CO.

2-26. Using Table 2.6 and the list of the elements on the inside front cover, we have

(a) barium hydride (b) lithium sulfide
(c) beryllium oxide (d) methane (not named according to Table 2.6)

MOLECULAR MASSES

We shall report the molecular masses to five significant figures in each case, which is sufficient for most of our purposes.

2-28. (a) The molecular mass of $CaWO_4$ is

$$\begin{aligned}\text{molecular mass} &= (\text{atomic mass of Ca}) + (\text{atomic mass of W}) + 4 \times (\text{atomic mass of O}) \\ &= (40.078) + (183.84) + (4 \times 15.9994) = 287.92\end{aligned}$$

(b) The molecular mass of Fe_3O_4 is

$$\begin{aligned}\text{molecular mass} &= 3 \times (\text{atomic mass of Fe}) + 4 \times (\text{atomic mass of O}) \\ &= (3 \times 55.845) + (4 \times 15.9994) = 231.53\end{aligned}$$

(c) The molecular mass of Na_3AlF_6 is

$$\begin{aligned}\text{molecular mass} &= 3 \times (\text{atomic mass of Na}) + (\text{atomic mass of Al}) + 6 \times (\text{atomic mass of F}) \\ &= (3 \times 22.98977) + (26.98154) + (6 \times 18.9984) = 209.94\end{aligned}$$

(d) The molecular mass of $Be_3Al_2Si_6O_{18}$ is

$$\begin{aligned}\text{molecular mass} &= 3 \times (\text{atomic mass of Be}) + 2 \times (\text{atomic mass of Al}) \\ &\quad + 6 \times (\text{atomic mass of Si}) + 18 \times (\text{atomic mass of O}) \\ &= (3 \times 9.01218) + (2 \times 26.98154) + (6 \times 28.0855) + (18 \times 15.9994) = 537.50\end{aligned}$$

(e) The molecular mass of Zn_2SiO_4 is

$$\begin{aligned}\text{molecular mass} &= 2 \times (\text{atomic mass of Zn}) + (\text{atomic mass of Si}) + 4 \times (\text{atomic mass of O}) \\ &= (2 \times 65.38) + (28.0855) + (4 \times 15.9994) = 222.84\end{aligned}$$

2-30. (a) The molecular mass of $C_{20}H_{30}O$ is

$$\begin{aligned}\text{molecular mass} &= 20 \times (\text{atomic mass of C}) + 30 \times (\text{atomic mass of H}) + (\text{atomic mass of O}) \\ &= (20 \times 12.0107) + (30 \times 1.00794) + (15.9994) = 286.45\end{aligned}$$

(b) The molecular mass of $C_{12}H_{17}ClN_4OS$ is

$$\begin{aligned}\text{molecular mass} &= 12 \times (\text{atomic mass of C}) + 17 \times (\text{atomic mass of H}) + (\text{atomic mass of Cl}) \\ &\quad + 4 \times (\text{atomic mass of N}) + (\text{atomic mass of O}) + (\text{atomic mass of S}) \\ &= (12 \times 12.0107) + (17 \times 1.00794) + (35.453) + (4 \times 14.0067) \\ &\quad + (15.9994) + (32.065) = 300.81\end{aligned}$$

(c) The molecular mass of $C_{17}H_{20}N_4O_6$ is

$$\begin{aligned}\text{molecular mass} &= 17 \times (\text{atomic mass of C}) + 20 \times (\text{atomic mass of H}) \\ &\quad + 4 \times (\text{atomic mass of N}) + 6 \times (\text{atomic mass of O}) \\ &= (17 \times 12.0107) + (20 \times 1.00794) + (4 \times 14.0067) + (6 \times 15.9994) = 376.36\end{aligned}$$

(d) The molecular mass of $C_{56}H_{88}O_2$ is

$$\begin{aligned}\text{molecular mass} &= 56 \times (\text{atomic mass of C}) + 88 \times (\text{atomic mass of H}) + 2 \times (\text{atomic mass of O}) \\ &= (56 \times 12.0107) + (88 \times 1.00794) + (2 \times 15.9994) = 793.30\end{aligned}$$

(e) The molecular mass of $C_6H_8O_6$ is

$$\begin{aligned}\text{molecular mass} &= 6 \times (\text{atomic mass of C}) + 8 \times (\text{atomic mass of H}) + 6 \times (\text{atomic mass of O}) \\ &= (6 \times 12.0107) + (8 \times 1.00794) + (6 \times 15.9994) = 176.12\end{aligned}$$

2-32. (a) The formula of dopamine is $C_8H_{11}NO_2$ and its molecular mass is

$$\begin{aligned}\text{molecular mass} &= 8 \times (\text{atomic mass of C}) + 11 \times (\text{atomic mass of H}) \\ &\quad + (\text{atomic mass of N}) + 2 \times (\text{atomic mass of O}) \\ &= (8 \times 12.0107) + (11 \times 1.00794) + 14.0067 + (2 \times 15.9994) \\ &= 153.18\end{aligned}$$

(b) The formula of nicotine is $C_{10}H_{14}N_2$ and its molecular mass is

$$\begin{aligned}\text{molecular mass} &= 10 \times (\text{atomic mass of C}) + 14 \times (\text{atomic mass of H}) + 2 \times (\text{atomic mass of N}) \\ &= (10 \times 12.0107) + (14 \times 1.00794) + (2 \times 14.0067) \\ &= 162.23\end{aligned}$$

MASS PERCENTAGES AND ATOMIC MASSES

2-34. The molecular mass of N_2O is

$$\begin{aligned}\text{molecular mass} &= 2 \times (\text{atomic mass of N}) + (\text{atomic mass of O}) \\ &= (2 \times 14.0067) + (15.9994) = 44.0128\end{aligned}$$

The mass percentage of nitrogen in N_2O is

$$\begin{aligned}\text{mass percentage of N} &= \frac{2 \times \text{atomic mass of N}}{\text{molecular mass of N}_2\text{O}} \times 100 \\ &= \frac{2 \times 14.0067}{44.0128} \times 100 = 63.6483\%\end{aligned}$$

The mass percentage of oxygen in N_2O is

$$\begin{aligned}\text{mass percentage of O} &= \frac{\text{atomic mass of O}}{\text{molecular mass of N}_2\text{O}} \times 100 \\ &= \frac{15.9994}{44.0128} \times 100 = 36.3517\%\end{aligned}$$

2-36. The molecular mass of Na_3AlF_6 is 209.9413. The mass percentage of sodium in Na_3AlF_6 is

$$\text{mass percentage of Na} = \frac{3 \times \text{atomic mass of Na}}{\text{molecular mass of } Na_3AlF_6} \times 100$$
$$= \frac{3 \times 22.98977}{209.9413} \times 100 = 32.85171\%$$

The mass percentage of aluminum in Na_3AlF_6 is

$$\text{mass percentage of Al} = \frac{\text{atomic mass of Al}}{\text{molecular mass of } Na_3AlF_6} \times 100$$
$$= \frac{26.98154}{209.9413} \times 100 = 12.85194\%$$

The mass percentage of fluorine in Na_3AlF_6 is

$$\text{mass percentage of F} = \frac{6 \times \text{atomic mass of F}}{\text{molecular mass of } Na_3AlF_6} \times 100$$
$$= \frac{6 \times 18.9984}{209.9413} \times 100 = 54.2963\%$$

2-38. The molecular mass of $SO_3 = 80.063$.

$$\text{mass percentage of S} = \frac{32.065}{80.063} \times 100 = 40.050\%$$

$$\text{mass of S} = \text{mass of } SO_3 \times \frac{\text{mass \% of S}}{100}$$
$$= 5.585 \text{ g} \times \frac{40.050}{100} = 2.237 \text{ g}$$

2-40. Take a 100-gram sample of the ore. The mass of $Cr_2O_3(s)$ in the sample is 42.7 grams. The molecular mass of Cr_2O_3 is 151.9904. The mass percentage of chromium in Cr_2O_3 is

$$\text{mass percentage of Cr in } Cr_2O_3 = \frac{2 \times 51.9961}{151.9904} \times 100 = 68.420\%$$

The mass of chromium in 42.7 grams of Cr_2O_3 is

$$\text{mass of Cr} = (42.7 \text{ g})\left(\frac{68.42}{100}\right) = 29.2 \text{ g}$$

The mass percentage of chromium in the ore is

$$\text{mass percentage Cr} = \frac{29.2 \text{ g Cr}}{100 \text{ g sample}} \times 100 = 29.2\%$$

PROTONS, NEUTRONS, AND ELECTRONS

2-42. From the atomic numbers, we find that

	Number of protons	Number of electrons	Number of neutrons
(a) phosphorus-30	15	15	$30 - 15 = 15$
(b) technetium-97	43	43	$97 - 43 = 54$
(c) iron-55	26	26	$55 - 26 = 29$
(d) americium-240	95	95	$240 - 95 = 145$

2-44. The atomic number determines the element. The mass number is the sum of the atomic number and the number of neutrons.

Symbol	Atomic number	Number of neutrons	Mass number
${}^{48}_{20}\text{Ca}$	20	28	48
${}^{90}_{40}\text{Zr}$	40	50	90
${}^{131}_{53}\text{I}$	53	78	131
${}^{99}_{42}\text{Mo}$	42	57	99

2-46. The atomic number determines the element. The mass number is the sum of the atomic number and the number of neutrons.

Symbol	Atomic number	Number of neutrons	Mass number
${}^{39}_{19}\text{K}$	19	20	39
${}^{56}_{26}\text{Fe}$	26	30	56
${}^{84}_{36}\text{Kr}$	36	48	84
${}^{120}_{50}\text{Sn}$	50	70	120

ISOTOPIC COMPOSITION

2-48. We shall show the multiplication steps to the proper number of significant figures and the final atomic mass to the correct number of significant figures. Using the data in Table 2.9, we have that

$$\begin{aligned} 23.985\,041\,90 \times 0.7899 &= 18.95 \\ +\,24.985\,837\,02 \times 0.1000 &= 2.499 \\ +\,25.982\,593\,04 \times 0.1101 &= 2.861 \\ \hline \text{atomic mass of magnesium} &= 24.31 \end{aligned}$$

2-50. We shall show the multiplication steps to the proper number of significant figures and the final atomic mass to the correct number of significant figures. The atomic mass of naturally occurring silicon is given by

$$\begin{aligned} 27.976\,926\,5327 \times 0.922\,297 &= 25.8030 \\ +\,28.976\,494\,72 \times 0.046\,832 &= 1.3570 \\ +\,29.973\,770\,22 \times 0.030\,872 &= 0.92535 \\ \hline \text{atomic mass of silicon} &= 28.0854 \end{aligned}$$

2-52. Because there are only two isotopes of gallium, the abundance of gallium-71 must equal 100% − 60.108%, the abundance of the other isotope of gallium, or 39.892%. The atomic mass of gallium is equal to the isotopic mass of the unknown gallium isotope times its abundance plus the isotopic mass of gallium-71 times its abundance. Let x be the isotopic mass of the gallium isotope and write

$$69.723 = \left(\frac{60.108}{100}\right) x + \left(\frac{39.892}{100}\right)(70.924\ 7050)$$

Collecting like terms yields

$$41.430 = 0.60108\,x$$

Solving for x gives $x = 68.925$.

2-54. Let x be the percentage of boron-10 in naturally occurring boron. The percentage of boron-11 must be $100 - x$. Now set up the equation

$$\text{atomic mass of B} = 10.811 = (10.013)\left(\frac{x}{100}\right) + (11.009)\left(\frac{100 - x}{100}\right)$$

Multiply this equation through by 100 to obtain

$$1081.1 = 10.013\,x + 1100.9 - 11.009\,x$$

Collecting like terms, we get

$$0.996\,x = 19.8$$

Solving for x yields

$$x = 19.9\% = \%\ \text{boron-10}$$

The percentage of boron-11 is

$$\%\ \text{boron-11} = 100 - x = 80.1\%$$

2-56. Let x be the percentage of europium-151 in naturally occurring europium. The percentage of europium-153 must be $100 - x$. Now set up the equation

$$\text{atomic mass of Eu} = 151.964 = (150.9199)\left(\frac{x}{100}\right) + (152.9212)\left(\frac{100 - x}{100}\right)$$

Multiply this equation through by 100 to obtain

$$15196.4 = 150.9199\,x + 15292.12 - 152.9212\,x$$

Collecting like terms, we get

$$2.0013\,x = 95.7$$

Solving for x yields

$$x = 47.8\% = \%\ \text{europium-151}$$

The percentage of europium-153 is

$$\%\ \text{europium-153} = 100.00 - x = 52.2\%$$

IONS

2-58. The number of electrons is equal to the atomic number minus the ionic charge. Thus,
(a) 36 (b) 18 (c) 46 (d) 18

2-60. The number of electrons is equal to the atomic number minus the ionic charge. Thus,
(a) 54 (b) 54 (c) 2 (d) 28

2-62. Some ions that are isoelectronic are as follows:

(a) Na^+, Mg^{2+}, O^{2-}, N^{3-} (b) Rb^+, Sr^{2+}, Br^-
(c) Cs^+, I^-, Te^{2-} (d) Cs^+, Ba^{2+}, I^-, Te^{2-}

2-64. (a) sulfide anion (b) aluminum cation (c) hydride anion (d) vanadium cation

2-66. (a) molecular mass of $NH_4^+ = 14.0067 + (4)(1.00794) = 18.038$
(b) molecular mass of $HO_2^- = 1.00794 + (2)(15.9994) = 33.007$
(c) molecular mass of $AgCl_2^- = 107.8682 + (2)(35.453) = 178.77$
(d) molecular mass of $PCl_6^- = 30.973\,76 + (6)(35.453) = 243.69$

ADDITIONAL PROBLEMS

2-68. The elements with names similar to the planets (or former planets) are mercury, uranium, neptunium, and plutonium. (See also Table A.5.)

2-70. Isotopes are atoms of an element that contain the same number of protons (same atomic number) but different numbers of neutrons. The chemical properties of the isotopes of an element are the same, although some physical properties such as mass and radioactivity may differ.

2-72.* The natural abundance of deuterium given in Table 2.9 is 0.0115%. Assuming that the abundance of deuterium in natural water is the same as the natural abundance of D, for DHO, we have

$$\text{percentage of D in DHO} = (2)(0.0115\%) = 0.023\%$$

where we have multiplied by 2 because either hydrogen atom in an H_2O molecule could be replaced by a deuterium atom. For D_2O, the probability that both hydrogen atoms are replaced by deuterium atoms is

$$\text{percentage of D in } D_2O = (0.0115\%)^2 = 0.00013\%$$

2-74. Let x be the molecular mass of the protein. Then

$$\text{mass \% of cobalt} = \frac{\text{atomic mass of Co}}{\text{molecular mass of protein}} \times 100 = \frac{58.93}{x} \times 100 = 0.168$$

Solving for x, we have $x = 35\,100$ or a multiple of this if there is more than one cobalt atom in the protein.

2-76. We shall show the multiplication steps to the proper number of significant figures and the final atomic mass to the correct number of significant figures. The atomic mass of naturally occurring tungsten is given by

$$\begin{aligned} 179.946\,706 \times 0.0012 &= 0.216 \\ +\,181.948\,206 \times 0.2650 &= 48.216 \\ +\,182.950\,2245 \times 0.1431 &= 26.180 \\ +\,183.950\,9326 \times 0.3064 &= 56.363 \\ +\,185.954\,362 \times 0.2843 &= 52.867 \\ \hline \text{atomic mass of tungsten} &= 183.84 \end{aligned}$$

2-78. Let's reduce the data to 1.00 gram of nitrogen. Then we have

$$\text{compound 1:} \quad \text{grams of oxygen combined} = (1.00\text{ g N})\left(\frac{0.703\text{ g O}}{0.615\text{ g N}}\right) = 1.14\text{ g O}$$

$$\text{compound 2:} \quad \text{grams of oxygen combined} = (1.00\text{ g N})\left(\frac{2.90\text{ g O}}{1.27\text{ g N}}\right) = 2.28\text{ g O}$$

The ratio of the masses of oxygen that combine with one gram of nitrogen is $2.28/1.14 = 2/1$. If we had based our calculation on one gram of oxygen, then we would have obtained $1/2$. In either case, the ratio comes out in terms of small whole numbers.

2-80. Atomic mass is the ratio of the mass of an atom to the mass of an atom of carbon-12. Thus we have

$$\frac{\text{atomic mass hydrogen}}{1} = \frac{1.008}{12}$$

$$\text{atomic mass of hydrogen} = 0.0840$$

$$\frac{\text{atomic mass of oxygen}}{1} = \frac{16.00}{12}$$

$$\text{atomic mass of oxygen} = 1.333$$

2-82. The ratio of the amount of oxygen that combines with hydrogen is $0.832\text{ g}/0.104\text{ g} = 8/1$, and so oxygen would have an atomic mass of 8 based upon an $\text{H} = 1$ scale. Similarly, the ratio of the amount of nitrogen that combines with hydrogen is $0.403\text{ g}/0.0864\text{ g} = 4.66/1$, and so nitrogen would have an atomic mass of 4.66 based upon an $\text{H} = 1$ scale. Chemists used these (incorrect) atomic masses for many years.

CHAPTER 3. The Periodic Table and Chemical Periodicity

BALANCING EQUATIONS

3-2. The procedure for balancing equations of the type considered in this chapter is the balancing by inspection method outlined in Section 3-2.

(a) We can start with any of the three kinds of atoms. Looking at the right side, we start with H. If we multiply KHF_2 by 2, then we have two H atoms on both sides of the equation.

$$2\,KHF_2 \longrightarrow KF + H_2 + F_2$$

We now multiply KF by 2 to have two K atoms on both sides of the equation:

$$2\,KHF_2(s) \longrightarrow 2\,KF(s) + H_2(g) + F_2(g)$$

The equation is now balanced.

(b) We can start with any of the three kinds of atoms. Let's start with C. If we multiply CO_2 by 3, then we have three C atoms on both sides of the equation.

$$C_3H_8 + O_2 \longrightarrow 3\,CO_2 + H_2O$$

If we multiply H_2O by 4, then we have eight H atoms on both sides of the equation.

$$C_3H_8 + O_2 \longrightarrow 3\,CO_2 + 4\,H_2O$$

We have 10 O atoms on the right side of the equation, so we must multiply O_2 by 5 to obtain the balanced equation

$$C_3H_8(g) + 5\,O_2(g) \longrightarrow 3\,CO_2(g) + 4\,H_2O(l)$$

(c) $P_4O_{10}(s) + 6\,H_2O(l) \longrightarrow 4\,H_3PO_4(l)$

(d) $3\,N_2H_4(g) \longrightarrow 4\,NH_3(g) + N_2(g)$

3-4. The procedure for balancing equations of the type considered in this chapter is the balancing by inspection method outlined in Section 3-2.

(a) $H_2SO_4(aq) + 2\,KOH(aq) \longrightarrow K_2SO_4(aq) + 2\,H_2O(l)$

(b) $Li_3N(s) + 3\,H_2O(l) \longrightarrow 3\,LiOH(aq) + NH_3(g)$

(c) $Al_4C_3(s) + 12\,HCl(aq) \longrightarrow 4\,AlCl_3(aq) + 3\,CH_4(g)$

(d) $ZnS(s) + 2\,HBr(aq) \longrightarrow ZnBr_2(aq) + H_2S(g)$

3-6. We have the following

(a) $\underset{\text{phosphorus trichloride}}{PCl_3(g)} + \underset{\text{chlorine}}{Cl_2(g)} \longrightarrow \underset{\text{phosphorus pentachloride}}{PCl_5(s)}$

(b) $\underset{\text{antimony}}{2\,Sb(s)} + \underset{\text{chlorine}}{3\,Cl_2(g)} \longrightarrow \underset{\text{antimony trichloride}}{2\,SbCl_3(s)}$

(c) $\underset{\text{gallium bromide}}{2\,GaBr_3(s)} + \underset{\text{chlorine}}{3\,Cl_2(g)} \longrightarrow \underset{\text{gallium chloride}}{2\,GaCl_3(s)} + \underset{\text{bromine}}{3\,Br_2(l)}$

(d) $\underset{\text{magnesium nitride}}{Mg_3N_2(s)} + \underset{\text{hydrogen chloride}}{6\,HCl(g)} \longrightarrow \underset{\text{magnesium chloride}}{3\,MgCl_2(s)} + \underset{\text{ammonia}}{2\,NH_3(g)}$

3-8. See Section 3-3 for the representative reactions.

(a) $Sr(s) + S(s) \longrightarrow \underset{\text{strontium sulfide}}{SrS(s)}$

(b) $2\,K(s) + 2\,H_2O(l) \longrightarrow \underset{\text{potassium hydroxide}}{2\,KOH(aq)} + \underset{\text{hydrogen}}{H_2(g)}$

(c) $Ca(s) + 2\,H_2O(l) \longrightarrow \underset{\text{calcium hydroxide}}{Ca(OH)_2(aq)}$

(d) $2\,Al(s) + 3\,Cl_2(g) \longrightarrow \underset{\text{aluminum chloride}}{2\,AlCl_3(s)}$

PERIODIC TABLE

3-10. By analogy with the properties of the other noble gases, we predict the following:

(a) colorless; (b) odorless; (c) $Ra(g)$; (d) no reaction

3-12. Se, a main-group (16) nonmetal; As, a main-group (15) semimetal; Mo, a transition (6) metal; Rn, a main-group (18) nonmetal; Ta, a transition (5) metal; Bi, a main-group (15) metal; In, a main-group (13) metal

3-14. The Group-17 elements are all very reactive. They all react with potassium, strontium, and aluminum according to

$$X_2 + 2\,K(s) \longrightarrow 2\,KX(s)$$
$$X_2 + 2\,Sr(s) \longrightarrow SrX_2(s)$$
$$3\,X_2 + 2\,Al(s) \longrightarrow 2\,AlX_3(s)$$

where X stands for the Group-17 elements. Fluorine is the most reactive halogen. Fluorine is the only halogen that forms compounds with xenon and krypton.

3-16. The Group-2 metals are fairly reactive. Although reactive, they are not as reactive as the Group-1 metals. They react with iodine, water, and oxygen according to

$$M(s) + I_2(s) \longrightarrow MI_2(s)$$
$$M(s) + 2\,H_2O(l) \longrightarrow M(OH)_2(s) + H_2(g)$$
$$2\,M(s) + O_2(g) \longrightarrow 2\,MO(s)$$

where M stands for the Group-2 metals. Barium is the most reactive nonradioactive (stable) Group-2 metal, and is also the most metallic.

3-18. Francium is a Group-1 metal, and thus, we predict the reactions of francium by analogy with the other Group-1 metals

(a) $2\,Fr(s) + Br_2(l) \longrightarrow 2\,FrBr(s)$

(b) $2\,Fr(s) + H_2(g) \longrightarrow 2\,FrH(s)$

(c) $2\,Fr(s) + 2\,H_2O(l) \longrightarrow 2\,FrOH(s) + H_2(g)$

(d) $2\,Fr(s) + S(s) \longrightarrow Fr_2S(s)$

3-20. Using the symbol for the element makes it easier to find the element in the periodic table. Palladium (Pd), silver (Ag), and zinc (Zn) are transition metals.

3-22. (a) Ge, germanium, is a semimetal (b) Ga, gallium, is a metal (c) Gd, gadolinium, is a metal (d) Al, aluminum, is a metal (e) At, astatine, is a semimetal

ADDITIONAL PROBLEMS

3-24. Group 11 is sometimes referred to as the "coinage metals" because the group contains the metals copper, silver, and gold, all of which have been used as coins in various cultures. However, it is not an IUPAC approved name because other metals, such as iron, lead, nickel, and zinc are also used in coins.

3-26. Because argon was the first and at that time the only noble gas, Mendeleev had no group to place argon in. Furthermore, because there was no gap between the elements chlorine and potassium, which both fit well into their respective groups, there was no logical place to insert argon. Thus, he reasoned (incorrectly) that argon was probably not a new element, but rather a compound or new form of some other element. Only after other noble gases were discovered, was a new family officially added to the periodic table.

3-28. We first write down the chemical formulas of the reactants and products and then balance the equation by inspection.

(a) $2\,K(s) + 2\,H_2O(l) \longrightarrow 2\,KOH(s) + H_2(g)$

(b) $KH(s) + H_2O(l) \longrightarrow KOH(s) + H_2(g)$

(c) $SiO_2(s) + 3\,C(s) \longrightarrow SiC(s) + 2\,CO(g)$

(d) $SiO_2(s) + 4\,HF(g) \longrightarrow SiF_4(g) + 2\,H_2O(l)$

(e) $2\,P(s) + 3\,Cl_2(g) \longrightarrow 2\,PCl_3(l)$

3-30. The equation for the reaction between lithium nitride and water is

$$Li_3N(s) + 3\,H_2O(l) \longrightarrow NH_3(g) + 3\,LiOH(aq)$$

By analogy, we have

$$Na_3P(s) + 3\,H_2O(l) \longrightarrow PH_3(g) + 3\,NaOH(aq)$$

3-32. We would predict that LiF would dissolve the least in water because the first member of a group does not always behave chemically as the other members of the group.

3-34. (a) $6\,F_2(g) + 2\,Al_2O_3(s) \longrightarrow 4\,AlF_3(s) + 3\,O_2(g)$

(b) $4\,NH_3(g) + 5\,O_2(g) \longrightarrow 4\,NO(g) + 6\,H_2O(l)$

(c) $2\,C_6H_6(l) + 15\,O_2(g) \longrightarrow 12\,CO_2(g) + 6\,H_2O(l)$

(d) $3\,H_2SO_4(aq) + Al_2(CO_3)_3(s) \longrightarrow Al_2(SO_4)_3(aq) + 3\,CO_2(g) + 3\,H_2O(l)$

3-36. (a) $2\,Na(s) + S(l) \longrightarrow Na_2S(l)$

(b) $Ca(s) + Br_2(l) \longrightarrow CaBr_2(s)$

(c) $2\,Ba(s) + O_2(g) \longrightarrow 2\,BaO(s)$

(d) $2\,SO_2(g) + O_2(g) \longrightarrow 2\,SO_3(g)$

(e) $4\,Mg(s) + 3\,N_2(g) \longrightarrow 2\,Mg_2N_3(s)$

3-38. (a) The melting point is 630.63°C.

(b) Antimony is a silvery white solid.

(c) Its density is 6.697 $gm{\cdot}cm^{-3}$.

(d) Some uses of antimony are in diodes, alloys with other metals, in matches, soldering, and as a source of neutrons in nuclear reactors.

3-40. The chemical equations for the preparations of copper, tin, and iron are

$$2\,CuO(s) + C(s) \longrightarrow 2\,Cu(s) + CO_2(g)$$
$$SnO_2(s) + C(s) \longrightarrow Sn(s) + CO_2(g)$$
$$2\,Fe_2O_3(s) + 3\,C(s) \longrightarrow 4\,Fe(s) + 3\,CO_2(g)$$

3-42. By inspection,

$$2\,HCl(aq) + 2\,NaI(aq) + NaClO(aq) \longrightarrow I_2(s) + 3\,NaCl(aq) + H_2O(l)$$

3-44.* Here is a systematic way to do this problem. Let the balanced equation be

$$a\,HCl(aq) + b\,K_2Cr_2O_7(aq) + c\,C_2H_5OH(aq) \longrightarrow d\,CrCl_3(aq) + e\,CO_2(g) + f\,KCl(aq) + g\,H_2O(l)$$

Imposing an atom balance on each side of the equation gives the following simultaneous algebraic equations:

$$\begin{aligned} H:&\ a + 6c = 2g \\ Cl:&\ a = 3d + f \\ K:&\ 2b = f \\ Cr:&\ 2b = d \\ O:&\ 7b + c = 2e + g \\ C:&\ 2c = e \end{aligned}$$

The equations for K and Cr give $f = d = 2b$. We are free to choose any one of a through g arbitrarily because stoichiometric coefficients are relative values. Let's choose $b = 1$, which gives $f = d = 2$, and $b = 1$. Substituting these values into the above equations gives

$$\begin{aligned} H:&\ a + 6c = 2g \\ Cl:&\ a = 8 \\ O:&\ 7 + c = 2e + g \\ C:&\ 2c = e \end{aligned}$$

Substituting the C equation and $a = 8$ into the H and O equations gives

$$8 = 2g - 6c$$
$$7 = 3c + g$$

Multiplying the second of these two equations by 2 and adding gives $4g = 22$, or $g = 11/2$. Therefore, $c = 1/2$ and $e = 1$. Thus the balanced equation is

$$8\,HCl(aq) + K_2Cr_2O_7(aq) + \tfrac{1}{2}\,C_2H_5OH(aq) \longrightarrow 2\,CrCl_3(aq) + CO_2(g) + 2\,KCl(aq) + \tfrac{11}{2}\,H_2O(l)$$

Multiplying through by 2 gives

$$16\,HCl(aq) + 2\,K_2Cr_2O_7(aq) + C_2H_5OH(aq) \longrightarrow 4\,CrCl_3(aq) + 2\,CO_2(g) + 4\,KCl(aq) + 11\,H_2O(l)$$

3-46.** Here is a systematic way to do this problem. Let the balanced equation be

$$a\,As_2S_5(s) + b\,NaNO_3(aq) + c\,HCl(aq) \longrightarrow d\,H_3AsO_4(aq) + e\,NaHSO_4(aq) + f\,NO_2(g) + g\,H_2O(l) + h\,NaCl(aq)$$

Imposing an atom balance on each side of the equation gives the following simultaneous algebraic equations:

$$\begin{aligned} \text{As}: &\quad 2a = d \\ \text{S}: &\quad 5a = e \\ \text{Na}: &\quad b = e + h \\ \text{N}: &\quad b = f \\ \text{O}: &\quad 3b = 4d + 4e + 2f + g \\ \text{H}: &\quad c = 3d + e + 2g \\ \text{Cl}: &\quad c = h \end{aligned}$$

We are free to choose any one of a through h arbitrarily, and we choose $a = 1$. This choice gives $a = 1$, $d = 2$, and $e = 5$ and

$$b = h + 5 \qquad (1)$$
$$b = f \qquad (2)$$
$$3b = 28 + 2f + g \qquad (3)$$
$$c = 11 + 2g \qquad (4)$$
$$c = h \qquad (5)$$

Substituting (2) into (3) and (5) into (4) gives

$$b = h + 5 \qquad (6)$$
$$b = 28 + g \qquad (7)$$
$$h = 11 + 2g \qquad (8)$$

Substitute (8) into (6) and equate the result to (7) to get $b = 11 + 2g + 5 = 28 + g$ or $g = 12$. Equation (7) now gives $b = 40$, Equation (2) gives $f = 40$, Equation (8) gives $h = 35$, and Equation (15) gives $c = 35$.
Summing up, the balanced equation is

$$\mathrm{As_2S_5}(s) + 40\,\mathrm{NaNO_3}(aq) + 35\,\mathrm{HCl}(aq) \longrightarrow$$
$$2\,\mathrm{H_3AsO_4}(aq) + 5\,\mathrm{NaHSO_4}(aq) + 40\,\mathrm{NO_2}(g) + 12\,\mathrm{H_2O}(l) + 35\,\mathrm{NaCl}(aq)$$

3-48. Thorium, protactinium, and uranium (90 to 92) are the only naturally occurring actinides.
Thorium (90) is the most abundant actinide and is as common as lead.
Americium (95) is used in smoke detectors.
Californium (98) is used in neutron activation analysis.

CHAPTER 4. Early Quantum Theory

IONIZATION ENERGIES

4-2. Ionization energies decrease with increasing atomic size within a family. The farther away an electron is from the nucleus, the easier it is to remove the electron from the atom. Thus we write Ba < Sr < Ca < Mg.

4-4. The energy required is the sum of the first three ionization energies of a boron atom. Therefore,

$$E = (1.33 + 4.04 + 6.08)\ \text{aJ} = 11.45\ \text{aJ}$$

4-6. The energy required is given by

$$E = (6.022 \times 10^{23}\ \text{atom})\,[(3.94 + 8.72)\ \text{aJ}\cdot\text{atom}^{-1}] = 7.624 \times 10^{24}\ \text{aJ} = 7.624 \times 10^{6}\ \text{J} = 7.624\ \text{MJ}$$

4-8. Using the data in Table 4.1, we have for a beryllium atom

n	I_n/aJ	$\ln(I_n/\text{aJ})$
1	1.49	0.399
2	2.92	1.07
3	24.7	3.21
4	39.9	3.69

The following plot suggests that the four electrons are arranged in two shells, with two electrons in an inner, tightly held shell and two in an outer shell.

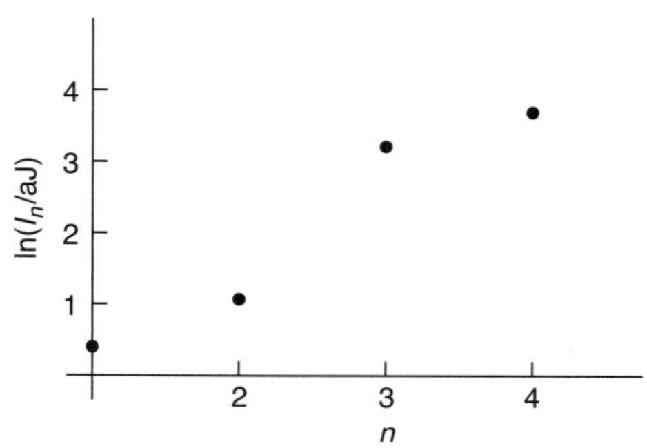

LEWIS ELECTRON-DOT FORMULAS

4-10. The Group-16 elements have six outer-shell electrons, and so their Lewis electron-dot formulas are

$$\cdot\ddot{\underset{..}{\text{O}}}\cdot \quad \cdot\ddot{\underset{..}{\text{S}}}\cdot \quad \cdot\ddot{\underset{..}{\text{Se}}}\cdot \quad \cdot\ddot{\underset{..}{\text{Te}}}\cdot$$

4-12. Boron (Group 3) has three outer electrons; B^{3+} has no outer electrons, and so we write B^{3+}.
Nitrogen (Group 15) has five outer electrons; N^{3-} has eight outer electrons, and so we write $:\ddot{\underset{..}{\text{N}}}:^{3-}$.
Fluorine (Group 17) has seven outer electrons; F^- has eight outer electrons, and so we write $:\ddot{\underset{..}{\text{F}}}:^{-}$.
Oxygen (Group 16) has six outer electrons; O^{2-} has eight outer electrons, and so we write $:\ddot{\underset{..}{\text{O}}}:^{2-}$.
The Na^+ ion has no valence electrons, and so we write Na^+.

ELECTROMAGNETIC RADIATION

4-14. (a) Yellow light has a longer wavelength than green light.
(b) Green light has a greater frequency.
(c) Green light has a greater energy.

4-16. Solve Equation 4.1 for ν.

$$\nu = \frac{c}{\lambda} = \frac{2.998 \times 10^8\ \text{m}\cdot\text{s}^{-1}}{589 \times 10^{-9}\ \text{m}} = 5.088 \times 10^{14}\ \text{s}^{-1}$$

4-18. The range of wavelength is between

$$\lambda = \frac{c}{\nu} = \frac{2.998 \times 10^8\ \text{m}\cdot\text{s}^{-1}}{34.2 \times 10^9\ \text{s}^{-1}} = 8.77 \times 10^{-3}\ \text{m} = 8.77\ \text{mm}$$

and

$$\lambda = \frac{c}{\nu} = \frac{2.998 \times 10^8\ \text{m}\cdot\text{s}^{-1}}{35.2 \times 10^9\ \text{s}^{-1}} = 8.52 \times 10^{-3}\ \text{m} = 8.52\ \text{mm}$$

4-20. The energy that corresponds to a wavelength of 80 nanometers is given by Equation 4.5:

$$E = \frac{hc}{\lambda} = \frac{(6.626 \times 10^{-34}\ \text{J}\cdot\text{s})(2.998 \times 10^8\ \text{m}\cdot\text{s}^{-1})}{80 \times 10^{-9}\ \text{m}}$$
$$= 2.5 \times 10^{-18}\ \text{J} = 2.5\ \text{aJ}$$

Thus, the energy of 80 nm radiation is not quite sufficient to ionize an argon atom.

PHOTONS

4-22. The energy per photon of X-rays of wavelength 210 pm is given by Equation 4.5.

$$E = \frac{hc}{\lambda} = \frac{(6.626 \times 10^{-34}\ \text{J}\cdot\text{s})(2.998 \times 10^8\ \text{m}\cdot\text{s}^{-1})}{210 \times 10^{-12}\ \text{m}}$$
$$= 9.46 \times 10^{-16}\ \text{J}\cdot\text{photon}^{-1}$$

The energy of 6.022×10^{23} photons is given by

$$E = (9.46 \times 10^{-16}\ \text{J}\cdot\text{photon}^{-1})(6.022 \times 10^{23}\ \text{photons}) = 5.70 \times 10^8\ \text{J}$$

4-24. The energy of each photon is given by

$$E = \frac{hc}{\lambda} = \frac{(6.626 \times 10^{-34}\ \text{J}\cdot\text{s})(2.998 \times 10^{8}\ \text{m}\cdot\text{s}^{-1})}{193 \times 10^{-9}\ \text{m}}$$
$$= 1.03 \times 10^{-18}\ \text{J}\cdot\text{photon}^{-1}$$

The energy of a pulse of these photons is given by

$$E_{\text{pulse}} = (1.03 \times 10^{-18}\ \text{J}\cdot\text{photon}^{-1})(2.5 \times 10^{16}\ \text{photon}) = 0.026\ \text{J} = 26\ \text{mJ}$$

4-26. The energy of a single photon is given by

$$E_{\text{photon}} = \frac{hc}{\lambda} = \frac{(6.626 \times 10^{-34}\ \text{J}\cdot\text{s})(2.998 \times 10^{8}\ \text{m}\cdot\text{s}^{-1})}{1.54 \times 10^{-10}\ \text{m}}$$
$$= 1.29 \times 10^{-15}\ \text{J}\cdot\text{photon}^{-1}$$

The energy of 6.022×10^{23} of these photons is

$$E = (1.29 \times 10^{-15}\ \text{J}\cdot\text{photon}^{-1})(6.022 \times 10^{23}\ \text{photons})$$
$$= 7.77 \times 10^{8}\ \text{J}$$

PHOTOELECTRIC EFFECT

4-28. The frequency corresponding to a wavelength of 840 nanometers is given by

$$\nu = \frac{c}{\lambda} = \frac{2.998 \times 10^{8}\ \text{m}\cdot\text{s}^{-1}}{840 \times 10^{-9}\ \text{m}} = 3.57 \times 10^{14}\ \text{s}^{-1} = 3.57 \times 10^{14}\ \text{Hz}$$

This frequency is not large enough to eject an electron from a copper surface.

4-30. The kinetic energy of an ejected electron is given by

$$E_{\text{k}} = \frac{1}{2}mv^2 = \frac{(9.1094 \times 10^{-31}\ \text{kg})(5.00 \times 10^{5}\ \text{m}\cdot\text{s}^{-1})^2}{2} = 1.14 \times 10^{-19}\ \text{J}$$

The energy of the incident radiation is given by

$$E_{\text{rad}} = \frac{hc}{\lambda} = \frac{(6.626 \times 10^{-34}\ \text{J}\cdot\text{s})(2.998 \times 10^{8}\ \text{m}\cdot\text{s}^{-1})}{390.0 \times 10^{-9}\ \text{m}} = 5.094 \times 10^{-19}\ \text{J}$$

The difference $E_{\text{rad}} - E_{\text{k}}$ is the threshold energy, and so the threshold frequency is given by

$$\nu = \frac{E_{\text{rad}} - E_{\text{k}}}{h} = \frac{5.094 \times 10^{-19}\ \text{J} - 1.14 \times 10^{-19}\ \text{J}}{6.626 \times 10^{-34}\ \text{J}\cdot\text{s}} = 5.96 \times 10^{14}\ \text{s}^{-1}$$

DE BROGLIE WAVELENGTH

4-32. We first convert the speed in miles per hour to meters per second.

$$v = (1250\ \text{mile}\cdot\text{h}^{-1})\left(\frac{1.609\ \text{km}}{1\ \text{mile}}\right)\left(\frac{10^3\ \text{m}}{1\ \text{km}}\right)\left(\frac{1\ \text{hr}}{3600\ \text{s}}\right) = 559\ \text{m}\cdot\text{s}^{-1}$$

The de Broglie wavelength is given by Equation 4.7

$$\lambda = \frac{h}{mv} = \frac{6.626 \times 10^{-34}\ \text{J}\cdot\text{s}}{(5.00 \times 10^{-3}\ \text{kg})(559\ \text{m}\cdot\text{s}^{-1})} = 2.37 \times 10^{-34}\ \text{m}$$

an undetectably small wavelength.

4-34. We can use Equation 4.7 to write

$$v = \frac{h}{m\lambda} = \frac{6.626 \times 10^{-34}\ \text{J}\cdot\text{s}}{(9.1094 \times 10^{-31}\ \text{kg})(96.0 \times 10^{-12}\ \text{m})} = 7.58 \times 10^{6}\ \text{m}\cdot\text{s}^{-1}$$

4-36. We solve Equation 4.7 for m to get

$$m = \frac{h}{\lambda v} = \frac{6.626 \times 10^{-34}\ \text{J}\cdot\text{s}}{(0.10 \times 10^{-15}\ \text{m})(100\ \text{m}\cdot\text{s}^{-1})} = 6.6 \times 10^{-20}\ \text{kg}$$

which is the maximum mass because λ is the smallest measureable wavelength.

HYDROGEN ATOMIC SPECTRUM

4-38. Farther. It takes less energy to ionize the hydrogen atom in the $n = 3$ state because the electron has already gained energy to reach the excited state.

4-40. We use Equation 4.12 with $n_i = 3$ and $n_f = 1$.

$$E = (2.1799\ \text{aJ})\left(\frac{1}{1^2} - \frac{1}{3^2}\right) = 1.9377\ \text{aJ} = 1.9377 \times 10^{-18}\ \text{J}$$

The frequency of a photon with this energy is obtained from Equation 4.3.

$$\nu = \frac{E}{h} = \frac{1.9377 \times 10^{-18}\ \text{J}}{6.626 \times 10^{-34}\ \text{J}\cdot\text{s}} = 2.9244 \times 10^{15}\ \text{s}^{-1}$$

This line appears in the Lyman series in the ultraviolet region.

4-42. To find the frequency of the lines, we first use Equaton 4.10 to find the energy of the photon emitted using $n_f = 3$ and $n_i = 4, 5, 6, \ldots$.

$$E = (2.1799\ \text{aJ})\left(\frac{1}{3^2} - \frac{1}{n_i^2}\right) = (2.1799\ \text{aJ})\left(\frac{1}{9} - \frac{1}{n_i^2}\right)$$

and then use $\lambda = hc/E$ to find the wavelengths of the lines. The results are

n_f	4	5	6	7	8
$\lambda/\mu\text{m}$	1.875	1.282	1.094	1.005	0.9544

These wavelengths are in the infrared region.

4-44. The wavelength of the given line is

$$\lambda = \frac{1}{9140\ \text{cm}^{-1}} = 1.094 \times 10^{-4}\ \text{cm} = 1.094\mu\text{m}$$

and the energy of the given line corresponds to (Equation 4.5)

$$E = \frac{hc}{\lambda} = \frac{(6.626 \times 10^{-34}\ \text{J·s})(2.998 \times 10^{8}\ \text{m·s}^{-1})}{1.094 \times 10^{-6}\ \text{m}} = 1.816 \times 10^{-19}\ \text{J}$$

We use Equation 4.10 with $n_\text{f} = 3$ and n_i being as yet unknown.

$$E = 1.816 \times 10^{-19}\ \text{J} = (2.1799\ \text{aJ})\left(\frac{1}{3^2} - \frac{1}{n_\text{i}^2}\right)$$

which yields

$$0.08330 = \frac{1}{9} - \frac{1}{n_\text{i}^2}$$

Solving for n_i, we have that

$$n_\text{i} = (35.96)^{1/2} = 5.996 \approx 6$$

4-46. The energy of a photon of wavelength 97.2 nm is (Equation 4.5)

$$E = \frac{hc}{\lambda} = \frac{(6.626 \times 10^{-34}\ \text{J·s})(2.998 \times 10^{8}\ \text{m·s}^{-1})}{97.2 \times 10^{-9}\ \text{m}} = 2.044 \times 10^{-18}\ \text{J}$$

We use Equation 4.10 with $n_\text{i} = 1$ and n_f being as yet unknown.

$$E = 2.044 \times 10^{-18}\ \text{J} = (2.1799\ \text{aJ})\left(\frac{1}{1^2} - \frac{1}{n_\text{f}^2}\right)$$

which yields

$$0.9375 = 1 - \frac{1}{n_\text{f}^2}$$

Solving for n_f, we have that $n_\text{f}^2 = 16$. Thus, $n_\text{f} = 4$. The electron then makes a transition from the $n = 4$ state to some lower state and emits a photon of wavelength 486 nm. The energy of this second photon is

$$E = \frac{hc}{\lambda} = \frac{(6.626 \times 10^{-34}\ \text{J·s})(2.998 \times 10^{8}\ \text{m·s}^{-1})}{486 \times 10^{-9}\ \text{m}} = 4.087 \times 10^{-19}\ \text{J}$$

Now using Equation 4.10 with $n_\text{i} = 4$ and n_f being as yet unknown,

$$E = 4.087 \times 10^{-19}\ \text{J} = (2.1799\ \text{aJ})\left(\frac{1}{n_\text{f}^2} - \frac{1}{4^2}\right)$$

which yields

$$0.1875 = \frac{1}{n_\text{f}^2} - \frac{1}{16}$$

or $n^2 = 4.00$. Thus, the final state is $n = 2$.

4-48. We would expect to see lines in the emission spectrum from transitions for the $n = 3$ state to the $n = 2$ state and from the $n = 3$ state to the $n = 1$ state. We would also expect to see a line in the emission spectrum from the $n = 2$ state to the $n = 1$ state resulting from electrons that first decayed from the $n = 3$ state to the $n = 2$ state. The energies of these three lines are $E_{3\to2} = 5$, $E_{3\to1} = 16$, and $E_{2\to1} = 11$.

4-50. To find the wavelengths of the lines, we first use the equation given in the previous problem to find the energies of the transition, writing an equation analogous to Equation 4.12.

$$E = (2.1799 \text{ aJ})(Z^2)\left(\frac{1}{n_\mathrm{i}^2} - \frac{1}{n_\mathrm{f}^2}\right)$$

Let $Z = 2$ and $n_\mathrm{f} = 4$ to obtain

$$E = (2.1799 \text{ aJ})(4)\left(\frac{1}{n_\mathrm{i}^2} - \frac{1}{16}\right)$$

and then use

$$\lambda = \frac{hc}{E} = \frac{(6.626 \times 10^{-34}\text{ J·s})(2.998 \times 10^{8}\text{ m·s}^{-1})}{E}$$

to find the wavelengths of the lines. The results are

n_f	5	6	7	8	9
E/aJ	0.1962	0.3028	0.3670	0.4087	0.4373
$\lambda/\mu\text{m}$	1.012	0.65560	0.5413	0.4860	0.4542

These wavelengths occur in the infrared-visible region and correspond to visible emission lines in stars.

ADDITIONAL PROBLEMS

4-52. The term quantum refers to the discrete packages of energy that are central to quantum theory. In contrast to classical physics where energy is regarded as a continuum, in quantum theory, energy is restricted to discrete values called quanta.

4-54. She could place the sample in a flame and look for a blue flame as shown in Figure 4.27. Measuring the emission lines with a spectrometer and comparing frequencies of the radiation to those listed in a handbook would confirm whether the sample contained cesium ions.

4-56. Although everyday objects, such as a moving baseball, do have an associated wavelength as found from the de Broglie equation, the wavelength is far too small to be observable.

4-58. When $n = \infty$ and $E = 0$, the hydrogen atom has been ionized; that is, the electron has been removed from the atom and the proton and electron are far apart.

4-60. A plot of $\ln(I_n/\text{aJ})$ against n shows a sizeable break between $n = 2$ and $n = 3$, so this new element probably belongs to Group 2.

4-62. The data are plotted below. Equation 4.4, $E_\mathrm{k} = h\nu - h\nu_0$, tells us that a plot of E_k versus ν is a straight line whose slope is the Planck constant. The intercept of the straight line with the horizontal axis (where $E_\mathrm{k} = 0$) gives the minimum energy required to eject an electron, $h\nu_0$.

From the graph we find that $h\nu_0 = 7.4 \times 10^{-19}$ J, and so the threshold frequency $\nu_0 = 7.4 \times 10^{-19}\ \text{J}/6.626 \times 10^{-34}\ \text{J·s} = 1.1 \times 10^{15}\ \text{s}^{-1}$. The slope is given by (for example, if we take the second and fourth set of data)

$$\text{slope} = h = \frac{(15.84 - 9.21) \times 10^{-19}\ \text{J}}{(3.50 - 2.50) \times 10^{15}\ \text{s}^{-1}} = 6.63 \times 10^{-34}\ \text{J·s}$$

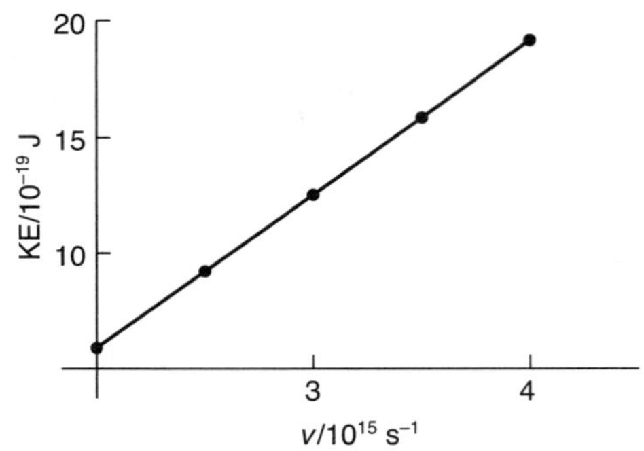

The intercept of the straight line with the horizontal axis (where $E_k = 0$) gives the frequency of light that is just sufficient to eject an electron, which is $1.1 \times 10^{15}\ \text{s}^{-1}$.

4-64. Combining Equations 4.3 and 4.10 with $n_f = 4$ and n_i being as yet unknown, we have that

$$\nu = \left(\frac{2.1799 \times 10^{-18}\ \text{J}}{6.626 \times 10^{-34}\ \text{J·s}}\right)\left(\frac{1}{4^2} - \frac{1}{n_i^2}\right) = 1.141 \times 10^{14}\ \text{s}^{-1}$$

or

$$0.03468 = \frac{1}{16} - \frac{1}{n_i^2}$$

Solving for n_i, we find that

$$n_i = (35.97)^{1/2} = 5.998 \approx 6$$

4-66. The frequencies of the lines in the Lyman series of the hydrogen atom are given by combining Equations 4.3 and 4.10 with $n_f = 1$,

$$\nu_{n\to 1} = (3.29 \times 10^{15}\ \text{s}^{-1})\left(\frac{1}{1^2} - \frac{1}{n^2}\right) \qquad n = 2,\ 3,\ \ldots$$

The values of ν and $1/n^2$ are given by

Transition	$\nu_{n\to 1}/10^{15}\ \text{s}^{-1}$	$1/n^2$
$2 \to 1$	2.47	0.2500
$3 \to 1$	2.92	0.1110
$4 \to 1$	3.08	0.0625
$5 \to 1$	3.16	0.0400
$6 \to 1$	3.20	0.0278

These data are plotted below. Note that a plot of $\nu_{n\to 1}$ versus $1/n^2$ is a straight line.

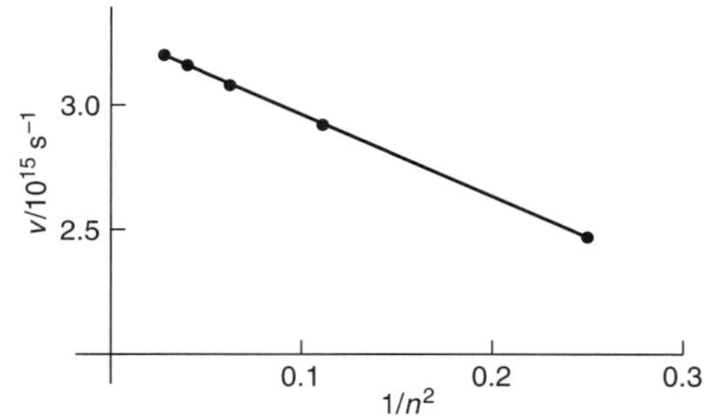

The slope of the line is given by (taking the first and fourth points arbitrarily)

$$\text{slope} = \frac{3.16 \times 10^{15}\ \text{s}^{-1} - 2.47 \times 10^{15}\ \text{s}^{-1}}{0.040 - 0.250} = -3.29 \times 10^{15}\ \text{s}^{-1}$$

4-68 The energy per photon is given by Equations 4.5.

$$E = \frac{hc}{\lambda} = \frac{(6.626 \times 10^{-34}\ \text{J}\cdot\text{s})(2.998 \times 10^{8}\ \text{m}\cdot\text{s}^{-1})}{510 \times 10^{-9}\ \text{m}}$$
$$= 3.90 \times 10^{-19}\ \text{J}\cdot\text{photon}^{-1}$$

The number of photons striking one square meter per second is

$$\text{number of photons} = \frac{1.0 \times 10^{3}\ \text{J}\cdot\text{s}^{-1}\cdot\text{m}^{-2}}{3.90 \times 10^{-19}\ \text{J}\cdot\text{photon}^{-1}} = 2.6 \times 10^{21}\ \text{photons}\cdot\text{s}^{-1}\cdot\text{m}^{-2}$$

The number of photons striking one square centimeter per second is

$$(2.6 \times 10^{21}\ \text{photons}\cdot\text{s}^{-1}\cdot\text{m}^{-2})\left(\frac{1\ \text{m}}{100\ \text{cm}}\right)^2 = 2.6 \times 10^{17}\ \text{photons}\cdot\text{s}^{-1}\cdot\text{cm}^{-2}$$

4-70. Equation 4.8 is $2\pi r = n\lambda$. If we substitute Equation 4.7 for λ into this equation, then we obtain

$$2\pi r = n\left(\frac{h}{mv}\right)$$

Multiply both sides by mv and divide both sides by 2π and get

$$mvr = n\frac{h}{2\pi}$$

CHAPTER 5. Quantum Theory and Atomic Structure

HEISENBERG UNCERTAINTY PRINCIPLE

5-2. The uncertainty in its position is ± 1 nm or $\Delta x = 2$ nm, and so the uncertainty in its momentum is given by

$$\Delta p \approx \frac{h}{4\pi\Delta x} = \frac{6.626 \times 10^{-34}\ \text{J}\cdot\text{s}}{(4\pi)(2 \times 10^{-9}\ \text{m})} = 2.6 \times 10^{-26}\ \text{kg}\cdot\text{m}\cdot\text{s}^{-1}$$

The uncertainty in the speed of the alpha particle is given by

$$\Delta v = \frac{\Delta p}{m} = \frac{2.6 \times 10^{-26}\ \text{kg}\cdot\text{m}\cdot\text{s}^{-1}}{6.65 \times 10^{-27}\ \text{kg}} = 4\ \text{m}\cdot\text{s}^{-1}$$

5-4. We must first convert the given quantities to SI units:

$$m = (2750\ \text{lb})\left(\frac{0.4536\ \text{kg}}{1\ \text{lb}}\right) = 1247\ \text{kg}$$

$$v = (85.5\ \text{miles}\cdot\text{h}^{-1})\left(\frac{1.609\ \text{km}}{1\ \text{mile}}\right)\left(\frac{10^3\ \text{m}}{1\ \text{km}}\right)\left(\frac{1\ \text{h}}{3600\ \text{s}}\right) = 38.2\ \text{m}\cdot\text{s}^{-1}$$

$$\Delta x = (5.0\ \text{ft})\left(\frac{12\ \text{in}}{1\ \text{ft}}\right)\left(\frac{2.54\ \text{cm}}{1\ \text{in}}\right)\left(\frac{1\ \text{m}}{100\ \text{cm}}\right) = 1.5\ \text{m}$$

The uncertainty in the car's momentum is given by

$$\Delta p \approx \frac{h}{4\pi\Delta x} = \frac{6.626 \times 10^{-34}\ \text{J}\cdot\text{s}}{(4\pi)(1.5\ \text{m})} = 3.5 \times 10^{-35}\ \text{kg}\cdot\text{m}\cdot\text{s}^{-1}$$

and the uncertainty in its speed is given by

$$\Delta v = \frac{\Delta p}{m} = \frac{3.5 \times 10^{-35}\ \text{kg}\cdot\text{m}\cdot\text{s}^{-1}}{1247\ \text{kg}} = 2.8 \times 10^{-38}\ \text{m}\cdot\text{s}^{-1}$$

This is a ridiculously small uncertainty.

QUANTUM NUMBERS AND ORBITALS

5-6. See the text.

5-8. (a) For $n = 3$ and $l = 1$, the orbital is a $3p$ orbital.

(b) For $n = 5$ and $l = 0$, the orbital is a $5s$ orbital.

(c) For $n = 2$ and $l = 1$, the orbital is a $2p$ orbital.

(d) For $n = 4$ and $l = 3$, the orbital is a $4f$ orbital.

5-10. (a) Allowed. (b) Not allowed, m_l cannot equal $+1$ if $l = 0$. (c) Allowed.

(d) Not allowed, l cannot equal 1 if $n = 1$. (e) Not allowed, m_s cannot equal 0.

ORBITALS AND ELECTRONS

5-12. For a $4f$ orbital, $n = 4$ and $l = 3$. The possible sets of quantum numbers are

n	l	m_l	m_s
4	3	−3	+1/2 or − 1/2
4	3	−2	+1/2 or − 1/2
4	3	−1	+1/2 or − 1/2
4	3	0	+1/2 or − 1/2
4	3	1	+1/2 or − 1/2
4	3	2	+1/2 or − 1/2
4	3	3	+1/2 or − 1/2

5-14. Note that for each value of m_l, m_s can have the values $+\frac{1}{2}$ or $-\frac{1}{2}$. Thus, the maximum number of electrons in a subshell is equal to two times the number of possible values of m_l.

				Maximum number of electrons	Total
$n = 1$	$l = 0$	$m_l = 0$	$m_s = +\frac{1}{2}, -\frac{1}{2}$	$2 \times 1 = 2$ electrons	2
$n = 2$	$l = 0$	$m_l = 0$	$m_s = +\frac{1}{2}, -\frac{1}{2}$	$2 \times 1 = 2$ electrons	
	$l = 1$	$m_l = -1, 0, 1$	$m_s = +\frac{1}{2}, -\frac{1}{2}$	$2 \times 3 = 6$ electrons	8
$n = 3$	$l = 0$	$m_l = 0$	$m_s = +\frac{1}{2}, -\frac{1}{2}$	$2 \times 1 = 2$ electrons	
	$l = 1$	$m_l = -1, 0, 1$	$m_s = +\frac{1}{2}, -\frac{1}{2}$	$2 \times 3 = 6$ electrons	
	$l = 2$	$m_l = -2, -1, 0, 1, 2$	$m_s = +\frac{1}{2}, -\frac{1}{2}$	$2 \times 5 = 10$ electrons	18
$n = 4$	$l = 0$	$m_l = 0$	$m_s = +\frac{1}{2}, -\frac{1}{2}$	$2 \times 1 = 2$ electrons	
	$l = 1$	$m_l = -1, 0, 1$	$m_s = +\frac{1}{2}, -\frac{1}{2}$	$2 \times 3 = 6$ electrons	
	$l = 2$	$m_l = -2, -1, 0, 1, 2$	$m_s = +\frac{1}{2}, -\frac{1}{2}$	$2 \times 5 = 10$ electrons	
	$l = 3$	$m_l = -3, -2, -1, 0, 1, 2, 3$	$m_s = +\frac{1}{2}, -\frac{1}{2}$	$2 \times 7 = 14$ electrons	32

Note that the maximum number of electrons with the same value of n in an atom is equal to $2n^2$.

5-16. In an f-transition series, the seven f orbitals are being filled. A set of seven f orbitals can hold up to 14 electrons.

ELECTRON CONFIGURATIONS OF ATOMS

5-18. (a) The $2s$ orbital can hold a maximum of two electrons.

(b) The $3s$ orbital can hold a maximum of two electrons; also the $2p$ orbitals must be filled before the $3s$ orbital.

(c) The $2p$ and $3p$ orbitals can each hold a maximum of six electrons.

(d) The $3d$ orbitals can hold a maximum of 10 electrons; also the $3p$ orbitals must be filled before the $4s$ and $3d$ orbitals.

5-20. We have the following:

(a) $1s^22s^22p^63s^23p^1$ — 13 electrons, aluminum
(b) $1s^22s^22p^63s^23p^64s^23d^3$ — 23 electrons, vanadium
(c) $1s^22s^22p^5$ — 9 electrons, fluorine
(d) $1s^22s^22p^63s^23p^64s^23d^{10}4p^1$ — 31 electrons, gallium
(e) $1s^22s^22p^4$ — 8 electrons, oxygen

5-22. Consult Figure 5.15 to confirm your answers.

(a) Si: $[Ne]3s^23p^2$ (b) Ni: $[Ar]4s^23d^8$ (c) Se: $[Ar]4s^23d^{10}4p^4$
(d) Cd: $[Kr]5s^24d^{10}$ (e) Mg: $[Ne]3s^2$

5-24. Consult Figure 5.15 to confirm your answers.

(a) Ba: $[Xe]6s^2$ (b) Ag: $[Kr]5s^14d^{10}$ (c) Gd: $[Xe]4f^75d^16s^2$
(d) Pd: $[Kr]4d^{10}$ (e) Sn: $[Kr]5s^24d^{10}5p^2$

5-26. The electron configuration for sulfur is

$1s$	$2s$	$2p$	$3s$	$3p$
↑↓	↑↓	↑↓ ↑↓ ↑↓	↑↓	↑↓ ↑ ↑

The electron configuration for chromium is

$1s$	$2s$	$2p$	$3s$	$3p$	$4s$	$3d$
↑↓	↑↓	↑↓ ↑↓ ↑↓	↑↓	↑↓ ↑↓ ↑↓	↑	↑ ↑ ↑ ↑ ↑

ELECTRON CONFIGURATIONS AND THE PERIODIC TABLE

5-28. The ground-state electron configurations are as follows:

(a) Y: $[Kr]5s^24d^1$ (b) Po: $[Xe]6s^24f^{14}5d^{10}6p^4$ (c) Co: $[Ar]4s^23d^7$
(d) Es: $[Rn]7s^25f^{11}$ (e) Pb: $[Xe]6s^24f^{14}5d^{10}6p^2$

5-30. (a) The first inner transition metal series to begin to fill f orbitals is the lanthanide series; thus the element is lanthanum, La. Note, however, that La is an exception and is actually a $5d^1$ metal. The element with the lowest atomic number whose ground state contains an f electron is cerium, Ce.

(b) The elements in the fourth row of the periodic table fill the $3d$ orbitals. Thus, the element is the third member of the transition metal series, which is vanadium, V.

(c) The next to last element in the $3d$ transition metals series has a filled d orbital; the element is copper, Cu. (Its outer configuration is $4s^13d^{10}$.)

(d) Elements in the third row of the periodic table have a filled $2p$ orbital. The element that has four electrons in the $3p$ orbital is sulfur, S.

5-32. We write the ground-state electron configurations and apply Hund's rule to each case.

(a) Al: $[\mathrm{Ne}]3s^2 3p^1$; one unpaired electron

(b) Cr: $[\mathrm{Ar}]4s^1 3d^5$; six unpaired electrons

(c) S: $[\mathrm{Ne}]3s^2 3p^4$; two unpaired electrons

(d) Hg: $[\mathrm{Xe}]6s^2 4f^{14} 5d^{10}$; no unpaired electrons

GROUND-STATE ELECTRON CONFIGURATIONS OF IONS

5-34. In the formation of positive ions, the corresponding noble-gas core is that for the noble gas in the preceding row of the periodic table.

(a) 2 electrons; Ca^{2+}, argon

(b) 1 electron, Li^+, helium

(c) 1 electron, Na^+, neon

(d) 2 electrons, Mg^{2+}, neon

5-36. First we determine the number of electrons in the ion, and then we write the ground-state electron configuration.

(a) O^+: $8 - 1 = 7$ electrons: $1s^2 2s^2 2p^3$ isoelectronic with nitrogen

(b) C^-: $6 + 1 = 7$ electrons: $1s^2 2s^2 2p^3$ isoelectronic with nitrogen

(c) F^+: $9 - 1 = 8$ electrons: $1s^2 2s^2 2p^4$ isoelectronic with oxygen

(d) O^{2+}: $8 - 2 = 6$ electrons: $1s^2 2s^2 2p^2$ isoelectronic with carbon

5-38. We have the following:

(a) O^{2-}: [Ne], 0 unpaired electrons

(b) Ca^+: $[\mathrm{Ar}]4s^1$, 1 unpaired electron

(c) He^+: $1s^1$, 1 unpaired electron

(d) Pb^{2+}: $[\mathrm{Xe}]5d^{10}6s^2$, 0 unpaired electrons

(e) N^{3-}: [Ne], 0 unpaired electrons

5-40. (a) Ne; (b) K; (c) H; (d) Hg; (e) Ne

5-42. In this problem we work out the electronic configurations of the reactants and the products.

(a) $\mathrm{I}(g) + \mathrm{e}^- \longrightarrow \mathrm{I}^-(g)$

$$[\mathrm{Kr}]5s^2 4d^{10} 5p^5 + \mathrm{e}^- \longrightarrow [\mathrm{Kr}]5s^2 4d^{10} 5p^6 \text{ or } [\mathrm{Xe}]$$

(b) $\mathrm{K}(g) + \mathrm{F}(g) \longrightarrow \mathrm{K}^+(g) + \mathrm{F}^-(g)$

$$[\mathrm{Ar}]4s^1 + [\mathrm{He}]2s^2 2p^5 \longrightarrow [\mathrm{Ar}] + [\mathrm{He}]2s^2 2p^6 \text{ or } [\mathrm{Ne}]$$

EXCITED-STATE ELECTRON CONFIGURATIONS

5-44. Count the total number of electrons and then determine the ground-state electron configuration. Compare your result with the configuration given.

(a) excited state of Ne (10 electrons)

(b) excited state of Na (11 electrons)

(c) excited state of He (2 electrons)

(d) ground state of K (19 electrons)

VALENCE ELECTRONS

5-46. We have the following:

(a) 0 valence electrons He (b) 8 valence electrons $:\ddot{\underset{..}{N}}:^{3-}$

(c) 7 valence electrons $\cdot\ddot{\underset{..}{F}}\cdot^{+}$ (d) 1 valence electrons Na·

(e) 0 valence electrons K^+

ATOMIC RADII

5-48. Recall that atomic radii increase as we move down a column of the periodic table, whereas atomic radii decrease as we move from left to right across a row of the periodic table.

(a) O > F; same row, but oxygen has the smaller nuclear charge

(b) Xe > Kr; xenon has more electronic shells than krypton

(c) Cl > F; chlorine has more electronic shells than fluorine

(d) Ca > Mg; calcium has more electronic shells than magnesium

5-50. Recall that atomic radii increase as we move down a column of the periodic table, whereas atomic radii decrease as we move from left to right across a row of the periodic table.

(a) Li < Na < Rb < Cs (b) P < Al < Mg < Na (c) Mg < Ca < Sr < Ba

5-52. $Mg^{2+} < Na^+ < F^- < O^{2-} < N^{3-}$. These ions are all isoelectronic with Ne, so the size depends upon only their respective ionic charge.

ADDITIONAL PROBLEMS

5-54. The term Δx is the uncertainty associated with a particular measurement of position and Δp is the uncertainty in the momentum of an object. If we know the momentum of an electron with 100% certainty, that is, $\Delta p = 0$, then the uncertainty in its position is infinity, or $\Delta x = \infty$; that is, we have no idea where the electron is.

5-56. Recall that ionization energies increase as we move from left to right across the periodic table and they decrease as we move down a group.

(a) B < O < F < Ne (b) Sn < Te < I < Xe

(c) Cs < Rb < K < Ca (d) Na < Al < S < Ar

5-58. The electrons in the outermost occupied shell (those beyond the noble-gas core) of a neutral atom or a monatomic ion are called valence electrons.

5-60. (a) We see that the $3d$ orbitals are partially occupied, so that the element must be a $3d$ transition metal. There are eight electrons in the $3d$ orbitals; thus the element is the eighth member of the series, or nickel.

(b) The outer electron configuration is $5s^1 4d^{10}$, which corresponds to that of the ninth member of the $4d$ transition metal series, or silver.

(c) The outer electron configuration is $3s^2 3p^4$, which corresponds to that of the member of Group 16 located in the third row of the periodic table, or sulfur.

(d) The outer electron configuration is $6s^2 5d^{10} 6p^2$, which corresponds to that of the second element after the $5d$ transition metal series, or lead.

5-62. We refer to the periodic table and find that

(a) $3s^2 3p^1$—third row, first *p*-block element, which is aluminum

(b) $2s^2 2p^4$—second row, fourth *p*-block element, which is oxygen

(c) $4s^2 3d^{10}$—fourth row, last transition metal, which is zinc

(d) $4s^2 3d^{10} 4p^6$—fourth row, sixth *p*-block element, which is krypton

5-64. See Figure 5.15.

(a) ground state of V (b) excited state of He (c) excited state of Ca

(d) ground state of Mo (e) excited state of La

5-66. We first determine the total number of electrons in the ion, then we determine the ground-state electron configuration, and finally we count the number of unpaired electrons using Hund's rule.

	Species	Ground state configuration	Number of unpaired electrons	Isoelectronic neutral atom
(a)	F^+	$[He]2s^2 2p^4$	2	O
(b)	Sn^{2+}	$[Kr]5s^2 4d^{10}$	0	Cd
(c)	Bi^{3+}	$[Xe]6s^2 4f^{14} 5d^{10}$	0	Hg
(d)	Ar^+	$[Ne]3s^2 3p^5$	1	Cl

5-68. If the 2*s* and 2*p* orbitals had the same energy, the electrons would half fill all the degenerate orbitals before the pairing of any electrons occurs, yielding the ground-state outer electron configurations of the second-row elements

$$2s^1,\ 2s^1 2p_x^1,\ 2s^1 2p_x^1 2p_y^1,\ 2s^1 2p_x^1 2p_y^1 2p_z^1,\ 2s^2 2p_x^1 2p_y^1 2p_z^1,\ 2s^2 2p_x^2 2p_y^1 2p_z^1,$$
$$2s^2 2p_x^2 2p_y^2 2p_z^1,\ 2s^2 2p_x^2 2p_y^2 2p_z^2$$

5-70.* The ground-state electron configurations of the elements beryllium and boron are:

$$\text{Be: [He]}2s^2 \qquad \text{and} \qquad \text{B: [He]}2s^2 2p^1$$

Because the energy of the 2*p* orbital is slightly greater than that of the 2*s* orbital, it is easier to remove an electron from a 2*p* orbital than from a 2*s* orbital. Even though the magnitude of the nuclear charge increases in going from beryllium to boron, due to shielding of the nucleus by the inner electrons and the average increased distance from the nucleus of the 2*p* electrons compared with the 2*s* electrons, it is slightly easier to remove a 2*p* electron from a boron atom than a 2*s* electron from a beryllium atom.

The ground-state electron configurations of the elements oxygen and nitrogen are

$$\text{O: [He]}2s^2 2p^3 \qquad \text{and} \qquad \text{N: [He]}2s^2 2p^4$$

We see that an oxygen atom has three unpaired 2*p* electrons, whereas two of the 2*p* electrons in a nitrogen atom are paired. Because the pairing of electrons requires energy, the removal of a paired electron requires less energy than the removal of an unpaired electron. Even though the magnitude of the nuclear charge increases in going from oxygen to nitrogen, this pairing energy in nitrogen is greater than the additional effective nuclear charge and so it is easier to remove a paired 2*p* electron from a nitrogen atom than an unpaired 2*p* electron from an oxygen atom.

CHAPTER 6. Ionic Bonds and Compounds

NOMENCLATURE AND CHEMICAL FORMULAS

6-2. We first determine the number of electrons in the ion and then compare the result with the Z value for the noble gas.

(a) yes; Xe (b) no (c) yes; Kr (d) yes; Ne (e) yes; Kr (f) no

6-4. We determine the charges on the atomic ions in ionic compounds using Figure 6.3.

(a) Li^+ and O^{2-} lithium oxide (b) Ca^{2+} and S^{2-} calcium sulfide

(c) Mg^{2+} and N^{3-} magnesium nitride (d) Al^{3+} and S^{2-} aluminum sulfide

6-6. The chemical formula and systematic name of each is

(a) Ga_2S_3 gallium sulfide (b) Fe_2Se_3 iron(III) selenide (c) PbO_2 lead(IV) oxide

(d) $BaAt_2$ barium astatide (e) Zn_3N_2 zinc nitride

6-8. To determine the chemical formula of an ionic compound, we use the procedure outlined in Example 6-4.

(a) Al_2S_3 (b) Na_2O (c) BaF_2 (d) LiH

6-10. (a) Cs_2O (b) Na_2Se (c) Li_2S (d) CaI_2

6-12. We designate the charge on the transition-metal ion by a Roman numeral in the name of the compound.

(a) tin(IV) oxide (b) iron(III) fluoride (c) lead(IV) oxide

(d) cobalt(III) nitride (e) mercury(II) selenide.

6-14. The systematic names are

(a) sodium hydride (b) tin(II) iodide (c) gold(I) sulfide

(d) cadmium sulfide (e) potassium oxide

6-16. To determine the chemical formula of an ionic compound formed from a positive ion and a negative ion, we use the fact that the compound has no net charge.

(a) Ru_2S_3 (b) ScF_3 (c) OsO_4 (d) MnS (e) $PtCl_4$

6-18. (a) $TlCl_3$ (b) CdI_2 (c) Zn_3As_2 (d) $LuBr_3$ (e) Ga_2O_3

6-20. The older names are

(a) stannic oxide (b) ferric fluoride (c) plumbic oxide

(d) cobaltic nitride (e) mercuric selenide

6-22. The chemical formula and systematic name of each is as follows

(a) PbF_4 lead(IV) fluoride (b) Hg_2Cl_2 mercury(I) chloride

(c) PbS lead(II) sulfide (d) HgO mercury(II) oxide

6-24. (a) $2\,CO(g) + O_2(g) \longrightarrow 2\,CO_2(g)$

(b) $2\,Cs(s) + Br_2(l) \longrightarrow 2\,CsBr(s)$

(c) $2\,NO(g) + O_2(g) \longrightarrow 2\,NO_2(g)$

(d) $4\,NH_3(g) + 5\,O_2(g) \longrightarrow 4\,NO(g) + 6\,H_2O(l)$

ELECTRON CONFIGURATIONS AND CHEMICAL REACTIONS

6-26. In order to emphasize the attainment of noble-gas configurations for ions, we write the electron configurations in terms of noble-gas configurations.

(a) $Ga([Ar]4s^23d^{10}4p^1) + 3\,F([He]2s^22p^5) \longrightarrow Ga^{3+}([Ar]3d^{10}) + 3\,F^-([Ne]) \longrightarrow GaF_3(g)$

(b) $Ag([Kr]5s^14d^{10}) + Cl([Ne]3s^23p^5) \longrightarrow Ag^+([Kr]4d^{10}) + Cl^-([Ar]) \longrightarrow AgCl(g)$

(c) $3\,Li([He]2s^1) + N([He]2s^22p^3) \longrightarrow 3\,Li^+([He]) + N^{3-}([Ne]) \longrightarrow Li_3N(g)$

6-28. (a) $3\ \cdot\text{Ca}\cdot + 2\ \cdot\ddot{\text{N}}\cdot \longrightarrow \underbrace{3\,\text{Ca}^{2+} + 2\ :\ddot{\underset{\cdot\cdot}{\text{N}}}:^{3-}}_{\text{Ca}_3\text{N}_2}$

(b) $\cdot\dot{\text{Al}}\cdot + 3\ :\ddot{\underset{\cdot\cdot}{\text{Cl}}}\cdot \longrightarrow \underbrace{\text{Al}^{3+} + 3\ :\ddot{\underset{\cdot\cdot}{\text{Cl}}}:^{-}}_{\text{AlCl}_3}$

(c) $2\,\text{Li}\cdot + \cdot\ddot{\underset{\cdot\cdot}{\text{O}}}\cdot \longrightarrow \underbrace{2\,\text{Li}^+ + :\ddot{\underset{\cdot\cdot}{\text{O}}}:^{2-}}_{\text{Li}_2\text{O}}$

ELECTRON CONFIGURATIONS OF IONS

6-30. Recall that for neutral atoms, the $(n+1)s$ orbitals are of lower energy than the nd orbitals, for example $4s < 3d$. However, for ions, the reverse is true; that is, $4s > 3d$. Thus, we first determine the number of electrons in the ion and then write the electron configuration filling the orbitals in the order $1s$, $2s$, $2p$, $3s$, $3p$, $4s$, and so forth.

(a) $Ru^{2+}([Kr]4d^6)$ (b) $W^{3+}([Xe]4f^{14}5d^3)$ (c) $Pd^{2+}([Kr]4d^8)$ (d) $Ti^+([Ar]3d^3)$

6-32. (a) $Fe^{2+}([Ar]3d^6)$, 6 d electrons

(b) $Zn^{2+}([Ar]3d^{10})$, 10 d electrons

(c) $V^{2+}([Ar]3d^3)$, 3 d electrons

(d) $Ni^{2+}([Ar]3d^8)$, 8 d electrons

For d transition series +2 ions, the number of d electrons is equal to the position of the metal in the series found by counting off from the start of the series.

6-34. For $n = 3$, an 18-electron outer configuration is one of the type $3s^2 3p^6 3d^{10}$, and the corresponding noble gas is argon.

(a) $Cu^+(3s^2 3p^6 3d^{10})$ or $Cu^+([Ar]3d^{10})$

(b) $Ga^{3+}(3s^2 3p^6 3d^{10})$ or $Ga^{3+}([Ar]3d^{10})$

In this case, $n = 5$ and the corresponding 18-electron outer configuration is of the type $(n-1)f^{14} ns^2 np^6 nd^{10}$ or $4f^{14} 5s^2 5p^6 5d^{10}$. The corresponding noble gas is Xe:

(c) $Hg^{2+}(4f^{10} 5s^2 5p^6 5d^{10})$ or $Hg^{2+}([Xe]5d^{10})$

(d) $Au^+(4f^{10} 5s^2 5p^6 5d^{10})$ or $Au^+([Xe]4f^{14} 5d^{10})$

6-36. We simply determine the number of electrons in the positive ion and in the negative ion. If the numbers are the same, then the ions are isoelectronic.

(a) $Na^+(10)$, $Cl^-(18)$; not isoelectronic

(b) $Rb^+(36)$, $Br^-(36)$; isoelectronic

(c) $Sr^{2+}(36)$, $Cl^-(18)$; not isoelectronic

(d) $SrBr_2$; isoelectronic ions, each has 36 electrons

(e) MgF_2; isoelectronic ions, each has 10 electrons

(f) $K^+(18)$, $I^-(54)$; not isoelectronic ions

IONIC RADII

6-38. (a) H A positive ion is smaller than the parent atom.

(b) Fe^{2+} The higher positively charged ion is smaller because of the larger nuclear attraction for the same number of electrons.

(c) S^{2-} A negative ion is larger than the parent atom.

(d) O^{2-} The higher negatively charged ion is larger because of the greater repulsion of the outer electrons.

6-40. (a) Cl^- A negative ion is larger than a positive ion with the same number of electrons.

(b) Au^+ The higher positively charged ion is smaller because of the larger nuclear attraction for the same number of electrons.

(c) Cr^+ Same reason as (b).

(d) P^{3-} The higher the negative charge for the same number of electrons, the larger the ion is because the electrons repel each other to a greater extent.

6-42. These ions are all isoelectronic. Thus, the relative sizes are

$$Mo^{6+} < Y^{3+} < Rb^+ < Br^- < Se^{2-}$$

IONIZATION ENERGIES AND ELECTRON AFFINITIES

6-44. See Figure 4.1 for ionization energies. The lower the ionization energy of an atom, the easier it is to remove an electron.

$$K > Na > B > H > He$$

6-46. The magnitude of the electron affinity of a chlorine atom is greater that that of a sulfur atom because a chlorine atom completely fills its outer shell with the addition of one electron.

6-48. (a) $Na(g) + H(g) \longrightarrow Na^+(g) + H^-(g) \quad E_{rxn} = 0.83\text{ aJ} - 0.12\text{ aJ} = 0.71\text{ aJ}$

(b) $Mg(g) + O(g) \longrightarrow Mg^{2+}(g) + O^{2-}(g)$

$$\begin{aligned} E_{rxn} &= \text{sum of the first two ionization energies of Mg}(g) \\ &\quad + \text{the sum of the first two electron affinities of O}(g) \\ &= (1.23 + 2.41)\text{ aJ} + (-0.234 + 1.30)\text{ aJ} = 4.71\text{ aJ} \end{aligned}$$

(c) $Mg(g) + 2\,Br(g) \longrightarrow Mg^{2+}(g) + 2\,Br^-(g)$

$$\begin{aligned} E_{rxn} &= \text{the sum of the first two ionization energies of Mg}(g) \\ &\quad + 2 \times \text{the electron affinitiy of Br}(g) \\ &= (1.23 + 2.41)\text{ aJ} + (2)(-0.540\text{ aJ}) = 2.56\text{ aJ} \end{aligned}$$

CALCULATIONS INVOLVING COULOMB'S LAW

6-50. The equation for Coulomb's law is

$$E_{\text{coulomb}} = \frac{(231\text{ aJ}\cdot\text{pm})(Q_1 Q_2)}{d}$$

From Table 6.4, the radius of a Na^+ ion is 102 picometers and the radius of a F^- ion is 133 picometers, and $Q_1 = +1$, and $Q_2 = -1$. Thus, we have

$$E_{\text{coulomb}} = \frac{(231\text{ aJ}\cdot\text{pm})(1)(-1)}{102\text{ pm} + 133\text{ pm}} = -0.983\text{ aJ}$$

6-52. (a) For the ionization of Mg, we have

$$\begin{aligned} Mg(g) &\longrightarrow Mg^+(g) + e^- \qquad I_1 = 1.23\text{ aJ} \\ Mg^+(g) &\longrightarrow Mg^{2+}(g) + e^- \qquad I_2 = 2.41\text{ aJ} \end{aligned}$$

Adding the above two equations, we obtain

$$Mg(g) \longrightarrow Mg^{2+}(g) + 2\,e^- \qquad I = 3.64\text{ aJ}$$

(b) For the addition of two electrons to O (Table 6.5),

$$\begin{aligned} O(g) + e^- &\longrightarrow O^-(g) \qquad EA_1 = -0.234\text{ aJ} \\ O^-(g) + e^- &\longrightarrow O^{2-}(g) \qquad EA_2 = 1.30\text{ aJ} \end{aligned}$$

Adding the above two equations, we obtain

$$O(g) + 2\,e^- \longrightarrow O^{2-}(g) \qquad EA = 1.07\text{ aJ}$$

If we add steps a and b, then we have

$$Mg(g) + O(g) \longrightarrow Mg^{2+}(g) + O^{2-}(g) \qquad E_{(1+2)} = 3.64\text{ aJ} + 1.07\text{ aJ} = 4.71\text{ aJ}$$

(c) We now bring the Mg^{2+} ion and the O^{2-} ion to their ion-pair separation distance. According to Table 6.4, the radius of a Mg^{2+} ion is 72 pm and the radius of a O^{2-} ion is 140 pm. Their separation as an ion pair is 212 pm. We now use Coulomb's law to calculate the energy released by the formation of one ion pair from the separated ions:

$$\mathrm{Mg^{2+}}(g) + \mathrm{O^{2-}}(g) \longrightarrow \mathrm{Mg^{2+}O^{2-}}(g)$$

$$E_{\text{coulomb}} = \frac{(231\ \text{aJ}\cdot\text{pm})(Q_1 Q_2)}{d} = \frac{(231\ \text{aJ}\cdot\text{pm})(+2)(-2)}{212\ \text{pm}} = -4.36\ \text{aJ}$$

Thus, we have

$$\mathrm{Mg^{2+}}(g) + \mathrm{O^{2-}}(g) \longrightarrow \mathrm{Mg^{2+}O^{2-}}(g) \qquad E_{\text{coulomb}} = -4.36\ \text{aJ}$$
$$d = 212\ \text{pm}$$

We combine this result with the result

$$\mathrm{Mg}(g) + \mathrm{O}(g) \longrightarrow \mathrm{Mg^{2+}}(g) + \mathrm{O^{2-}}(g) \qquad E_{(1+2)} = 4.71\ \text{aJ}$$

to obtain the final result

$$\mathrm{Mg^{2+}}(g) + \mathrm{O^{2-}}(g) \longrightarrow \mathrm{Mg^{2+}O^{2-}}(g) \qquad E_{\text{rxn}} = 0.35\ \text{aJ}$$
$$d = 212\ \text{pm}$$

6-54. The energy involved in forming the separated ions is given by

$$\begin{aligned} E_{(1+2)} &= \text{the sum of the first two ionization energies of Zn}(g) \\ &\quad - \text{the sum of the first two electron affinities of S}(g) \\ &= 4.38\ \text{aJ} + (-0.332\ \text{aJ} + 0.980\ \text{aJ}) = 5.03\ \text{aJ} \end{aligned}$$

The energy involved in bringing the two ions together (Table 6.4) is given by Coulomb's law

$$E_{\text{coulomb}} = \frac{(231\ \text{aJ}\cdot\text{pm})(Q_1 Q_2)}{d} = \frac{(231\ \text{aJ}\cdot\text{pm})(+2)(-2)}{74\ \text{pm} + 184\ \text{pm}} = -3.58\ \text{aJ}$$

The total energy is

$$E_{\text{rxn}} = 5.03\ \text{aJ} - 3.58\ \text{aJ} = 1.45\ \text{aJ}$$

6-56. The diagram is shown below.

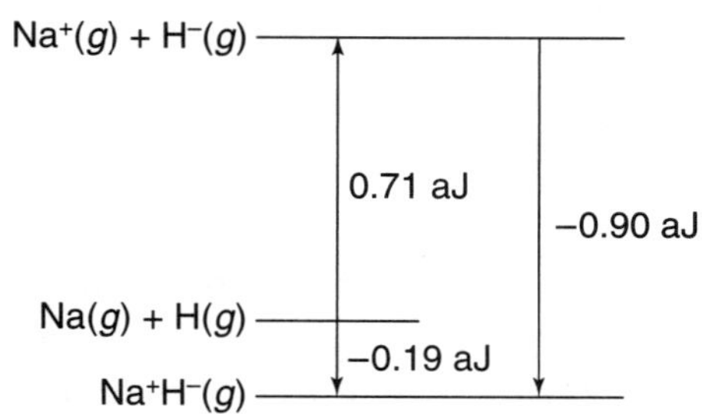

ADDITIONAL PROBLEMS

6-58. Calcium belongs to Group 2 of the periodic table, whose outer electron configuration is ns^2. By losing two electrons, Group 2 elements can achieve a noble-gas-like electron configuration. Thus, calcium (and all the Group 2 elements) forms a stable +2 cation.
Fluorine belongs to Group 17 of the periodic table, whose outer electron configuration is ns^2np^5. By gaining one electron, Group 17 elements can achieve a noble-gas-like electron configuration. Thus, fluorine (and all the Group 17 elements) forms a stable −1 anion.

6-60. Cations have fewer electrons than their neutral parent atoms, so that the electrostatic attraction between the nucleus and the electrons is greater in the cations than the neutral parent atoms. The greater nuclear-electron attraction leads to a smaller cation. Anions have more electrons than their neutral parent atoms so the the electrostatic attraction between the nucleus and the electrons is less in the anions than the neutral parent atoms. The lower nuclear-electron attraction leads to a larger anion than the parent atom.

6-62. (a) The ground-state electron configuration of a lutetium atom is $[Xe]6s^24f^{14}5d^1$, and so we predict that the ionic charge of a lutetium ion is +3 ($[Xe]4f^{14}$).

(b) The ground-state electron configuration of a lawrencium atom is $[Rn]7s^25f^{14}6d^1$, and so we predict that the ionic charge of a lawrencium ion is +3 ($[Rn]5f^{14}$).

6-64. (a) You do not need the II because +2 is the only common oxidation state of zinc. The correct name is zinc oxide.

(b) You do not need to say monohydride because NaH is the only hydride of sodium. The correct name is sodium hydride.

(c) Stannous(II) is a hybrid of systematic and older nomenclature. The correct name is tin(II) nitride.

(d) You do not need to say dichloride because +2 is the only oxidation state of cadmium. The correct name is cadmium chloride.

(e) The III designation dictates the oxidation state of iron, so the tri is redundant. The correct name is iron(III) chloride.

6-66. (a) The +2 transition metal ions that are d^{10} ions are Zn^{2+}, Cd^{2+}, and Hg^{2+}.

(b) The +4 transition metal ions that are d^0 ions are those for which the +2 ions are d^2 (Ti^{2+}, Zr^{2+}, Hf^{2+}). Thus, we see that Ti^{4+}, Zr^{4+}, Hf^{4+} are d^0 ions.

6-68. All the outer electrons of K^+ and Cu^+ are in the $n = 3$ shell, but the nucleus of a copper atom has a greater charge than that of a potassium atom and thus a greater attraction for the electrons.

6-70. The anions are P^{3-}, S^{2-}, and Cl^-. For the same row of the periodic table, the ion with the greatest negative charge will have the largest radius because the extra electrons will cause both a greater electron-electron repulsion and a greater electron cloud expansion. Thus, the smallest ion is the chloride ion because it has the smallest negative charge.

6-72. (a) $LiCl(s)$ because chloride ions are smaller than bromide ions.

(b) $MgO(s)$ because the ionic charges are +2 and −2.

(c) $CaO(s)$ because the ionic charges are +2 and −2.

6-74. The fluoride ion is so small that the coulombic force between the fluoride ion and a metal ion is greater than between any other anion and the metal ion.

6-76.* The bond energy of this ionic species is given by

$$E_{\text{ionic}} = 2\, E_{\text{ionic}}(\text{two Ba}^{2+}\text{Cl}^{-} \text{ pair energies}) + E_{\text{ionic}}(\text{Cl}^{-}\ \text{Cl}^{-} \text{ pair energy})$$

$$= (2)\left[\frac{(231\ \text{aJ}\cdot\text{pm})(+2)(-1)}{(135+181)\ \text{pm}}\right] + \frac{(231\ \text{aJ}\cdot\text{pm})(-1)(-1)}{(2)(135+181)\ \text{pm}}$$

$$= -2.92\ \text{aJ} + 0.36\ \text{aJ} = -2.55\ \text{aJ}$$

6-78.* Choosing the sodium ions, for example, we see from Figure 6.13 that while each sodium ion is surrounded by six chloride anions, which must necessarily increase the magnitude of the total lattice energy, there are twelve next-nearest-neighbor positively charged sodium ions that likewise decrease the total lattice energy. It is the sum of these alternating positive and negative terms over successive neighbors that determines the value of the lattice energy and not just the terms due to the six nearest neighbors to each ion.

CHAPTER 7. Lewis Formulas

COMPOUNDS INVOLVING SINGLE BONDS

7-2. Because there is no unique central atom in the molecule, we shall assume that the two nitrogen atoms are bonded together and that the fluorine atoms are attached to the two nitrogen atoms.

F N N F

F F

The total number of valence electrons is $(2 \times 5) + (4 \times 7) = 38$. We use two valence electrons to form the N–N bond and eight valence electrons to form the N–F bonds. We place 24 valence electrons as lone pairs on the F atoms and the remaining four as a lone pairs on each of the two N atoms. The Lewis formula is

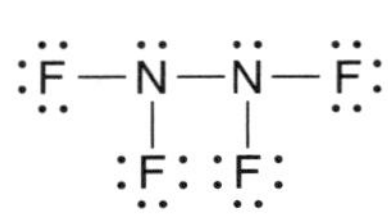

7-4. (a) We shall assume that the phosphorous atom is central and that each bromine atom is attached to it. The total number of valence electrons is $(1 \times 5) + (3 \times 7) = 26$. We use six of the valence electrons to form the P–Br bonds. We place 18 of the valence electrons as lone pairs on the three bromine atoms and the remaining two as a lone pair on the P atom. The Lewis formula is

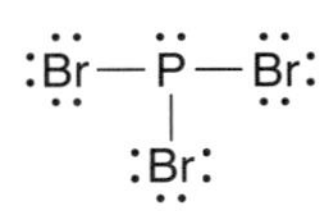

(b) Because silicon is the unique atom in this molecule, we shall assume that it is central and that each fluorine atom is attached to it. The total number of valence electrons is $(1 \times 4) + (4 \times 7) = 32$. We use eight of the valence electrons to form Si–F bonds, which satisfies the octet rule about the Si atom. We place the remaining 24 as lone pairs on the F atoms. The completed Lewis formula is

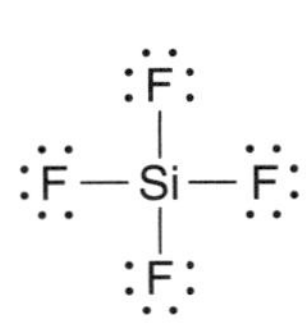

(c) We shall assume that the nitrogen atom is central and that each iodine atom is attached to it. The total number of valence electrons is $(1 \times 5) + (3 \times 7) = 26$. We use six of the valence electrons to form the N–I bonds. We place 18 of the valence electrons as lone pairs on the three iodine atoms and the remaining two as a lone pair on the N atom. The Lewis formula is

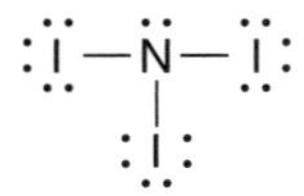

(d) The total number of valence electrons is $(1 \times 6) + (2 \times 1) = 8$. We use four of the valence electrons to form the Se–H bonds. We place the remaining four valence electrons as two lone pairs on the Se atom. The Lewis formula is

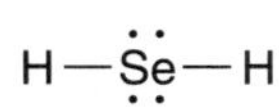

7-6. (a) Six of the 14 valence electrons are used to form the C–H bonds, two to form the C–S bond, two to form the S–H bond, and the four remaining valence electrons are placed as lone pairs on the S atom.

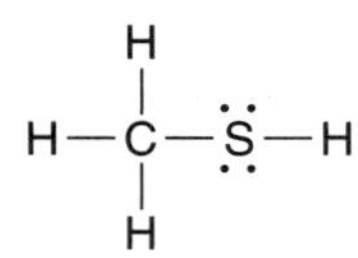

(b) We shall assume that the unique oxygen atom is central and that each carbon atom is attached to it. The hydrogen atoms must be terminal atoms, and so we attach these to the two carbon atoms. Twelve of the 20 valence electrons are used to form the C–H bonds, four to form the C–O bonds, and the four remaining valence electrons are placed as lone pairs on the O atom. The Lewis formula is

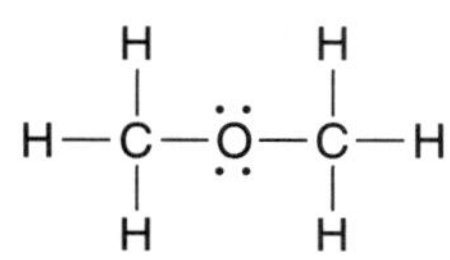

(c) We shall assume that the unique nitrogen atom is central and that each carbon atom is attached to it. The hydrogen atoms must be terminal atoms, and so we attach these to the three carbon atoms. Eighteen of the 26 valence electrons are used to form the C–H bonds, six to form the C–N bonds, and the two remaining valence electrons are placed as a lone pair on the N atom. The Lewis formula is

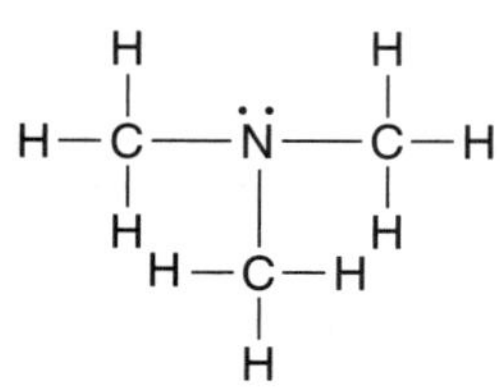

MULTIPLE BONDS

7-8. (a) The hydrogen atom must be terminal and the nitrogen atom is unique, so we arrange the atoms as

O N O H

There is a total of $(2 \times 6) + (1 \times 5) + (1 \times 1) = 18$ valence electrons. If we add one bond between each O atom and the N atom and one between the O atom and the H atom, we cannot satisfy the octet rule around each atom. Thus we add one more bond between the terminal O atom and the N atom. We place the remaining ten valence electrons as lone pairs to satisfy the octet rule about each atom. The Lewis formula is

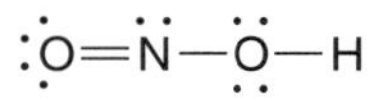

Notice that each atom in this formula has a zero formal charge, which would not be the case had we connected the H atom directly to the N atom in the molecule.

(b) Make the unique silicon atom central. There is a total of $(2 \times 6) + (1 \times 4) = 16$ valence electrons. If we add one bond between each atom and then try to satisfy the octet rule about each of the atoms, we find that we are four electrons short. Thus we add two more bonds between the silicon and oxygen atoms and obtain

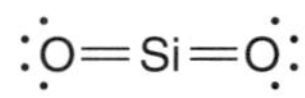

(c) We arrange the atoms as suggested by the chemical formula. There is a total of 18 valence electrons. If we add one bond between each of the atoms, we cannot satisfy the octet rule around each atom. Thus we add one more bond between two of the carbon atoms to obtain

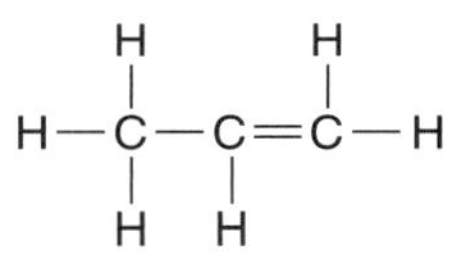

(d) Because there is no unique central atom in the molecule, we shall assume that the two carbon atoms are bonded together and that the chlorine atoms are attached to the two carbon atoms as suggested by the chemical formula. The total number of valence electrons is 36. If we add four bonds between the carbon and chlorine atoms, and one bond between the two carbon atoms, we cannot satisfy the octet rule on each atom. Thus we add one more bond between the two carbon atoms to obtain

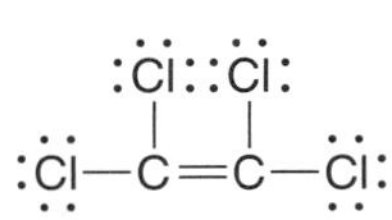

Notice that each atom in this formula has a zero formal charge, which would not be the case had we located the double bond anywhere else in the formula.

7-10. We arrange the three nitrogen atoms in a row. There are $(3 \times 5) + 1 = 16$ valence electrons. If we add one bond between each atom and then try to satisfy the octet rule about the three nitrogen atoms, we find that we are four electrons short. Thus we add two more bonds between the nitrogen atoms and obtain

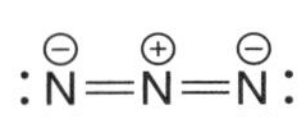

7-12. We arrange the atoms as suggested by the chemical formula. There is a total of 24 valence electrons, or 12 electron pairs in an acetone molecule. The only way to satisfy the octet rule for

the carbon and oxygen atoms by using 12 electron pairs is to write

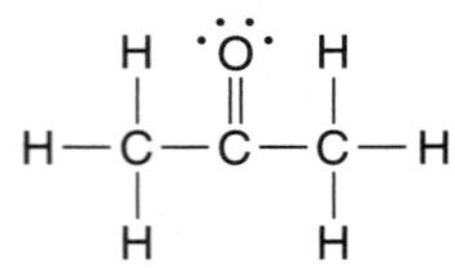

FORMAL CHARGE

7-14. The Lewis formula for the arrangement NOCl is

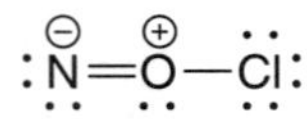

where we have assigned formal charges according to Equation 7.1

$$\text{formal charge on N} = 5 - 4 - \frac{1}{2}(4) = -1$$

$$\text{formal charge on O} = 6 - 2 - \frac{1}{2}(6) = +1$$

$$\text{formal charge on Cl} = 7 - 6 - \frac{1}{2}(2) = 0$$

The Lewis formula for the arrangement NClO is

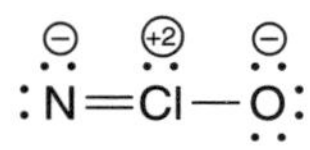

where we have assigned formal charges according to Equation 7.1

$$\text{formal charge on N} = 5 - 4 - \frac{1}{2}(4) = -1$$

$$\text{formal charge on O} = 6 - 6 - \frac{1}{2}(2) = -1$$

$$\text{formal charge on Cl} = 7 - 2 - \frac{1}{2}(6) = +2$$

The Lewis formula for the arrangement ONCl has lowest formal charge (all zeros)

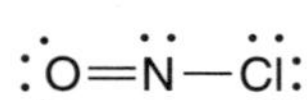

Thus we predict that the arrangement ONCl is the most likely.

7-16. The Lewis formula for the arrangement N–O–O is

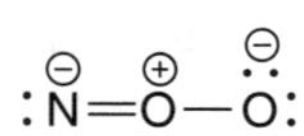

where we have assigned formal charges according to Equation 7.1. The arrangement where nitrogen is the central atom has the Lewis formulas

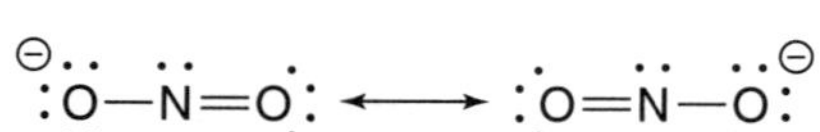

where we have assigned formal charges to each of the atoms in the two resonance formulas in accordance with Equation 7.1. Thus we see that the Lewis formula with the nitrogen atom in the center is the preferred formula of the ion.

RESONANCE

7-18. There are 24 valence electrons, or 12 electron pairs in a CH_3COO^- ion. Two resonance forms are

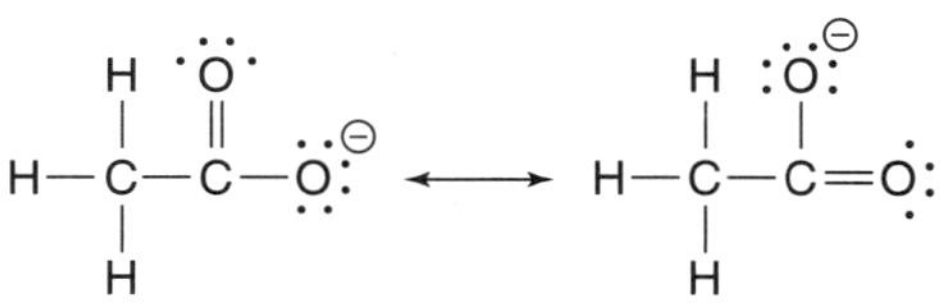

where we have assigned formal charges according to Equation 7.1. The superposition of these two resonance forms gives the resonance hybrid

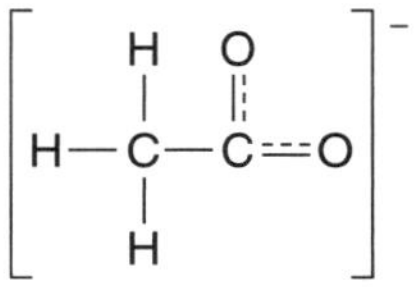

The two carbon–oxygen bonds in a CH_3COO^- ion are equivalent; they have the same bond length and the same bond energy.

7-20. There are 18 valence electrons, or nine electron pairs in an ozone molecule. The two resonance forms are

The superposition of these two resonance forms gives the resonance hybrid

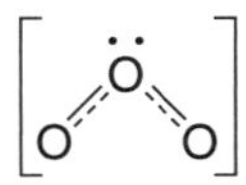

Both oxygen–oxygen bonds in an ozone molecule are equivalent; they have the same bond length and the same bond energy.

7-22. The two resonance forms of a naphthalene molecule are

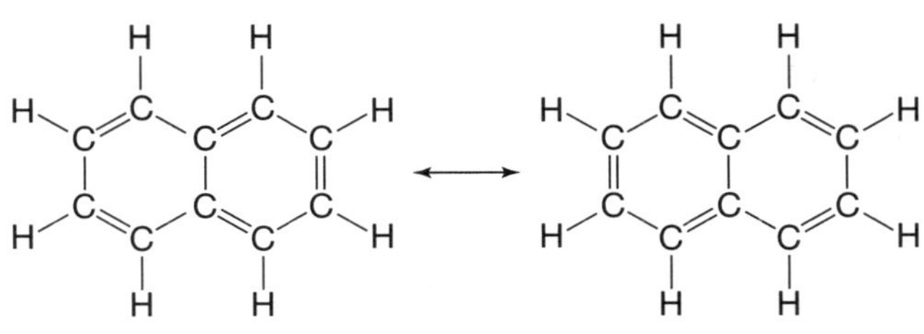

or simply,

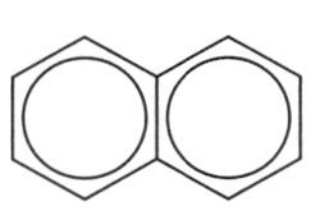

using the abbreviation for the superimposed formula of benzene.

FREE RADICALS

7-24. (a) A BrO_3 molecule contains a total of 59 electrons and 25 valence electrons. The Lewis formulas for a BrO_3 molecule are

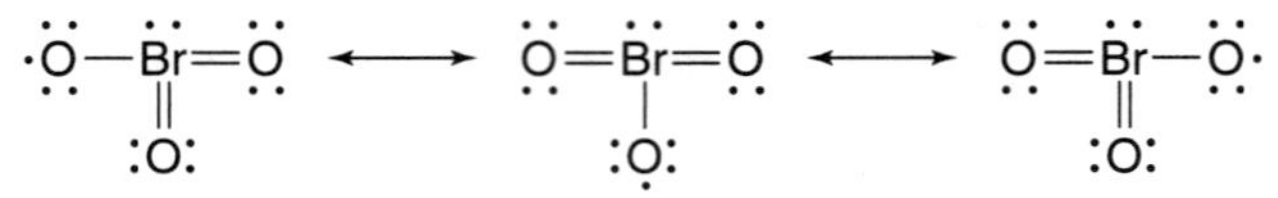

odd electron

(b) A SO_3 molecule contains a total of 40 electrons and 24 valence electrons. The Lewis formula of a SO_3 molecule is

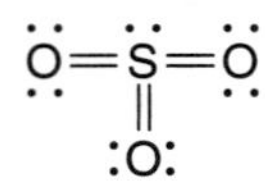

(c) A HNO molecule contains a total of 16 electrons and 12 valence electrons. The Lewis formula of a HNO molecule is

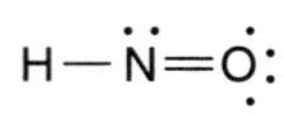

(d) A HO_2 molecule contains a total of 17 electrons and 13 valence electrons. The Lewis formula of a HO_2 molecule is

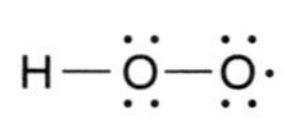

odd electron

7-26. (a)

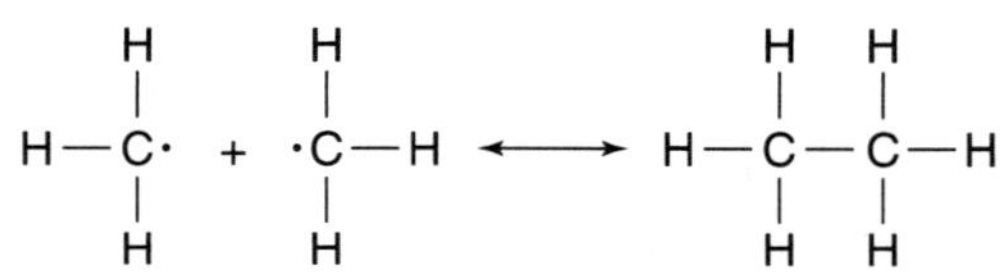

(b)

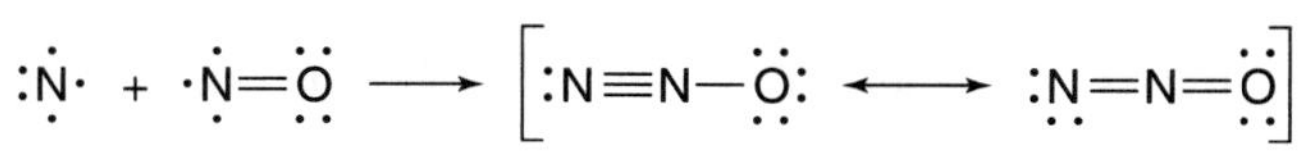

(c)

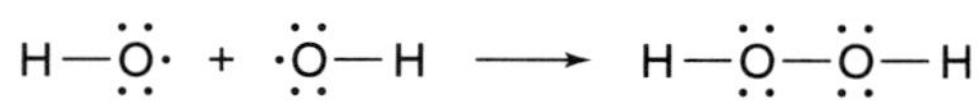

EXPANDED OCTETS

7-28. (a) A XeF_2 molecule has 22 valence electrons. We take Xe as the central atom. If we add one bond between each atom and then try to satisfy the octet rule about each of the atoms, we find that we have two additional electrons to account for. Because xenon is located in the fifth row of the periodic table ($n = 5$) it can expand its octet by using its $5d$ orbitals and so we place these additional two electrons on the xenon atom by expanding its octet, to obtain

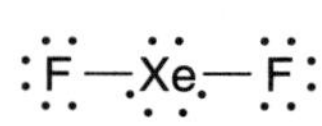

(b) A XeF_4 molecule has 36 valence electrons or 18 electron pairs. Placing the Xe atom as the central atom, we have

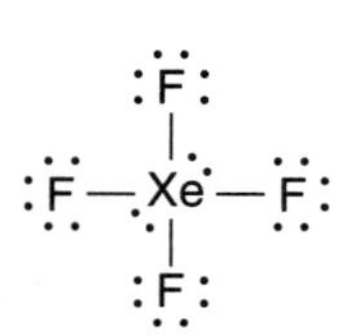

(c) A XeF_6 molecule has 25 electron pairs. Placing the Xe atom as the central atom, we have

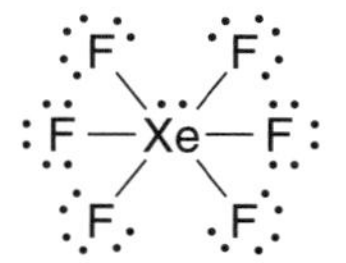

(d) A $XeOF_4$ molecule has 21 electron pairs. Placing the Xe atom in the center and forming a double bond to the oxygen atom to give it a zero formal charge, we have

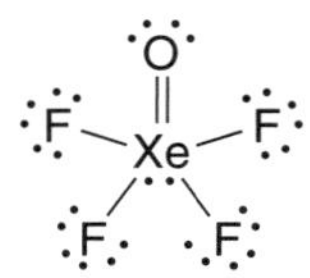

(e) A XeO_2F_2 molecule has 17 electron pairs. Placing the Xe atom in the center and forming a double bond to each of the oxygen atoms to give them each a zero formal charge, we have

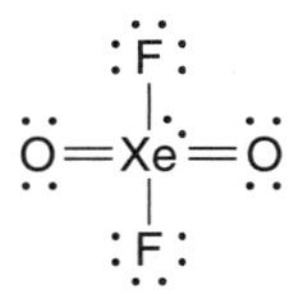

7-30. There are 36 valence electrons or 18 electron pairs in an ICl_4^- ion. Placing the iodine atom as the central atom, we have

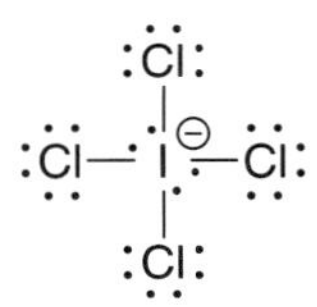

There are 34 valence electrons or 17 electron pairs in an IF_4^+ ion. Placing the iodine atom as the central atom, we have

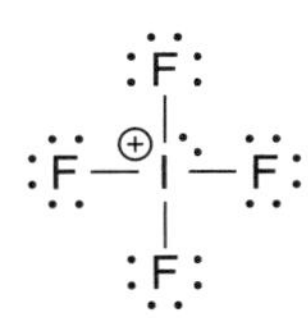

There are 11 electron pairs in an IF_2^- ion. Placing the iodine atom as the central atom, we have

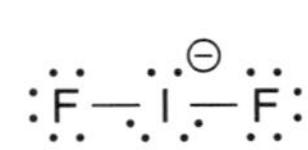

7-32. There are 26 valence electrons or 13 electron pairs in a SOF_2 molecule. Placing the sulfur atom as the central atom, we have

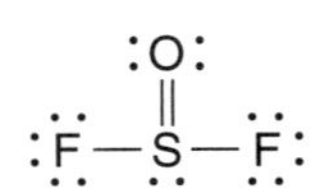

There are 32 valence electrons or 16 electron pairs in a SO_2F_2 molecule. Placing the sulfur atom as the central atom, we have

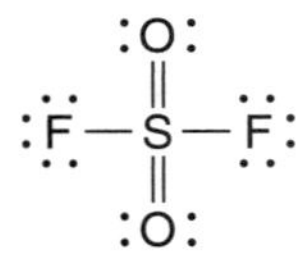

ELECTRONEGATIVITY AND DIPOLE MOMENTS

7-34. Electronegativities increase as you go from left to right across a row in the periodic table and decrease as you go down a column in the periodic table. Thus we have

$$Cl > S > Se > Sb > In$$

7-36. Electronegativities increase as you go from left to right across a row in the periodic table and decrease as you go down a column in the periodic table. Because each molecule in a given set involves one atom that is different and one or more atoms that are the same, and because the geometry of each molecule in each set is the same, we can rank the electronegativity difference based on the electronegativity of the atoms that are different in each member of the set. The rankings are as follows:

(a) $IF_3 < BrF_3 < ClF_3$ (electronegativity increases from I to Cl)
(b) $H_2Te < H_2Se < H_2S < H_2O$ (electronegativity increases from Te to O)
(c) $H_2S < SO_2 < O_3$ (electronegativity increases with charge separation)

Note that the Lewis formula for O_3 is

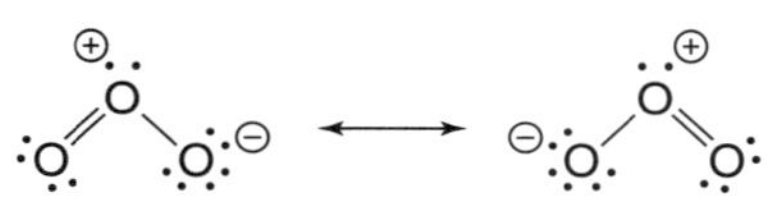

7-38. (a) Fluorine is more electronegative than hydrogen, and so we have

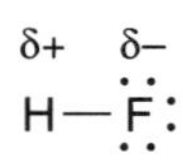

(b) Phosphorous is more electronegative than hydrogen, and so we have

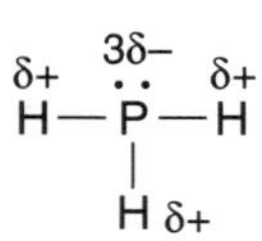

(c) Sulfur is more electronegative than hydrogen, and so we have

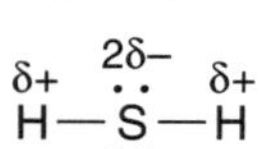

ADDITIONAL PROBLEMS

7-40. A chemical bond is what holds the atoms in a molecule together. So far we have learned about two kinds of chemical bonds, ionic and covalent bonds. An ionic bond is formed from positively and negatively charged ions. See Table 6.1 for some differences between ionically bonded and covalently bonded compounds.

7-42. When a charge is shared by two or more atoms in a molecule, rather than tending to be located on just one atom, we say that the charge is delocalized. Resonance structures are the method by which we portray delocalized charge using Lewis formulas.

7-44. (a) Some resonance forms are

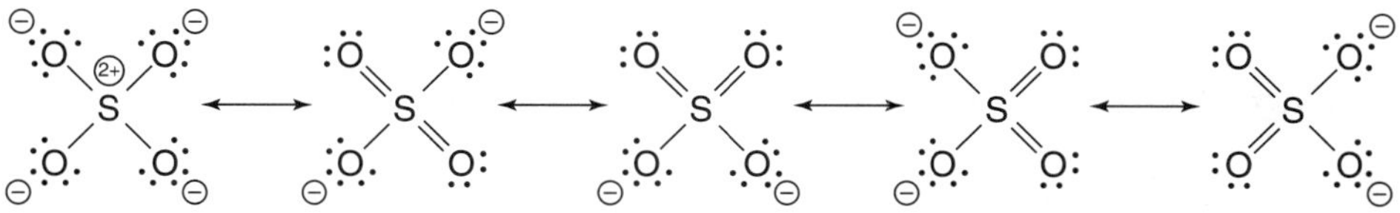

The superposition of these resonance forms gives

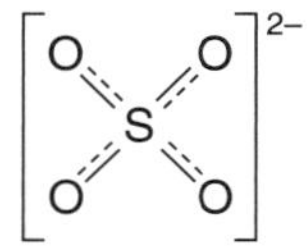

All four bonds are equivalent; they have the same bond lengths and the same bond energy.

(b) Some resonance forms are

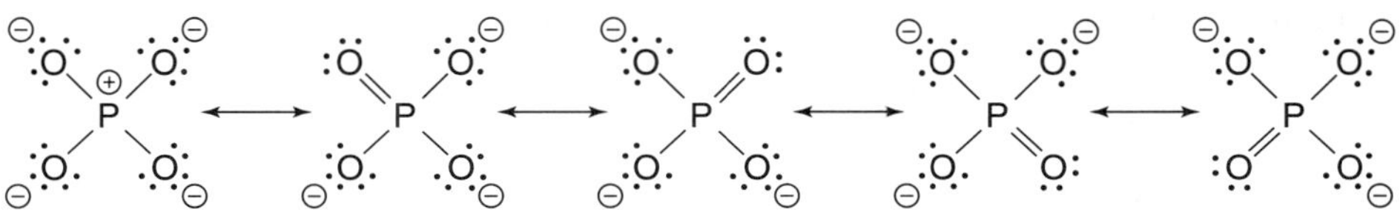

The superposition of these resonance forms gives

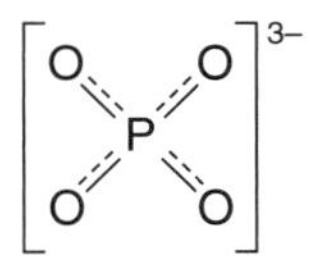

All four bonds are equivalent; they have the same bond lengths and the same bond energy.

(c) Two resonance forms are

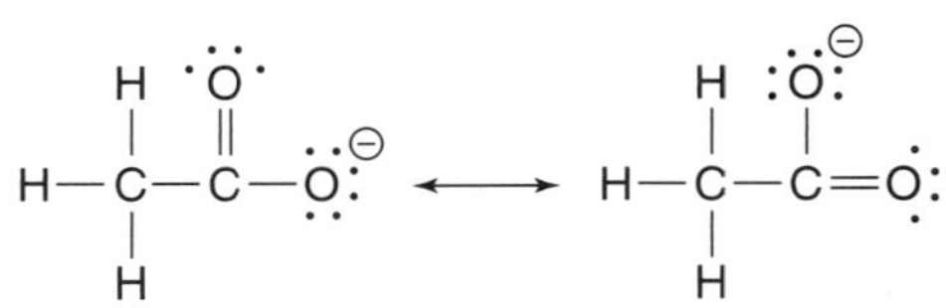

The superposition of these resonance forms is

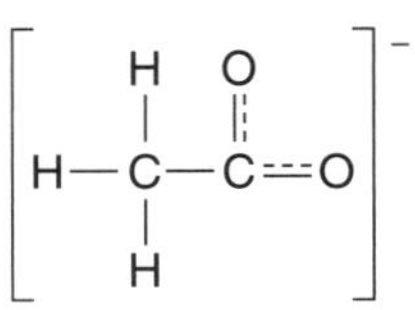

The two C–O bonds are equivalent; they have the same bond lengths and the same bond energy.

7-46. (a) 32 valence electrons, 16 electron pairs, nitrogen atom central

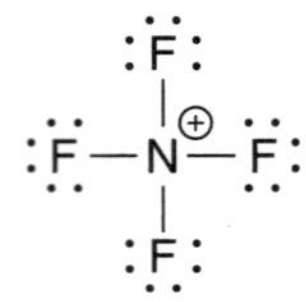

(b) 34 valence electrons, 17 electron pairs, chlorine atom central

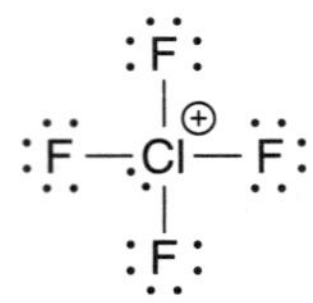

(c) 8 valence electrons, 4 electron pairs, phosphorous atom central

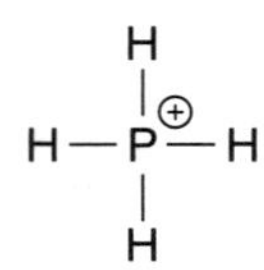

(d) 48 valence electrons, 24 electron pairs, arsenic atom central

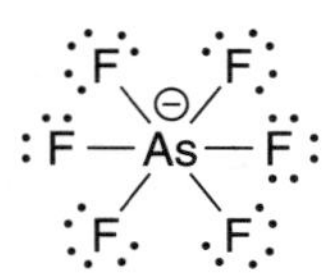

(e) 36 valence electrons, 18 electron pairs, bromine atom central

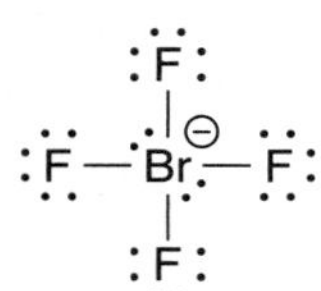

7-48. (a) 26 valence electrons, 13 electron pairs, chlorine atom central

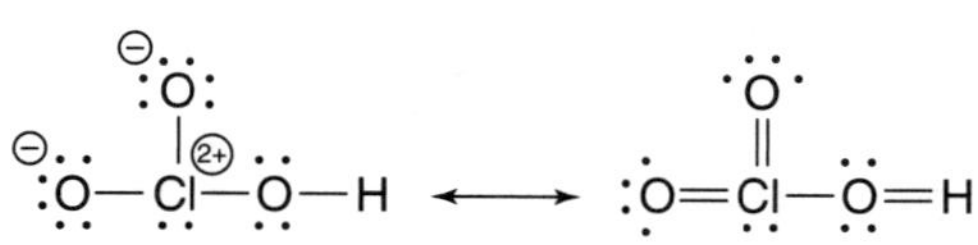

+ other resonance forms with an expanded valence shell

(b) 18 valence electrons, 9 electron pairs, nitrogen atom central

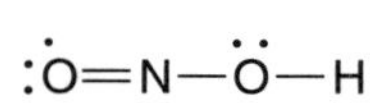

(c) 32 valence electrons, 16 electron pairs, iodine atom central

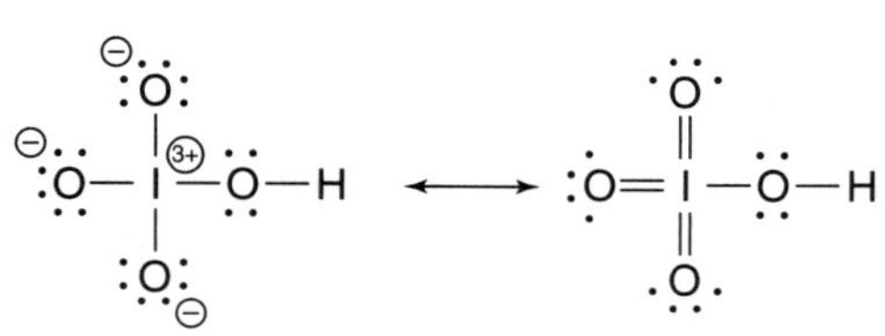

+ other resonance forms with an expanded valence shell

(d) 20 valence electrons, 10 electron pairs, bromine atom central

7-50. (a) 20 valence electrons, 10 electron pairs, sulfur atom central

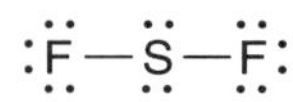

(b) 34 valence electrons, 17 electron pairs, sulfur atom central

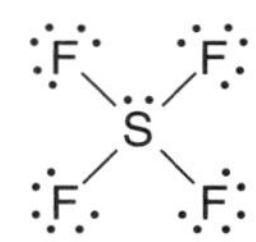

(c) 82 valence electrons, 41 electron pairs, sulfur atoms central

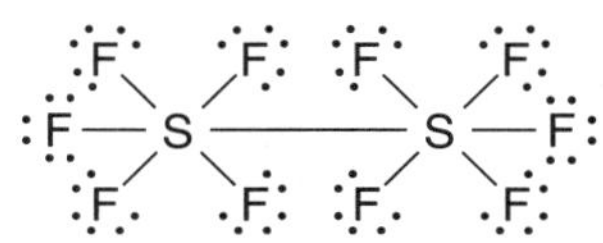

(d) 26 valence electrons, 13 electron pairs, sulfur atoms central

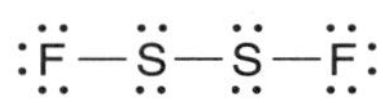

7-52. The Lewis formula is

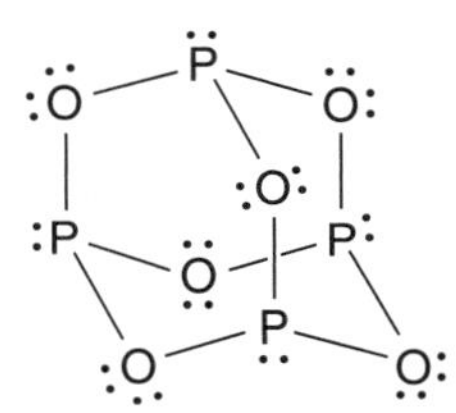

7-54. The Lewis formula of N_2F_2 with the arrangement FNNF is

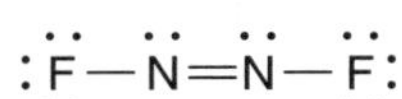

The Lewis formula of N_2F_2 with the arrangement FFNN is

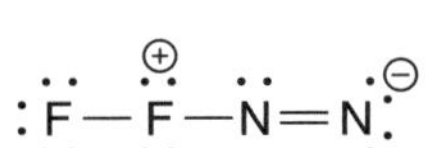

The Lewis formula of N_2F_2 with the arrangement NFFN is

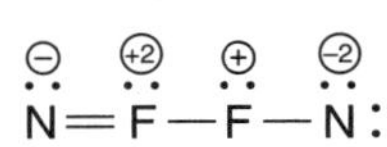

where the formal charges in each formula have been assigned according to Equation 7.1. The arrangement FNNF has the lowest formal charge, and so we predict that the structure of a N_2F_2 is molecule FNNF.

7-56. Line up the carbon and oxygen atoms as they appear in the chemical formula and place the hydrogen atoms in terminal positions to get

(a) 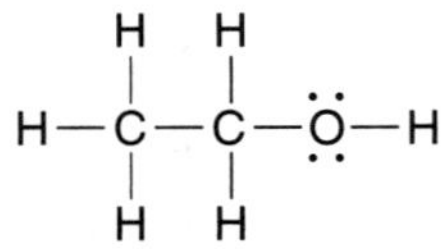

(b) 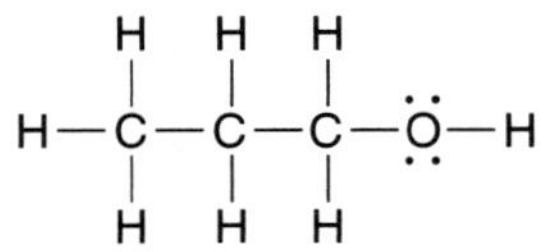

(c) 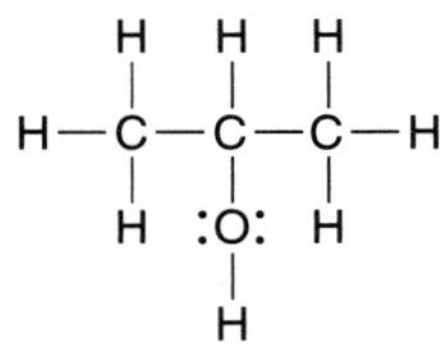

7-58. The Lewis formula of a BrCl molecule is

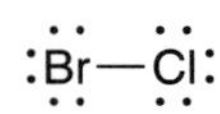

The electronegativity of a chlorine atom is greater than that of a bromine atom, and so we have

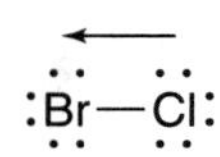

7-60. The Lewis formula for a CH_3OH molecule is

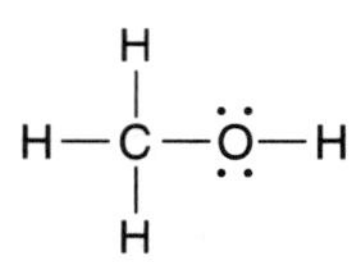

When a proton is removed, the electron remains with the oxygen atom; thus the Lewis formula for the methoxide ion is

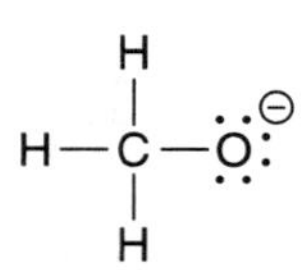

The names are

$KOCH_3$	potassium methoxide
$Al(OCH_3)_3$	aluminum methoxide

7-62. The Lewis formula of a dinitrogen oxide molecule is

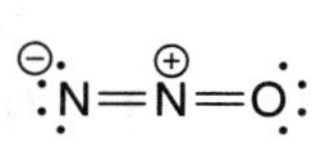

The Lewis formulas of the odd-electron nitrogen oxide molecule are

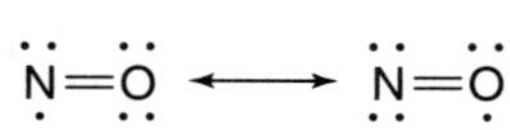

The Lewis formulas of a dinitrogen trioxide molecule are

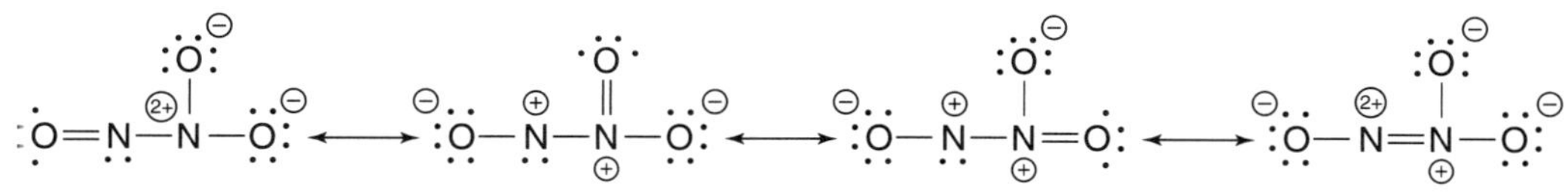

Their superposition is

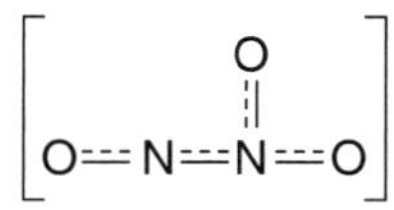

The Lewis formulas of the odd-electron nitrogen dioxide molecule are

The Lewis formulas of the dinitrogen tetroxide molecule are

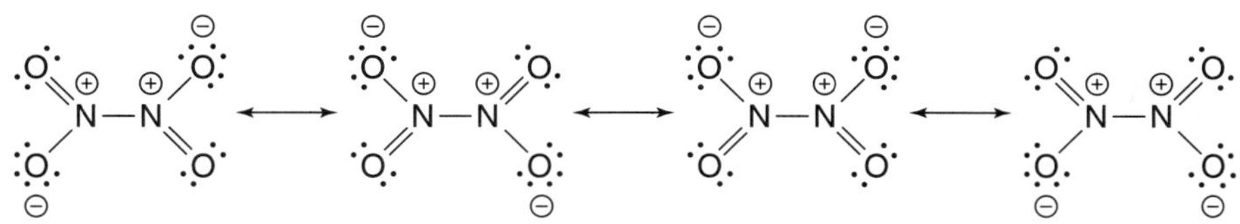

Their superposition is

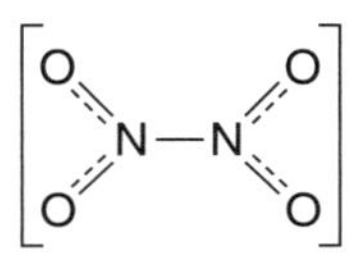

The Lewis formulas of a dinitrogen pentoxide molecule are similar to those of a dinitrogen tetroxide molecule, but with a central oxygen atom

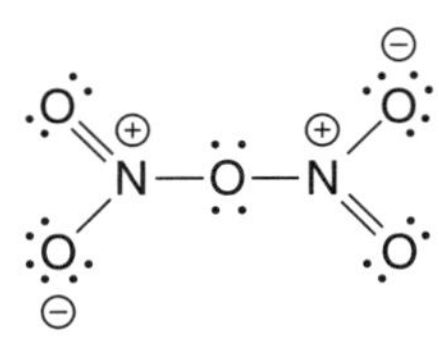

Their superposition is

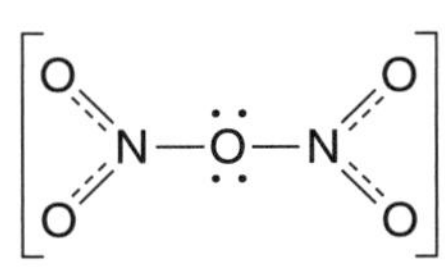

7-64. The Lewis formulas of the thiosulfate ion are

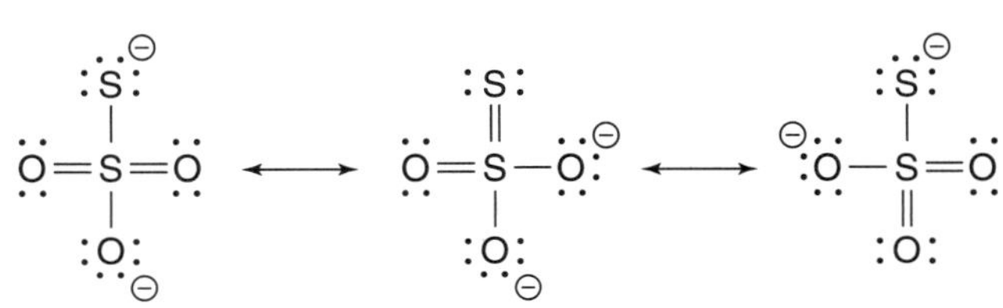

Their superposition is

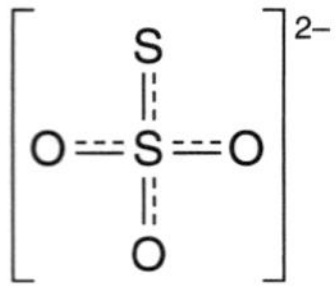

Each of the sulfur–oxygen bonds in the thiosulfate ion are equivalent; they have the same length and the same bond energy.

CHAPTER 8. Predictions of Molecular Geometries

MOLECULES AND IONS INVOLVING ONLY SINGLE BONDS

8-2. (a) An NH_4^+ ion has $(1 \times 5) + (4 \times 1) = 10$ valence electrons. Its Lewis formula is

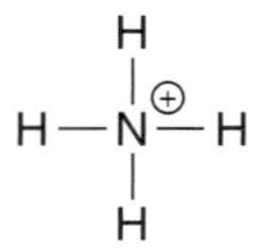

NH_4^+ is an AX_4 tetrahedral molecule (see Table 8.2) and thus has no 90° bond angles.

(b) A PF_5 molecule has 40 valence electrons. Its Lewis formula is

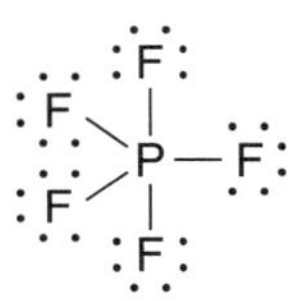

PF_5 is an AX_5 trigonal bipyramidal molecule and thus has some 90° bond angles.

(c) An AlF_6^{3-} ion has 48 valence electrons. Its Lewis formula is

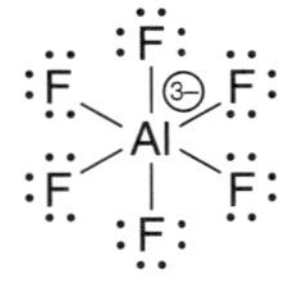

AlF_6^{3-} is an AX_6 octahedral molecule and thus has 90° bond angles.

(d) A $SiCl_4$ molecule has 32 valence electrons. Its Lewis formula is

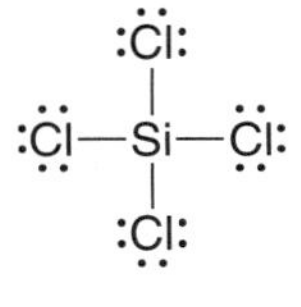

$SiCl_4$ is an AX_4 tetrahedral molecule and thus has no 90° bond angles.

8-4. (a) SeF_6 is an AX_6 octahedral molecule and thus has 180° bond angles.

(b) BrF_2^- is an AX_2E_3 linear molecule and thus has a 180° bond angle.

(c) SCl_2 is an AX_2E_2 bent molecule and thus has no 180° bond angles.

(d) $SiBr_4$ is an AX_4 tetrahedral molecule and thus has no 180° bond angles.

8-6. See Table 8.2.

(a) 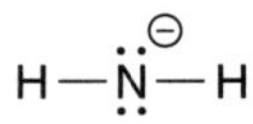 AX_2E_2 bent

(b) 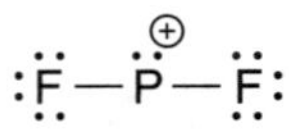 AX_2E bent

(c) 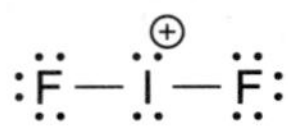 AX_2E_2 bent

(d) 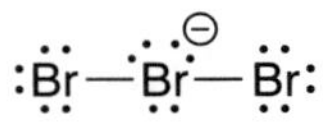 AX_2E_3 linear

8-8. See Table 8.2.

(a) 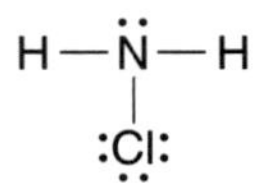 AX_3E trigonal pyramidal

(b) 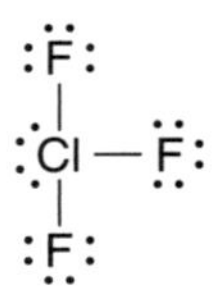 AX_3E_2 T-shaped

(c) 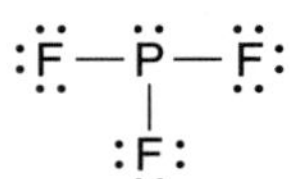 AX_3E trigonal pyramidal

(d) 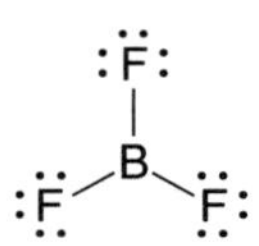 AX_3 trigonal planar

8-10. See Table 8.2.

(a) 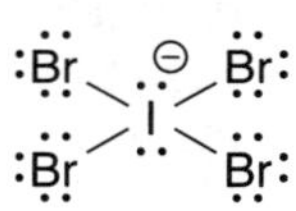 AX_4E_2 square planar

(b) 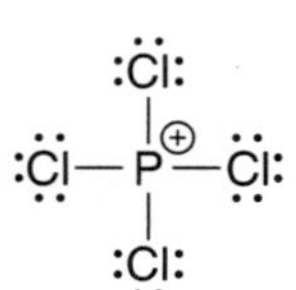 AX_4 tetrahedral

(c) 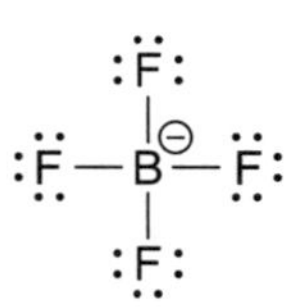 AX_4 tetrahedral

(d) 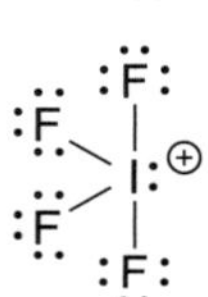 AX_4E seesaw-shaped

8-12. We have for the given fluorides

(a) 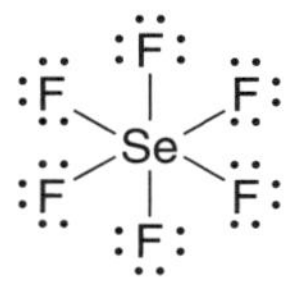 AX_6 octahedral 90°

(b) 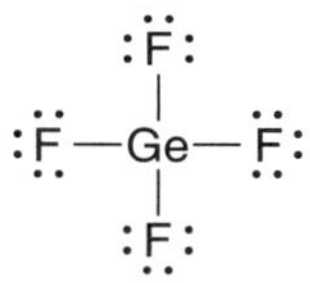 AX_4 tetrahedral 109.5°

(c) 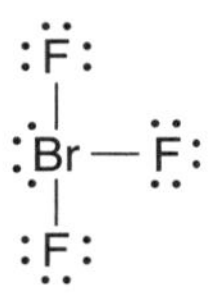 AX_3E_2 T-shaped 90°

(d) 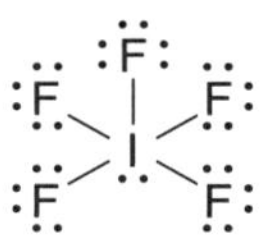 AX_5E square pyramidal < 90°

8-14. We have for the given ions

(a) 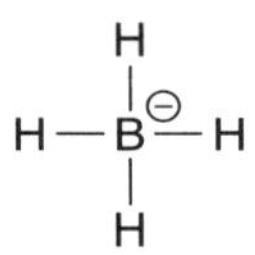 AX_4 tetrahedral 109.5°

(b) 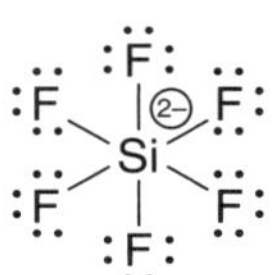 AX_6 octahedral 90°

(c) 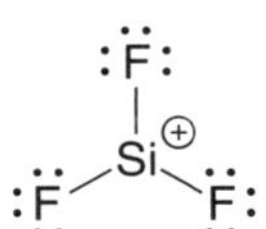 AX_3 trigonal planar 120°

(d) 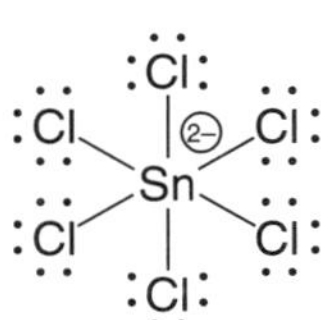 AX_6 octahedral 90°

8-16. We have for the given molecules

(a) 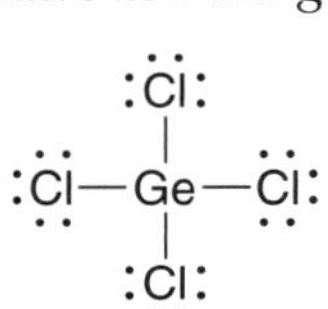 AX_4 tetrahedral 109.5°

(b) 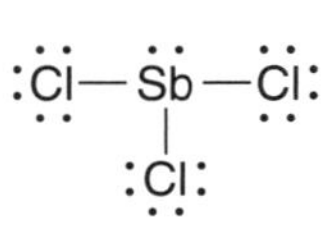 AX_3E trigonal pyramidal < 120°

(c) 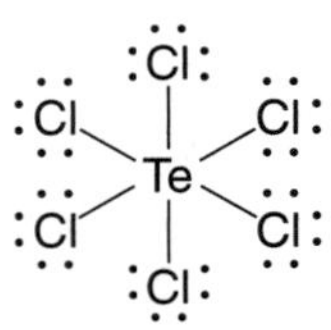 AX_6 octahedral 90°

(d) 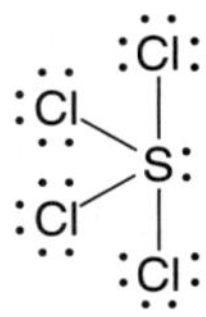 AX_4E seesaw-shaped < 120° and < 180°

8-18. See Table 8.2.

(a) 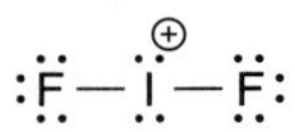 AX_2E_2 bent 109.5°

(b) 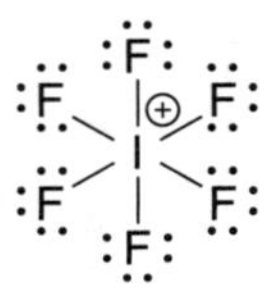 AX_6 octahedral 90°

(c) 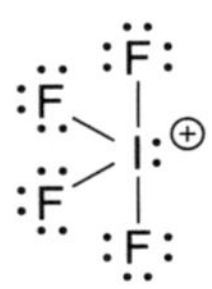 AX_4E seesaw-shaped < 120°, < 180°

(d) 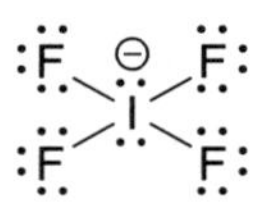 AX_4E_2 square planar 90°

MOLECULES OR IONS THAT MAY INVOLVE MULTIPLE BONDS

8-20. We have the following

(a) 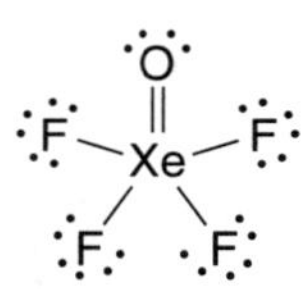 AX_5E square pyramidal < 90°

(b) 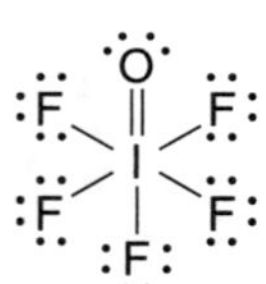 AX_6 octahedral 90°

(c) 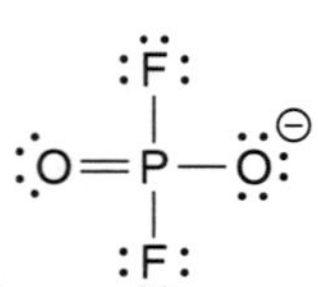 AX_4 tetrahedral ~ 109.5°

(d) 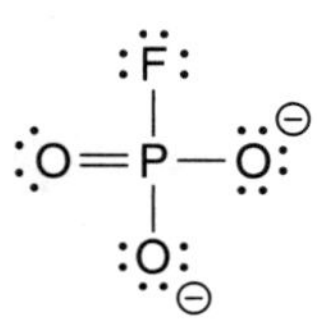 AX_4 tetrahedral ~ 109.5°

8-22. We have the following

(a) 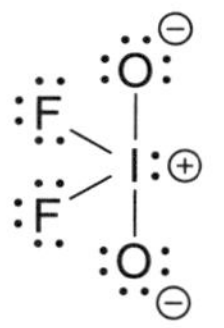 AX_4E seesaw-shaped $< 90^\circ$, $< 120^\circ$

(b) 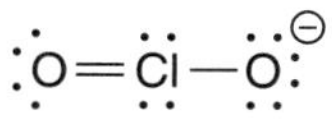 AX_2E_2 bent $< 109.5^\circ$

(c) 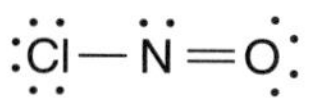 AX_2E bent $< 120^\circ$

(d) 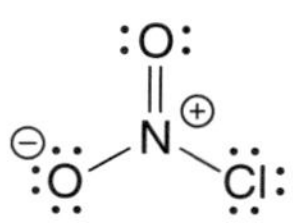 AX_3 trigonal planar $< 120^\circ$

8-24. We have the following

(a) 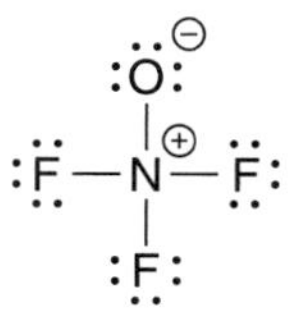 AX_4 tetrahedral $\sim 109.5^\circ$

(b) 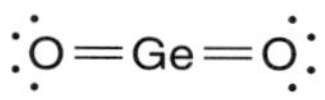 AX_2 linear 180°

(c) 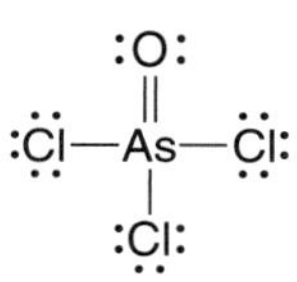 AX_4 tetrahedral $< 109.5^\circ$

(d) 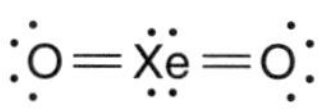 AX_2E_2 bent $< 109.5^\circ$

8-26. We have the following

(a) 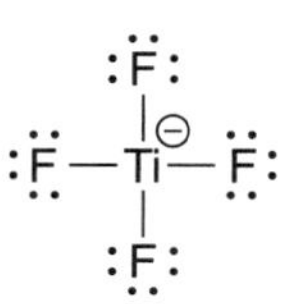 AX_4 tetrahedral 109.5°

(b) 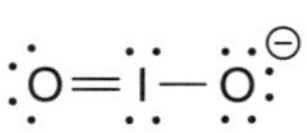 AX_2E_2 bent $< 109.5^\circ$

(c) 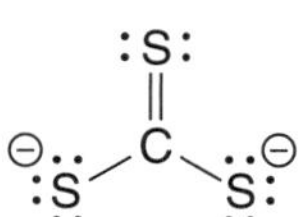 AX_3 trigonal planar 120°

(d) 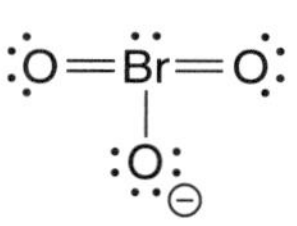 AX_3E trigonal pyramidal $< 109.5^\circ$

8-28. We have the following

(a)		AX_2E	a bent ion,	$< 120°$
(b)	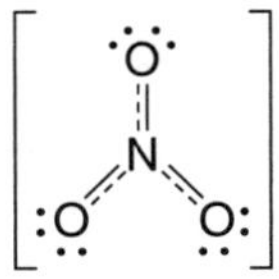	AX_3	a trigonal planar ion,	120°
(c)	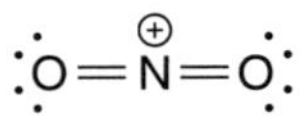	AX_2	a linear ion,	180°
(d)	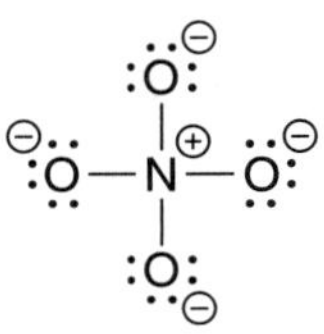	AX_4	a tetrahedral ion,	109.5°

MOLECULES WITH MORE THAN ONE CENTRAL ATOM

8-30. The bonding around the nitrogen atom that is bonded to two hydrogen atoms is trigonal pyramidal and the bonding in NCN is linear. A sketch of the cyanamide molecule is

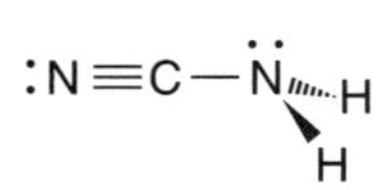

8-32. The Lewis formula for acetic acid is

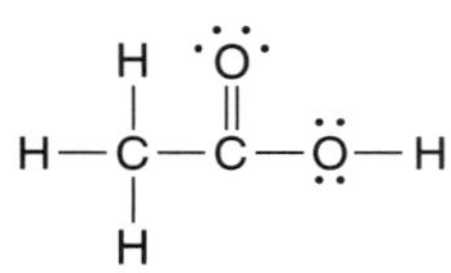

The bonding around the first carbon atom is tetrahedral, trigonal planar around the C=O bond, and bent around the oxygen atom that is bonded to the hydrogen atom. A sketch of the acetic acid molecule is

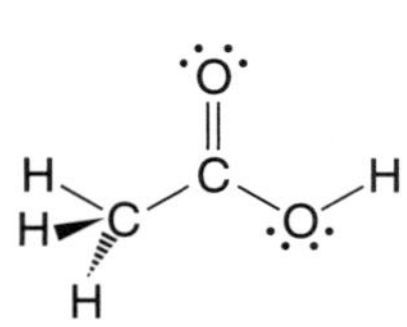

MOLECULAR SHAPES AND DIPOLE MOMENTS

8-34. (a) CCl_4 is an AX_4 molecule. Chlorine is more electronegative than carbon, and so we have

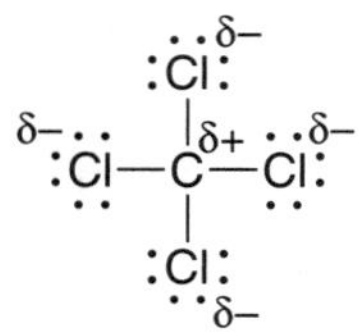

Because the molecule is tetrahedral, it has no net dipole.

(b) PCl_3 is an AX_3E molecule. Chlorine is more electronegative than phosphorus, and so we have

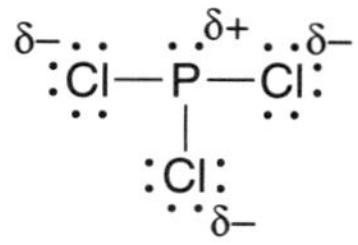

Because the molecule has a trigonal pyramidal shape, it has a net dipole moment and so is polar.

(c) ClF_3 is an AX_3E_2 molecule. Fluorine is more electronegative than chlorine, and so we have

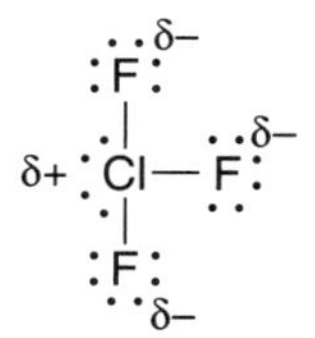

Because the molecule is T-shaped, it has a net dipole moment and so is polar.

(d) CCl_2F_2 is an AX_4 molecule. Both chlorine and fluorine are more electronegative than carbon, and so we have

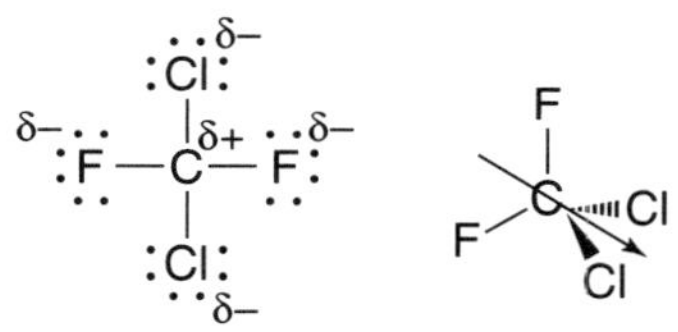

Because fluorine is more electronegative than chlorine and because the molecule is tetrahedral, the two chlorine and two fluorine atoms are not opposite to one another (as can be seen from the projection drawing or building a model); it has a net dipole moment and so is polar.

8-36. We have the following

(a) 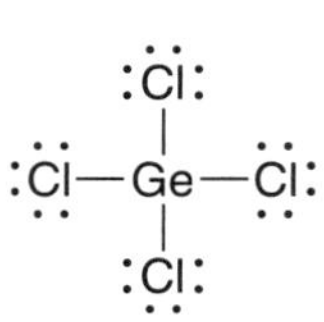 AX_4 tetrahedral (no dipole moment)

(b) 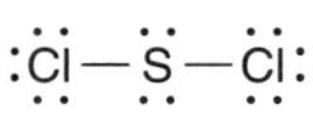 AX_2E_2 bent (dipole moment)

(c) 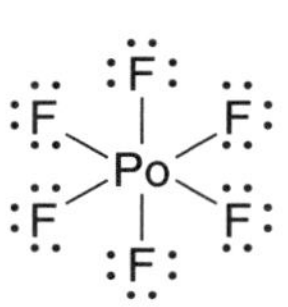 AX_6 octahedral (no dipole moment)

(d) 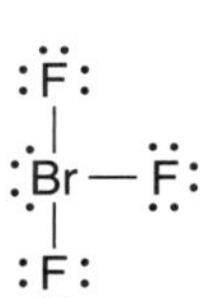 AX_3E_2 T-shaped (dipole moment)

8-38. We have the following

(a) 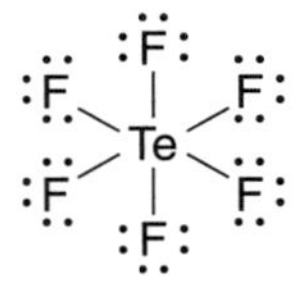 AX_6 octahedral (no dipole moment)

(b) 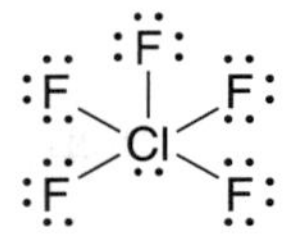 AX_5E square pyramidal (dipole moment)

(c) 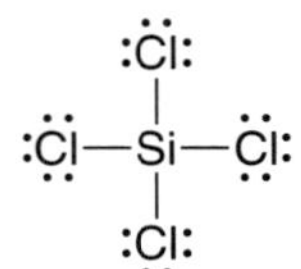 AX_4 tetrahedral (no dipole moment)

(d) 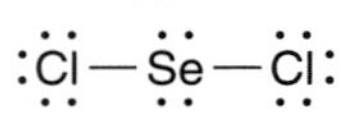 AX_2E_2 bent (dipole moment)

8-40. We have the following

(a) 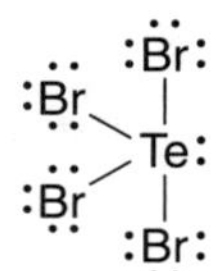 AX_4E seesaw-shaped, polar

(b) 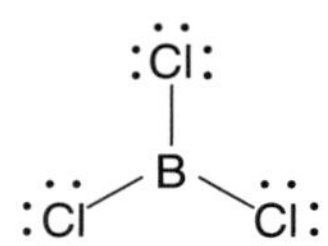 AX_3 trigonal planar, nonpolar

(c) 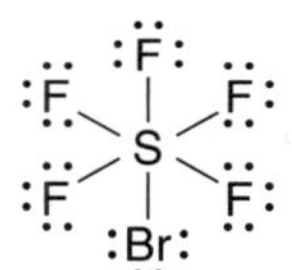 AX_6 octahedral, polar

(d) 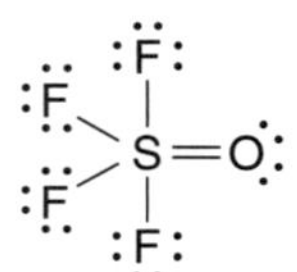 AX_5 trigonal bipyramidal, polar

ISOMERS

8-42. The five structural formulas are

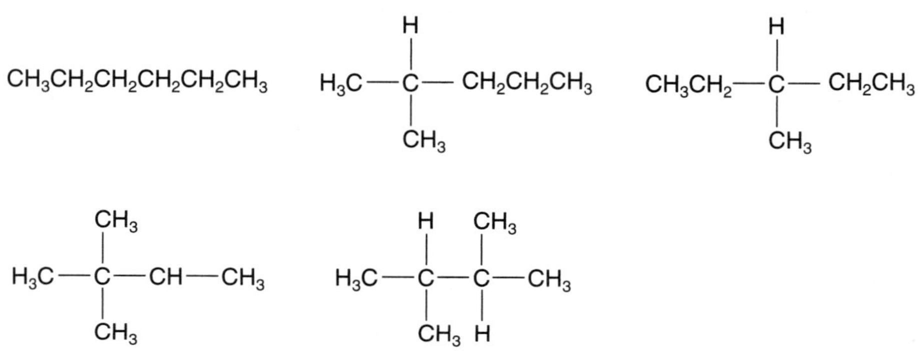

8-44. (a) All octahedral vertices are equivalent. Thus, there is only one possible arrangement of the ligands for octahedral AX_5Y molecules.

(b) Two isomers, one with the two Y's opposite and one with the two Y's adjacent:

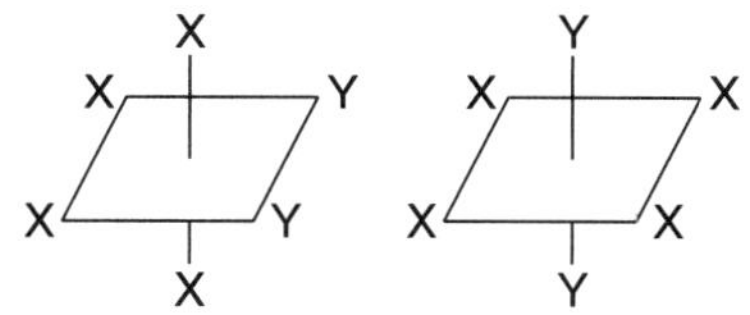

(c) Two isomers, one with the three Y's on an octahedral face and one with the three Y's along three of the four vertices:

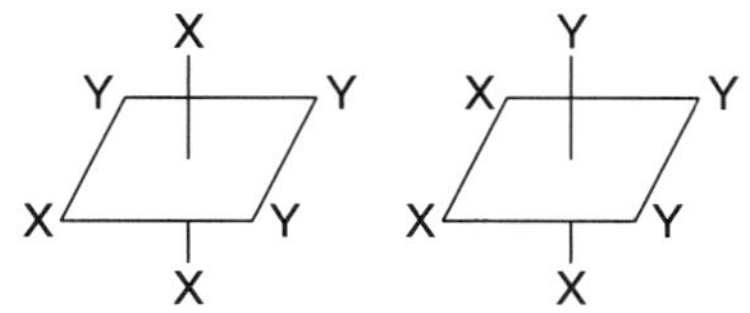

8-46. The geometric isomers of $N_2H_2F_2$ are

8-48. (a) Optical isomers (The four substituents are different.)

(b) No optical isomers (The four substituents must be different.)

(c) No optical isomers (The four substituents must be different.)

(d) Optical isomers (The four substituents are different.)

ADDITIONAL PROBLEMS

8-50. The ligands in molecules tend to move as far apart from each other as possible about the central atom because of bond-pair bond-pair repulsions. The bond angles in a regular tetrahedron are 109.5°, while those in a square are only 90°. The tetrahedron is the preferred geometry for an AX_4 molecule because a tetrahedron places the ligands farther apart than a square.

8-52. In order to be polar, a molecule must have an electronegativity difference between at least two of its atoms and the atoms in the molecule must be arranged asymmetrically so that the dipole moments within the molecule do not cancel each other out.

8-54. (a) H_2O (AX_2E_2)

(b) NH_2^- (AX_2E_2)

(c) AlH_4^- (AX_4)

(d) SF_6 (AX_6)

8-56. We have the following

(a) 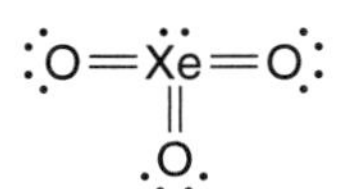 AX_3E trigonal pyramidal

(b) 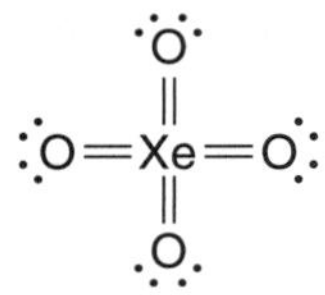 AX_4 tetrahedral

(c) 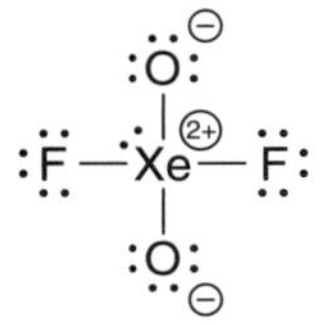 AX_4E seesaw-shaped

(d) 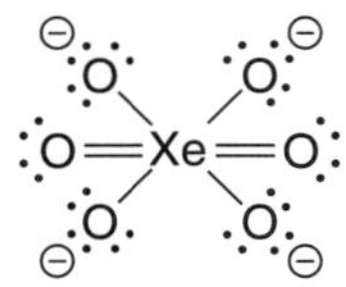 AX_6 octahedral

8-58. We have the following

(a) 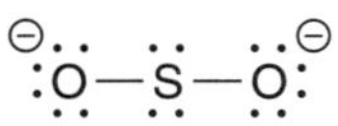 AX_2E_2 bent

(b) 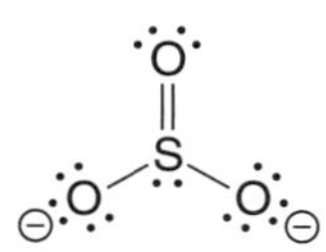 + resonance forms AX_3E trigonal pyramidal

(c) 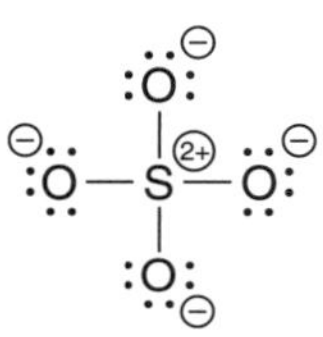 + resonance forms AX_4 tetrahedral

8-60. The PCl_4^+ ions are tetrahedral and the PI_6^- ions are octahedral.

8-62. Using the AX_mE_n notation, we find that

(a) SOF_2 is AX_3E (b) ClF_3 is AX_3E_2 (c) NO_2Cl is AX_3 (d) BF_3 is AX_3

Thus, only SOF_2 is trigonal pyramidal.

8-64. Using the AX_mE_n notation, we find that

(a) $BeCl_2$ is AX_2 (b) O_3 is AX_2E (c) OCl_2 is AX_2E_2 (d) NOF is AX_2E

Thus, only $BeCl_2$ is linear.

8-66. We have the following

(a) 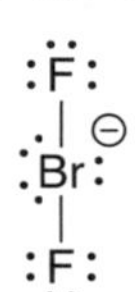 AX_2E_3 linear

(b) 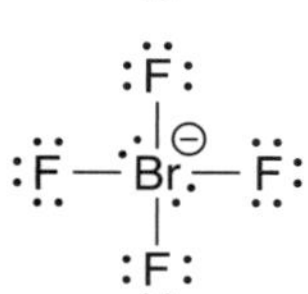 AX_4E_2 square planar

(c) 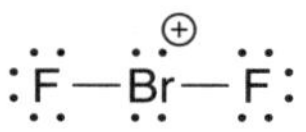 AX_2E_2 bent

(d) 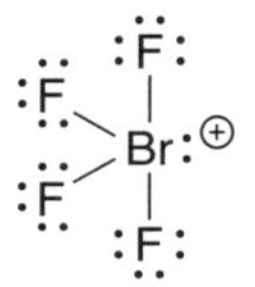 AX_4E seesaw-shaped

8-68. The Lewis formulas are

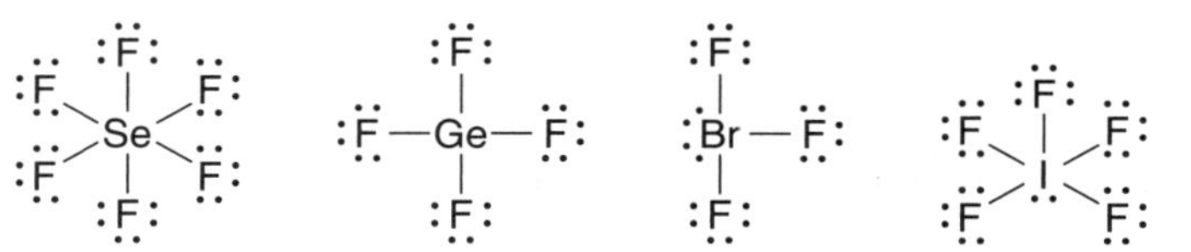

SeF_6 is AX_6 and octahedral; GeF_4 is AX_4 and tetrahedral; BrF_3 is AX_3E_2 and T-shaped; and IF_5 is AX_5E and square pyramidal. Thus, we have

(1) a	(2) c, d	(3) b	(4) none	(5) none
(6) none	(7) none	(8) a	(9) c, d	

8-70. The Lewis formulas are

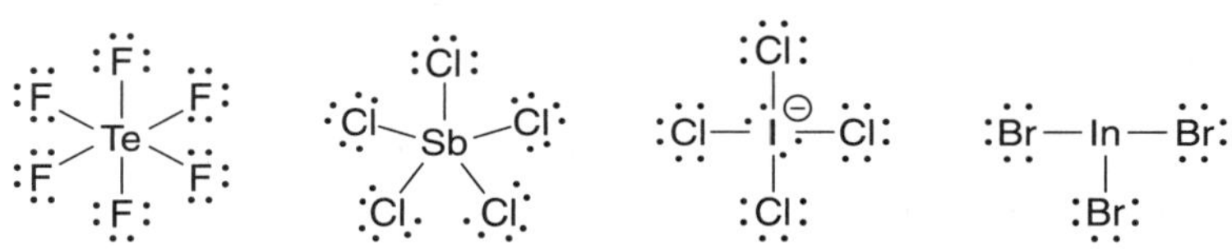

TeF_6 is AX_6 and octahedral; $SbCl_5$ is AX_5 and trigonal bipyramidal; ICl_4^- is AX_4E_2 and square planar; and $InBr_3$ is AX_3 and trigonal planar. Thus, we have

(1) a, b, c (2) none (3) b, d

8-72. See the text.

8-74. The Lewis formula for a dimethyl sulfide molecule is

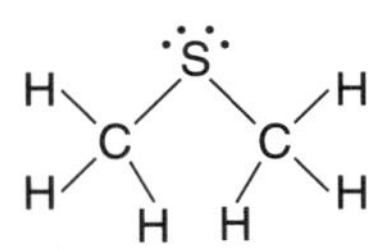

The bonding around the central sulfur atom has an AX_2E_2 configuration and so has a bent shape with C–S–C bond angles of less than 109.5°. Each of the two carbon atoms has an AX_4 configuration and so has a tetrahedral shape with S–C–H bond angles of about 109.5°. Because the molecule is asymmetric and because a sulfur atom is slightly more electronegative than a carbon atom, the molecule has a net dipole moment (shown below) and therefore is polar. A perspective drawing of the Lewis formula of a dimethyl sulfide molecule using wedge-shaped bonds is

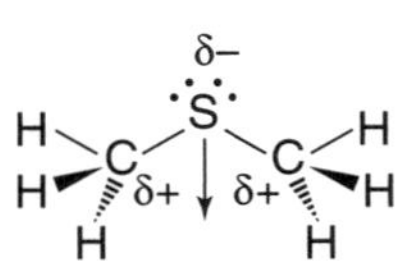

8-76. (a) All the vertices of a tetrahedron are equivalent, and thus it makes no difference where the ligand Y is placed. Therefore, there is only one possible arrangement of the ligands for tetrahedral AX_3Y molecules.

(b) We can form two geometric isomers of AX_2Y_2 square planar molecules—one with identical ligands adjacent to each other, and one with them opposite to each other.

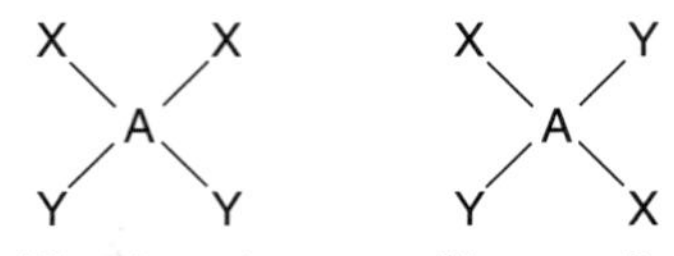

(c) Because each of the four ligands bonded to the central atom in a tetrahedral AHXYZ molecule is different, it is chiral, and so we can form two optical isomers.

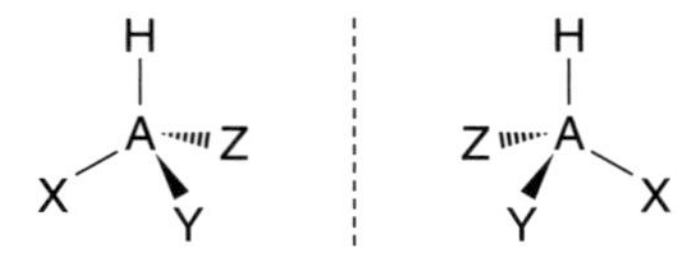

(d) All the vertices of a square are equivalent, and thus it makes no difference where the ligand Y is placed. Therefore, there is only one possible arrangement of the ligands for square planar AX_3Y molecules.

8-78. The chiral center is the carbon atom attached to $-NH_2$, $-COOH$, and $-CH_2OH$ (four different substituents). See the structure of alanine on page 261.

8-80. The triangle formed by the center sphere and any two others in Figure 8.2b is given by θ in

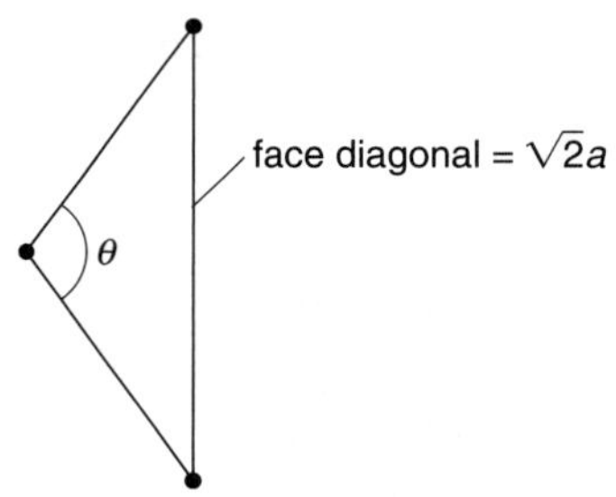

If we now draw a line from the center dot perpendicular to the face diagonal, then we bisect θ and the face diagonal

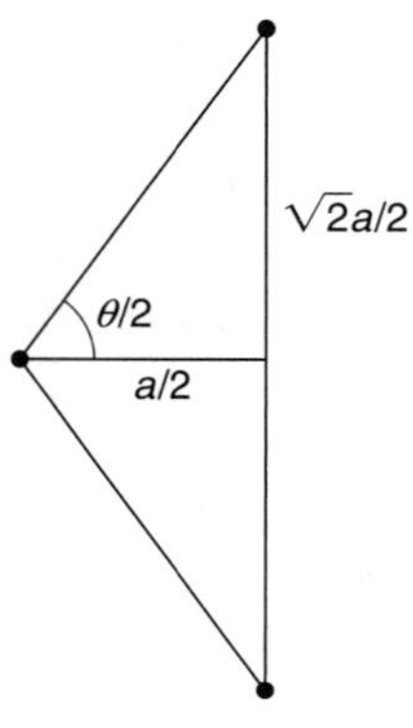

The length of this new line is $a/2$ because it is one half the length of an edge (see Figure 8.2b). Using trigonometry, we write

$$\tan\frac{\theta}{2} = \frac{\sqrt{2}\,a/2}{a/2} = \sqrt{2}$$

or

$$\frac{\theta}{2} = \tan^{-1}\sqrt{2}$$

Using the $\tan^{-1}$ key on your calculator, you find that

$$\frac{\theta}{2} = 54.74°$$

or $\theta = 109.5°$

CHAPTER 9. Covalent Bonding

DIATOMIC MOLECULES

9-2. There are ten electrons in a diatomic boron molecule. Using Figure 9.13, we place electrons into the molecular orbitals according to the Pauli exclusion principle; $(\sigma_{1s})^2(\sigma_{1s}^*)^2(\sigma_{2s})^2(\sigma_{2s}^*)^2(\pi_{2p_x})^1(\pi_{2p_y})^1$ is the ground-state electron configuration of a diatomic boron molecule. The bond order is given by Equation 9.1

$$\text{bond order} = \frac{6 \text{ bonding electrons} - 4 \text{ antibonding electrons}}{2} = 1$$

9-4. There are ten electrons in a diatomic boron molecule. Using Figure 9.13, we place electrons into the molecular orbitals according to the Pauli exclusion principle; $(\sigma_{1s})^2(\sigma_{1s}^*)^2(\sigma_{2s})^2(\sigma_{2s}^*)^2(\pi_{2p_x})^1(\pi_{2p_y})^1$ is the ground-state electron configuration of a diatomic boron molecule. Hund's rule states that we should place one electron in each of the two π orbitals with unpaired spins. Thus, there are two unpaired electrons in a B_2 molecule, and so B_2 is paramagnetic.

9-6. Using Figure 9.13 and Equation 9.1, we find that the ground-state electron configurations and bond orders of an O_2 molecule and an O_2^+ ion are

	Ground-state electron configuration	Bond order
O_2	$(\sigma_{1s})^2(\sigma_{1s}^*)^2(\sigma_{2s})^2(\sigma_{2s}^*)^2(\sigma_{2p})^2(\pi_{2p})^4(\pi_{2p}^*)^2$	2
O_2^+	$(\sigma_{1s})^2(\sigma_{1s}^*)^2(\sigma_{2s})^2(\sigma_{2s}^*)^2(\sigma_{2p})^2(\pi_{2p})^4(\pi_{2p}^*)^1$	$2\frac{1}{2}$

We find that the bond order of an O_2 molecule is 2 while the bond order of an O_2^+ ion is $2\frac{1}{2}$. The bond energy increases as the bond order increases. Therefore, we predict (correctly) that the bond energy of an O_2 molecule is less than that of an O_2^+ ion.

9-8. Using Figure 9.13 and Equation 9.1, we find that the ground-state electron configurations and bond orders of a C_2 molecule and a C_2^{2-} ion are

	Ground-state electron configuration	Bond order
C_2	$(\sigma_{1s})^2(\sigma_{1s}^*)^2(\sigma_{2s})^2(\sigma_{2s}^*)^2(\pi_{2p})^4$	2
C_2^{2-}	$(\sigma_{1s})^2(\sigma_{1s}^*)^2(\sigma_{2s})^2(\sigma_{2s}^*)^2(\pi_{2p})^4(\sigma_{2p})^2$	3

Because the bond order of a C_2^{2-} ion is greater than the bond order of a C_2 molecule, we predict (correctly) that a C_2^{2-} ion has a larger bond energy and a shorter bond length than a C_2 molecule.

9-10. Using Figure 9.15 and Equation 9.1, we find that the ground-state electron configurations and bond orders of the diatomic species NO, NO^+, and NO^- are

	Ground-state electron configuration	Bond order
NO	$(\sigma_{1s})^2(\sigma_{1s}^*)^2(\sigma_{2s})^2(\sigma_{2s}^*)^2(\pi_{2p})^4(\sigma_{2p})^2(\pi_{2p}^*)^1$	$2\frac{1}{2}$
NO^+	$(\sigma_{1s})^2(\sigma_{1s}^*)^2(\sigma_{2s})^2(\sigma_{2s}^*)^2(\pi_{2p})^4(\sigma_{2p})^2$	3
NO^-	$(\sigma_{1s})^2(\sigma_{1s}^*)^2(\sigma_{2s})^2(\sigma_{2s}^*)^2(\pi_{2p})^4(\sigma_{2p})^2(\pi_{2p}^*)^2$	2

We predict that a NO^- ion has the longest bond length because it has the smallest bond order of the three species.

9-12. Determine the total number of electrons and use Figure 9.13 and Equation 9.1.

(a) 13 electrons
$(\sigma_{1s})^2(\sigma_{1s}^*)^2(\sigma_{2s})^2(\sigma_{2s}^*)^2(\pi_{2p})^4(\sigma_{2p})^1$
bond order $= 2\frac{1}{2}$

(b) 9 electrons
$(\sigma_{1s})^2(\sigma_{1s}^*)^2(\sigma_{2s})^2(\sigma_{2s}^*)^2(\pi_{2p})^1$
bond order $= \frac{1}{2}$

(c) 3 electrons
$(\sigma_{1s})^2(\sigma_{1s}^*)^1$
bond order $= \frac{1}{2}$

(d) 20 electrons
$(\sigma_{1s})^2(\sigma_{1s}^*)^2(\sigma_{2s})^2(\sigma_{2s}^*)^2(\sigma_{2p})^2(\pi_{2p})^4(\pi_{2p}^*)^4(\sigma_{2p}^*)^2$
bond order $= 0$

9-14. We use the value of the bond order to predict the stability.

(a) 1/2, stable

(b) 0, unstable

(c) 1 1/2, stable

(d) −1/2, unstable

9-16. If the electron that is removed occupied a bonding orbital, then a weaker net bonding will result. For example, the ground-state electron configuration of a B_2 molecule (10 electrons) is $(\sigma_{1s})^2(\sigma_{1s}^*)^2(\sigma_{2s})^2(\sigma_{2s}^*)^2(\pi_{2p})^2$. The addition of an electron will produce $(\sigma_{1s})^2(\sigma_{1s}^*)^2(\sigma_{2s})^2(\sigma_{2s}^*)^2(\pi_{2p})^3$ with one additional bonding electron, and thus a stronger net bonding.

9-18. The ground-state electron configuration of a P_2 molecule is

$$(\sigma_{1s})^2(\sigma_{1s}^*)^2(\sigma_{2s})^2(\sigma_{2s}^*)^2(\pi_{2p})^4(\sigma_{2p})^2(\pi_{2p}^*)^4(\sigma_{2p}^*)^2(\sigma_{3s})^2(\sigma_{3s}^*)^2(\pi_{3p})^4(\sigma_{3p})^2$$

The bond order is 3 and there are no unpaired electrons (diamagnetic). The Lewis formula of P_2 is :P≡P:, which agrees with the calculated bond order of 3.

POLYATOMIC MOLECULES

9-20. A mercury atom has two valence electrons and each chlorine atom has seven, for a total of 16 valence electrons. The Lewis formula of a $HgCl_2$ molecule is

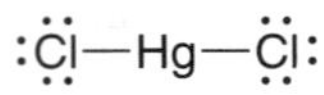

Because $HgCl_2$ is a linear molecule, it is appropriate to use sp orbitals on the mercury atom. We can form two localized bond orbitals by combining each mercury sp orbital with a chlorine $3p$ orbital, oriented 180° from each other. Four of the valence electrons occupy the two localized $Hg(sp) + Cl(3p_z)$ bond orbitals pairwise to form the two localized covalent bonds. The remaining 12 valence electrons occupy the remaining $3p$ orbitals on the chlorine atoms as three lone pairs on each atom.

9-22. There are six valence electrons from the sulfur atom and seven from each fluorine atom for a total of 20 valence electrons. The Lewis formula of a SF_2 molecule is

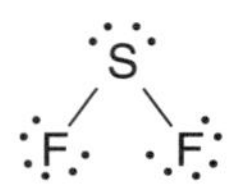

SF_2 is an AX_2E_2 molecule, and so we use sp^3 hybrid orbitals to describe the bonding. We can form the two localized bond orbitals that point toward the vertices of a regular tetrahedron by combining each sulfur sp^3 orbital with a fluorine $2p$ orbital. Four of the valence electrons occupy these two localized bond orbitals pairwise to form the two localized σ bonds. Four of the valence electrons occupy pairwise the two remaining sp^3 orbitals as lone pairs on the sulfur atom. The remaining 12 valence electrons are lone pairs on the two fluorine atoms.

9-24. The Lewis formula of a NF_3 molecule is

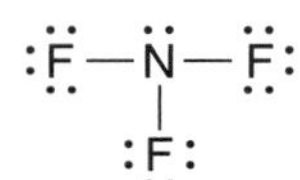

Because NF_3 is a trigonal pyramidal molecule, it is appropriate to use sp^3 orbitals on the nitrogen atom. The three localized bond orbitals are formed by combining an sp^3 orbital on the nitrogen atom with a fluorine $2p$ orbital. Six of the 26 valence electrons occupy these three σ-bond orbitals pairwise. The lone electron pair on the nitrogen atom occupies the remaining sp^3 orbital and the remaining 18 valence electrons are lone pairs on the fluorine atoms.

9-26. The Lewis formula for a CH_3^+ ion is

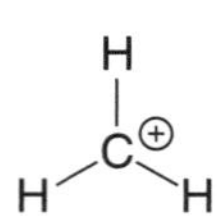

Because there are three pairs of electrons about the carbon atom, it is appropriate to use sp^2 hydrid orbitals on the carbon atom. The three localized bond orbitals are formed by combining one of the sp^2 orbitals on the carbon atom with a hydrogen $1s$ orbital to form the three σ bonds of the methyl cation.

9-28. There are 26 valence electrons in a $HCCl_3$ molecule. Its Lewis formula is

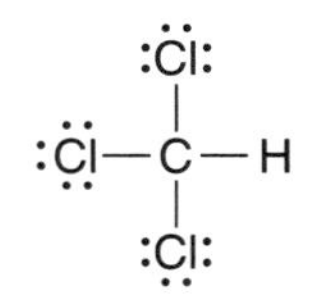

Because a $HCCl_3$ molecule is tetrahedral, it is appropriate to use sp^3 orbitals on the carbon atom. We can form four localized σ-bond orbitals by combining a carbon sp^3 orbital with a $3p$ orbital on each of the three chlorine atoms and one with a $1s$ orbital on the hydrogen atom. There are 26 valence electrons in a $HCCl_3$ molecule. Eight of the 26 valence electrons occupy the four bond orbitals pairwise to form four σ-bond orbitals and the remaining 18 valence electrons are lone-pair electrons on the three chlorine atoms.

9-30. 22 valence electrons. The Lewis formula of a XeF_2 molecule is

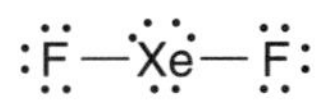

XeF_2 is an AX_2E_3 molecule, and so we use sp^3d hybrid orbitals to describe the bonding. Four of the 22 valence electrons form two localized $Xe(sp^3d)$–$F(2p_z)\sigma$ bonds. Six of the valence electrons occupy the remaining three sp^3d orbitals as three lone pairs on the xenon atom. The remaining 12 valence electrons are lone-pair electrons on the four fluorine atoms.

9-32. 42 valence electrons. The Lewis formula of a TeF_5^- ion is

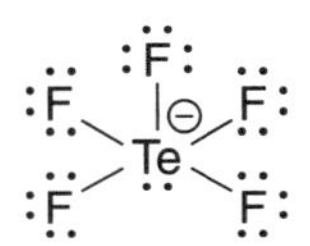

TeF_5^- is an AX_5E ion, and so we use sp^3d^2 hybrid orbitals to describe the bonding. Ten of the 42 valence electrons form five localized $Te(sp^3d^2)$–$F(2p_z)$ σ bonds. Two of the valence electrons occupy the remaining sp^3d^2 orbital as a lone pair on the tellurium atom. The remaining 30 valence electrons are lone-pair electrons on the five fluorine atoms.

MOLECULES WITH NO UNIQUE CENTRAL ATOM

9-34. Because there are four pairs of electrons about each of the carbon atoms and about the oxygen atom, it is appropriate to use sp^3 orbitals on the two carbon atoms and the oxygen atom. The four σ-bond orbitals on the carbon atom on the left are formed by combining three of the sp^3 orbitals on the carbon atom with three hydrogen $1s$ orbitals and one with an sp^3 orbital on the second carbon atom. The three remaining σ bonds on the second carbon atom are formed by combining two of the sp^3 orbitals on the carbon atom with two hydrogen $1s$ orbitals and one with an sp^3 orbital on the oxygen atom. The remaining σ-bond orbital on the oxygen atom is formed by combining one of the sp^3 orbitals on the oxygen atom with a $1s$ hydrogen orbital. Sixteen of the 20 valence electrons occupy the eight σ-bond orbitals pairwise to form the eight σ bonds. Two electron pairs occupy the remaining two sp^3 orbitals on the oxygen atom. The shape is tetrahedral about each carbon atom and the oxygen–hydrogen bond is bent and is about 105°.

9-36. Because there are four pairs of electrons about each of the carbon atoms and the nitrogen atom, it is appropriate to use sp^3 orbitals on both carbon atoms and the nitrogen atom. The formation of the localized bond orbitals can be illustrated by

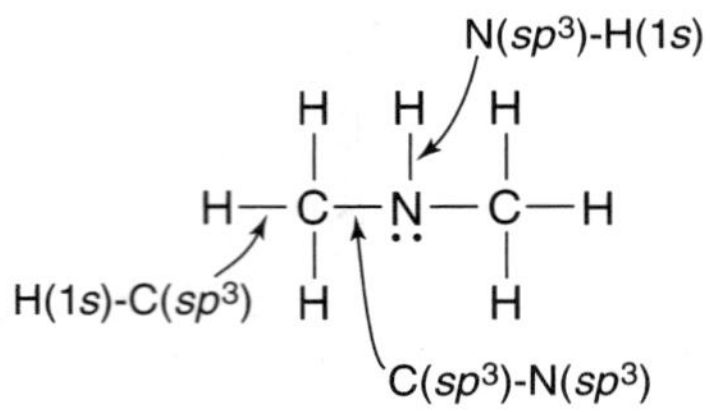

Eighteen of the 20 valence electrons occupy the nine σ-bond orbitals and the remaining two occupy an sp^3 orbital on the nitrogen atom as a lone pair. The shape around each carbon atom is tetrahedral and the shape around the nitrogen atom is trigonal pyramidal.

9-38. We use sp^3 orbitals on the oxygen atom and all three carbon atoms. The formation of the localized bond orbitals can be illustrated by

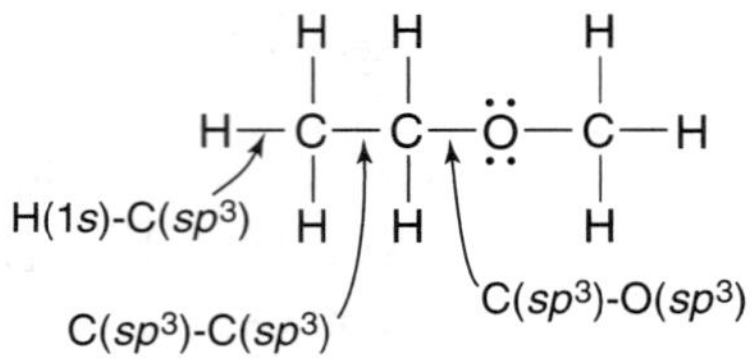

Twenty-two of the 26 valence electrons occupy the 11 localized bond orbitals and the remaining four occupy two of the oxygen sp^3 orbitals as lone pairs.

MULTIPLE BONDS

9-40. (a) The Lewis formula for $CH_2{=}CCl_2$ is

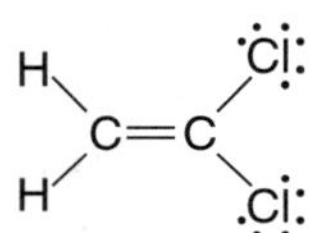

There are five σ bonds and one π bond.

(b) The Lewis formula for HOOCCOOH is

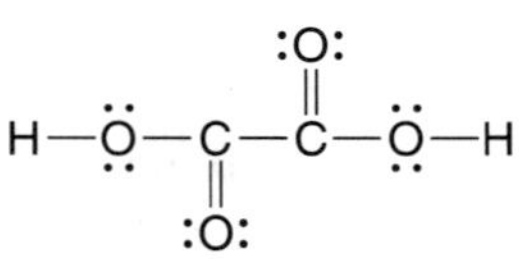

There are seven σ bonds and two π bonds.

(c) The Lewis formula for FHC=C=CHF is

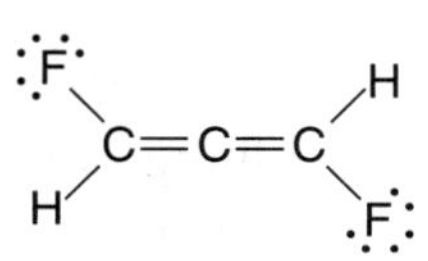

There are six σ bonds and two π bonds.

(d) There are eight σ bonds and two π bonds.

9-42. The Lewis formula for a methyl cyanide molecule is

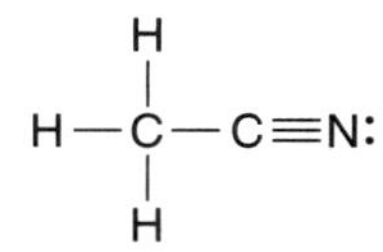

There are five σ bonds and two π bonds. There are $(2 \times 4) + (1 \times 5) + (3 \times 1) = 16$ valence electrons, 14 of which occupy the seven bond orbitals. The other two valence electrons occupy the lone pair on the nitrogen atom. The bond order of the carbon–carbon bond is 1 and that of the carbon–nitrogen bond is 3.

9-44. The Lewis formula of an acetylide ion is

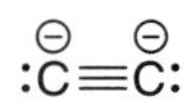

Because the molecule is linear, it is appropriate to use sp orbitals on each of the carbon atoms. The σ-bond orbital between the two carbon atoms is formed by combining an sp orbital on each of the carbon atoms. Each of the two π-bond orbitals is formed by combining a $2p$ orbital perpendicular to the bonding axis on each of the carbon atoms. Two of the 10 valence electrons occupy the σ-bond orbital, four occupy the two π-bond orbitals, and the remaining four valence electrons are lone-pair electrons. Each lone pair occupies an sp orbital on a carbon atom. The C≡C bond has a bond order of 3.

Using molecular orbital theory to describe the bonding in a C_2^{2-} ion, we find that the ground-state electron configuration according to Figure 9.13 is $(\sigma_{1s})^2(\sigma_{1s}^*)^2(\sigma_{2s})^2(\sigma_{2s}^*)^2(\pi_{2p})^4(\sigma_{2p})^2$. Thus a C_2^{2-} ion has no unpaired electrons and is not paramagnetic.

GEOMETRIC ISOMERS

9-46. The two isomers of 1,2-dibromo-1,2-dichloroethene are

trans-isomer *cis*-isomer

9-48. Only FHC=C=CHF exhibits *cis-trans* isomerization. The two isomers are

trans-isomer *cis*-isomer

DELOCALIZED BONDS

9-50. The Lewis formula for an anthracene molecule is

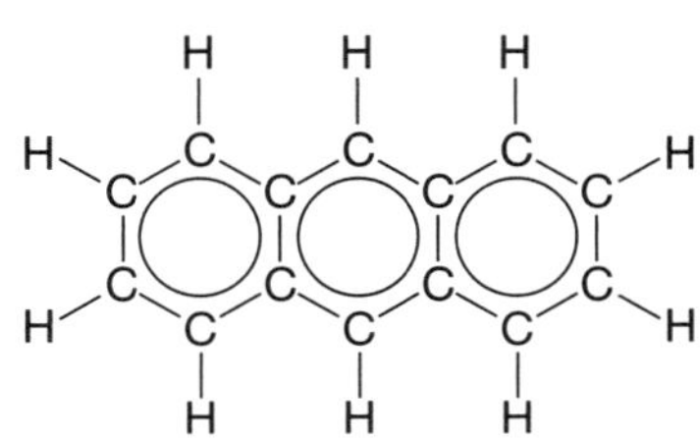

Because the geometry about each carbon atom is trigonal planar, it is appropriate to use sp^2 hybrid orbitals on each of the carbon atoms. The σ-bond framework is

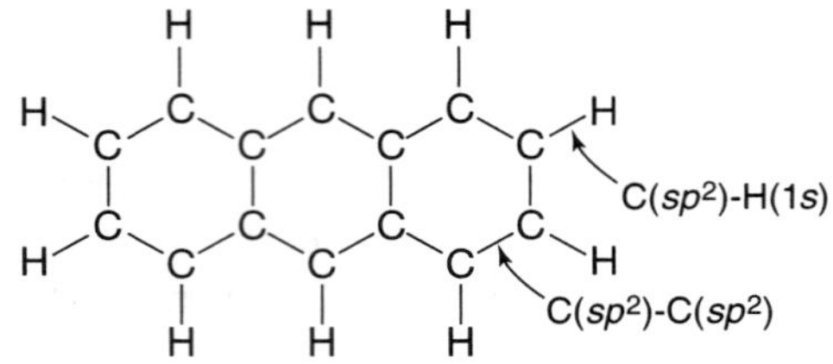

The remaining $2p$ orbitals on each carbon atom combine to form fourteen π orbitals (conservation of orbitals) that are delocalized over the three rings as indicated by the circles in the Lewis formula. Fifty-two of the 66 valence electrons occupy the 26 σ-bond orbitals and the remaining 14 valence electrons occupy the seven delocalized π-bond orbitals with the lowest energy.

ADDITIONAL PROBLEMS

9-52. A bonding molecular orbital formed from the combination of atomic orbitals concentrates the electronic charge between the nuclei and tends to draw the nuclei together. The energy of the bonding molecular orbital is lower than that of the atomic orbitals on the atoms from which it is formed.

In an antibonding molecular orbital formed from the combination of atomic orbitals, the electronic charge is concentrated on the far side of the nuclei and tends to draw the nuclei apart. The energy of the antibonding molecular orbital is greater than that of the atomic orbitals on the atoms from which it is formed.

9-54. Bond lengths decrease and bond strengths (bond energy) increase with increasing bond order.

9-56. A carbon–carbon double bond consists of one σ bond and one π bond. Because energy is required to break the π bond, essentially no rotation occurs about a double bond at room temperature, thereby locking the π bond into a planar shape. Thus, it is possible to form *cis-trans* isomers of the molecule if there are two substituents at each end of the carbon–carbon π-bond.

A carbon–carbon single bond consists of only one σ bond. The carbon atoms are free to rotate about the σ bond so that the substituents are not locked into place. Thus, it is not possible to form *cis-trans* isomers about a carbon–carbon single bond.

A carbon–carbon triple bond consists of one σ bond and two π bonds. There is only one substituent at each end of the triple bond and so it is not possible to form *cis-trans* isomers about a triple bond.

9-58. Antibonding orbitals have a much higher energy than bonding orbitals and are generally unoccupied, so we may ignore the antibonding orbitals when considering molecules in their ground state.

9-60. Because boron and oxygen are close to one another in the periodic table, we shall use Figure 9.15 to place the electrons in the molecular bond orbitals. There are 13 valence electrons in a BO molecule. The ground-state electron configuration is

$$(\sigma_{1s})^2(\sigma_{1s}^*)^2(\sigma_{2s})^2(\sigma_{2s}^*)^2(\pi_{2p})^4(\sigma_{2p})^1$$

The bond order is

$$\text{bond order} = \frac{9-4}{2} = 2\tfrac{1}{2}$$

9-62. Because carbon and nitrogen are close to one another in the periodic table, we shall use Figure 9.15 to place the electrons in the molecular bond orbitals. There are 14 valence electrons in a CN^- ion. The ground-state electron configuration is

$$(\sigma_{1s})^2(\sigma_{1s}^*)^2(\sigma_{2s})^2(\sigma_{2s}^*)^2(\pi_{2p})^4(\sigma_{2p})^2$$

$$\text{bond order} = \frac{10-4}{2} = 3$$

Both a N_2 molecule and a CO molecule have a bond order of 3.

9-64. Table 9.2 shows that the bond order for N_2 in the ground state is 3. The bond order for nitrogen in the given excited state is $(9-5)/2 = 2$. Therefore, the bond length of nitrogen in the excited state is greater than that of a nitrogen molecule in the ground state.

9-66. The ground-electronic state of a F_2 molecule is $(\sigma_{1s})^2(\sigma_{1s}^*)^2(\sigma_{2s})^2(\sigma_{2s}^*)^2(\pi_{2p})^4(\pi_{2p}^*)^4(\sigma_{2p})^2$. The excited state of a F_2 molecule can electronically relax to the ground state by undergoing a $\sigma_{2p}^* \to \sigma_{2p}$ transition, or to a less excited state by transitions such as $\pi_{2p} \to \sigma_{2p}$ or $\pi_{2p}^* \to \sigma_{2p}$. Because the bond order of this excited state is 0, some of the molecules may also undergo photodissociation.

9-68. The central carbon atom in an acetone molecule forms two σ bonds to the two other carbon atoms and a σ bond and a π bond to the oxygen atom. Each carbon–carbon σ-bond orbital is formed by combining an sp^2 orbital on the central carbon atom with an sp^3 orbital on the other carbon atom. The oxygen–carbon σ bond is formed by combining the remaining sp^2 orbital on the central carbon atom with an sp^2 orbital on the oxygen atom. The π-bond orbital is formed by combining the remaining $2p$ orbital on the central carbon atom with the remaining $2p$ orbital on the oxygen atom. Eighteen of the 24 valence electrons occupy the nine σ-bond orbitals and two of them occupy the π-bond orbital. The two lone electron pairs on the oxygen atom occupy the remaining two sp^2 orbitals. The shape of an acetone molecule is trigonal planar around the central carbon atom.

9-70. The Lewis formula of an ethyl cyanide molecule is

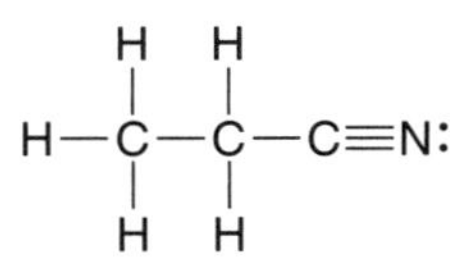

We use sp^3 orbitals on the carbon atoms bonded to the hydrogen atoms and sp orbitals on the cyanide carbon atom and the nitrogen atom. The σ-bond framework in an ethyl cyanide molecule is

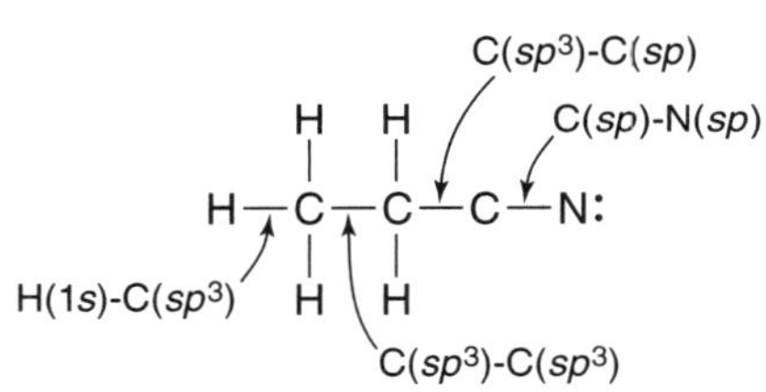

Sixteen of the 22 valence electrons occupy the eight σ-bond orbitals and two others occupy the second nitrogen sp orbital as a lone pair. The two carbon $2p$ orbitals and the two nitrogen $2p$ orbitals combine to form the two carbon–nitrogen π-bond orbitals that are occupied by the remaining four valence electrons to form two π bonds. The two carbon–nitrogen π bonds and the carbon–nitrogen σ bond constitute the carbon–nitrogen triple bond. The use of sp orbitals on the cyanide carbon atom produces a 180° bond angle.

9-72. (a) The Lewis formula is 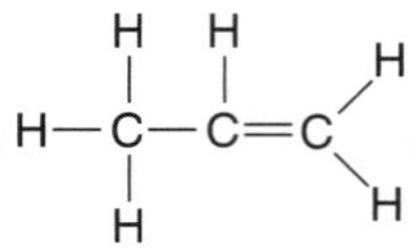 . Therefore, there are 8 σ bonds and 1 π bond.

(b) The Lewis formula is 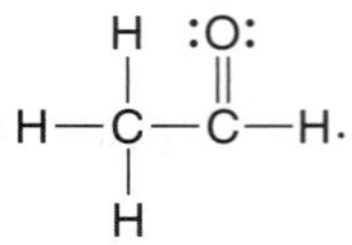. Therefore, there are 6 σ bonds and 1 π bond.

(c) The Lewis formula is 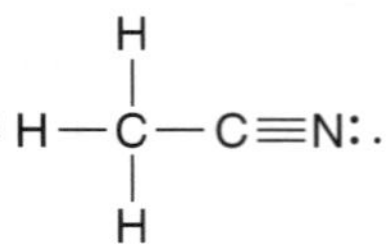. Therefore, there are 5 σ bonds and 2 π bonds.

(d) The Lewis formula is 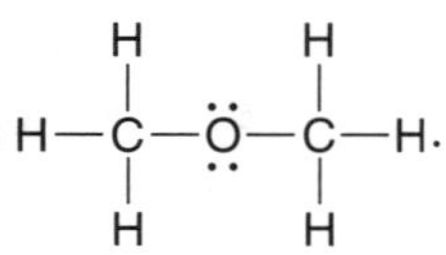. Therefore, there are 8 σ bonds.

9-74. The Lewis formulas of the three different isomers of a dibromoethene molecule are

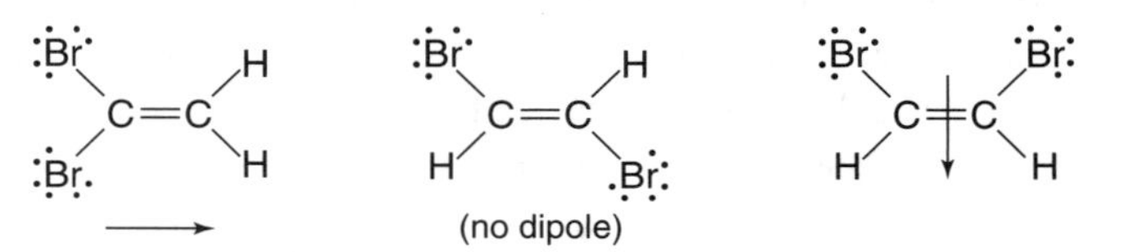

(no dipole)

1,2-dibromoethene *cis*-1,2-dibromoethene *trans*-1,2-dibromoethene

The first and second isomers are structural isomers. The second and third isomers are *cis-trans* isomers. Only the *cis*-1,2-dibromoethane is nonpolar.

9-76. The sigma bond structure of cyanazine is

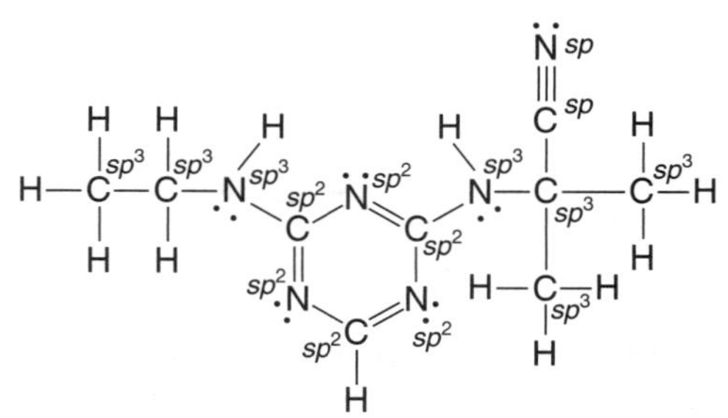

9-78. There are 24 valence electrons in a NO_3^- ion. Because there are three sets of electrons about the nitrogen atom, it is appropriate to use sp^2 hybrid orbitals on the nitrogen atom. We shall also use sp^2 hybrid orbitals on each of the three oxygen atoms. The σ-bond framework of a NO_3^- ion is

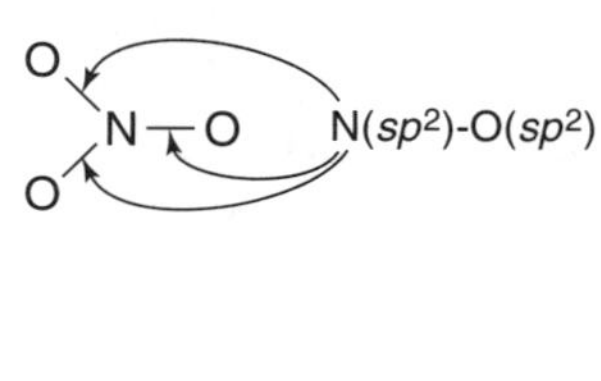

The σ-bond framework accounts for 18 of the 24 valence electrons. The remaining six valence electrons occupy the bonding π orbitals (two electrons) and the two nonbonding π orbitals (four electrons) shown in the diagram. The bond order of each nitrogen–oxygen bond is $1\frac{1}{3}$, 1 from the σ bond and $\frac{1}{3}$ from the delocalized electrons in a bonding orbital that extends over the 3 nitrogen–oxygen bonds in the ion. This picture of the bonding is consistent with the Lewis resonance formula of a NO_3^- ion.

9–80.* The two carbon–oxygen π bonds in a CO_2 molecule are formed from the $2p_x$ and $2p_y$ orbitals on the central carbon atom and the $2p_x$ and $2p_y$ orbitals on each of the two oxygen atoms. Take the z axis to be the bonding axis. Because the $2p_x$ and $2p_y$ orbitals are perpendicular to one another, the two π-bonding orbitals formed from these orbitals are also perpendicular to one another, and so they do not combine effectively.

CHAPTER 10. Chemical Reactivity

CHEMICAL NOMENCLATURE AND POLYATOMIC ANIONS

10-2. In order to name the following compounds, we need to know the names and formulas of the polyatomic ions given in Table 10.1.

(a) sodium acetate (b) calcium chlorate

(c) ammonium carbonate (d) barium nitrate

10-4. In order to name the following compounds, we need to know the names and formulas of the polyatomic ions given in Table 10.1.

(a) ammonium thiosulfate (b) sodium sulfite

(c) potassium carbonate (d) sodium thiosulfate

10-6. The following compounds contain transition-metal ions. The ionic charge of the metal ion is indicated by a Roman numeral. See Section 6-4 of the text.

(a) chromium(II) sulfate (b) cobalt(II) cyanide

(c) tin(II) nitrate (d) copper(I) carbonate

10-8. In order to write the chemical formula for each of the following compounds, we need to know the formula and ionic charge of the polyatomic ions given in Table 10.1. Recall that the net charge on an ionic compound must be zero.

(a) $HC_2H_3O_2$ (b) $HClO_3$ (c) H_2CO_3 (d) $HClO_4$

10-10. See the solution to Problem 10-8.

(a) $NaClO_4$ (b) $KMnO_4$ (c) $CaSO_3$ (d) $LiCN$

10-12. The Roman numeral following the name of the transition-metal ion indicates the ionic charge of the metal ion.

(a) $Hg_2(C_2H_3O_2)_2$ (b) $Hg(CN)_2$ (c) $Fe(ClO_4)_2$ (d) $CrSO_3$

10-14. (a) $NaClO(s)$ (b) $H_2O_2(l)$ (c) $KOH(s)$ (d) $CH_3COOH(aq)$

ACIDS AND BASES

10-16. (a) basic (b) acidic (c) acidic (d) basic (e) basic

10-18. (a) LiOH(aq) dissociates in aqueous solution to produce $OH^-(aq)$ ions and so is a base.

(b) Na_2O is an oxide of a Group 1 metal, and so reacts with water according to

$$Na_2O(s) + H_2O(l) \xrightarrow[H_2O(l)]{} 2\,Na^+(aq) + 2\,OH^-(aq)$$

Thus, Na_2O forms a basic solution.

(c) Lead is not one of the metals listed in Figure 10.7 whose oxides form basic solutions. In addition, from the solubility Rules, PbO(s) is insoluble in water. Thus, we conclude that PbO(s) will not form a basic solution.

(d) $HNO_3(aq)$ dissociates in aqueous solution to produce $H^+(aq)$ ions and so is an acid, not a base.

OXYACIDS AND OXYANIONS

10-20. (a) organic acid (b) oxyacid (c) organic acid (d) oxyacid

10-22. (a) NO_2^- is a nitrite anion. Because the name of the anion ends in *-ite*, the acid is nitrous acid.

(b) SO_2^{2-} contains one less oxygen atom than the SO_3^{2-} anion and so is called a hyposulfite anion. Because the name of the anion ends in *-ite*, the acid is hyposulfurous acid.

(c) ClO_2^- is a chlorite anion, and so the acid is chlorous acid.

(d) IO_3^- is an iodate anion, and so the acid is iodic acid.

10-24. The names are

(a) copper(II) hypochlorite (b) scandium(III) iodate

(c) iron(III) bromate (d) ruthenium(III) periodate

10-26. The Lewis formulas are

(a) Perchlorate ion: 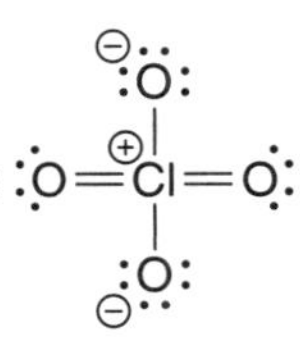 + other resonance forms

(b) Carbonate ion: 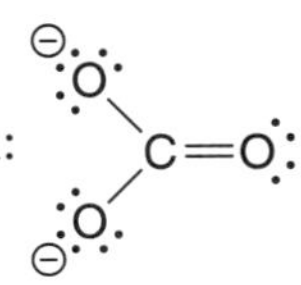 + other resonance forms

(c) Phosphate ion: 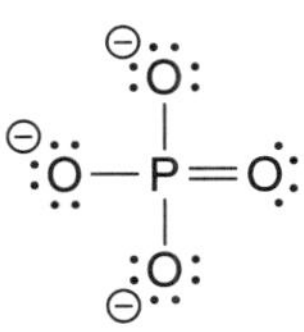 + other resonance forms

(d) Acetate ion: 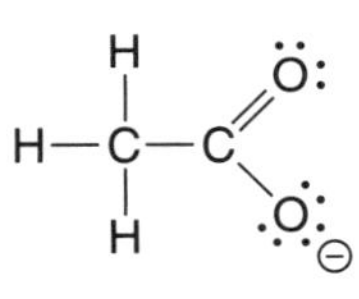 + other resonance forms

10-28. The Lewis formulas of the corresponding acids of the anions in Problem 10-26 are

(a) Perchloric acid: 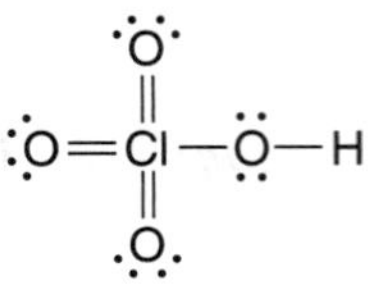

(b) Carbonic acid: 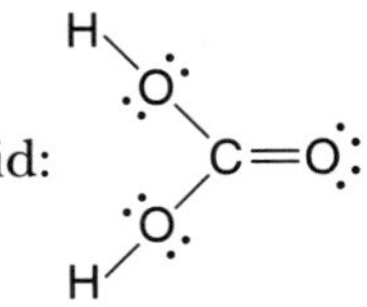

(c) Phosphoric acid: 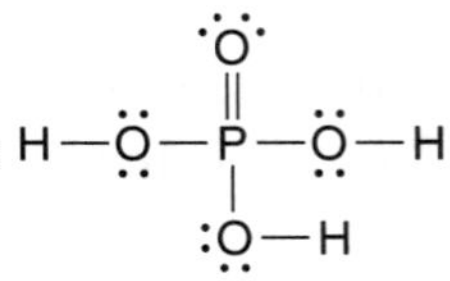

(d) Acetic acid: 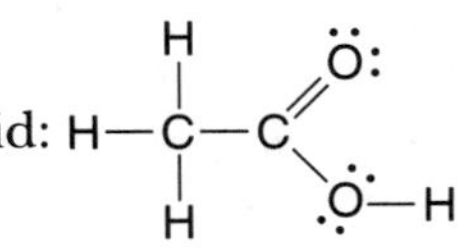

HYDRATES

10-30. The names of the hydrates are

(a) barium hydroxide octahydrate

(b) lead(II) chloride dihydrate

(c) lithium hydroxide monohydrate

(d) lithium chromate dihydrate

10-32. (a) $H_2C_2O_4 \cdot 2\,H_2O$ (b) $Al_2(SO_4)_3 \cdot 8\,H_2O$ (c) $NdI_3 \cdot 9\,H_2O$ (d) $Na_2HPO_4 \cdot 7\,H_2O$

CLASSIFICATION OF REACTIONS AND PREDICTION OF REACTION PRODUCTS

10-34. We classify the reactions as follows

(a) decomposition

(b) single replacement

(c) double replacement

(d) single replacement

10-36. (a) decomposition

$$2\,NaClO_3(s) \longrightarrow 2\,NaCl(s) + 3\,O_2(g)$$

(b) combination; already balanced

(c) single replacement

$$H_2(g) + 2\,AgCl(s) \xrightarrow[H_2O(l)]{} 2\,Ag(s) + 2\,HCl(aq)$$

(d) double replacement

$$Hg_2(NO_3)_2(aq) + 2\,CH_3COOH(aq) \longrightarrow Hg_2(CH_3COO)_2(s) + 2\,HNO_3(aq)$$

10-38. (a) $4\,Li(s) + O_2(g) \longrightarrow 2\,Li_2O(s)$

(b) $MgO(s) + CO_2(g) \longrightarrow MgCO_3(s)$

(c) $2\,H_2(g) + O_2(g) \longrightarrow 2\,H_2O(l)$

(d) $N_2(g) + 3\,H_2(g) \longrightarrow 2\,NH_3(g)$

10-40. See Table 10.8 to see which metals react with water or an acid to produce hydrogen gas.

(a) $Ba(s) + 2\,H_2O(l) \longrightarrow Ba(OH)_2(aq) + H_2(g)$

(b) $Fe(s) + H_2SO_4(aq) \longrightarrow 2\,FeSO_4(aq) + H_2(g)$

(c) $Ca(s) + 2\,HBr(aq) \longrightarrow CaBr_2(aq) + H_2(g)$

(d) $Pb(s) + 2\,HCl(aq) \longrightarrow PbCl_2(aq) + H_2(g)$

NET IONIC EQUATIONS

10-42. To determine the net ionic equation that corresponds to the complete ionic equation, we determine the precipitate or molecular product formed. The ions that are used to form this species are the ones in the net ionic equation.

(a) $H^+(aq) + OH^-(aq) \longrightarrow H_2O(l)$

(b) $Pb^{2+}(aq) + CO_3^{2-}(aq) \longrightarrow PbCO_3(s)$

(c) $2\,Ag^+(aq) + SO_4^{2-}(aq) \longrightarrow Ag_2SO_4(s)$

(d) $S^{2-}(aq) + Zn^{2+}(aq) \longrightarrow ZnS(s)$

10-44. To determine the net ionic equation that corresponds to the complete ionic equation, we determine the precipitate or molecular product formed. The ions that are used to form this species are the ones in the net ionic equation.

(a) $2\,AgNO_3(aq) + Na_2S(aq) \longrightarrow Ag_2S(s) + 2\,NaNO_3(aq)$
$2\,Ag^+(aq) + S^{2-}(aq) \longrightarrow Ag_2S(s)$

(b) $H_2SO_4(aq) + Pb(NO_3)_2(aq) \longrightarrow PbSO_4(s) + 2\,HNO_3(aq)$
$SO_4^{2-}(aq) + Pb^{2+}(aq) \longrightarrow PbSO_4(s)$

(c) $Hg(NO_3)_2(aq) + 2\,NaI(aq) \longrightarrow HgI_2(s) + 2\,NaNO_3(aq)$
$Hg^{2+}(aq) + 2\,I^-(aq) \longrightarrow HgI_2(s)$

(d) $CdCl_2(aq) + 2\,AgClO_4(aq) \longrightarrow 2\,AgCl(s) + Cd(ClO_4)_2(aq)$
$Cl^-(aq) + Ag^+(aq) \longrightarrow AgCl(s)$

SOLUBILITY RULES

10-46. (a) insoluble, Rule 5 (b) soluble, Rule 4 (c) soluble, Rule 1
(d) insoluble, Rule 3 (e) insoluble, Rule 3

10-48. (a) insoluble, Rule 3 (b) soluble, Rule 2 (c) insoluble, Rule 3
(d) soluble, Rule 2 (e) insoluble, Rule 3

10-50. (a) soluble; $FeBr_3(s) \xrightarrow[H_2O(l)]{} Fe^{3+}(aq) + 3\,Br^-(aq)$

(b) insoluble

(c) soluble; $(NH_4)_2CO_3(s) \xrightarrow[H_2O(l)]{} 2\,NH_4^+(aq) + CO_3^{2-}(aq)$

(d) soluble; $K_2S(s) \xrightarrow[H_2O(l)]{} 2\,K^+(aq) + S^{2-}(aq)$

DOUBLE-REPLACEMENT REACTIONS

10-52. (a) $CaSO_4$ is insoluble by Rule 6.

$$H_2SO_4(aq) + Ca(ClO_4)_2(aq) \longrightarrow CaSO_4(s) + 2\,HClO_4(aq)$$
$$SO_4^{2-}(aq) + Ca^{2+}(aq) \longrightarrow CaSO_4(s)$$

(b) All compounds are soluble; thus there is no reaction.

(c) $Hg_2(C_7H_5O_2)_2$ is insoluble by Rule 3.

$$Hg_2(NO_3)_2(aq) + 2\,NaC_7H_5O_2(aq) \longrightarrow Hg_2(C_7H_5O_2)_2(s) + 2\,NaNO_3(aq)$$
$$Hg_2^{2+}(aq) + 2\,C_7H_5O_2^-(aq) \longrightarrow Hg_2(C_7H_5O_2)_2(s)$$

(d) $PbBr_2$ is insoluble by Rule 3.

$$Pb(CH_3COO)_2(aq) + 2\,KBr(aq) \longrightarrow PbBr_2(s) + 2\,KCH_3COO(aq)$$

$$Pb^{2+}(aq) + 2\,Br^-(aq) \longrightarrow PbBr_2(s)$$

10-54. (a) $NH_4NO_3(aq) + NaOH(aq) \longrightarrow NaNO_3(aq) + H_2O(l) + NH_3(aq)$

$$NH_4^+(aq) + OH^-(aq) \longrightarrow H_2O(l) + NH_3(aq)$$

(b) $2\,HNO_3(aq) + BaCO_3(s) \longrightarrow Ba(NO_3)_2(aq) + H_2O(l) + CO_2(g)$

$$2\,H^+(aq) + BaCO_3(s) \longrightarrow Ba^{2+}(aq) + H_2O(l) + CO_2(g)$$

(c) $2\,H_2O_2(aq) \longrightarrow 2\,H_2O(l) + O_2(g)$

The reaction equation is the same as the complete equation.

10-56. (a) $2\,HClO_4(aq) + Ca(OH)_2(aq) \longrightarrow \underset{\text{calcium perchlorate}}{Ca(ClO_4)_2(aq)} + 2\,H_2O(l)$

$$H^+(aq) + OH^-(aq) \longrightarrow H_2O(l)$$

(b) $2\,HCl(aq) + CaCO_3(s) \longrightarrow \underset{\text{calcium chloride}}{CaCl_2(aq)} + \underset{\text{carbon dioxide}}{CO_2(g)} + H_2O(l)$

$$2\,H^+(aq) + CaCO_3(s) \longrightarrow Ca^{2+}(aq) + CO_2(g) + H_2O(l)$$

(c) $6\,HNO_3(aq) + Al_2O_3(s) \longrightarrow \underset{\text{aluminum nitrate}}{2\,Al(NO_3)_3(aq)} + 3\,H_2O(l)$

$$6\,H^+(aq) + Al_2O_3(s) \longrightarrow 2\,Al^{3+}(aq) + 3\,H_2O(l)$$

(d) $H_2SO_4(aq) + Cu(OH)_2(s) \longrightarrow \underset{\text{copper(II) sulfate}}{CuSO_4(aq)} + 2\,H_2O(l)$

$$2\,H^+(aq) + Cu(OH)_2(s) \longrightarrow Cu^{2+}(aq) + 2\,H_2O(l)$$

10-58. (a) $K_2CrO_4(aq) + Pb(NO_3)_2(aq) \longrightarrow 2\,KNO_3(aq) + PbCrO_4(s)$

(b) $2\,HCl(aq) + Na_2S(aq) \longrightarrow 2\,NaCl(aq) + H_2S(g)$

(c) $Ba(OH)_2(aq) + ZnSO_4(aq) \longrightarrow Zn(OH)_2(s) + BaSO_4(s)$

(d) $2\,HNO_3(aq) + CaO(s) \longrightarrow Ca(NO_3)_2(aq) + H_2O(l)$

10-60. (a) precipitation reaction

(b) gas forming reaction

(c) precipitation reaction

(d) acid-base reaction

OXIDATION-REDUCTION

10-62. (a) The lithium atoms in Li(s) are neutral, and so have an ionic charge of zero. The selenium atoms in Se(s) are neutral, and so have an ionic charge of zero. In one Li_2Se formula unit, the ionic charge of the lithium atom (Group 1 metal) is $+1$ and that of each selenium atom (a Group 16 nonmetal) is -2. Because the lithium atom loses an electron in going from an ionic charge of 0 to $+1$, the lithium atom is oxidized and Li(s) is the reducing agent. Because each selenium atom gains two electrons in going from an ionic charge of 0 to -2, each selenium atom is reduced and Se(s) is the oxidizing agent.

(b) The scandium atoms in Sc(s) are neutral, and so have an ionic charge of zero. An iodine molecule is neutral, and so each iodine atom in $I_2(g)$ has an ionic charge of zero. In one ScI_3 formula unit, the ionic charge of the scandium atom (Group 3 metal) is +3 and that of each iodine atom (a Group 17 nonmetal) is −1. Because the scandium atom loses three electrons in going from an ionic charge of 0 to +3, the scandium atom is oxidized and Sc(s) is the reducing agent. Because each iodine atom gains an electron in going from an ionic charge of 0 to −1, each iodine atom is reduced and $I_2(g)$is the oxidizing agent.

(c) The gallium atoms in Ga(s) are neutral, and so have an ionic charge of zero. A phosphorus molecule is neutral, and so each phosphorus atom in $P_4(s)$ has an ionic charge of zero. In one GaP formula unit, the ionic charge of the gallium atom (Group 3 metal) is +3 and that of each phosphorus atom (a Group 15 nonmetal) is −3. Because the gallium atom loses three electrons in going from an ionic charge of 0 to +3, the gallium atom is oxidized and Ga(s) is the reducing agent. Because each phosphorus atom gains three electrons in going from an ionic charge of 0 to −3, each phosphorus atom is reduced and $P_4(s)$is the oxidizing agent.

(d) The potassium atoms in K(s) are neutral, and so have an ionic charge of zero. A fluorine molecule is neutral, and so each fluorine atom in $F_2(g)$ has an ionic charge of zero. In one KF formula unit, the ionic charge of the potassium atom (Group 1 metal) is +1 and that of each fluorine atom (a Group 17 nonmetal) is −1. Because the potassium atom loses one electron in going from an ionic charge of 0 to +1, the potassium atom is oxidized and K(s) is the reducing agent. Because each fluorine atom gains an electron in going from an ionic charge of 0 to −1, each fluorine atom is reduced and $F_2(g)$is the oxidizing agent.

10-64. (a) In the chemical equation, two lithium atoms are oxidized, $2 \times (0 \rightarrow +1)$, and one selenium atom is reduced, $1 \times (0 \rightarrow -2)$, and so there is a transfer of two electrons in this equation.

(b) In the chemical equation, two scandium atoms are oxidized, $2 \times (0 \rightarrow +3)$, and six iodine atoms are reduced, $6 \times (0 \rightarrow -2)$, and so there is a transfer of 12 electrons in this equation.

(c) In the chemical equation, four gallium atoms are oxidized, $4 \times (0 \rightarrow +3)$, and four phosphorus atoms are reduced, $4 \times (0 \rightarrow -3)$, and so there is a transfer of 12 electrons in this equation.

(d) In the chemical equation, two potassium atoms are oxidized, $2 \times (0 \rightarrow +1)$, and two fluoride atoms are reduced, $2 \times (0 \rightarrow -1)$, and so there is a transfer of two electrons in this equation.

10-66. The balanced chemical equation that describes the reaction in the fuel cell is

$$CH_4(g) + 2\,O_2(g) \longrightarrow CO_2(g) + 2\,H_2O(l)$$

In this equation, the ionic charge on each oxygen atom goes from 0 in $O_2(g)$ to −2 in $CO_2(g)$ and $H_2O(l)$. Therefore, the oxygen atoms are reduced and $O_2(g)$ is the oxidizing agent. Because each of the four hydrogen atoms in a $CH_4(g)$ molecule has an ionic charge of +1, we conclude that the ionic charge of the carbon atom in $CH_4(g)$ is −4. Because each of the two oxygen atoms in a $CO_2(g)$ molecule has an ionic charge of −2, we conclude that the ionic charge of the carbon atom in $CO_2(g)$ is +4. Therefore, the carbon atoms are oxidized $(-4 \rightarrow +4)$ and $CH_4(g)$ is the reducing agent. In the chemical equation, one carbon atom is oxidized, $1 \times (-4 \rightarrow +4)$, and four oxygen atoms are reduced, $4 \times (0 \rightarrow -2)$, and so there is a transfer of eight electrons in this equation.

ADDITIONAL PROBLEMS

10-68. In a single replacement reaction, one species replaces another in a compound. In a double-replacement reaction, two ionic species exchange cations and anions.

10-70. The two products are carbon dioxide and water.

10-72. (a) $2\,Na(s) + H_2(g) \longrightarrow 2\,NaH(s)$

(b) $2\,Al(s) + 3\,S(s) \longrightarrow Al_2S_3(s)$

(c) $H_2O(g) + C(s) \longrightarrow CO(g) + H_2(g)$

(d) $PCl_3(l) + Cl_2(g) \longrightarrow PCl_5(s)$

10-74. $HCHO_2(aq) + NH_3(aq) \longrightarrow NH_4CHO_2(aq)$

10-76. (a) $Cl_2(g) + 2\,NaI(aq) \longrightarrow 2\,NaCl(aq) + I_2(s)$

(b) $Br_2(l) + 2\,NaI(aq) \longrightarrow 2\,NaBr(aq) + I_2(s)$

(c) no reaction

(d) no reaction

10-78. (a) $ZnS(s) + 2\,HCl(aq) \longrightarrow ZnCl_2(aq) + H_2S(g)$

(b) $2\,PbO_2(s) \longrightarrow 2\,PbO(s) + O_2(g)$

(c) $3\,CaCl_2(aq) + 2\,H_3PO_4(aq) \longrightarrow Ca_3(PO_4)_2(s) + 6\,HCl(aq)$

10-80. (a) $C_{12}H_{22}O_{11}(s) \longrightarrow 12\,C(s) + 11\,H_2O(l)$

(b) $Cl_2(g) + 2\,NaBr(aq) \longrightarrow 2\,NaCl(aq) + Br_2(l)$

(c) $Li_2O(s) + H_2O(l) \longrightarrow 2\,LiOH(aq)$

10-82. (a) $Na_2CO_3 \cdot 10\,H_2O(s) \longrightarrow Na_2CO_3(s) + 10\,H_2O(g)$

(b) $Pb(NO_3)_2(aq) + Na_2SO_4(aq) \longrightarrow PbSO_4(s) + 2\,NaNO_3(aq)$

(c) $2\,Fe(s) + 3\,Pb(NO_3)_2(aq) \longrightarrow 2\,Fe(NO_3)_3(aq) + 3\,Pb(s)$

10-84. Powdered calcium carbonate is best. It will neutralize the hydrochloric acid and you can use an excess, unlike sodium hydroxide, where an excess would lead to a problem similar to the original problem. Water simply dilutes the acid, but does not neutralize it.

10-86. $2\,Pb(l) + O_2(g) \longrightarrow 2\,PbO(s)$

$Ag(l) + O_2(g) \longrightarrow$ no reaction

10-88. $2\,HgS(s) + 2\,O_2(g) \xrightarrow{\text{heat}} HgO(s) + SO_2(g)$

$HgO(s) + HgS(s) \xrightarrow{\text{heat}} Hg(g) + SO_2(g)$

$Hg(g) \xrightarrow{\text{cold}} Hg(l)$

10-90.* 1. Add $Cl^-(aq)$, for example by adding $NaCl(aq)$. If $Pb^{2+}(aq)$ is present, it will precipitate as $PbCl_2(s)$. None of the other possible cations will form an insoluble chloride precipitate.

2. To the resulting solution from step 1, add $SO_4^{2-}(aq)$, for example by adding $Na_2SO_4(aq)$. If $Ca^{2+}(aq)$ is present, it will precipitate as $CaSO_4(s)$. None of the remaining cations will form an insoluble sulfate precipitate.

3. To the resulting solution, add $NaOH(aq)$. If $Fe^{3+}(aq)$ is present, it will precipitate as $Fe(OH)_3(s)$. The ammonium ion does not form an insoluble hydroxide precipitate.

4. Now heat the solution and smell to test for ammonia. Although ammonia gas is very soluble in water, some ammonia gas will be released upon heating. In basic solution, some ammonia will be present due to the reaction described by

$$NH^{+}(aq) + OH^{-}(aq) \longrightarrow NH_3(aq) + H_2O(l)$$

10-92.* $a\,CH_3CH_2OH + b\,K_2Cr_2O_7 + c\,H_2SO_4 \longrightarrow d\,CH_3COOH + e\,Cr_2(SO_4)_3 + f\,K_2SO_4 + g\,H_2O$
The number of each atom must be the same on both sides of the equation, and so we write

$$\begin{aligned} &C: && 2a = 2d \\ &K: && 2b = 2f \\ &O: && a + 7b + 4c = 2d + 12e + 4f + g \\ &Cr: && 2b = 2e \\ &S: && c = 3e + f \\ &H: && 6a + 2c = 4d + 2g \end{aligned}$$

Take $b = 1$ (arbitrarily). Then $e = 1$, $f = 1$, and $c = 4$. Now we have the equations

$$\begin{aligned} &C: && a = d \\ &O: && 23 = d + 16 + g \\ &H: && 2d + 8 = 2g \quad (g = d + 4) \end{aligned}$$

Therefore, $d = 3/2 = a$ and $g = 11/2$. Putting this all together and eliminating the fractional balancing coefficients by multiplying by 2, we write

$$3\,CH_3CH_2OH(aq) + 2\,K_2Cr_2O_7(aq) + 8\,H_2SO_4(aq) \longrightarrow$$
$$3\,CH_3COOH(aq) + 2\,Cr_2(SO_4)_3(aq) + 2\,K_2SO_4(aq) + 11\,H_2O(l)$$

CHAPTER 11. Chemical Calculations

MOLES AND MASS PERCENT

11-2. (a) $\text{mass of Hg} = (3.00\text{ mol Hg})\left(\frac{200.59\text{ g Hg}}{1\text{ mol Hg}}\right) = 602\text{ g}$

(b) $\text{mass of Fe(OH)}_3 = (1.872\times10^{24}\text{ molecules Fe(OH)}_3)\left(\frac{1\text{ mol Fe(OH)}_3}{6.022\times10^{23}\text{ molecules Fe(OH)}_3}\right)$

$$\times\left(\frac{106.87\text{ g Fe(OH)}_3}{1\text{ mol Fe(OH)}_3}\right) = 332.2\text{ g}$$

(c) $\text{mass of }^{18}\text{O atoms} = (1.00\text{ mol O}_2)\left(\frac{18.00\text{ g }^{18}\text{O}}{1\text{ mol }^{18}\text{O}}\right) = 18\text{ g}$

(d) $\text{mass of N}_2 = (2.0\text{ mol N}_2)\left(\frac{28.01\text{ g N}_2}{1\text{ mol N}_2}\right) = 56\text{ g}$

11-4. $\text{mass \% C} = \frac{\text{mass C}}{\text{mass C}_{12}\text{H}_{17}\text{NO}}\times100 = \frac{144.1284}{191.2695}\times100 = 75.35\%$

$$\text{mass \% H} = \frac{\text{mass H}}{\text{mass C}_{12}\text{H}_{17}\text{NO}}\times100 = \frac{17.13498}{191.2695}\times100 = 8.959\%$$

$$\text{mass \% N} = \frac{\text{mass N}}{\text{mass C}_{12}\text{H}_{17}\text{NO}}\times100 = \frac{14.0067}{191.2695}\times100 = 7.323\%$$

$$\text{mass \% O} = \frac{\text{mass O}}{\text{mass C}_{12}\text{H}_{17}\text{NO}}\times100 = \frac{15.9994}{191.2695}\times100 = 8.365\%$$

Note that the sum of the percentages is 100.00%.

11-6. (a) $\text{mass \% C} = \frac{\text{mass C}}{\text{mass CH}_3\text{OH}}\times100 = \frac{12.0107}{32.0419}\times100 = 37.48\%$

$$\text{mass \% O} = \frac{\text{mass O}}{\text{mass CH}_3\text{OH}}\times100 = \frac{15.9994}{32.0419}\times100 = 49.93\%$$

$$\text{mass \% H} = \frac{\text{mass H}}{\text{mass CH}_3\text{OH}}\times100 = \frac{4.03176}{32.0419}\times100 = 12.58\%$$

Note that the sum of the percentages is 100.00%.

(b) $$\text{mass \% O} = \frac{\text{mass O}}{\text{mass H}_2\text{O}} \times 100 = \frac{15.9994}{18.0153} \times 100 = 88.81\%$$

$$\text{mass \% H} = \frac{\text{mass H}}{\text{mass H}_2\text{O}} \times 100 = \frac{2.0159}{18.0153} \times 100 = 11.19\%$$

Note that the sum of the percentages is 100.00%.

(c) $$\text{mass \% O} = \frac{\text{mass O}}{\text{mass H}_2\text{O}_2} \times 100 = \frac{31.9988}{34.00147} \times 100 = 94.07\%$$

$$\text{mass \% H} = \frac{\text{mass H}}{\text{mass H}_2\text{O}_2} \times 100 = \frac{2.01588}{34.00147} \times 100 = 5.926\%$$

Note that the sum of the percentages is 100.00%.

(d) $$\text{mass \% Mg} = \frac{\text{mass Mg}}{\text{mass MgSO}_4\cdot 7\,\text{H}_2\text{O}} \times 100 = \frac{24.3050}{246.475} \times 100 = 9.861\%$$

$$\text{mass \% S} = \frac{\text{mass S}}{\text{mass MgSO}_4\cdot 7\,\text{H}_2\text{O}} \times 100 = \frac{32.065}{246.475} \times 100 = 13.01\%$$

$$\text{mass \% O} = \frac{\text{mass O}}{\text{mass MgSO}_4\cdot 7\,\text{H}_2\text{O}} \times 100 = \frac{175.9934}{246.475} \times 100 = 71.40\%$$

$$\text{mass \% H} = \frac{\text{mass H}}{\text{mass MgSO}_4\cdot 7\,\text{H}_2\text{O}} \times 100 = \frac{14.1112}{246.475} \times 100 = 5.725\%$$

Note that the sum of the percentages is 100.00%.

11-8. (a) $$\text{mass of O} = (13.0\text{ g})\left(\frac{\text{mass O}}{\text{mass Ca(CH}_3\text{COO)}_2}\right) = (13.0\text{ g})\left(\frac{64.00}{158.17}\right) = 5.26\text{ g}$$

(b) $$\text{mass of F} = (0.22\text{ g})\left(\frac{\text{mass F}}{\text{mass XeF}_4}\right) = (0.22\text{ g})\left(\frac{76.0}{207.3}\right) = 0.081\text{ g}$$

(c) $$\text{mass of H}_2\text{O} = (25.0\text{ g})\left(\frac{\text{mass H}_2\text{O}}{\text{mass BaCl}_2\cdot 2\,\text{H}_2\text{O}}\right) = (25.0\text{ g})\left(\frac{36.03}{244.26}\right) = 3.69\text{ g}$$

(d) $$\text{mass of C} = (2.00\text{ g})\left(\frac{\text{mass C}}{\text{mass C}_6\text{H}_{14}}\right) = (2.00\text{ g})\left(\frac{72.06}{86.18}\right) = 1.67\text{ g}$$

SIMPLEST OR EMPIRICAL FORMULAS

11-10. Take a 100-gram sample and write

$$69.9\text{ g Fe} \Leftrightarrow 30.1\text{ g O}$$

Divide each quantity by its corresponding atomic mass to get

$$(69.9\text{ g})\left(\frac{1\text{ mol Fe}}{55.85\text{ g Fe}}\right) = 1.25\text{ mol Fe} \Leftrightarrow (30.1\text{ g})\left(\frac{1\text{ mol O}}{16.00\text{ g O}}\right) = 1.88\text{ mol O}$$

Divide both sides by the smaller quantity (1.25) to obtain

$$1.00\text{ mol Fe} \Leftrightarrow 1.50\text{ mol O}$$

Multiply both sides by 2 to obtain

$$2.00\text{ mol Fe} \Leftrightarrow 3.00\text{ mol O}$$

The empirical formula of rust is Fe_2O_3.

11-12. The mass percentage of Fe in the compound is

$$\text{mass \% Fe} = \frac{3.78 \text{ g Fe}}{5.95 \text{ g iron sulfide}} \times 100 = 63.5\%$$

Therefore, mass % S = 36.5%. Assume a 100-gram sample and write

$$63.5 \text{ g Fe} \approx 36.5 \text{ g S}$$

Divide each quantity by its corresponding atomic mass to get

$$(63.5 \text{ g})\left(\frac{1 \text{ mol Fe}}{55.85 \text{ g Fe}}\right) = 1.14 \text{ mol Fe} \approx (36.5 \text{ g})\left(\frac{1 \text{ mol S}}{32.07 \text{ g S}}\right) = 1.14 \text{ mol S}$$

Divide both sides by 1.14 to obtain

$$1.00 \text{ mol Fe} \approx 1.00 \text{ mol S}$$

The empirical formula is FeS.

11-14. We are given that 5.00 grams of aluminum react with oxygen to yield 9.45 grams of aluminum oxide. Therefore, the mass of oxygen that reacted is 9.45 g − 5.00 g = 4.45 g. Thus, the mass of oxygen in the compound is 4.45 grams, and so we have the stoichiometric correspondence

$$5.00 \text{ g Al} \approx 4.45 \text{ g O}$$

Divide each quantity by its corresponding atomic mass to get

$$0.185 \text{ mol Al} \approx 0.278 \text{ mol O}$$

or

$$1.00 \text{ mol Al} \approx 1.50 \text{ mol O}$$

Multiply both sides by 2 to obtain

$$2.00 \text{ mol Al} \approx 3.00 \text{ mol O}$$

The empirical formula is Al_2O_3.

11-16. (a) Take a 100-gram sample and write

$$71.89 \text{ g Tl} \approx 28.11 \text{ g Br}$$

$$\frac{71.89 \text{ g}}{204.4 \text{ g}\cdot\text{mol}^{-1}} = 0.3517 \text{ mol Tl} \approx \frac{28.11 \text{ g}}{79.90 \text{ g}\cdot\text{mol}^{-1}} = 0.3518 \text{ mol Br}$$

$$1.00 \text{ mol Tl} \approx 1.00 \text{ mol Br}$$

The empirical formula is TlBr, thallium(I) bromide.

(b) Take a 100-gram sample and write

$$74.51 \text{ g Pb} \approx 25.49 \text{ g Cl}$$

$$\frac{74.51 \text{ g}}{207.2 \text{ g}\cdot\text{mol}^{-1}} = 0.3596 \text{ mol Pb} \approx \frac{25.49 \text{ g}}{35.45 \text{ g}\cdot\text{mol}^{-1}} = 0.7190 \text{ mol Cl}$$

$$1.00 \text{ mol Pb} \approx 2.00 \text{ mol Cl}$$

The empirical formula is $PbCl_2$, lead(II) chloride.

(c) Take a 100-gram sample and write

$$82.24 \text{ g N} \approx 17.76 \text{ g H}$$

$$\frac{82.24 \text{ g}}{14.01 \text{ g}\cdot\text{mol}^{-1}} = 5.870 \text{ mol N} \approx \frac{17.76 \text{ g}}{1.008 \text{ g}\cdot\text{mol}^{-1}} = 17.62 \text{ mol H}$$

$$1.00 \text{ mol N} \approx 3.00 \text{ mol H}$$

The empirical formula is NH_3, ammonia.

(d) Take a 100-gram sample and write

$$72.24 \text{ g Mg} \approx 27.76 \text{ g N}$$

$$\frac{72.24 \text{ g}}{24.31 \text{ g}\cdot\text{mol}^{-1}} = 2.972 \text{ mol Mg} \approx \frac{27.76 \text{ g}}{14.01 \text{ g}\cdot\text{mol}^{-1}} = 1.981 \text{ mol N}$$

$$1.50 \text{ mol Mg} \approx 1.00 \text{ mol N}$$

$$3.00 \text{ mol Mg} \approx 2.00 \text{ mol N}$$

The empirical formula is Mg_3N_2, magnesium nitride.

DETERMINATION OF ATOMIC MASS

11-18. The mass percentage of the element X is

$$\text{mass \% X} = 100\% - 74.8\% = 25.2\%$$

Assuming a 100-gram sample, we have

$$74.8 \text{ g Cl} \approx 25.2 \text{ g X}$$

We do not know the atomic mass of X, but we can divide 74.8 g Cl by the atomic mass of Cl, 35.45, to obtain

$$2.11 \text{ mol Cl} \approx 25.2 \text{ g X}$$

We know from the given empirical formula that

$$4 \text{ mol Cl} \approx 1 \text{ mol X}$$

Thus, we have

$$(2.11 \text{ mol Cl})\left(\frac{1 \text{ mol X}}{4 \text{ mol Cl}}\right) \approx 25.2 \text{ g X}$$

or

$$0.528 \text{ mol X} \approx 25.2 \text{ g X}$$

Divide both sides by 0.528 to get

$$1.00 \text{ mol X} \approx 47.7 \text{ g X}$$

The atomic mass of X is 47.7, which corresponds to titanium (Ti).

11-20. Let x be the atomic mass of element X. The formula mass of HXO_3 is given by

$$\text{formula mass of } HXO_3 = (1 \times 1.008) + x + (3 \times 16.00) = 49.01 + x$$

We have the stoichiometric correspondence

$$\begin{aligned} 0.0133 \text{ mol HXO}_3 &\Bumpeq 1.123 \text{ g HXO}_3 \\ 1 \text{ mol HXO}_3 &\Bumpeq 84.44 \text{ g HXO}_3 \end{aligned}$$

Thus,

$$49.01 + x = 84.44$$

Solving for x, we have $x = 35.4$. The atomic mass of X is 35.4, which corresponds to chlorine (Cl).

MOLECULAR FORMULAS

11-22. Take a 100-gram sample and write

$$40.0 \text{ g C} \Bumpeq 6.71 \text{ g H} \Bumpeq 53.3 \text{ g O}$$

Divide by the corresponding atomic masses to get

$$3.33 \text{ mol C} \Bumpeq 6.66 \text{ mol H} \Bumpeq 3.33 \text{ mol O}$$

Now divide by the smallest quantity (3.33) to get

$$1.00 \text{ mol C} \Bumpeq 2.00 \text{ mol H} \Bumpeq 1.00 \text{ mol O}$$

The simplest formula is CH_2O, whose formula mass is 30.03. Given that the molecular mass is 180.2, which is six times the formula mass of the simplest formula, the molecular formula is $C_6H_{12}O_6$.

11-24. Take a 100-gram sample and write

$$0.373 \text{ g Fe} \Bumpeq 99.627 \text{ g hemoglobin}$$

Divide the mass of iron by its atomic mass to get

$$0.00668 \text{ mol Fe} \Bumpeq 99.627 \text{ g hemoglobin}$$

We also have that

$$4 \text{ atoms Fe} \Bumpeq 1 \text{ hemoglobin molecule}$$

Multiplying each side by Avogadro's number gives

$$4 \text{ mol Fe} \Bumpeq 1 \text{ mol hemoglobin}$$

so that

$$\begin{aligned} (0.00668 \text{ mol Fe})\left(\frac{1 \text{ mol hemoglobin}}{4 \text{ mol Fe}}\right) &\Bumpeq 99.627 \text{ g hemoglobin} \\ 0.00167 \text{ mol hemoglobin} &\Bumpeq 99.627 \text{ g hemoglobin} \end{aligned}$$

Dividing by 0.00167, we have that

$$1 \text{ mol hemoglobin} \Bumpeq 59\,700 \text{ g hemoglobin}$$

The molecular mass of hemoglobin is 59 700.

COMBUSTION ANALYSIS

11-26. The masses of carbon and hydrogen in the original sample are

$$\text{mass of C} = (2.367\text{ g CO}_2)\left(\frac{12.01\text{ g C}}{44.01\text{ g CO}_2}\right) = 0.6460\text{ g}$$

$$\text{mass of H} = (0.4835\text{ g H}_2\text{O})\left(\frac{2\times 1.008\text{ g H}}{18.02\text{ g H}_2\text{O}}\right) = 0.05409\text{ g}$$

The mass percentages of C and H are

$$\text{mass \% of C} = \frac{0.6460\text{ g}}{1.000\text{ g}}\times 100 = 64.60\%$$

$$\text{mass \% of H} = \frac{0.05410\text{ g}}{1.000\text{ g}}\times 100 = 5.409\%$$

The mass percentage of iron is obtained by the difference

$$\text{mass \% Fe} = 100.00\% - 64.60\% - 5.409\% = 29.99\%$$

Take a 100-gram sample and write

$$64.60\text{ g C} \Bumpeq 5.409\text{ g H} \Bumpeq 29.99\text{ g Fe}$$

$$5.379\text{ mol C} \Bumpeq 5.367\text{ mol H} \Bumpeq 0.5370\text{ mol Fe}$$

$$10.0\text{ mol C} \Bumpeq 9.99\text{ mol H} \Bumpeq 1.00\text{ mol Fe}$$

The empirical formula is $C_{10}H_{10}Fe$.

11-28. The masses of carbon and hydrogen in the original sample are

$$\text{mass of C} = (46.20\text{ mg CO}_2)\left(\frac{12.01\text{ mg C}}{44.0095\text{ mg CO}_2}\right) = 12.61\text{ mg}$$

$$\text{mass of H} = (15.13\text{ mg H}_2\text{O})\left(\frac{2\times 1.008\text{ mg H}}{18.02\text{ mg H}_2\text{O}}\right) = 1.693\text{ mg}$$

The mass percentages of C and H are

$$\text{mass \% of C} = \frac{12.61\text{ mg}}{15.42\text{ mg}}\times 100 = 81.78\%$$

$$\text{mass \% of H} = \frac{1.693\text{ mg}}{15.42\text{ mg}}\times 100 = 10.98\%$$

The mass percentage of oxygen is obtained by the difference

$$\text{mass \% O} = 100.00\% - 81.78\% - 10.98\% = 7.24\%$$

Take a 100-gram sample and write

$$81.78\text{ g C} \Bumpeq 10.98\text{ g H} \Bumpeq 7.24\text{ g O}$$

$$6.809\text{ mol C} \Bumpeq 10.89\text{ mol H} \Bumpeq 0.453\text{ mol O}$$

$$15.0\text{ mol C} \Bumpeq 24.0\text{ mol H} \Bumpeq 1.00\text{ mol O}$$

The empirical formula is $C_{15}H_{24}O$.

11-30. The mass percentages of carbon, hydrogen, and lead are

$$\text{mass \% of C} = \frac{(6.34\text{ g CO}_2)\left(\dfrac{12.01\text{ g C}}{44.01\text{ g CO}_2}\right)}{5.83\text{ g}} \times 100 = 29.7\%$$

$$\text{mass \% of H} = \frac{(3.26\text{ g H}_2\text{O})\left(\dfrac{2 \times 1.008\text{ g H}}{18.02\text{ g H}_2\text{O}}\right)}{5.83\text{ g}} \times 100 = 6.26\%$$

$$\text{mass \% Pb} = 100.00\% - 29.7\% - 6.26\% = 64.0\%$$

Take a 100-gram sample and write

$$29.7\text{ g C} \simeq 6.26\text{ g H} \simeq 64.0\text{ g Pb}$$

$$2.47\text{ mol C} \simeq 6.21\text{ mol H} \simeq 0.309\text{ mol Pb}$$

$$7.99\text{ mol C} \simeq 20.1\text{ mol H} \simeq 1.00\text{ mol Pb}$$

The empirical formula is $C_8H_{20}Pb$.

CALCULATIONS INVOLVING CHEMICAL REACTIONS

11-32. The number of moles that correspond to 50.0 grams of iodine, $I_2(s)$, is

$$\text{moles of I}_2 = (50.0\text{ g})\left(\frac{1\text{ mol I}_2}{253.8\text{ g I}_2}\right) = 0.197\text{ mol}$$

We see from the reaction that two moles of sodium iodide $NaI(s)$ are required to produce one mole of $I_2(s)$. Thus,

$$\text{moles of NaI} = (0.197\text{ mol I}_2)\left(\frac{2\text{ mol NaI}}{1\text{ mol I}_2}\right) = 0.394\text{ mol}$$

The mass of $NaI(s)$ required is

$$\text{mass of NaI} = (0.394\text{ mol})\left(\frac{149.9\text{ g NaI}}{1\text{ mol NaI}}\right) = 59.1\text{ g}$$

11-34. The number of moles of $KClO_3(s)$ reacted is

$$\text{moles of KClO}_3 = (10.0\text{ g})\left(\frac{1\text{ mol KClO}_3}{122.55\text{ g KClO}_3}\right) = 0.0816\text{ mol}$$

We see from the equation for the reaction that three moles of $O_2(g)$ are produced from two moles of $KClO_3(s)$. Therefore,

$$\text{moles of O}_2 = (0.0816\text{ mol KClO}_3)\left(\frac{3\text{ mol O}_2}{2\text{ mol KClO}_4}\right) = 0.1224\text{ mol}$$

The mass of $O_2(g)$ prepared is

$$\text{mass of O}_2 = (0.1224\text{ mol})\left(\frac{32.00\text{ g O}_2}{1\text{ mol O}_2}\right) = 3.92\text{ g}$$

11-36. The number of moles of phosphate rock reacted is

$$\text{moles of } Ca_{10}(OH)_2(PO_4)_6 = (100 \text{ metric tons})\left(\frac{10^3 \text{ kg}}{1 \text{ metric ton}}\right)\left(\frac{10^3 \text{ g}}{1 \text{ kg}}\right) \times \left(\frac{1 \text{ mol } Ca_{10}(OH)_2(PO_4)_6}{1004.48 \text{ g } Ca_{10}(OH)_2(PO_4)_6}\right)$$
$$= 9.955 \times 10^4 \text{ mol}$$

We see from the equation for the reaction that six moles of $H_3PO_4(l)$ are produced from one mole of $Ca_{10}(OH)_2(PO_4)_6(s)$. Therefore,

$$\text{moles of } H_3PO_4 = [9.955 \times 10^4 \text{ mol } Ca_{10}(OH)_2(PO_4)_6]\left(\frac{6 \text{ mol } H_3PO_4}{1 \text{ mol } Ca_{10}(OH)_2(PO_4)_6}\right)$$
$$= 5.973 \times 10^5 \text{ mol}$$

The mass of $H_3PO_4(l)$ produced is

$$\text{mass of } H_3PO_4 = (5.973 \times 10^5 \text{ mol})\left(\frac{97.995 \text{ g } H_3PO_4}{1 \text{ mol } H_3PO_4}\right)\left(\frac{1 \text{ kg}}{10^3 \text{ g}}\right)\left(\frac{1 \text{ metric ton}}{10^3 \text{ kg}}\right)$$
$$= 58.5 \text{ metric ton}$$

11-38. The number of moles of $TiO_2(s)$ is

$$\text{moles of } TiO_2 = (4.10 \times 10^3 \text{ kg})\left(\frac{10^3 \text{ g}}{1 \text{ kg}}\right)\left(\frac{1 \text{ mol } TiO_2}{79.87 \text{ g } TiO_2}\right) = 5.13 \times 10^4 \text{ mol}$$

Note that one mole of $Ti(s)$ results from each mole of $TiO_2(s)$; thus,

$$\text{moles of Ti} = 5.13 \times 10^4 \text{ mol}$$

The mass of $Ti(s)$ produced is

$$\text{mass of Ti} = (5.13 \times 10^4 \text{ mol})\left(\frac{47.867 \text{ g Ti}}{1 \text{ mol Ti}}\right) = 2.46 \times 10^6 \text{ g} = 2.46 \times 10^3 \text{ kg}$$

CALCULATIONS WITHOUT THE CHEMICAL EQUATION

11-40. The number of moles of $H_2S(g)$ in 10.0 metric tons is given by

$$\text{moles of } H_2S = (10.0 \text{ metric ton})\left(\frac{10^3 \text{ kg}}{1 \text{ metric ton}}\right)\left(\frac{10^3 \text{ g}}{1 \text{ kg}}\right)\left(\frac{1 \text{ mol } H_2S}{34.08 \text{ g } H_2S}\right)$$
$$= 2.934 \times 10^5 \text{ mol}$$

The number of moles of S is given by

$$\text{moles of S} = (2.934 \times 10^5 \text{ mol } H_2S)\left(\frac{1 \text{ mol S}}{1 \text{ mol } H_2S}\right) = 2.934 \times 10^5 \text{ mol}$$

Because all the sulfur in H_2S ends up as sulfur, the number of metric tons of sulfur is given by

$$\text{mass of S} = (2.934 \times 10^5 \text{ mol})\left(\frac{32.065 \text{ g S}}{1 \text{ mol S}}\right)\left(\frac{1 \text{ kg}}{10^3 \text{ g}}\right)\left(\frac{1 \text{ metric ton}}{10^3 \text{ kg}}\right)$$
$$= 9.41 \text{ metric ton}$$

11-42. The number of moles of sodium is given by

$$\text{moles of Na} = (10.0\text{ g})\left(\frac{1\text{ mol Na}}{22.99\text{ g Na}}\right) = 0.4350\text{ mol}$$

Because all the sodium ends up in $NaNH_2(s)$, we have that

$$\text{moles of NaNH}_2 = (0.4350\text{ mol Na})\left(\frac{1\text{ mol NaNH}_2}{1\text{ mol Na}}\right) = 0.4350\text{ mol}$$

The number of grams of $NaNH_2(s)$ produced is given by

$$\text{mass of NaNH}_2 = (0.4350\text{ mol})\left(\frac{39.02\text{ g NaNH}_2}{1\text{ mol NaNH}_2}\right) = 17.0\text{ g}$$

MIXTURES

11-44. The mass of $KClO_3$ is 14.17 grams and the mass % is 78.0%.

11-46. The number of moles of $BaSO_4(s)$ precipitated is given by

$$\text{moles of BaSO}_4 = (4.37\text{ g BaSO}_4)\left(\frac{1\text{ mol BaSO}_4}{233.4\text{ g BaSO}_4}\right) = 0.01872\text{ mol}$$

Let

$$x\text{ g} = \text{number of grams of K}_2\text{SO}_4$$

and

$$3.00\text{ g} - x\text{ g} = \text{number of grams of MnSO}_4$$

Then

$$\text{moles of K}_2\text{SO}_4 = (x\text{ g})\left(\frac{1\text{ mol K}_2\text{SO}_4}{174.27\text{ g K}_2\text{SO}_4}\right) = \frac{x\text{ mol}}{174.27}$$

and

$$\text{moles of MnSO}_4 = (3.00\text{ g} - x\text{ g})\left(\frac{1\text{ mol MnSO}_4}{151.0\text{ g MnSO}_4}\right) = \frac{(3.00 - x)\text{ mol}}{151.0}$$

Because

$$\text{K}_2\text{SO}_4(aq) + \text{Ba(NO}_3)_2(aq) \longrightarrow 2\,\text{KNO}_3(aq) + \text{BaSO}_4(s)$$

and

$$\text{MnSO}_4(aq) + \text{BaNO}_3(aq) \longrightarrow \text{Mn(NO}_3)_2(aq) + \text{BaSO}_4(s)$$

the total number of moles of $K_2SO_4(s)$ and $MnSO_4(s)$ in the sample is equal to the number of moles of $BaSO_4(s)$ precipitated or

$$\text{moles of K}_2\text{SO}_4 + \text{moles of MnSO}_4 = \text{moles of BaSO}_4$$

Thus, we have that

$$\frac{x\text{ mol}}{174.27} + \frac{(3.00 - x)\text{ mol}}{151.0} = 0.01872\text{ mol}$$

Solving for x, we have

$$5.7382 \times 10^{-3}\,x + 0.01987 - 6.6225 \times 10^{-3}\,x = 0.01872$$
$$0.884 \times 10^{-3}\,x = 0.00115$$
$$x = 1.30$$

The mass of $K_2SO_4(s)$ is 1.30 g.
The mass percentage of $K_2SO_4(s)$ in the sample is

$$\text{mass \% } K_2SO_4 = \frac{1.30\text{ g}}{3.00\text{ g}} \times 100 = 43.3\%$$

and

$$\text{mass \% MnSO}_4 = 100.0\% - 43.3\% = 56.7\%$$

11-48. Let x g = number of grams of Al(s) and 9.87 g − x g = number of grams of Mg(s). The number of moles of aluminum and of magnesium are given by

$$\text{moles of Al} = (x\text{ g})\left(\frac{1\text{ mol Al}}{26.98\text{ g Al}}\right) = \frac{x\text{ mol}}{26.98}$$
$$\text{moles of Mg} = (9.87\text{ g} - x\text{ g})\left(\frac{1\text{ mol Mg}}{24.31\text{ g Mg}}\right) = \frac{(9.87 - x)\text{ mol}}{24.31}$$

The equations for the reactions are

$$2\,\text{Al}(s) + 6\,\text{HCl}(aq) \longrightarrow 2\,\text{AlCl}_3(aq) + 3\,\text{H}_2(g)$$
$$\text{Mg}(s) + 2\,\text{HCl}(aq) \longrightarrow \text{MgCl}_2(aq) + \text{H}_2(g)$$

The number of moles of hydrogen produced is given by

$$\text{moles of H}_2 = (0.998\text{ g})\left(\frac{1\text{ mol H}_2}{2.016\text{ g H}_2}\right) = 0.4950\text{ mol}$$

We have that the total number of moles of hydrogen produced is equal to the number of moles of hydrogen produced by the reaction of aluminum plus the number of moles of hydrogen produced by the reaction of magnesium:

$$0.4950\text{ mol H}_2 = \left(\frac{x\text{ mol Al}}{26.98}\right)\left(\frac{3\text{ mol H}_2}{2\text{ mol Al}}\right) + \left(\frac{(9.87 - x)\text{ mol Mg}}{24.31}\right)\left(\frac{1\text{ mol H}_2}{1\text{ mol Mg}}\right)$$
$$0.4950 = 0.05560\,x - 0.4060 - 0.04114\,x$$

Solving for x, we have that

$$x = 6.15$$

Thus, the mass of aluminum is 6.15 grams and the mass of magnesium is 9.87 g − 6.15 g = 3.72 grams. The mass percentage of each is

$$\text{mass \% Al} = 62.3\% \qquad \text{and} \qquad \text{mass \% Mg} = 37.7\%$$

LIMITING REACTANT

11-50. Because we are given the quantities of two reactants, we must check to see if one of them is a limiting reactant. The number of moles of $P_4(s)$ is

$$\text{moles of } P_4 = (20.0\text{ g})\left(\frac{1\text{ mol }P_4}{123.90\text{ g }P_4}\right) = 0.1614\text{ mol}$$

and the number of moles of $NaOH(s)$ is

$$\text{moles of NaOH} = (50.0\text{ g})\left(\frac{1\text{ mol NaOH}}{40.00\text{ g NaOH}}\right) = 1.250\text{ mol}$$

Because one mole of $P_4(s)$ reacts with three moles of $NaOH(aq)$, we see that 0.1614 moles of $P_4(s)$ require 0.4842 moles of $NaOH(aq)$. Thus, $NaOH(aq)$ is in excess and $P_4(s)$ is the limiting reactant. The mass of $PH_3(g)$ produced is

$$\text{mass of } PH_3 = (0.1614\text{ mol }P_4)\left(\frac{1\text{ mol }PH_3}{1\text{ mol }P_4}\right)\left(\frac{33.99\text{ g }PH_3}{1\text{ mol }PH_3}\right) = 5.49\text{ g}$$

11-52. Because we are given the quantities of two reactants, we must check to see if one of them is a limiting reactant. The number of moles of NaBr is

$$\text{moles of NaBr} = (25.0\text{ g})\left(\frac{1\text{ mol NaBr}}{102.89\text{ g NaBr}}\right) = 0.243\text{ mol}$$

and the number of moles of $Cl_2(g)$ is

$$\text{moles of } Cl_2 = (25.0\text{ g})\left(\frac{1\text{ mol }Cl_2}{70.90\text{ g }Cl_2}\right) = 0.353\text{ mol}$$

Each mole of Cl_2 requires two moles of NaBr, or 0.353 moles of Cl_2 require 0.706 moles of NaBr. Thus, Cl_2 is in excess and NaBr is the limiting reactant. The mass of $Br_2(l)$ produced is

$$\text{mass of } Br_2 = (0.243\text{ mol NaBr})\left(\frac{1\text{ mol }Br_2}{2\text{ mol NaBr}}\right)\left(\frac{159.8\text{ g }Br_2}{1\text{ mol }Br_2}\right)$$
$$= 19.4\text{ g}$$

11-54. (a) The balanced equation is

$$CaCO_3(s) + 2\,HCl(aq) \longrightarrow CaCl_2(aq) + H_2O(l) + CO_2(g)$$

(b) Because we are given the quantities of two reactants, we must check to see if one of them is a limiting reactant. We have the stoichiometric correspondences

$$25.0\text{ g }CaCO_3 \approx 0.250\text{ mol }CaCO_3$$
$$15.0\text{ g HCl} \approx 0.411\text{ mol HCl}$$

Therefore, HCl is the limiting reactant because 0.411 moles of HCl require only 0.205 moles of $CaCO_3$. The mass of calcium chloride produced is

$$\text{mass of } CaCl_2 = (0.411\text{ mol HCl})\left(\frac{1\text{ mol }CaCl_2}{2\text{ mol HCl}}\right)\left(\frac{110.98\text{ g }CaCl_2}{1\text{ mol }CaCl_2}\right) = 22.8\text{ g}$$

(c) Only 0.205 moles of $CaCO_3$ will react, and so there is an excess of (0.250 – 0.205) moles = 0.045 moles of $CaCO_3$, or 4.50 grams remaining.

11-56. (a) The balanced equation is

$$CdCl_2(aq) + 2\,AgClO_4(aq) \longrightarrow Cd(ClO_4)_2(aq) + 2\,AgCl(s)$$

(b) Because we are given the quantities of two reactants, we must check to see if one of them is a limiting reactant. We have the stoichiometric correspondences

$$17.5\text{ g } CdCl_2 \Leftrightarrow 0.0955\text{ mol } CdCl_2$$
$$35.5\text{ g } AgClO_4 \Leftrightarrow 0.171\text{ mol } AgClO_4$$

Therefore, $CdCl_2$ is in excess because the 0.171 moles of $AgClO_4$ require only 0.0855 moles of $CdCl_2$. The mass of silver chloride produced is

$$\text{mass of AgCl} = (0.171\text{ mol } AgClO_4)\left(\frac{2\text{ mol AgCl}}{2\text{ mol } AgClO_4}\right)\left(\frac{143.32\text{ g AgCl}}{1\text{ mol AgCl}}\right) = 24.5\text{ g}$$

(c) There is an excess of (0.0955 − 0.0855) moles = 0.010 moles of $CdCl_2$ or 1.8 grams.

11-58. The equation for the reaction is

$$Hg(NO_3)_2(aq) + 2\,NaBr(aq) \longrightarrow HgBr_2(s) + 2\,NaNO_3(aq)$$

Because we are given the quantities of two reactants, we must check to see if one of them is a limiting reactant. We have the stoichiometric correspondences

$$25.0\text{ g } Hg(NO_3)_2 \Leftrightarrow 0.0770\text{ mol } Hg(NO_3)_2$$
$$15.0\text{ g NaBr} \Leftrightarrow 0.1458\text{ mol NaBr}$$

Therefore, $Hg(NO_3)_2$ is slightly in excess because the 0.146 moles of NaBr require only 0.0729 moles of $Hg(NO_3)_2$. The mass of mercury(II) bromide produced is

$$\text{mass of } HgBr_2 = (0.1458\text{ mol NaBr})\left(\frac{1\text{ mol } HgBr_2}{2\text{ mol NaBr}}\right)\left(\frac{360.40\text{ g } HgBr_2}{1\text{ mol } HgBr_2}\right) = 26.3\text{ g}$$

There is an excess of (0.0770 − 0.0729) moles = 0.0041 moles of $Hg(NO_3)_2$ or 1.3 grams.

PERCENTAGE YIELD

11-60. The theoretical yield of antimony is

$$\begin{aligned}\text{theoretical yield} &= (60.0\text{ g } Sb_4O_6)\left(\frac{1\text{ mol } Sb_4O_6}{583.04\text{ g } Sb_4O_6}\right)\left(\frac{4\text{ mol Sb}}{1\text{ mol } Sb_4O_6}\right)\left(\frac{121.76\text{ g Sb}}{1\text{ mol Sb}}\right)\\ &= 50.1\text{ g}\end{aligned}$$

The percent yield is

$$\begin{aligned}\%\text{ yield} &= \frac{\text{actual yield}}{\text{theoretical yield}} \times 100\\ &= \frac{49.0\text{ g}}{50.1\text{ g}} \times 100 = 97.8\%\end{aligned}$$

11-62. The theoretical yield of $CH_3CH_2OH(l)$ is

$$\text{theoretical yield} = (10.0 \times 10^3 \text{ g } C_2H_4)\left(\frac{1 \text{ mol } C_2H_4}{28.05 \text{ g } C_2H_4}\right)\left(\frac{1 \text{ mol } CH_3CH_2OH}{1 \text{ mol } C_2H_4}\right)$$

$$\times \left(\frac{46.07 \text{ g } CH_3CH_2OH}{1 \text{ mol } CH_3CH_2OH}\right)$$

$$= 16.4 \times 10^3 \text{ g} = 16.4 \text{ kg}$$

The percent yield is

$$\% \text{ yield} = \frac{\text{actual yield}}{\text{theoretical yield}} \times 100$$

$$= \frac{13.5 \text{ kg}}{16.4 \text{ kg}} \times 100 = 82.3\%$$

11-64. (a) The balanced equation is

$$Fe(s) + H_2SO_4(aq) \longrightarrow FeSO_4(aq) + H_2(g)$$

(b) The theoretical mass of iron required is

$$\text{mass of Fe} = (1.00 \text{ kg } H_2)\left(\frac{10^3 \text{ g}}{1 \text{ kg}}\right)\left(\frac{1 \text{ mol } H_2}{2.016 \text{ g } H_2}\right)\left(\frac{1 \text{ mol Fe}}{1 \text{ mol } H_2}\right)\left(\frac{55.845 \text{ g Fe}}{1 \text{ mol Fe}}\right)$$

$$= 27\,700 \text{ g} = 27.7 \text{ kg}$$

But the yield of the reaction is 95% so that the actual mass of iron required is

$$\text{mass of iron} = \frac{27.7 \text{ kg}}{0.95} = 29.2 \text{ kg}$$

ADDITIONAL PROBLEMS

11-66. We cannot use the same procedure to determine the mass of oxygen present in the original sample because we don't know what fraction of the oxygen atoms in the products came from the substance being analyzed. Consider the combustion of methanol, $CH_3OH(l)$, which is described by the chemical equation

$$2\, CH_3OH(l) + 3\, O_2(g) \longrightarrow 2\, CO_2(g) + 4\, H_2O(g)$$

All the carbon atoms in the $CO_2(g)$ produced and all the hydrogen atoms in the $H_2O(l)$ in this reaction come from the $CH_3OH(l)$; however, not all the oxygen atoms in these two products do. The oxygen atoms in the products come from *both* the $CH_3OH(l)$ and the $O_2(g)$. In addition, we use an excess of oxygen gas to ensure complete combustion to the substance being analyzed. Thus, we cannot determine the mass of oxygen originating from the substance being analyzed in the same manner that we use to find the masses of carbon and hydrogen.

11-68. The reaction may fail to go to completion; there may be other reactions that give rise to undesired products; some of the desired product may not be readily recoverable or may be lost in the purification process; or the original reactants may be impure.

11-70. The number of nitrogen atoms in 2.0 liters of $N_2(l)$ is given by

$$\text{number of N atoms} = (2.0\text{ L N}_2)\left(\frac{1000\text{ mL}}{1\text{ L}}\right)\left(\frac{0.808\text{ g}}{1\text{ mL}}\right)\left(\frac{1\text{ mol N}_2}{28.01\text{ g N}_2}\right) \times \left(\frac{6.022\times10^{23}\text{ molecule}}{1\text{ mol}}\right)\left(\frac{2\text{ atoms}}{1\text{ molecule}}\right) = 6.95\times10^{25}$$

Because there are the same number of nitrogen atoms in $N_2(g)$ as in $NH_4NO_3(s)$, we require 6.95×10^{25} formula units of $NH_4NO_3(s)$, or 115.4 moles, or 9240 grams, or 9.24 kilograms.

11-72. Because we are given the quantities of two reactants, we must check to see if one of them is a limiting reactant. The number of moles of the reactants are

$$\text{moles of CS}_2 = (1.00\times10^3\text{ g})\left(\frac{1\text{ mol CS}_2}{76.15\text{ g CS}_2}\right) = 13.1\text{ mol}$$

$$\text{moles of NaOH} = (1.00\times10^3\text{ g})\left(\frac{1\text{ mol NaOH}}{40.00\text{ g NaOH}}\right) = 25.0\text{ mol}$$

Because 25.0 moles of $NaOH(aq)$ require only 12.5 moles of $CS_2(g)$, we see that $CS_2(g)$ is in excess and $NaOH(aq)$ is the limiting reactant. The mass of each product that is produced is

$$\text{mass of Na}_2\text{CS}_3 = (25.0\text{ mol NaOH})\left(\frac{2\text{ mol Na}_2\text{CS}_3}{6\text{ mol NaOH}}\right)\left(\frac{154.19\text{ g Na}_2\text{CS}_3}{1\text{ mol Na}_2\text{CS}_3}\right) = 1280\text{ g}$$

$$\text{mass of Na}_2\text{CO}_3 = (25.0\text{ mol NaOH})\left(\frac{1\text{ mol Na}_2\text{CO}_3}{6\text{ mol NaOH}}\right)\left(\frac{105.99\text{ g Na}_2\text{CO}_3}{1\text{ mol Na}_2\text{CO}_3}\right) = 442\text{ g}$$

$$\text{mass of H}_2\text{O} = (25.0\text{ mol NaOH})\left(\frac{3\text{ mol H}_2\text{O}}{6\text{ mol NaOH}}\right)\left(\frac{18.02\text{ g H}_2\text{O}}{1\text{ mol H}_2\text{O}}\right) = 225\text{ g}$$

11-74. (a) The mass of $Na_2O_2(s)$ produced is

$$\text{mass of Na}_2\text{O}_2 = (0.600\text{ g Na})\left(\frac{1\text{ mol Na}}{22.99\text{ g Na}}\right)\left(\frac{1\text{ mol Na}_2\text{O}_2}{2\text{ mol Na}}\right)\left(\frac{77.98\text{ g Na}_2\text{O}_2}{1\text{ mol Na}_2\text{O}_2}\right) = 1.02\text{ g}$$

(b) The mass of $KO_2(s)$ produced is

$$\text{mass of KO}_2 = (0.600\text{ g K})\left(\frac{1\text{ mol K}}{39.10\text{ g K}}\right)\left(\frac{1\text{ mol KO}_2}{1\text{ mol K}}\right)\left(\frac{71.10\text{ g KO}_2}{1\text{ mol KO}_2}\right) = 1.09\text{ g}$$

(c) The mass of $RbO_2(s)$ produced is

$$\text{mass of RbO}_2 = (0.600\text{ g Rb})\left(\frac{1\text{ mol Rb}}{85.47\text{ g Rb}}\right)\left(\frac{1\text{ mol RbO}_2}{1\text{ mol Rb}}\right)\left(\frac{117.47\text{ g RbO}_2}{1\text{ mol RbO}_2}\right) = 0.825\text{ g}$$

(d) The mass of $CsO_2(s)$ produced is

$$\text{mass of } CsO_2 = (0.600\text{ g Cs})\left(\frac{1\text{ mol Cs}}{132.91\text{ g Cs}}\right)\left(\frac{1\text{ mol } CsO_2}{1\text{ mol Cs}}\right)\left(\frac{164.91\text{ g } CsO_2}{1\text{ mol } CsO_2}\right)$$
$$= 0.744\text{ g}$$

11-76. The number of moles of glucose is

$$\text{moles of } C_6H_{12}O_6 = (28.0\text{ g } C_6H_{12}O_6)\left(\frac{1\text{ mol } C_6H_{12}O_6}{180.16\text{ g } C_6H_{12}O_6}\right) = 0.155\text{ mol}$$

The number of moles of oxygen required is

$$\text{moles of } O_2 = (0.155\text{ mol } C_6H_{12}O_6)\left(\frac{6\text{ mol } O_2}{1\text{ mol } C_6H_{12}O_6}\right) = 0.930\text{ mol}$$

The mass of $O_2(g)$ required is

$$\text{mass of } O_2 = (0.930\text{ mol } O_2)\left(\frac{32.00\text{ g } O_2}{1\text{ mol } O_2}\right) = 29.8\text{ g}$$

The number of moles of $CO_2(g)$ produced is

$$\text{moles of } CO_2 = (0.155\text{ mol } C_6H_{12}O_6)\left(\frac{6\text{ mol } CO_2}{1\text{ mol } C_6H_{12}O_6}\right) = 0.930\text{ mol}$$

The mass of $CO_2(g)$ produced is

$$\text{mass of } CO_2 = (0.930\text{ mol } CO_2)\left(\frac{44.01\text{ g } CO_2}{1\text{ mol } CO_2}\right) = 40.9\text{ g}$$

11-78. Let x g = mass of NaCl(s) and $(12.42 - x)$ g = mass of $CaCl_2(s)$ in the mixture. We have that the number of moles of AgCl(s) precipitated is equal to

$$\text{moles of NaCl} + 2\,(\text{moles of } CaCl_2) = \text{moles of AgCl}$$

Thus, we have that

$$\frac{x\text{ g}}{58.44\text{ g}\cdot\text{mol}^{-1}} + \frac{(2)(12.42 - x)\text{ g}}{111.0\text{ g}\cdot\text{mol}^{-1}} = \frac{31.70\text{ g}}{143.3\text{ g}\cdot\text{mol}^{-1}}$$
$$0.01711\,x + 0.2238 - 0.01802\,x = 0.2212$$

Solving for x, we have that $x = 2.9$. Thus, the mass of NaCl(s) is 2.9 grams and the mass of $CaCl_2(s)$ is 12.42 g − 2.9 g = 9.5 grams. The mass percentage of each in the mixture is

$$\text{mass \% NaCl} = \frac{2.9\text{ g}}{12.42\text{ g}} \times 100 = 23\%$$

$$\text{mass \% } CaCl_2 = \frac{9.5\text{ g}}{12.42\text{ g}} \times 100 = 77\%$$

11-80. The mass of water in the nickel sulfate hydrate is

$$\text{mass of } H_2O = 12.060\text{ g} - 7.101\text{ g} = 4.959\text{ g}$$

The number of moles of nickel sulfate hydrate is given by

$$\text{moles of NiSO}_4\cdot x\,\text{H}_2\text{O} = (7.101\text{ g NiSO}_4)\left(\frac{1\text{ mol NiSO}_4}{154.76\text{ g NiSO}_4}\right)\left(\frac{1\text{ mol NiSO}_4\cdot x\,\text{H}_2\text{O}}{1\text{ mol NiSO}_4}\right)$$
$$= 0.04588\text{ mol}$$

The number of moles of water in the nickel sulfate hydrate sample is given by

$$\text{moles of H}_2\text{O} = (4.959\text{ g H}_2\text{O})\left(\frac{1\text{ mol H}_2\text{O}}{18.02\text{ g H}_2\text{O}}\right) = 0.2752\text{ mol}$$

We have the stoichiometric correspondence

$$x\text{ mol H}_2\text{O} \Leftrightarrow x\text{ mol NiSO}_4\cdot x\,\text{H}_2\text{O}$$

Thus, we have that

$$0.2752 = 0.04588\ x$$

Solving for x, we have $x = 6.00$. There are 6 water molecules per formula unit of nickel sulfate hydrate.

11-82. Take a 100-gram sample and so we have the stoichiometric correspondence

$$85.63\text{ g B} \Leftrightarrow 14.37\text{ g H}$$
$$7.921\text{ mol B} \Leftrightarrow 14.26\text{ mol H}$$
$$1\text{ mol B} \Leftrightarrow 1.800\text{ mol H}$$
$$5\text{ mol B} \Leftrightarrow 9.00\text{ mol H}$$

The empirical formula is B_5H_9.

11-84. Let x g be the mass of $Na_2SO_4(s)$. Therefore, the mass of $NaHSO_4(s)$ is $2.606\text{ g} - x\text{ g}$. We have that

$$\text{moles of Na}_2\text{SO}_4 = (x\text{ g})\left(\frac{1\text{ mol Na}_2\text{SO}_4}{142.04\text{ g Na}_2\text{SO}_4}\right) = 0.0070403\ x\text{ mol}$$

$$\text{moles of NaHSO}_4 = (2.606\text{ g} - x\text{ g})\left(\frac{1\text{ mol NaHSO}_4}{120.06\text{ g NaHSO}_4}\right) = 0.021706\text{ mol} - 0.0083292\ x\text{ mol}$$

$$\text{moles of BaSO}_4 = (4.688\text{ g})\left(\frac{1\text{ mol BaSO}_4}{233.4\text{ g BaSO}_4}\right) = 0.020086\text{ mol}$$

Now, from the given chemical equations, we have that

$$\text{moles of Na}_2\text{SO}_4 + \text{moles of NaHSO}_4 = \text{moles of BaSO}_4$$
$$0.0070403\ x\text{ mol} + 0.021706\text{ mol} - 0.0083292\ x\text{ mol} = 0.020086\text{ mol}$$
$$0.0012889\ x = 0.00162$$
$$x = 1.257$$

The mass of $Na_2SO_4(s)$ is 1.257 grams. The mass percentages of $Na_2SO_4(s)$ and $NaHSO_4(s)$ are

$$\text{mass \% } Na_2SO_4 = \frac{1.257 \text{ g } Na_2SO_4}{2.606 \text{ g sample}} \times 100 = 48.2\%$$
$$\text{mass \% } NaHSO_4 = 100\% - 48.2\% = 51.8\%$$

11-86. The number of moles of ammonia, $NH_3(g)$, is given by

$$\text{moles of } NH_3 = (6.40 \times 10^4 \text{ kg})\left(\frac{10^3 \text{ g}}{1 \text{ kg}}\right)\left(\frac{1 \text{ mol } NH_3}{17.03 \text{ g } NH_3}\right) = 3.76 \times 10^6 \text{ mol}$$

From the set of three equations, we see that

$$\text{moles of } HNO_3 = (1 \text{ mol } NH_3)\left(\frac{4 \text{ mol NO}}{4 \text{ mol } NH_3}\right)\left(\frac{2 \text{ mol } NO_2}{2 \text{ mol NO}}\right)\left(\frac{2 \text{ mol } HNO_3}{3 \text{ mol } NO_2}\right) = \frac{2}{3} \text{ mol}$$

Therefore,

$$\text{moles of } HNO_3 = (3.76 \times 10^6 \text{ mol } NH_3)\left(\frac{2 \text{ mol } HNO_3}{3 \text{ mol } NH_3}\right) = 2.51 \times 10^6 \text{ mol}$$

The mass of nitric acid produced is

$$\text{mass of } HNO_3 = (2.51 \times 10^6 \text{ mol } HNO_3)\left(\frac{63.01 \text{ g } HNO_3}{1 \text{ mol } HNO_3}\right) = 1.58 \times 10^8 \text{ g} = 1.58 \times 10^5 \text{ kg}$$

In practice, the $NO(g)$ produced in Step 3 is cycled back into Step 2, and so the theoretical yield is greater.

11-88. Take a 100-gram sample and write

$$67.31 \text{ g C} \mathrel{\hat{=}} 6.978 \text{ g H} \mathrel{\hat{=}} 4.618 \text{ g N} \mathrel{\hat{=}} 21.10 \text{ g O}$$
$$5.604 \text{ mol C} \mathrel{\hat{=}} 6.923 \text{ mol H} \mathrel{\hat{=}} 0.3297 \text{ mol N} \mathrel{\hat{=}} 1.319 \text{ mol O}$$
$$17.00 \text{ mol C} \mathrel{\hat{=}} 21.00 \text{ H} \mathrel{\hat{=}} 1.000 \text{ mol N} \mathrel{\hat{=}} 4.001 \text{ mol O}$$

Thus, the empirical formula of the white powder is $C_{17}H_{21}NO_4(s)$. The powder may be cocaine but further analysis is required to confirm it.

11-90. The reaction of the burning of gasoline (represented as octane) can be described by the equation

$$2\,C_8H_{18}(l) + 25\,O_2(g) \longrightarrow 16\,CO_2(g) + 18\,H_2O(g)$$

The 500 gallons of gasoline will yield

$$\begin{aligned}\text{mass of } CO_2 &= (500 \text{ gallon octane})\left(\frac{3785.4 \text{ mL}}{1 \text{ gallon}}\right)\left(\frac{0.7 \text{ g}}{1 \text{ mL}}\right)\left(\frac{1 \text{ mol } C_8H_{18}}{114.2 \text{ g } C_8H_{18}}\right)\\ &\times \left(\frac{16 \text{ mol } CO_2}{2 \text{ mol } C_8H_{18}}\right)\left(\frac{44.01 \text{ g } CO_2}{1 \text{ mol } CO_2}\right)\left(\frac{1 \text{ kg}}{1000 \text{ g}}\right)\left(\frac{2.2046 \text{ lb}}{1 \text{ kg}}\right)\left(\frac{1 \text{ ton}}{2000 \text{ lb}}\right)\\ &= 4.5 \text{ ton}\end{aligned}$$

which is about five tons, as the ad claims.

11-92. The masses of carbon and hydrogen in the original sample are

$$\text{mass of C} = (54.246\text{ mg CO}_2)\left(\frac{12.0107\text{ mg C}}{44.0095\text{ mg CO}_2}\right) = 14.804\text{ mg}$$

$$\text{mass of H} = (22.206\text{ mg H}_2\text{O})\left(\frac{2\times 1.00794\text{ g H}}{18.0153\text{ g H}_2\text{O}}\right) = 2.4848\text{ mg}$$

The mass of sulfur in the 23.725-milligram sample is given by

$$\text{mass of S} = (10.255\text{ mg SO}_2)\left(\frac{32.065\text{ g S}}{64.0638\text{ g SO}_2}\right) = 5.1328\text{ mg}$$

The mass percentages of each of these elements in the compound are given by

$$\text{mass \% of C} = \frac{14.804\text{ mg}}{30.450\text{ mg}} \times 100 = 48.617\%$$

$$\text{mass \% of H} = \frac{2.4848\text{ mg}}{30.450\text{ mg}} \times 100 = 8.160\%$$

$$\text{mass \% of S} = \frac{5.1328\text{ mg}}{23.725\text{ mg}} \times 100 = 21.635\%$$

The mass percentage of oxygen is obtained by the difference

$$\text{mass \% O} = 100.000\% - 48.617\% - 8.160\% - 21.635\% = 21.588\%$$

Take a 100-gram sample and write

$$48.617\text{ g C} \approx 8.160\text{ g H} \approx 21.635\text{ g S} \approx 21.588\text{ g O}$$

$$4.4078\text{ mol C} \approx 8.0957\text{ mol H} \approx 0.67473\text{ mol S} \approx 1.3493\text{ mol O}$$

$$5.999\text{ mol C} \approx 12.00\text{ mol H} \approx 1\text{ mol S} \approx 2.000\text{ mol O}$$

The empirical formula of the compound is $C_6H_{12}SO_2$.

11-94.* The mass of the oil is

$$m = dV = (0.96\text{ g}\cdot\text{mL}^{-1})(1\,ts)\left(\frac{4.93\text{ mL}}{1\,ts}\right) = 4.73\text{ g}$$

The number of moles in the oil is

$$\text{moles of oil} = (4.73\text{ g})\left(\frac{1\text{ mol}}{180\text{ g}}\right) = 0.0263\text{ mol}$$

The area of the oil layer on the pond is

$$\text{area} = (0.5\text{ acre})\left(\frac{4050\text{ m}^2}{1\text{ acre}}\right) = 2025\text{ m}^2$$

The volume of the oil in the teaspoon in cubic meters is

$$V = (4.93\text{ cm}^3)\left(\frac{1\text{ m}}{10^2\text{ cm}}\right)^3 = 4.93\times 10^{-6}\text{ m}^3$$

The thickness of the oil layer is given by

$$\text{volume of one teaspoon of oil} = \text{volume of the oil layer}$$
$$4.93 \times 10^{-6}\ \text{m}^3 = \text{area} \times \text{thickness} = (2025\ \text{m}^2)(d)$$

Solving for the thickness, d, we have

$$d = \frac{4.93 \times 10^{-6}\ \text{m}^3}{2025\ \text{m}^2} = 2.43 \times 10^{-9}\ \text{m}$$

For convenience, we shall assume that the oil molecule is a cube with the length of each side equal to 2.43×10^{-9} m. We are assuming that the layer is one molecule in thickness. If we take the layer to be a square whose length is 2.43×10^{-9} m times the number of cubes or molecules along each side, n, then the area of the layer is given by

$$(2.43 \times 10^{-9}\ \text{m})\,n \times (2.43 \times 10^{-9}\ \text{m})\,n = 2025\ \text{m}^2$$

or

$$n^2 = \frac{2025\ \text{m}^2}{5.905 \times 10^{-18}\ \text{m}^2} = 3.429 \times 10^{20}$$

The number of molecules or cubes, N, in the layer is equal to n^2. Thus we have

$$N = 3.429 \times 10^{20}\ \text{molecules} \Bumpeq 0.0263\ \text{moles}$$

or

$$1.3 \times 10^{22}\ \text{molecules} \Bumpeq 1.00\ \text{mole}$$

Thus we estimate Avogadro's number to be 1×10^{22} molecules·mol^{-1}.

We could have assumed other shapes for the layer and oil molecule, but a square and cube are the most convneient choice.

CHAPTER 12. Chemical Calculations for Solutions

Many of the calculations in this chapter require the volume of a solution to be expressed in liters. Often the volume is given in milliliters. The unit conversion factor is

$$1 = \frac{10^3 \text{ mL}}{1 \text{ L}}$$

You should become proficient in converting from milliliters to liters and liters to milliliters. The conversion from one unit to the other will not be shown in many of the following solutions.

PREPARATIONS OF SOLUTIONS

12-2. The number of moles of NaOH is

$$\text{moles of NaOH} = (572 \text{ g NaOH})\left(\frac{1 \text{ mol NaOH}}{40.00 \text{ g NaOH}}\right) = 14.3 \text{ mol}$$

Because the 14.3 moles are dissolved in 1 liter of solution, the molarity is 14.3 M.

12-4. The number of moles of caffeine is

$$\text{moles of } C_8H_{10}N_4O_2 = (300 \text{ mg } C_8H_{10}N_4O_2)\left(\frac{1 \text{ g}}{10^3 \text{ mg}}\right)\left(\frac{1 \text{ mol } C_8H_{10}N_4O_2}{194.19 \text{ g } C_8H_{10}N_4O_2}\right)$$
$$= 1.545 \times 10^{-3} \text{ mol}$$

The molarity is calculated by using Equation 12.2:

$$M = \frac{n}{V} = \left(\frac{1.545 \times 10^{-3} \text{ mol}}{1 \text{ cup}}\right)\left(\frac{4 \text{ cup}}{0.946 \text{ L}}\right) = 0.00653 \text{ M}$$

12-6. (a) The number of moles of NaCl in the solution is

$$\text{moles of NaCl} = MV = (0.200 \text{ mol}\cdot\text{L}^{-1})(50.0 \ \mu\text{L})\left(\frac{1 \text{ L}}{10^6 \ \mu\text{L}}\right)$$
$$= 1.00 \times 10^{-5} \text{ mol}$$

12-8. From Equation 12.4, we have

$$M_1 V_1 = M_2 V_2$$

Thus,

$$(12.0\ \text{mmol}\cdot\text{mL}^{-1})\, V_1 = (1.0\ \text{mmol}\cdot\text{mL}^{-1})(250\ \text{mL})$$

Solving for V_1, we have

$$V_1 = \frac{(1.0)(250\ \text{mL})}{12.0} = 21\ \text{mL}$$

12-10. The number of moles of $CuSO_4$ in 50.0 milliliters of a 0.200 molar solution is

$$\text{moles of CuSO}_4 = MV = (0.200\ \text{mol}\cdot\text{L}^{-1})(0.0500\ \text{L}) = 0.0100\ \text{mol}$$

The number of moles of $CuSO_4\cdot 5\,H_2O$ required is

$$\text{moles of CuSO}_4\cdot 5\,\text{H}_2\text{O} = (0.0100\ \text{mol CuSO}_4)\left(\frac{1\ \text{mol CuSO}_4\cdot 5\,\text{H}_2\text{O}}{1\ \text{mol CuSO}_4}\right) = 0.0100\ \text{mol}$$

The mass of $CuSO_4\cdot 5\,H_2O$ required is

$$\text{mass of CuSO}_4\cdot 5\,\text{H}_2\text{O} = (0.0100\ \text{mol})\left(\frac{249.69\ \text{g CuSO}_4\cdot 5\,\text{H}_2\text{O}}{1\ \text{mol CuSO}_4\cdot 5\,\text{H}_2\text{O}}\right) = 2.50\ \text{g}$$

Dissolve 2.50 grams in about 40 mL of water in a 50-mL volumetric flask and dilute the solution to the 50-mL mark and shake to make the solution homogeneous.

12-12. Take a 100.0-gram sample of the concentrated phosphoric acid solution. We have 85 grams of H_3PO_4 in the sample. The number of moles of H_3PO_4 in the sample is

$$\text{moles of H}_3\text{PO}_4 = (85\ \text{g H}_3\text{PO}_4)\left(\frac{1\ \text{mol H}_3\text{PO}_4}{98.00\ \text{g H}_3\text{PO}_4}\right) = 0.867\ \text{mol}$$

From the definition of molarity, we have that

$$M = \frac{n}{V} = \frac{0.867\ \text{mol}}{V} = 15\ \text{M}$$

Solving for V, we have

$$V = \frac{0.867\ \text{mol}}{15\ \text{mol}\cdot\text{L}^{-1}} = 0.058\ \text{L} = 58\ \text{mL}$$

The density of the solution is given by

$$d = \frac{m}{V} = \frac{100.0\ \text{g}}{58\ \text{mL}} = 1.7\ \text{g}\cdot\text{mL}^{-1}$$

CONCENTRATION OF IONS

12-14. $C_{12}H_{22}O_{11}(aq)$, nonelectrolyte; $NaCl(aq)$, strong electrolyte; $NaHCO_3(aq)$, strong electrolyte; $NH_3(aq)$, weak electrolyte; $CH_3COCH_3(aq)$, nonelectrolyte

12-16. Nickel(III) chloride dissociates in aqueous solution according to

$$NiCl_3(s) \xrightarrow[H_2O]{} Ni^{3+}(aq) + 3\,Cl^-(aq)$$

Therefore, the solution is 0.050 M in $Ni^{3+}(aq)$ ions and 3×0.050 M $= 0.150$ M in $Cl^-(aq)$ ions.

REACTIONS IN SOLUTION

12-18. The number of moles of BaO_2 reacted is

$$\text{moles of } BaO_2 = (17.6\text{ g})\left(\frac{1\text{ mol } BaO_2}{169.3\text{ g } BaO_2}\right) = 0.1040\text{ mol}$$

The number of moles of H_2SO_4 required to react with the BaO_2 is given by

$$\text{moles of } H_2SO_4 = (0.1040\text{ mol } BaO_2)\left(\frac{1\text{ mol } H_2SO_4}{1\text{ mol } BaO_2}\right) = 0.1040\text{ mol}$$

The volume of sulfuric acid required is given by

$$V = \frac{n}{M} = \frac{0.1040\text{ mol}}{3.75\text{ mol}\cdot\text{L}^{-1}} = 0.0277\text{ L} = 27.7\text{ mL}$$

12-20. We must first check to see which, if either, of the reactants is a limiting reactant. The number of moles of $CoCO_3(s)$ is given by

$$\text{moles of } CoCO_3 = (2.17\text{ g})\left(\frac{1\text{ mol } CoCO_3}{118.94\text{ g } CoCO_3}\right) = 0.0182\text{ mol}$$

The number of moles of $HCl(aq)$ is

$$\text{moles of HCl} = (100.0\text{ mL})\left(\frac{1\text{ L}}{10^3\text{ mL}}\right)(0.375\text{ M}) = 0.0375\text{ mol}$$

Because 0.0375 moles of $HCl(aq)$ react with $0.0375/2 = 0.0188$ moles of $CoCO_3$, we see that $HCl(aq)$ is in excess and that $CoCO_3(s)$ is the limiting reactant. The mass of $CoCl_2\cdot 6\,H_2O(s)$ that can be produced is given by

$$\text{mass of } CoCl_2\cdot 6\,H_2O = (0.0182\text{ mol } CoCO_3)\left(\frac{1\ CoCl_2\cdot 6\,H_2O}{1\text{ mol } CoCO_3}\right)\left(\frac{237.9\text{ g } CoCl_2\cdot 6\,H_2O}{1\text{ mol } CoCl_2\cdot 6\,H_2O}\right)$$
$$= 4.33\text{ g}$$

12-22. We must first check to see which reactant is the limiting reactant. The number of moles of each reactant is given by

$$\text{moles of } Mg_2Si = (1.09\text{ g } Mg_2Si)\left(\frac{1\text{ mol } Mg_2Si}{76.70\text{ g } Mg_2Si}\right) = 0.0142\text{ mol}$$
$$\text{moles of HCl} = MV = (1.25\text{ M})(50.0\text{ mL}) = 62.5\text{ mmol} = 0.0625\text{ mol}$$

If we divide the number of moles of each reactant by its respective stoichiometric coefficient, then we see that moles of $HCl/4 = 0.0156$ moles is greater than moles of $Mg_2Si/1 = 0.0142$

moles. Therefore, Mg_2Si is the limiting reactant.

$$\text{mass of SiH}_4 = (0.0142\text{ mol Mg}_2\text{Si})\left(\frac{1\text{ SiH}_4}{1\text{ mol Mg}_2\text{Si}}\right)\left(\frac{32.12\text{ g SiH}_4}{1\text{ mol SiH}_4}\right)$$
$$= 0.456\text{ g}$$

12-24. The number of moles of NaBr per cubic meter is given by

$$\text{moles of NaBr} = MV = \left(\frac{4.00\times10^{-3}\text{ mol}}{1\text{ L}}\right)\left(\frac{10^3\text{ L}}{1\text{ m}^3}\right)(1\text{ m}^3) = 4.00\text{ mol}$$

The number of moles of $Br_2(l)$ produced is

$$\text{moles of Br}_2 = (4.00\text{ mol NaBr})\left(\frac{1\text{ mol Br}_2}{2\text{ mol NaBr}}\right) = 2.00\text{ mol}$$

and the number of grams of $Br_2(l)$ is

$$\text{mass of Br}_2 = (2.00\text{ mol Br}_2)\left(\frac{159.81\text{ g Br}_2}{1\text{ mol Br}_2}\right) = 320\text{ g}$$

The number of moles of $Cl_2(g)$ required is

$$\text{moles of Cl}_2 = (4.00\text{ mol NaBr})\left(\frac{1\text{ mol Cl}_2}{2\text{ mol NaBr}}\right) = 2.00\text{ mol}$$

and the number of grams of $Cl_2(g)$ required is

$$\text{mass of Cl}_2 = (2.00\text{ mol Cl}_2)\left(\frac{70.90\text{ g Cl}_2}{1\text{ mol Cl}_2}\right) = 142\text{ g}$$

12-26. The number of moles of $AgCl(s)$ is

$$\text{moles of AgCl} = (0.231\text{ g AgCl})\left(\frac{1\text{ mol AgCl}}{143.3\text{ g AgCl}}\right) = 0.00161\text{ mol}$$

The number of moles of $NH_3(aq)$ required is

$$\text{moles of NH}_3 = (0.00161\text{ mol AgCl})\left(\frac{2\text{ mol NH}_3}{1\text{ mol AgCl}}\right) = 0.00322\text{ mol}$$

The volume of the $NH_3(aq)$ solution required can be found using Equation 12.3.

$$V = \frac{n}{M} = \frac{0.00322\text{ mol}}{0.100\text{ mol}\cdot\text{L}^{-1}} = 0.0322\text{ L} = 32.2\text{ mL}$$

12-28. The equation for the reaction is

$$2\,\text{AgNO}_3(aq) + \text{CaI}_2(aq) \longrightarrow \text{Ca(NO}_3)_2(aq) + 2\,\text{AgI}(s)$$

We must first check to see which reactant is the limiting reactant. The number of moles of each reactant is given by

$$\text{moles of AgNO}_3 = MV = (0.850\text{ mol}\cdot\text{L}^{-1})(0.175\text{ L}) = 0.1488\text{ mol}$$
$$\text{moles of CaI}_2 = (0.765\text{ mol}\cdot\text{L}^{-1})(0.125\text{ L}) = 0.0956\text{ mol}$$

If we divide the number of moles of each reactant by its respective stoichiometric coefficient, then we see that moles of $AgNO_3/2 = 0.07440$ moles is less than moles of $CaI_2/1 = 0.0956$ moles. Therefore, $AgNO_3(aq)$ is the limiting reactant.

$$\text{mass of AgI} = (0.1488 \text{ mol AgNO}_3)\left(\frac{2 \text{ mol AgI}}{2 \text{ mol AgNO}_3}\right)\left(\frac{234.8 \text{ g AgI}}{1 \text{ mol AgI}}\right) = 34.9 \text{ g}$$

The total volume of the combined solutions is 300 milliliters. Because $AgNO_3(aq)$ is the limiting reactant, after the reaction has taken place, we have $Ca^{2+}(aq)$ ions, $NO_3^-(aq)$, and excess $I^-(aq)$ ions remaining in the solution. The number of moles of $Ca^{2+}(aq)$ is equal to the number of moles of the $CaI_2(aq)$ reactant, or 0.0956 moles. Therefore, the concentration of $Ca^{2+}(aq)$ ions remaining in the solution is given by

$$\text{concentration of Ca}^{2+}(aq) = \frac{0.0956 \text{ mol}}{0.300 \text{ L}} = 0.319 \text{ M}$$

The number of moles of $NO_3^-(aq)$ is equal to the number of moles of the $AgNO_3(aq)$ reactant, or 0.1488 moles. Therefore, the concentration of $NO_3^-(aq)$ ions remaining in the solution is given by

$$\text{concentration of NO}_3^-(aq) = \frac{0.1488 \text{ mol}}{0.300 \text{ L}} = 0.496 \text{ M}$$

Because $CaI_2(aq)$ is in excess, there are $I^-(aq)$ ions remaining in the solution following the reaction. The number of moles of $I^-(aq)$ ions before the reaction is given by

$$\text{moles of I}^- \text{ initially} = (0.0956 \text{ mol CaI}_2)\left(\frac{2 \text{ mol I}^-}{1 \text{ Mol CaI}_2}\right) = 0.1921 \text{ mol}$$

The number of moles of $I^-(aq)$ that reacted is

$$\text{moles I}^- \text{ reacted} = (0.1488 \text{ mol AgNO}_3)\left(\frac{2 \text{ mol AgI}}{2 \text{ mol AgNO}_3}\right)\left(\frac{1 \text{ mol I}^-}{1 \text{ mol AgI}}\right) = 0.1488 \text{ mol}$$

Therefore, the number of moles of $I^-(aq)$ remaining in the solution is 0.1912 moles − 0.1488 moles = 0.0422 moles, and the concentration of $I^-(aq)$ following the reaction is

$$\text{concentration of I}^-(aq) = \frac{0.0422 \text{ mol}}{0.300 \text{ L}} = 0.141 \text{ M}$$

CALCULATIONS INVOLVING ACID-BASE REACTIONS

12-30. The number of milliliters of $H_2SO_4(aq)$ required to neutralize the $NaOH(aq)$ solution is

$$\text{millimoles of H}_2\text{SO}_4 = MV = (0.300 \text{ mmol}\cdot\text{mL}^{-1})(24.6 \text{ mL}) = 7.38 \text{ mmol}$$

We see from the equation of the neutralization reaction

$$\text{H}_2\text{SO}_4(aq) + 2\,\text{NaOH}(aq) \longrightarrow \text{Na}_2\text{SO}_4(aq) + 2\,\text{H}_2\text{O}(l)$$

that it requires one mole of $H_2SO_4(aq)$ to neutralize two moles of $NaOH(aq)$. Thus, we have

$$\text{millimoles of NaOH} = (7.38\text{ mmol H}_2\text{SO}_4)\left(\frac{2\text{ mol NaOH}}{1\text{ mol H}_2\text{SO}_4}\right) = 14.76\text{ mmol}$$

The concentration of the $NaOH(aq)$ solution is

$$M = \frac{n}{V} = \frac{14.76\text{ mmol}}{20.0\text{ mL}} = 0.738\text{ M}$$

12-32. We first must determine the number of moles of $Mg(OH)_2(s)$ and $Al(OH)_3(s)$ in 500 milligrams.

$$\text{moles of Mg(OH)}_2 = [0.500\text{ g Mg(OH)}_2]\left(\frac{1\text{ mol Mg(OH)}_2}{58.32\text{ g Mg(OH)}_2}\right) = 8.57\times10^{-3}\text{ mol}$$

$$\text{moles of Al(OH)}_3 = [0.500\text{ g Al(OH)}_3]\left(\frac{1\text{ mol Al(OH)}_3}{78.00\text{ g Al(OH)}_3}\right) = 6.41\times10^{-3}\text{ mol}$$

We see from the equation of the neutralization reaction

$$\text{Mg(OH)}_2(s) + 2\,\text{HCl}(aq) \longrightarrow \text{MgCl}_2(aq) + 2\,\text{H}_2\text{O}(l)$$

that it requires two moles of $HCl(aq)$ to neutralize one mole of $Mg(OH)_2(s)$. Thus, we have

$$\text{moles of HCl} = [8.57\times10^{-3}\text{ moles of Mg(OH)}_2]\left(\frac{2\text{ mol HCl}}{1\text{ mol Mg(OH)}_2}\right) = 1.71\times10^{-2}\text{ mol}$$

The volume of the $HCl(aq)$ solution neutralized by the $Mg(OH)_2(s)$ is

$$V = \frac{n}{M} = \frac{1.71\times10^{-2}\text{ mol}}{0.10\text{ mol}\cdot\text{L}^{-1}} = 0.170\text{ L} = 170\text{ mL}$$

We see from the equation of the neutralization reaction

$$\text{Al(OH)}_3(s) + 3\,\text{HCl}(aq) \longrightarrow \text{AlCl}_3(aq) + 3\,\text{H}_2\text{O}(l)$$

that it requires three moles of $HCl(aq)$ to neutralize one mole of $Al(OH)_3(s)$. Thus, we have

$$\text{moles of HCl} = [6.41\times10^{-3}\text{ moles of Al(OH)}_3]\left(\frac{3\text{ mol HCl}}{1\text{ mol Al(OH)}_3}\right) = 1.92\times10^{-2}\text{ mol}$$

The volume of the $HCl(aq)$ solution neutralized by the $Al(OH)_3(s)$ is

$$V = \frac{n}{M} = \frac{1.92\times10^{-2}\text{ mol}}{0.10\text{ mol}\cdot\text{L}^{-1}} = 0.192\text{ L} = 190\text{ mL}$$

12-34. The equation for the neutralization reaction is

$$\text{NaOH}(aq) + \text{HBr}(aq) \longrightarrow \text{NaBr}(aq) + \text{H}_2\text{O}(l)$$

The number of millimoles of $NaOH(aq)$ added to the $HBr(aq)$ solution is

$$\text{millimoles of NaOH} = (500\text{ mL})(0.200\text{ mmol}\cdot\text{mL}^{-1}) = 100\text{ mmol}$$

The number of millimoles of HBr(aq) initially present is

$$\text{millimoles of HBr} = MV = (0.100\ \text{mmol}\cdot\text{mL}^{-1})(200\ \text{mL}) = 20.0\ \text{mmol}$$

Thus, an amount of NaOH(aq) *in excess* of that required to neutralize all the HBr(aq) was added. The final concentration of HBr(aq) is thus zero. The final number of moles of NaOH(aq) equals the initial number of moles of NaOH(aq) minus the number of moles of NaOH(aq) used to neutralize the HBr(aq). The volume of the final solution is

$$\begin{aligned}\text{final volume} &= \text{initial volume} + \text{volume added}\\ &= 500\ \text{mL} + 200\text{mL} = 700\ \text{mL}\end{aligned}$$

The final concentration of NaOH(aq) is given by

$$\text{molarity of NaOH} = \frac{n}{V} = \frac{100\ \text{mmol} - 20.0\ \text{mmol}}{700\ \text{mL}} = 0.11\ \text{M}$$

The number of moles of NaBr(aq) is equal to the number of moles of HBr(aq) neutralized. Thus,

$$\text{molarity of NaBr} = \frac{n}{V} = \frac{20.0\ \text{mmol}}{700\ \text{mL}} = 0.0286\ \text{M}$$

12-36. The number of moles of NaOH(aq) in the 25.0 milliliter sample is

$$\text{moles of NaOH} = (23.2\ \text{mL HCl})\left(\frac{1\ \text{L}}{10^3\ \text{mL}}\right)\left(\frac{0.100\ \text{mol HCl}}{1\ \text{L}}\right)\left(\frac{1\ \text{mol NaOH}}{1\ \text{mol HCl}}\right) = 0.00232\ \text{mol}$$

Therefore, the number of moles of NaOH(aq) in the 100.0 milliliter sample is

$$\text{moles of NaOH} = (0.00232\ \text{mol})\left(\frac{100.0\ \text{mL}}{25.0\ \text{mL}}\right) = 0.00928\ \text{mol}$$

The mass of NaOH in the 100 milliliters of NaOH(aq) solution is

$$\text{mass of NaOH} = (0.0928\ \text{mol})\left(\frac{40.00\ \text{g NaOH}}{1\ \text{mol NaOH}}\right) = 0.371\ \text{g}$$

The percentage of NaOH(s) in the sample is equal to the purity of the NaOH

$$\text{purity} = \frac{0.371\ \text{g NaOH}}{0.400\ \text{g sample}} \times 100 = 92.8\%$$

We assumed that any impurities do not react with HCl(aq).

12-38. The number of moles of base required to neutralize the acid is

$$\text{moles of NaOH} = MV = (0.250\ \text{mol}\cdot\text{L}^{-1})(0.0666\ \text{L}) = 0.01665\ \text{mol}$$

The number of moles of acid present in the original 100.0 milliliter sample was 0.01665 moles. Thus, we have the stoichiometric correspondence

$$1.00\ \text{g acid} \approx 0.01665\ \text{mol acid}$$

Dividing by 0.01665, we obtain

$$60.1\ \text{g acid} \approx 1.00\ \text{mol acid}$$

The formula mass of the acid is 60.1.

ADDITIONAL PROBLEMS

12-40. Water is a poor electrolyte, and so a solution of pure water should not conduct an electric current. Acetic acid is a weak electrolyte, and so an aqueous solution of acetic acid should weakly conduct an electric current. Potassium chloride is a strong electrolyte, and so an aqueous solution of potassium chloride should be a good conductor of an electric current.

12-42. The flow diagram would look like the following:

Given a volume and concentration of one of the reactants or products

↓

Multiply the volume of the solution in liters by the molarity of the solution

↓

Number of moles of the reactant or product

↓

Use the balancing coefficients in the chemical equation to form the stoichiometric conversion factor

↓

Number of moles of some other reactant or product

↓

Multiply by the atomic mass or molecular mass

↓

Mass of the other reactant or product

12-44. Take a one-liter (1000 mL) sample of the stock solution. The mass of the sample is

$$\text{mass of sample} = dV = (1.011\ \text{g}\cdot\text{mL}^{-1})(1000\ \text{mL}) = 1011\ \text{g}$$

The mass of hydrazine in the sample is

$$\text{mass of N}_2\text{H}_4 = (1011\ \text{g})\left(\frac{95.0\%}{100\%}\right) = 960.5\ \text{g}$$

The number of moles of hydrazine in the sample is

$$\text{moles of N}_2\text{H}_4 = (960.5\ \text{g N}_2\text{H}_4)\left(\frac{1\ \text{mol N}_2\text{H}_4}{32.05\ \text{g N}_2\text{H}_4}\right) = 30.0\ \text{mol}$$

The concentration of $N_2H_4(aq)$ is

$$M = \frac{n}{V} = \frac{30.0\ \text{mol}}{1.00\ \text{L}} = 30.0\ \text{M}$$

12-46. Take a one-liter (1000 milliliters) sample of the solution. The mass of HCl in the sample is

$$\text{mass of HCl} = (1.20\ \text{g}\cdot\text{mL}^{-1})(1000\ \text{mL})\left(\frac{40.0\%}{100\%}\right) = 480\ \text{g}$$

The number of moles of hydrochloric acid in the sample is

$$\text{moles of HCl} = (480\text{ g HCl})\left(\frac{1\text{ mol HCl}}{36.46\text{ g HCl}}\right) = 13.2\text{ mol}$$

The molarity of HCl(aq) is

$$M = \frac{n}{V} = \frac{13.2\text{ mol}}{1.00\text{ L}} = 13.2\text{ M}$$

12-48. The number of moles of NaCl(aq) in the one-liter solution is

$$\text{moles of NaCl} = (75.6\text{ g NaCl})\left(\frac{1\text{ mol NaCl}}{58.44\text{ g NaCl}}\right) = 1.29\text{ mol}$$

Thus, the molarity of NaCl(aq) is

$$M = \frac{1.29\text{ mol}}{1\text{ L}} = 1.29\text{ M}$$

The number of moles of Cl^-(aq) in 500 milliliters of the 1.29 M NaCl(aq) solution is

$$\text{moles of Cl}^- = (0.500\text{ L})(1.29\text{ M}) = 0.645\text{ mol}$$

Thus,

$$\text{moles of CaCl}_2\cdot 6\,\text{H}_2\text{O required} = (0.645\text{ mol Cl}^-)\left(\frac{1\text{ mol CaCl}_2\cdot 6\,\text{H}_2\text{O}}{2\text{ mol Cl}^-}\right) = 0.323\text{ mol}$$

$$\text{mass of CaCl}_2\cdot 6\,\text{H}_2\text{O} = (0.323\text{ mol CaCl}_2\cdot 6\,\text{H}_2\text{O})\left(\frac{219.1\text{ g CaCl}_2\cdot 6\,\text{H}_2\text{O}}{1\text{ mol CaCl}_2\cdot 6\,\text{H}_2\text{O}}\right) = 70.8\text{ g}$$

12-50. From Equation 12.4, we have $M_1 V_1 = M_2 V_2$. Thus,

$$V_1 = \frac{M_2 V_2}{M_1} = \frac{(0.500\text{ mmol}\cdot\text{mL}^{-1})(50.0\text{ mL})}{6.00\text{ mmol}\cdot\text{mL}^{-1}} = 4.2\text{ mL}$$

12-52. We use Equation 12.4 in the form

$$V_1 = \frac{M_2 V_2}{M_1}$$

Thus,

$$V_1 = \frac{(0.12\text{ mol}\cdot\text{L}^{-1})(250\text{ mL})}{8.0\text{ mol}\cdot\text{L}^{-1}} = 3.75\text{ mL}$$

We add 3.75 mL of 8.0 M H_2SO_4(aq) to a 250-mL volumetric flask that is about half-filled with water, swirl the solution, and dilute with water to the 250-mL mark on the flask.

12-54. 4.0 ppm of NaF corresponds to 4.0 grams of NaF(s) per 10^6 grams of water. This corresponds to

$$\text{moles of NaF per }10^6\text{ grams of water} = (4.0\text{ g NaF})\left(\frac{1\text{ mol NaF}}{41.99\text{ g NaF}}\right) = 0.095\text{ mol}$$

The solution is so dilute that the 10^6 grams of water corresponds to 10^6 milliliters of solution, or to 10^3 liters. Therefore, the molarity of the solution is

$$M = \frac{0.095\text{ mol}}{10^3\text{ L}} = 9.5 \times 10^{-5}\text{ M}$$

12-56. We have that

$$\text{mass \%} = \frac{\text{mass CaCl}_2}{\text{mass solution}} \times 100$$

Thus

$$\begin{aligned}\text{mass solution} &= \frac{\text{mass CaCl}_2}{\text{mass \%}} \times 100 \\ &= \frac{3.25\text{ g}}{14} \times 100 = 23\text{ g}\end{aligned}$$

12-58. The number of moles of $Al_2(SO_4)_3$ in 500 milliliters of the solution is

$$\text{moles of Al}_2(\text{SO}_4)_3 = (3.00\text{ mol}\cdot\text{L}^{-1})(0.500\text{ L}) = 1.50\text{ mol}$$

The number of grams of $Al_2(SO_4)_3\cdot 18\,H_2O(s)$ required is

$$\begin{aligned}\text{mass of Al}_2(\text{SO}_4)_3\cdot 18\,\text{H}_2\text{O} &= [1.50\text{ mol Al}_2(\text{SO}_4)_3]\left(\frac{1\text{ mol Al}_2(\text{SO}_4)_3\cdot 18\,\text{H}_2\text{O}}{1\text{ mol Al}_2(\text{SO}_4)_3}\right) \\ &\times \left(\frac{666.43\text{ g Al}_2(\text{SO}_4)_3\cdot 18\,\text{H}_2\text{O}}{1\text{ mol Al}_2(\text{SO}_4)_3\cdot 18\,\text{H}_2\text{O}}\right) \\ &= 1000\text{ g} = 1.00\text{ kg}\end{aligned}$$

12-60. The number of moles of $KOH(aq)$ in 500 milliliters of a 6.00 M solution is

$$\text{moles of KOH} = (6.00\text{ mol}\cdot\text{L}^{-1})(0.500\text{ L}) = 3.00\text{ mol}$$

$$\text{mass of KOH} = (3.00\text{ mol})\left(\frac{56.11\text{ g KOH}}{1\text{ mol KOH}}\right) = 168.3\text{ g}$$

The mass percentage of KOH in the $KOH(aq)$ sample is $100\% - 8.75\% = 91.25\%$. Using the formula

$$\text{mass \% KOH} = \frac{\text{mass KOH}}{\text{mass KOH solution}} \times 100$$

we have that

$$\begin{aligned}\text{mass KOH solution} &= \frac{\text{mass KOH}}{\text{mass \% KOH}} \times 100 \\ &= \frac{168.3\text{ g}}{91.25} \times 100 = 184\text{ g}\end{aligned}$$

12-62. The number of moles of $HCl(aq)$ required is

$$\text{moles of HCl} = (11.78\text{ g Na}_2\text{CO}_3)\left(\frac{1\text{ mol Na}_2\text{CO}_3}{106.0\text{ g Na}_2\text{CO}_3}\right)\left(\frac{2\text{ mol HCl}}{1\text{ mol Na}_2\text{CO}_3}\right) = 0.2223\text{ mol}$$

The volume of $HCl(aq)$ required is given by

$$V = \frac{n}{M} = \frac{0.2223\text{ mol}}{1.250\text{ mol}\cdot\text{L}^{-1}} = 0.1778\text{ L} = 177.8\text{ mL}$$

12-64. The number of moles of NaOH(aq) required to neutralize the acid is

$$\text{moles of NaOH} = (3.965\ \text{M})(25.95\ \text{mL})\left(\frac{1\ \text{L}}{10^3\ \text{mL}}\right) = 0.1029\ \text{mol}$$

The number of moles of $H_6C_4O_4(s)$ is

$$\text{moles of } H_6C_4O_4 = (6.076\ \text{g } H_6C_4O_4)\left(\frac{1\ \text{mol } H_6C_4O_4}{118.09\ \text{g } H_6C_4O_4}\right) = 0.05145\ \text{mol}$$

Because the number of moles of NaOH(aq) required is two times the number of moles of $H_6C_4O_4(s)$ present, each $H_6C_4O_4$ molecule must have two acidic protons.

12-66. We have

$$\text{millimoles of HCl} = (0.150\ \text{mmol}\cdot\text{mL}^{-1})(27.5\ \text{mL}) = 4.125\ \text{mmol}$$

$$\text{millimoles of } NH_3 = (4.125\ \text{mmol HCl})\left(\frac{1\ \text{mmol } NH_3}{1\ \text{mmol HCl}}\right) = 4.125\ \text{mmol}$$

$$\text{mass of nitrogen} = (4.125 \times 10^{-3}\ \text{mol } NH_3)\left(\frac{1\ \text{mol N}}{1\ \text{mol } NH_3}\right)\left(\frac{14.01\ \text{g N}}{1\ \text{mol N}}\right) = 0.05779\ \text{g}$$

$$\%\ \text{nitrogen} = \frac{0.05779\ \text{g}}{2.25\ \text{g}} \times 100 = 2.57\%$$

12-68. The equation for the reaction is

$$BaCl_2(aq) + K_2SO_4(aq) \longrightarrow BaSO_4(s) + 2\,KCl(aq)$$

We must first check to see which reactant, if either, is the limiting reactant.

$$\text{millimoles of } BaCl_2 = (0.450\ \text{mmol}\cdot\text{mL}^{-1})(20.0\ \text{mL}) = 9.00\ \text{mmol}$$

$$\text{millimoles of } K_2SO_4 = (0.250\ \text{mmol}\cdot\text{mL}^{-1})(36.0\ \text{mL}) = 9.00\ \text{mmol}$$

If we divide the number of millimoles of each reactant by its respective stoichiometric coefficient, then we see that the millimoles of $BaCl_2/1 = 9.00$ millimoles is equal to $K_2SO_4/1 = 9.00$ millimoles. Therefore, neither reactant is a limiting reactant. The number of moles of $BaSO_4$ precipitated is

$$\text{moles of } BaSO_4 = (9.00 \times 10^{-3}\ \text{mol } BaCl_2)\left(\frac{1\ \text{mol } BaSO_4}{1\ \text{mol } BaCl_2}\right) = 9.00 \times 10^{-3}\ \text{mol}$$

$$\text{mass of } BaSO_4 = (9.00 \times 10^{-3}\ \text{mol})\left(\frac{233.4\ \text{g } BaSO_4}{1\ \text{mol } BaSO_4}\right) = 2.10\ \text{g}$$

12-70. The equation for the reaction is

$$Zn(NO_3)_2(aq) + Na_2S(aq) \longrightarrow ZnS(s) + 2\,NaNO_3(aq)$$

We must first check to see which reactant, if either, is the limiting reactant.

$$\text{millimoles of } Zn(NO_3)_2 = (1.76\ \text{mmol}\cdot\text{mL}^{-1})(30.0\ \text{mL}) = 52.8\ \text{mmol}$$

$$\text{millimoles of } Na_2S = (2.18\ \text{mmol}\cdot\text{mL}^{-1})(30.0\ \text{mL}) = 65.4\ \text{mmol}$$

If we divide the number of millimoles of each reactant by its respective stoichiometric coefficient, then we see that the millimoles of $Zn(NO_2)_2/1 = 52.8$ millimoles is less than

$Na_2S/1 = 65.4$ millimoles. Therefore, $Zn(NO_3)_2(aq)$ is the limiting reactant. The number of moles of $ZnS(s)$ precipitated is

$$\text{moles of ZnS} = [0.0528\text{ mol Zn(NO}_3)_2]\left(\frac{1\text{ mol ZnS}}{1\text{ mol Zn(NO}_3)_2}\right) = 0.0528\text{ mol}$$

and so the mass of $ZnS(s)$ precipitated is

$$\text{mass of ZnS} = (0.0528\text{ mol})\left(\frac{97.45\text{ g ZnS}}{1\text{ mol ZnS}}\right) = 5.15\text{ g}$$

12-72. The number of moles of $PbCl_2(s)$ precipitated is given by

$$\text{moles of PbCl}_2 = (12.79\text{ g PbCl}_2)\left(\frac{1\text{ mol PbCl}_2}{278.1\text{ g PbCl}_2}\right) = 0.04599\text{ mol}$$

$$\text{moles of Pb(NO}_3)_2 = (0.04599\text{ mol PbCl}_2)\left(\frac{1\text{ mol Pb(NO}_3)_2}{1\text{ mol PbCl}_2}\right) = 0.04599\text{ mol}$$

The molarity of the solution is

$$M = \frac{n}{V} = \frac{0.04599\text{ mol}}{0.2000\text{ L}} = 0.2300\text{ M}$$

12-74. The equation for the precipitation of $AsO_4^{3-}(aq)$ is

$$AsO_4^{3-}(aq) + 3\,AgNO_3(aq) \longrightarrow Ag_3AsO_4(s) + 3\,NO_3^-(aq)$$

$$\text{millimoles of AgNO}_3 = (0.655\text{ mmol}\cdot\text{mL}^{-1})(37.5\text{ mL}) = 24.56\text{ mmol}$$

$$\text{millimoles of AsO}_4^{3-} = (24.56\text{ mmol AgNO}_3)\left(\frac{1\text{ mmol AsO}_4^{3-}}{3\text{ mmol AgNO}_3}\right) = 8.187\text{ mmol}$$

All the arsenic in $As_4O_6(s)$ ends up in AsO_4^{3-}; thus,

$$1\text{ mol As}_4\text{O}_6 \mathrel{\hat{=}} 4\text{ mol AsO}_4^{3-}$$

$$\text{moles of As}_4\text{O}_6 = (8.187\times10^{-3}\text{ mol AsO}_4^{3-})\left(\frac{1\text{ mol As}_4\text{O}_6}{4\text{ mol AsO}_4^{3-}}\right) = 2.047\times10^{-3}\text{ mol}$$

$$\text{mass of As}_4\text{O}_6 = (2.047\times10^{-3}\text{ mol})\left(\frac{395.68\text{ g As}_4\text{O}_6}{1\text{ mol As}_4\text{O}_6}\right) = 0.8099\text{ g}$$

The mass percentage of $As_4O_6(s)$ in the pesticide is

$$\text{mass \% As}_4\text{O}_6 = \frac{0.8099\text{ g As}_4\text{O}_6}{11.75\text{ g sample}}\times100\% = 6.89\%$$

12-76. We have that

$$\text{millimoles of NH}_3 = (3.52\text{ mg})\left(\frac{1\text{ mmol NH}_3}{17.03\text{ mg NH}_3}\right) = 0.2067\text{ mmol}$$

$$\text{millimoles of OBr}^- = (0.2067\text{ mmol NH}_3)\left(\frac{3\text{ mmol OBr}^-}{2\text{ mmol NH}_3}\right) = 0.3101\text{ mmol}$$

The molarity of the solution is

$$M = \frac{n}{V} = \frac{0.3101 \text{ mmol}}{10.00 \text{ mL}} = 0.0310 \text{ M}$$

12-78. The number of moles of $NaOH(aq)$ is given by

$$\text{moles of NaOH} = (0.1500 \text{ M})(0.02352 \text{ L}) = 0.003528 \text{ mol}$$

One mole of $NaOH(aq)$ will neutralize one mole of aspirin; thus

$$\text{moles of aspirin} = \text{moles of NaOH} = 0.003528 \text{ mol}$$

From the formula mass of aspirin, 180.16, we have the stoichiometric correspondence

$$0.003528 \text{ mol aspirin} \approx 0.6356 \text{ g aspirin}$$

The mass percentage of aspirin in a tablet is

$$\text{mass \% aspirin} = \frac{0.6356 \text{ g aspirin}}{1.00 \text{ g tablet}} \times 100 = 63.6\%$$

12-80. Take a 1.00 liter sample. The volume of ethanol in 1.00 liters of wine is

$$V = (1.00 \text{ L})\left(\frac{11\%}{100\%}\right) = 0.11 \text{ L} = 110 \text{ mL}$$

The mass of the ethanol is

$$\text{mass of ethanol} = (110 \text{ mL})(0.79 \text{ g}\cdot\text{mL}^{-1}) = 86.9 \text{ g}$$

The mass of glucose needed to produce 86.9 grams of ethanol is

$$\text{mass of glucose} = (86.9 \text{ g } CH_3CH_2OH)\left(\frac{1 \text{ mol } CH_3CH_2OH}{46.07 \text{ g } CH_3CH_2OH}\right)$$

$$\times \left(\frac{1 \text{ mol } C_6H_{12}O_6}{2 \text{ mol } CH_3CH_2OH}\right)\left(\frac{180.16 \text{ g } C_6H_{12}O_6}{1 \text{ mol } C_6H_{12}O_6}\right)$$

$$= 170 \text{ g}$$

The volume of water in a 1.00 liter sample is

$$V = 1000 \text{ mL} - 110 \text{ mL} = 890 \text{ mL}$$

The concentration of glucose in the grape juice is

$$\text{concentration of glucose} = \frac{170 \text{ g}}{0.890 \text{ L}} = 190 \text{ g}\cdot\text{L}^{-1}$$

12-82. We must first check to see which reactant, if either, is the limiting reactant.

$$\text{moles of } Cr_2O_7^{2-} = (0.560 \text{ mol}\cdot\text{L}^{-1})(0.0550 \text{ L}) = 0.0308 \text{ mol}$$

$$\text{moles of } CH_3CH_2OH = (0.963 \text{ mol}\cdot\text{L}^{-1})(0.100 \text{ L}) = 0.0963 \text{ mol}$$

If we divide the number of moles of each reactant by its respective stoichiometric coefficient, then we see that the moles of $Cr_2O_7^{2-}/2 = 0.0154$ moles is less than $CH_3CH_2OH/3 = 0.0321$

moles. Therefore, the limiting reactant is $Cr_2O_7^{2-}$. The number of grams of $HC_2H_3O_2$ produced is

$$\text{mass of } HC_2H_3O_2 = (0.0308 \text{ mol } Cr_2O_7^{2-}) \left(\frac{3 \text{ mol } HC_2H_3O_2}{2 \text{ mol } Cr_2O_7^{2-}}\right) \left(\frac{60.05 \text{ g } HC_2H_3O_2}{1 \text{ mol } HC_2H_3O_2}\right)$$

$$= 2.77 \text{ g}$$

12-84. The number of millimoles of $K_2Cr_2O_4$ that reacted is

$$\text{millimoles of } K_2Cr_2O_4 = (0.336 \text{ M})(50.0 \text{ mL}) = 16.80 \text{ mmol}$$

The number of millimoles of $KMnO_4$ required is

$$\text{millimoles of } KMnO_4 = (16.80 \text{ mmol } K_2Cr_2O_4) \left(\frac{2 \text{ mmol } KMnO_4}{5 \text{ mmol } K_2Cr_2O_4}\right) = 6.720 \text{ mmol}$$

The number of milliliters of $KMnO_4$ required is given by

$$V = \frac{n}{M} = \frac{6.720 \text{ mmol}}{0.475 \text{ mmol} \cdot \text{mL}^{-1}} = 14.1 \text{ mL}$$

CHAPTER 13. Properties of Gases

PRESSURE AND BOYLE'S LAW

13-2. The conversions from millibars to torr and to atmospheres are (Table 13.1)

$$(985\ \text{mbar})\left(\frac{1\ \text{bar}}{1000\ \text{mbar}}\right)\left(\frac{750.1\ \text{Torr}}{1\ \text{bar}}\right) = 739\ \text{Torr}$$

$$(985\ \text{mbar})\left(\frac{1\ \text{bar}}{1000\ \text{mbar}}\right)\left(\frac{0.9869\ \text{atm}}{1\ \text{bar}}\right) = 0.972\ \text{atm}$$

13-4. The volume will be larger than 10.0 mL because the pressure is decreased.

$$V = (10.0\ \text{mL})\left(\frac{1.20\ \text{bar}}{0.95\ \text{bar}}\right) = 12.6\ \text{mL}$$

13-6. We use Boyle's law in the form

$$P_\text{i}V_\text{i} = P_\text{f}V_\text{f} \qquad (\text{constant } T)$$

where i stands for initial and f for final. Thus, we have for the initial volume V_i

$$V_\text{i} = \frac{P_\text{f}V_\text{f}}{P_\text{i}} = \frac{(0.10\ \text{bar})(100\ \text{m}^3)}{100\ \text{bar}} = 0.10\ \text{m}^3$$

$$V_\text{i} = (0.10\ \text{m}^3)\left(\frac{100\ \text{cm}}{1\ \text{m}}\right)^3\left(\frac{1\ \text{mL}}{1\ \text{cm}^3}\right)\left(\frac{1\ \text{L}}{10^3\ \text{mL}}\right) = 100\ \text{L}$$

You need two cylinders.

TEMPERATURE AND CHARLES'S LAW

13-8. The volume will be smaller than 1.2 L because the temperature is decreased.

$$V = (1.2\ \text{L})\left(\frac{(-18 + 273)\ \text{K}}{(32 + 273)\ \text{K}}\right) = 1.0\ \text{L}$$

13-10. A gas thermometer is described in Example 13-3. We use Equation 13.7 in the form

$$\frac{V_\text{f}}{T_\text{f}} = \frac{V_\text{i}}{T_\text{i}}$$

We must convert 20.0°C to the Kelvin temperature scale

$$T_i = 20.0 + 273.15 = 293.2\ \text{K}$$

Thus, we have

$$\frac{8.4\ \text{mL}}{T_f} = \frac{12.6\ \text{mL}}{293.2\ \text{K}}$$

Solving for T_f, we have

$$T_f = \frac{(293.2\ \text{K})(8.4\ \text{mL})}{12.6\ \text{mL}} = 195\ \text{K} = -78°\text{C}$$

GAY-LUSSAC'S LAW

13-12. The equation for the reaction is

$$\text{X}_x(g) + 2\,\text{Y}_y(g) \longrightarrow 2\,\text{XY}_2(g)$$

Because the product has but one X per formula unit and we are given two of them, the original gas X must be diatomic. Because the formula of the product is XY_2 and we have two of them, we have four Y atoms. Thus, they must come in pairs and so Y is also diatomic.

13-14. The equation for the reaction is

$$2\,\text{H}_2(g) + \text{O}_2(g) \longrightarrow 2\,\text{H}_2\text{O}(g)$$

We see from Gay-Lussac's law that 1.0 liter of $O_2(g)$ and 2.0 liters of $H_2(g)$ produce 2.0 liters of $H_2O(g)$. The volume of oxygen needed is

$$\text{volume of O}_2 = (0.55\ \text{L H}_2)\left(\frac{1\ \text{L O}_2}{2\ \text{L H}_2}\right) = 0.28\ \text{L}$$

The volume of water produced is

$$\text{volume of H}_2\text{O} = (0.55\ \text{L H}_2)\left(\frac{2\ \text{L H}_2\text{O}}{2\ \text{L H}_2}\right) = 0.55\ \text{L}$$

IDEAL-GAS LAW

13-16. We first use the ideal-gas equation to compute the number of moles of propane.

$$n = \frac{PV}{RT}$$

Substituting the values for P, V, T in units of kelvin, and R in units of $\text{L·atm·mol}^{-1}\text{·K}^{-1}$ into the expression for n yields

$$n = \frac{(7.50\ \text{atm})(50.0\ \text{L})}{(0.08206\ \text{L·atm·mol}^{-1}\text{·K}^{-1})(298\ \text{K})} = 15.33\ \text{mol}$$

The number of grams of propane is given by

$$\text{mass of C}_3\text{H}_8 = (15.33\ \text{mol})\left(\frac{44.10\ \text{g C}_3\text{H}_8}{1\ \text{mol C}_3\text{H}_8}\right) = 676\ \text{g}$$

13-18. We have the stoichiometric correspondence

$$6.15 \times 10^{-3}\ \text{g CO}_2 \approx 1.397 \times 10^{-4}\ \text{mol CO}_2$$

The pressure exerted is given by

$$P = \frac{nRT}{V} = \frac{(1.397 \times 10^{-4}\ \text{mol})(8.3145\ \text{Pa}\cdot\text{m}^3\cdot\text{mol}^{-1}\cdot\text{K}^{-1})(348.2\ \text{K})}{(2.10\ \text{mL})\left(\dfrac{1\ \text{cm}^3}{1\ \text{mL}}\right)\left(\dfrac{1\ \text{m}}{100\ \text{cm}}\right)^3}$$
$$= 1.93 \times 10^5\ \text{Pa} = 193\ \text{kPa}$$

where we have used the fact that $1\ \text{Pa} = 1\ \text{N}\cdot\text{m}^{-2} = 1\ \text{J}\cdot\text{m}^{-3}$.

13-20. The number of moles is given by

$$n = \frac{PV}{RT} = \frac{(8.72 \times 10^4\ \text{Pa})(7.12\ \mu\text{L})\left(\dfrac{1\ \text{L}}{10^6\ \mu\text{L}}\right)\left(\dfrac{1\ \text{m}^3}{10^3\ \text{L}}\right)}{(8.3145\ \text{Pa}\cdot\text{m}^3\cdot\text{mol}^{-1}\cdot\text{K}^{-1})(295.2\ \text{K})}$$
$$= 2.53 \times 10^{-7}\ \text{mol}$$

The mass of the radon gas is

$$\text{mass of radon} = (2.53 \times 10^{-7}\ \text{mol Rn})\left(\frac{222\ \text{g Rn}}{1\ \text{mol Rn}}\right) = 5.62 \times 10^{-5}\ \text{g} = 56.2\ \mu\text{g}$$

13-22. We use the ideal-gas equation in the form

$$n = \frac{PV}{RT}$$

Substituting the values of P, V, R, and T into the expression for n, we have

$$n = \frac{(1.0 \times 10^{-3}\ \text{Torr})(1.00\ \text{mL})\left(\dfrac{1\ \text{L}}{10^3\ \text{mL}}\right)}{(62.3637\ \text{Torr}\cdot\text{L}\cdot\text{mol}^{-1}\cdot\text{K}^{-1})(293\ \text{K})} = 5.47 \times 10^{-11}\ \text{mol}$$

The number of molecules is given by

$$\text{number of molecules} = (5.47 \times 10^{-11}\ \text{mol})(6.022 \times 10^{23}\ \text{molecule}\cdot\text{mol}^{-1})$$
$$= 3.3 \times 10^{13}\ \text{molecule}$$

GAS DENSITY

13-24. We shall use Equation 13.12 to calculate the density of $CF_2Cl_2(g)$.

$$\rho = \frac{MP}{RT}$$

The molar mass of CF_2Cl_2 is $120.9\ \text{g}\cdot\text{mol}^{-1}$. The density at 0°C and 1.00 atm is given by

$$\rho = \frac{(120.9\ \text{g}\cdot\text{mol}^{-1})(1.00\ \text{atm})}{(0.08206\ \text{L}\cdot\text{atm}\cdot\text{mol}^{-1}\cdot\text{K}^{-1})(273\ \text{K})} = 5.40\ \text{g}\cdot\text{L}^{-1}$$

13-26. We first determine the empirical formula of the hydrocarbon. As usual, take a 100-gram sample and write

$$88.82 \text{ g C} \Bumpeq 11.18 \text{ g H}$$

Dividing the mass of each element by its atomic mass, we have

$$7.395 \text{ mol C} \Bumpeq 11.09 \text{ mol H}$$
$$1.00 \text{ mol C} \Bumpeq 1.50 \text{ mol H}$$

The empirical formula is C_2H_3. We now determine the molar mass of the hydrocarbon using the ideal-gas equation.

$$n = \frac{(772 \text{ Torr})(0.0349 \text{ L})}{(62.3637 \text{ Torr}\cdot\text{L}\cdot\text{mol}^{-1}\cdot\text{K}^{-1})(373.2 \text{ K})} = 0.001158 \text{ mol}$$

The molar mass of the hydrocarbon is

$$M = \frac{m}{n} = \frac{0.0626 \text{ g}}{0.001158 \text{ mol}} = 54.1 \text{ g}\cdot\text{mol}^{-1}$$

The molar mass of C_2H_3 is 27.04 $\text{g}\cdot\text{mol}^{-1}$. Therefore, the molecular formula of the hydrocarbon is C_4H_6.

13-28. We first determine the empirical formula of the compound. Taking a 100-gram sample, we have

$$85.60 \text{ g C} \Bumpeq 14.40 \text{ g H}$$

Dividing the mass of each element by its atomic mass, we have

$$7.127 \text{ mol C} \Bumpeq 14.29 \text{ mol H}$$
$$1.00 \text{ mol C} \Bumpeq 2.00 \text{ mol H}$$

Thus, the empirical formula is CH_2. The molar mass is computed from the gas density at a known temperature and pressure.

$$M = \frac{\rho RT}{P} = \frac{(0.9588 \text{ g}\cdot\text{L}^{-1})(62.3637 \text{ Torr}\cdot\text{L}\cdot\text{mol}^{-1}\cdot\text{K}^{-1})(298 \text{ K})}{635 \text{ Torr}}$$
$$= 28.1 \text{ g}\cdot\text{mol}^{-1}$$

The formula mass of CH_2 is 14.03. Thus, the molecular formula of ethylene is C_2H_4.

13-30. The number of moles is given by

$$n = \frac{PV}{RT} = \frac{(0.998 \text{ bar})(0.250 \text{ L})}{(0.083145 \text{ L}\cdot\text{bar}\cdot\text{mol}^{-1}\cdot\text{K}^{-1})(284.0 \text{ K})} = 1.06 \times 10^{-2} \text{ mol}$$

The molecular mass of the compound is given by

$$0.635 \text{ g} \Bumpeq 1.06 \times 10^{-2} \text{ mol}$$

or

$$59.9 \text{ g} \Bumpeq 1.00 \text{ mol}$$

From the combustion analysis data, we have that

$$\text{mass of C} = (2.637 \text{ g CO}_2)\left(\frac{12.01 \text{ g C}}{44.01 \text{ g CO}_2}\right) = 0.7197 \text{ g}$$

$$\text{mass of H} = (1.439 \text{ g H}_2\text{O})\left(\frac{2.016 \text{ g H}}{18.02 \text{ g H}_2\text{O}}\right) = 0.1610 \text{ g}$$

$$\text{mass \% C} = \frac{0.7197 \text{ g}}{1.200 \text{ g}} \times 100 = 59.98\%$$

$$\text{mass \% H} = \frac{0.1610 \text{ g}}{1.200 \text{ g}} \times 100 = 13.42\%$$

$$\text{mass \% O} = 100.00\% - 59.98\% - 13.42\% = 26.60\%$$

The empirical formula of the compound is given by

$$59.99 \text{ g C} \Bumpeq 13.42 \text{ g H} \Bumpeq 26.60 \text{ g O}$$

$$4.995 \text{ mol C} \Bumpeq 13.31 \text{ mol H} \Bumpeq 1.663 \text{ mol O}$$

$$3.000 \text{ mol C} \Bumpeq 8.004 \text{ mol H} \Bumpeq 1.000 \text{ mol O}$$

Therefore, the empirical formula of the compound is C_3H_8O. The molecular mass is 59.90, and so the molecular formula is C_3H_8O.

PARTIAL PRESSURES

13-32. The pressure of each gas is proportional to its number of moles, and so we can write

$$x_{H_2} = \frac{400.0 \text{ Torr}}{(400.0 + 355.1 + 75.2) \text{ Torr}} = \frac{400.0}{830.3} = 0.4818$$

$$x_{N_2} = \frac{355.1 \text{ Torr}}{830.3 \text{ Torr}} = 0.4277$$

$$x_{Ar} = \frac{75.2 \text{ Torr}}{830.3 \text{ Torr}} = 0.0906$$

Note that $x_{H_2} + x_{N_2} + x_{Ar} = 1.000$.

13-34. Let x be the mass fraction of nitrogen in air. Because we are assuming that air consists of only oxygen and nitrogen, the fraction of oxygen is $1 - x$. Then, using the formula masses of nitrogen and oxygen, we have

$$28.0134\,x + 31.9988(1 - x) = 29.0$$

or $x = 0.752$. The mass percentage of nitrogen is 75.2% and that of oxygen is 24.8%. The actual percentages are 78% nitrogen and 21% oxygen. This discrepancy suggests that air consists of more components than just oxygen and nitrogen.

13-36. The mixture is 75% $CO_2(g)$ and 25% $H_2O(g)$ by volume at 175°C. Because the volume is proportional to the number of moles at constant pressure and temperature, the volume fraction is equal to the mole fraction. Therefore, we have for the partial pressure of $CO_2(g)$ at 175°C

$$P_{CO_2} = (0.75)(225 \text{ kPa}) = 169 \text{ kPa}$$

Upon cooling, the volume remains constant because the container is rigid and the number of moles of $CO_2(g)$ in the gas phase also remains constant. Thus, we obtain from Charles's law or the ideal-gas equation

$$P_f = \frac{P_i T_f}{T_i} = P_{CO_2}$$

At 0°C, we have that

$$P_{CO_2} = \frac{(169 \text{ kPa})(273 \text{ K})}{448 \text{ K}} = 103 \text{ kPa}$$

13-38. After the valve has been opened, the volume of each gas is

$$\text{final volume} = 650.0 \text{ mL} + 500.0 \text{ mL} = 1150.0 \text{ mL}$$

We can use Boyle's law in the form $P_i V_i = P_f V_f$ to find the partial pressure of each gas (P_f).

$$\text{partial pressure of } N_2 = P_f = \frac{P_i V_i}{V_f} = \frac{(825 \text{ Torr})(650.0 \text{ mL})}{1150.0 \text{ mL}} = 466 \text{ Torr}$$

$$\text{partial pressure of } O_2 = P_f = \frac{P_i V_i}{V_f} = \frac{(732 \text{ Torr})(500.0 \text{ mL})}{1150.0 \text{ mL}} = 318 \text{ Torr}$$

$$\text{total pressure} = 466 \text{ Torr} + 318 \text{ Torr} = 784 \text{ Torr}$$

IDEAL-GAS LAW AND CHEMICAL REACTIONS

13-40. The number of moles of $NaCl(s)$ in 2.50 kilograms is

$$n = (2.50 \text{ kg})\left(\frac{10^3 \text{ g}}{1 \text{ kg}}\right)\left(\frac{1 \text{ mol NaCl}}{58.44 \text{ g NaCl}}\right) = 42.8 \text{ mol}$$

From the reaction stoichiometry, the number of moles of $H_2(g)$ produced is

$$n = (42.8 \text{ mol})\left(\frac{1 \text{ mol } H_2}{2 \text{ mol NaCl}}\right) = 21.4 \text{ mol}$$

The number of moles of $Cl_2(g)$ produced is equal to the number of moles of $H_2(g)$ produced. Thus, the number of moles of $Cl_2(g)$ is 21.4 moles. We now compute the volume of $Cl_2(g)$ or of $H_2(g)$ by using the ideal-gas equation

$$V = \frac{nRT}{P} = \frac{(21.4 \text{ mol})(0.08206 \text{ L·atm·mol}^{-1}\text{·K}^{-1})(298 \text{ K})}{10.0 \text{ atm}} = 52.3 \text{ L}$$

13-42. We must first check to see which reactant, if either, is the limiting reactant.

$$\text{moles of } CaC_2 = (100.0 \text{ g})\left(\frac{1 \text{ mol } CaC_2}{64.100 \text{ g } CaC_2}\right) = 1.560 \text{ mol}$$

$$\text{moles of } H_2O = (100.0 \text{ g})\left(\frac{1 \text{ mol } H_2O}{18.0153 \text{ g } H_2O}\right) = 5.551 \text{ mol}$$

We see that $CaC_2(s)$ is the limiting reactant. The number of moles of $C_2H_2(g)$ produced is

$$\text{moles of } C_2H_2 = (1.560 \text{ mol } CaC_2)\left(\frac{1 \text{ mol } C_2H_2}{1 \text{ mol } CaC_2}\right) = 1.560 \text{ mol}$$

We now use the ideal-gas equation to compute the volume V occupied by 1.560 moles of $C_2H_2(g)$ at the given values of T and P. At 0°C and 1.00 atm, we have

$$V = \frac{nRT}{P} = \frac{(1.560\ \text{mol})(0.08206\ \text{L·atm·mol}^{-1}\text{·K}^{-1})(273\ \text{K})}{1.00\ \text{atm}} = 34.9\ \text{L}$$

At 125°C and 1.00 atm, we have

$$V = \frac{nRT}{P} = \frac{(1.560\ \text{mol})(0.08206\ \text{L·atm·mol}^{-1}\text{·K}^{-1})(398\ \text{K})}{1.00\ \text{atm}} = 50.9\ \text{L}$$

13-44. The number of moles of $H_2S(g)$ that react is

$$n = \frac{PV}{RT} = \frac{(6.00\ \text{bar})(2.00 \times 10^6\ \text{L})}{(0.083145\ \text{L·bar·mol}^{-1}\text{·K}^{-1})(473\ \text{K})} = 3.05 \times 10^5\ \text{mol}$$

Adding the two chemical equations gives

$$4\,H_2S(g) + 3\,O_2(g) \longrightarrow SO_2(g) + 3\,S(l) + 4\,H_2O(g)$$

The mass of sulfur produced is

$$\text{mass of S} = (3.05 \times 10^5\ \text{mol H}_2\text{S})\left(\frac{3\ \text{mol S}}{4\ \text{mol H}_2\text{S}}\right)\left(\frac{32.065\ \text{g S}}{1\ \text{mol S}}\right)$$
$$= 7.33 \times 10^6\ \text{g} = 7.33 \times 10^3\ \text{kg}$$

Recall that 1 metric ton = 1000 kg, so the mass of sulfur produced is 7.43 metric tons.

13-46. The decomposition of two moles of TNT yields 12 + 5 + 3 = 20 moles of product gases. A 1000-gram sample of TNT yields the following number of moles of gases:

$$n = (1000\ \text{g TNT})\left(\frac{1\ \text{mol TNT}}{227.13\ \text{g TNT}}\right)\left(\frac{20\ \text{mol gas}}{2\ \text{mol TNT}}\right) = 44.0\ \text{mol}$$

The volume is calculated from the ideal-gas equation:

$$V = \frac{nRT}{P} = \frac{(44.0\ \text{mol})(0.08206\ \text{L·atm·mol}^{-1}\text{·K}^{-1})(273\ \text{K})}{1.00\ \text{atm}} = 986\ \text{L}$$

The pressure produced when the reaction is confined to 50 liters at 773 K is given by

$$P = \frac{nRT}{V} = \frac{(44.0\ \text{mol})(0.08206\ \text{L·atm·mol}^{-1}\text{·K}^{-1})(773\ \text{K})}{50\ \text{L}} = 56\ \text{atm}$$

13-48. The equation for the combustion reaction is

$$CH_4(g) + 2\,O_2(g) \longrightarrow CO_2(g) + 2\,H_2O(g)$$

Let the initial pressure of the O_2 be $P^0_{O_2}$. Because the pressure is proportional to the number of moles (at constant temperature and volume), we see that 125 Torr of the $CH_4(g)$ reacts to produce 125 Torr of $CO_2(g)$ and 250 Torr of $H_2O(g)$ and consumes 250 Torr of $O_2(g)$. The total pressure after the reaction is

$$P_{\text{after}} = (P^0_{O_2} - 250\ \text{Torr}) + 125\ \text{Torr} + 250\ \text{Torr} = P^0_{O_2} + 125\ \text{Torr}$$

The total pressure before the reaction is

$$P_{\text{before}} = P^0_{O_2} + 125\ \text{Torr}$$

Thus, $\Delta P_{\text{reaction}} = 0$.

MOLECULAR SPEEDS

13-50. Recall that in working kinetic theory problems, we use the value $R = 8.3145\ \text{J}\cdot\text{mol}^{-1}\cdot\text{K}^{-1}$. The root-mean-square speed in $\text{m}\cdot\text{s}^{-1}$ of a gas molecule is calculated from the equation

$$v_{\text{rms}} = \left(\frac{3RT}{M_{\text{kg}}}\right)^{1/2}$$

where M_{kg} is the molar mass in kilograms per mole. For N_2O, we have

$$M_{\text{kg}} = \frac{44.01\ \text{g}\cdot\text{mol}^{-1}}{1000\ \text{g}\cdot\text{kg}^{-1}} = 0.04401\ \text{kg}\cdot\text{mol}^{-1}$$

Thus, at the temperatures given, we have for the root-mean-square speed of a $N_2O(g)$ molecule at 20°C, 200°C, and 2000°C

$$v_{\text{rms}} = \left[\frac{(3)(8.3145\ \text{J}\cdot\text{mol}^{-1}\cdot\text{K}^{-1})(293\ \text{K})}{0.04401\ \text{kg}\cdot\text{mol}^{-1}}\right]^{1/2} = 408\ \text{m}\cdot\text{s}^{-1}$$

$$= \left[\frac{(3)(8.3145\ \text{J}\cdot\text{mol}^{-1}\cdot\text{K}^{-1})(473\ \text{K})}{0.04401\ \text{kg}\cdot\text{mol}^{-1}}\right]^{1/2} = 518\ \text{m}\cdot\text{s}^{-1}$$

$$= \left[\frac{(3)(8.3145\ \text{J}\cdot\text{mol}^{-1}\cdot\text{K}^{-1})(2273\ \text{K})}{0.04401\ \text{kg}\cdot\text{mol}^{-1}}\right]^{1/2} = 1140\ \text{m}\cdot\text{s}^{-1}$$

13-52. The ratio of the root-mean-square speed of $H_2(g)$ and $I_2(g)$ in a gas mixture is computed using Equation 13.29

$$\frac{v_{\text{rms},H_2}}{v_{\text{rms},I_2}} = \frac{\left(\frac{3RT}{M_{\text{kg},H_2}}\right)^{1/2}}{\left(\frac{3RT}{M_{\text{kg},I_2}}\right)^{1/2}} = \left(\frac{M_{I_2}}{M_{H_2}}\right)^{1/2} = \left(\frac{253.8}{2.016}\right)^{1/2} = 11.22$$

Note that $H_2(g)$ with a much lower mass than $I_2(g)$ has a much higher root-mean-square speed.

GRAHAM'S LAW OF EFFUSION

13-54. The ratio of the rates of diffusion of the two gases is given by Graham's law:

$$\frac{\text{rate}_A}{\text{rate}_B} = \left(\frac{M_B}{M_A}\right)^{1/2}$$

If we take A = hydrogen and B = carbon dioxide, then we have

$$\text{rate}_{H_2} = (\text{rate}_{CO_2})\left(\frac{M_{CO_2}}{M_{H_2}}\right)^{1/2}$$

Substituting in the values for rate_{CO_2}, M_{CO_2}, and M_{H_2} yields

$$\text{rate}_{H_2} = (1.50\ \text{mL}\cdot\text{day}^{-1})\left(\frac{44.01}{2.016}\right)^{1/2} = 7.01\ \text{mL}\cdot\text{day}^{-1}$$

In one day, 7.01 milliliters of hydrogen will leak out.

13-56. Graham's law gives

$$\frac{\text{rate}_{\text{A}}}{\text{rate}_{\text{B}}} = \left(\frac{M_{\text{B}}}{M_{\text{A}}}\right)^{1/2}$$

or

$$\frac{M_{\text{B}}}{M_{\text{A}}} = \left(\frac{\text{rate}_{\text{A}}}{\text{rate}_{\text{B}}}\right)^{2}$$

We can find the percentage of CO(g) in the gas mixture from the molecular mass calculated from Graham's law. If we let B = CO-CO_2 gas mixture and A = nitrogen, then we have

$$M_{\text{mixture}} = (M_{\text{N}_2})\left(\frac{\text{rate}_{\text{N}_2}}{\text{rate}_{\text{mixture}}}\right)^{2}$$

The rate of effusion of $N_2(g)$ is

$$\text{rate}_{\text{N}_2} = \frac{1.00\ \text{mL}}{175.0\ \text{s}} = 5.714 \times 10^{-3}\ \text{mL}\cdot\text{s}^{-1}$$

The rate of effusion of the CO-CO_2 gas mixture is

$$\text{rate}_{\text{mixture}} = \frac{1.00\ \text{mL}}{200.0\ \text{s}} = 5.00 \times 10^{-3}\ \text{mL}\cdot\text{s}^{-1}$$

Thus,

$$M_{\text{mixture}} = (28.02\ \text{g}\cdot\text{mol}^{-1})\left(\frac{5.714 \times 10^{-3}\ \text{mL}\cdot\text{s}^{-1}}{5.00 \times 10^{-3}\ \text{mL}\cdot\text{s}^{-1}}\right)^{2} = 36.59\ \text{g}\cdot\text{mol}^{-1}$$

If we let x be the fraction of the mixture that is CO(g), then $1.00 - x$ is the fraction that is $CO_2(g)$. The molecular mass of the mixture is given by

$$M_{\text{mixture}} = x\, M_{\text{CO}} + (1 - x)\, M_{\text{CO}_2}$$
$$36.59 = x(28.02) + (1 - x)(44.01) = 28.02\,x + 44.01 - 44.01\,x$$
$$15.99\,x = 7.42$$

Solving for x, we have that $x = 0.464$. The mixture is 46.4% CO(g).

MEAN FREE PATH

13-58. The mean free path is given by Equation 13.34

$$l = \frac{RT}{\pi\sqrt{2}\, d^2 N_{\text{A}} P}$$

Solving for P, we have

$$\begin{aligned} P &= \frac{RT}{(\pi\sqrt{2})\, N_{\text{A}} d^2 l} \\ &= \frac{(0.08206\ \text{L}\cdot\text{atm}\cdot\text{mol}^{-1}\cdot\text{K}^{-1})(293.2\ \text{K})}{\pi\sqrt{2}\,(6.022 \times 10^{23}\ \text{mol}^{-1})(270 \times 10^{-12}\ \text{m})^2(1.00 \times 10^{-6}\ \text{m})} \\ &= 0.123\ \text{atm} \end{aligned}$$

The other two values of l give 1.23×10^{-4} atm (for $l = 1.00$ mm) and 1.23×10^{-7} atm (for $l = 1.00$ m).

13-60. The number of collisions per second (collision frequency) is given by

$$z = \frac{v_{\text{rms}}}{l}$$

The root-mean-square speed of a $N_2(g)$ molecule at 20°C is

$$v_{\text{rms}} = \left(\frac{3RT}{M_{\text{kg}}}\right)^{1/2} = \left[\frac{(3)(8.3145\text{ J·mol}^{-1}\text{·K}^{-1})(293\text{ K})}{0.0280\text{ kg·mol}^{-1}}\right]^{1/2} = 511\text{ m·s}^{-1}$$

The mean free path for $N_2(g)$ molecule at 20°C and 1.0×10^{-3} Torr using the value $d_{N_2} = 370$ pm from Table 13.4,

$$\begin{aligned} l &= \frac{RT}{\pi\sqrt{2}\,N_A d^2 P} \\ &= \frac{(62.3637\text{ L·Torr·mol}^{-1}\text{·K}^{-1})\left(\frac{1\text{ m}^3}{10^3\text{ L}}\right)(293\text{ K})}{\pi\sqrt{2}\,(6.022 \times 10^{23}\text{ mol}^{-1})(370 \times 10^{-12}\text{ m})^2(1.0 \times 10^{-3}\text{ Torr})} \\ &= 5.0 \times 10^{-2}\text{ m} \end{aligned}$$

Thus,

$$z = \frac{511\text{ m·s}^{-1}}{5.0 \times 10^{-2}\text{ m}} = 1.0 \times 10^4\text{ collisions·s}^{-1}$$

VAN DER WAALS EQUATION

13-62. Start with

$$\left(P + \frac{n^2 a}{V^2}\right)(V - nb) = nRT$$

When V becomes large, the nb term can be neglected with respect to V and the n^2a/V^2 term can be neglected with respect to P to obtain $PV = nRT$.

13-64. We shall use Equation 13.39 to calculate the pressure

$$P = \frac{nRT}{V - nb} - \frac{n^2 a}{V^2}$$

The number of moles of propane is

$$n = (45\text{ g})\left(\frac{1\text{ mol } C_3H_8}{44.10\text{ g } C_3H_8}\right) = 1.02\text{ mol}$$

We obtain the values of a and b for C_3H_8 from Table 13.5. Thus, we have

$$\begin{aligned} P &= \frac{(1.02\text{ mol})(0.083145\text{ L·bar·mol}^{-1}\text{·K}^{-1})(573\text{ K})}{2.2\text{ L} - (1.02\text{ mol})(0.090494\text{ L·mol}^{-1})} - \frac{(1.02\text{ mol})^2(9.3919\text{ L}^2\text{·bar·mol}^{-2})}{(2.2\text{ L})^2} \\ &= 23.1\text{ bar} - 2.2\text{ bar} = 20.9\text{ bar} \end{aligned}$$

The pressure calculated by using the ideal-gas equation is

$$P = \frac{nRT}{V} = \frac{(1.02\text{ mol})(0.083145\text{ L}\cdot\text{bar}\cdot\text{mol}^{-1}\cdot\text{K}^{-1})(573\text{ K})}{2.2\text{ L}} = 22.1\text{ bar}$$

ADDITIONAL PROBLEMS

13-66. The density of mercury ($13.6\text{ g}\cdot\text{mL}^{-1}$) is much greater than that of most liquids and so the column of a mercury barometer needs to be only about a meter long. In contrast, the column of a water barometer would need to be at least 13.6 times longer or about 14 meters, which is almost the height of a three-story building. In addition, the vapor pressure of water varies with temperature and adds a significant error.

13-68. Start with 1 bar = 0.9869 atm (Table 13.1). Therefore,

$$(1.00\text{ bar})\left(\frac{0.9869\text{ atm}}{1\text{ bar}}\right)\left(\frac{760\text{ Torr}}{1\text{ atm}}\right)\left(\frac{1\text{ mm Hg}}{1\text{ Torr}}\right)\left(\frac{1\text{ cm}}{10\text{ mm}}\right)\left(\frac{1\text{ in}}{2.54\text{ cm}}\right) = 29.5\text{ in}$$

13-70. Mercury is $\dfrac{13.6\text{ g}\cdot\text{cm}^{-3}}{6.0\text{ g}\cdot\text{cm}^{-3}} = 2.267$ times as dense as gallium and so the column of gallium will be 2.267 times the height of the column of mercury. The column of mercury supported is 1300 mm and so

$$\text{column of gallium} = (1300\text{ mm})(2.267)\left(\frac{1\text{ m}}{10^3\text{ mm}}\right) = 3.0\text{ m}$$

13-72. The table of the data is as follows:

V/L	$1/P$/atm^{-1}	PV/L·atm
0.938	3.8	0.24
0.595	2.4	0.24
0.294	1.2	0.24
0.203	0.83	0.24
0.116	0.48	0.24
0.093	0.38	0.24
0.078	0.32	0.24

Boyle's law in an equation states that

$$V = \frac{c}{P}$$

Thus,

$$PV = \text{constant}$$

as the data show. A plot of V versus $1/P$ should be a straight line of the form $y = ax$, where $y = V$ and $x = 1/P$.

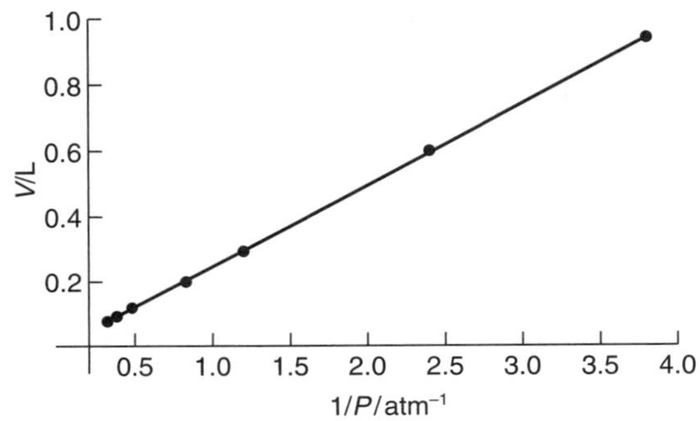

13-74. The equation for the reaction is

$$2\,C_8H_{18}(l) + 25\,O_2(g) \longrightarrow 16\,CO_2(g) + 18\,H_2O(l)$$

One gallon of gasoline corresponds to

$$\text{moles of } C_8H_{18} = (1 \text{ gal})\left(\frac{4 \text{ qt}}{1 \text{ gal}}\right)\left(\frac{1 \text{ L}}{1.0567 \text{ qt}}\right)\left(\frac{1000 \text{ mL}}{1 \text{ L}}\right)(0.70 \text{ g}\cdot\text{mL}^{-1}) \times \left(\frac{1}{114.2 \text{ g}\cdot\text{mol}^{-1}}\right) = 23.2 \text{ mol}$$

This quantity requires (23.2 mol) (25/2) = 290 moles of $O_2(g)$, which occupies

$$V_{O_2} = \frac{nRT}{P} = \frac{(290 \text{ mol})(0.083145 \text{ L}\cdot\text{bar}\cdot\text{mol}^{-1}\cdot\text{K}^{-1})(273 \text{ K})}{1 \text{ bar}} = 6580 \text{ L}$$

The corresponding volume of air is given by

$$V_{\text{air}} = V_{O_2}\left(\frac{100\% \text{ air}}{21\% \text{ } O_2}\right) = (6580 \text{ L})\left(\frac{100\% \text{ air}}{21\% \text{ } O_2}\right)\left(\frac{1 \text{ qt}}{0.9463 \text{ L}}\right)\left(\frac{1 \text{ gal}}{4 \text{ qt}}\right) = 8300 \text{ gallons}$$

The information on the website is approximately correct.

13-76. The number of moles of $H_2(g)$ produced is

$$n = \frac{PV}{RT} = \frac{(750 \text{ Torr})(0.150 \text{ L})}{(62.3637 \text{ L}\cdot\text{Torr}\cdot\text{mol}^{-1}\cdot\text{K}^{-1})(283 \text{ K})} = 6.374 \times 10^{-3} \text{ mol}$$

The number of moles of NaOH required is

$$n = (6.374 \times 10^{-3} \text{ mol } H_2)\left(\frac{2 \text{ mol NaOH}}{3 \text{ mol } H_2}\right) = 4.249 \times 10^{-3} \text{ mol}$$

The volume of NaOH(aq) required is

$$V = \frac{n}{M} = \frac{4.249 \times 10^{-3} \text{ mol}}{0.200 \text{ mol}\cdot\text{L}^{-1}} = 0.0212 \text{ L} = 21.2 \text{ mL}$$

13-78. The number of moles of $NH_3(g)$ added is

$$\text{moles of } NH_3 = (5.0 \text{ g})\left(\frac{1 \text{ mol } NH_3}{17.03 \text{ g } NH_3}\right) = 0.294 \text{ mol}$$

The number of moles of HCl(g) added is

$$\text{moles of HCl} = (10.0\text{ g})\left(\frac{1\text{ mol HCl}}{36.46\text{ g HCl}}\right) = 0.2743\text{ mol}$$

We see that HCl(g) is the limiting reactant. The number of moles of $NH_3(g)$ remaining is

$$\text{moles of NH}_3 = 0.294\text{ mol} - 0.2743\text{ mol} = 0.020\text{ mol}$$

The pressure of $NH_3(g)$ is

$$P = \frac{nRT}{V} = \frac{(0.020\text{ mol})(0.083145\text{ L}\cdot\text{bar}\cdot\text{mol}^{-1}\cdot\text{K}^{-1})(348\text{ K})}{1.00\text{ L}} = 0.58\text{ bar}$$

13-80. Of the gases listed in Table 13.5, the one that deviates most from ideal-gas behavior is the one with the largest values of the van der Waals constants a and b. Inspection of the data in the table shows that propane is the most nonideal of the gases listed.

13-82. The number of moles of sodium peroxide present is given by

$$\text{moles of Na}_2\text{O}_2 = (1000\text{ g})\left(\frac{1\text{ mol Na}_2\text{O}_2}{77.99\text{ g Na}_2\text{O}_2}\right) = 12.82\text{ mol}$$

and the number of moles of $CO_2(g)$ absorbed is given by

$$\text{moles of CO}_2 = (12.82\text{ mol Na}_2\text{O}_2)\left(\frac{2\text{ mol CO}_2}{2\text{ mol Na}_2\text{O}_2}\right) = 12.82\text{ mol}$$

The number of liters of carbon dioxide absorbed is given by

$$V = \frac{nRT}{P} = \frac{(12.82\text{ mol})(0.08206\text{ L}\cdot\text{atm}\cdot\text{mol}^{-1}\cdot\text{K}^{-1})(273\text{ K})}{1\text{ atm}} = 287\text{ L}$$

The number of moles of $O_2(g)$ produced is

$$\text{moles of O}_2 = (12.82\text{ mol Na}_2\text{O}_2)\left(\frac{1\text{ mol O}_2}{2\text{ mol Na}_2\text{O}_2}\right) = 6.410\text{ mol}$$

The number of liters of oxygen produced is given by

$$V = \frac{nRT}{P} = \frac{(6.410\text{ mol})(0.08206\text{ L}\cdot\text{atm}\cdot\text{mol}^{-1}\cdot\text{K}^{-1})(273\text{ K})}{1\text{ atm}} = 144\text{ L}$$

13-84. We must first check to see which reactant, if either, is the limiting reactant.

$$\text{moles of HCl} = MV = (0.1150\text{ mol}\cdot\text{L}^{-1})(0.3000\text{ L}) = 0.03450\text{ mol}$$

$$\text{moles of KMnO}_4 = (6.75\text{ g})\left(\frac{1\text{ mol KMnO}_4}{158.03\text{ g KMnO}_4}\right) = 0.04271\text{ mol}$$

If we divide the number of moles of each reactant by its respective stoichiometric coefficient, then we see that moles of HCl/16 = 0.0022 moles is less than moles of $KMnO_4/2 = 0.02136$ moles. Therefore, the limiting reactant is HCl(aq).

$$\text{moles of Cl}_2 = (0.03450\text{ mol HCl})\left(\frac{5\text{ mol Cl}_2}{16\text{ mol HCl}}\right) = 0.01078\text{ mol}$$

The volume of chlorine is

$$V = \frac{nRT}{P} = \frac{(0.01078\ \text{mol})(62.3637\ \text{L·Torr·mol}^{-1}\text{·K}^{-1})(288\ \text{K})}{815\ \text{Torr}} = 0.238\ \text{L} = 238\ \text{mL}$$

13-86. The volume of the room is

$$V = (3.0\ \text{m})(5.0\ \text{m})(6.0\ \text{m})\left(\frac{1000\ \text{L}}{1\ \text{m}^3}\right) = 9.0 \times 10^4\ \text{L}$$

The number of moles of water vapor in the room is

$$n = \frac{PV}{RT} = \frac{(35\ \text{Torr})(9.0 \times 10^4\ \text{L})}{(62.3637\ \text{L·Torr·mol}^{-1}\text{·K}^{-1})(308\ \text{K})} = 164\ \text{mol}$$

The mass of the water vapor is

$$\text{mass of H}_2\text{O} = (164\ \text{mol})\left(\frac{18.02\ \text{g H}_2\text{O}}{1\ \text{mol H}_2\text{O}}\right) = 2960\ \text{g}$$

Using the fact that the density of water is $1.00\ \text{g·mL}^{-1}$, we have that the volume of water obtained would be

$$V = \frac{m}{d} = \frac{2960\ \text{g}}{1.00\ \text{g·mL}^{-1}} = 3000\ \text{mL} = 3.0\ \text{L}$$

13-88.* Let x g equal the mass of NaH(s) and so $3.75\ \text{g} - x\,\text{g}$ is the mass of $CaH_2(s)$. The number of moles of NaH(s) and $CaH_2(s)$ are

$$\text{moles of NaH} = (x\ \text{g})\left(\frac{1\ \text{mol NaH}}{24.00\ \text{g NaH}}\right) = 0.04167\,x\ \text{mol}$$

$$\text{moles of CaH}_2 = (3.75\ \text{g} - x\,\text{g})\left(\frac{1\ \text{mol CaH}_2}{42.09\ \text{g CaH}_2}\right) = 0.08909\ \text{mol} - 0.02376\,x\ \text{mol}$$

$$\text{moles of H}_2 = \frac{PV}{RT} = \frac{(742\ \text{Torr})(4.12\ \text{L})}{(62.3637\ \text{L·Torr·mol}^{-1}\text{·K}^{-1})(290\ \text{K})} = 0.1690\ \text{mol}$$

The equations for the reactions are

$$\text{NaH}(s) + \text{H}_2\text{O}(l) \longrightarrow \text{NaOH}(aq) + \text{H}_2(g)$$
$$\text{CaH}_2(s) + 2\,\text{H}_2\text{O}(l) \longrightarrow \text{Ca(OH)}_2(aq) + 2\,\text{H}_2(g)$$

Thus, we have that one mole of $H_2(g)$ is produced for each mole of NaH(s) that reacts and two moles for each mole of $CaH_2(s)$ that reacts. Thus,

$$\begin{aligned}
\text{moles of H}_2 &= \text{moles of NaH} + 2\,(\text{moles of CaH}_2)\\
0.1690\ \text{mol} &= 0.04167\,x\ \text{mol} + (2)(0.08909\ \text{mol} - 0.02376\,x\ \text{mol})\\
0.1690 &= 0.04167\,x + 0.1782 - 0.04752\,x\\
5.85 \times 10^{-3}x &= 9.2 \times 10^{-3}\\
x &= 1.6
\end{aligned}$$

The mass of NaH(s) is 1.6 grams. The mass percentage of each in the mixture is

$$\text{mass \% NaH} = \frac{1.6\ \text{g}}{3.75\ \text{g}} \times 100 = 43\%$$
$$\text{mass \% CaH}_2 = 100\% - 43\% = 57\%$$

13-90. The balanced equation is

$$2\,\mathrm{N_xO_y}(g) + \mathrm{O_2}(g) \longrightarrow \mathrm{N_2O_4}(g)$$

We have $2x = 2$ and $2y + 2 = 4$. Thus, $x = 1$ and $y = 1$, which will balance the equation. Thus, the formula of the oxide is $\mathrm{NO}(g)$.

13-92.* The equation for the reaction that takes place when the mixture is heated is

$$2\,\mathrm{KClO_3}(s) \longrightarrow 2\,\mathrm{KCl}(s) + 3\,\mathrm{O_2}(g)$$

The pressure of $\mathrm{O_2}(g)$ is

$$P = 756\text{ Torr} - 15.5\text{ Torr} = 740\text{ Torr}$$

The number of moles of $\mathrm{O_2}(g)$ produced is

$$n = \frac{PV}{RT} = \frac{(740\text{ Torr})(0.0807\text{ L})}{(62.3637\text{ L}\cdot\text{Torr}\cdot\text{mol}^{-1}\cdot\text{K}^{-1})(291\text{ K})}$$
$$= 3.291 \times 10^{-3}\text{ mol}$$

The mass of $\mathrm{KClO_3}(s)$ that produced 3.291×10^{-3} moles of $\mathrm{O_2}(g)$ is

$$\text{mass of KClO}_3 = (3.291 \times 10^{-3}\text{ mol O}_2)\left(\frac{2\text{ mol KClO}_3}{3\text{ mol O}_2}\right)\left(\frac{122.55\text{ g KClO}_3}{1\text{ mol KClO}_3}\right)$$
$$= 0.2689\text{ g}$$

The mass percentage of $\mathrm{KClO_3}$ in the mixture is

$$\text{mass \% of KClO}_3 = \frac{0.2689\text{ g}}{0.428\text{ g}} \times 100 = 62.8\%$$

13-94. From the expressions for v_{rms} and v_{sound}, we have

$$\frac{v_{\text{rms}}}{v_{\text{sound}}} = \frac{\left(\dfrac{3RT}{M_{\text{kg}}}\right)^{1/2}}{\left(\dfrac{5RT}{3M_{\text{kg}}}\right)^{1/2}} = \frac{3}{\sqrt{5}}$$

Thus, at 293 K we have for the root-mean-square speed for an argon atom

$$v_{\text{rms}} = \left[\frac{(3)(8.3145\text{ J}\cdot\text{mol}^{-1}\cdot\text{K}^{-1})(293\text{ K})}{0.03995\text{ kg}\cdot\text{mol}^{-1}}\right]^{1/2} = 428\text{ m}\cdot\text{s}^{-1}$$

and the speed of sound in argon is

$$v_{\text{sound}} = \left(\frac{\sqrt{5}}{3}\right) v_{\text{rms}} = 319\text{ m}\cdot\text{s}^{-1}$$

CHAPTER 14. Thermochemistry

HEAT AND ENERGY

14-2. The work done on the system in decreasing the volume at constant pressure is given by

$$w = -P\Delta V = -(4.0\ \text{atm})(2.0\ \text{L} - 10.0\ \text{L}) = 32\ \text{L·atm}$$

Work is generally expressed in joules; therefore, we must convert L·atm to J.

$$w = (32\ \text{L·atm})\left(\frac{101.325\ \text{J}}{1\ \text{L·atm}}\right) = 3200\ \text{J} = 3.2\ \text{kJ}$$

14-4. The work done on the system in decreasing the volume at constant pressure is given by

$$w = -P\Delta V = -(350.0\times 10^3\ \text{Pa})(-250.0\ \text{cm}^3)\left(\frac{1\ \text{m}}{100\ \text{cm}}\right)^3 = +87.50\ \text{J}$$

14-6. The equation for the reaction is

$$\text{CH}_4(g) + 2\,\text{O}_2(g) \longrightarrow \text{CO}_2(g) + 2\,\text{H}_2\text{O}(l)$$

Let the initial volume of $O_2(g)$ be V_i. The initial volume of gases is $V_i + 2.0$ L. The final volume of gases, V_f, is 2.0 L of $CO_2(g)$ plus $V_i - 4.0$ L of $O_2(g)$, or $V_f = V_i - 2.0$ L. The change in volume is -2.0 L. The work done is given by

$$w = -P\Delta V = -(1.00\ \text{bar})\left(\frac{10^5\ \text{Pa}}{1\ \text{bar}}\right)(-2.0\ \text{L})\left(\frac{1\ \text{m}^3}{1000\ \text{L}}\right) = 200\ \text{J}$$

14-8. The equation for the reaction is

$$\text{Ba}(s) + \text{Cl}_2(g) \longrightarrow \text{BaCl}_2(s)$$

We are given the amount of heat evolved (-15.4 kilojoules) when 2.46 grams of barium reacts with chlorine. The amount of heat that is evolved when one mole of barium reacts to form one mole of $BaCl_2(s)$ is

$$q = \left(\frac{-15.4\ \text{kJ}}{2.46\ \text{g Ba}}\right)\left(\frac{137.3\ \text{g Ba}}{1\ \text{mol Ba}}\right) = -860\ \text{kJ·mol}^{-1}$$

14-10. The equation for the reaction is

$$2\,\mathrm{Mg}(s) + \mathrm{O_2}(g) \longrightarrow 2\,\mathrm{MgO}(s)$$

The number of moles of $MgO(g)$ produced is

$$\text{moles of MgO} = (0.165\text{ g Mg})\left(\frac{1\text{ mol Mg}}{24.31\text{ g Mg}}\right)\left(\frac{2\text{ mol MgO}}{2\text{ mol Mg}}\right) = 0.00679\text{ mol}$$

The heat evolved per mole of $MgO(s)$ is

$$\text{heat evolved per mole MgO} = \frac{-4.08\text{ kJ}}{0.00679\text{ mol}} = -601\text{ kJ}\cdot\text{mol}^{-1}$$

14-12. We have the stoichiometric correspondence

$$10.0\text{ g Fe} \approx 0.1791\text{ mol Fe}$$

The energy released per one mole of the stated reaction equation is $-1648\text{ kJ}\cdot\text{mol}^{-1}$, or -1648 kJ per four moles of $Fe(s)$. Therefore, the energy released when 0.1791 moles of Fe react is

$$\Delta H = (0.1791\text{ mol Fe})\left(\frac{-1648\text{ kJ}}{4\text{ mol Fe}}\right) = -73.8\text{ kJ}$$

14-14. We use Equation 14.10 in the form

$$\Delta U^\circ_{\mathrm{rxn}} = \Delta H^\circ_{\mathrm{rxn}} - P\Delta V_{\mathrm{rxn}}$$

For this system

$$P\Delta V = \Delta\nu_{\mathrm{gas}} RT$$

with $\Delta\nu_{\mathrm{gas}} = -5$ moles. Thus,

$$\begin{aligned}\Delta U^\circ_{\mathrm{rxn}} &= \Delta H^\circ_{\mathrm{rxn}} - \Delta\nu_{\mathrm{gas}} RT \\ &= -1166\text{ kJ}\cdot\text{mol}^{-1} - (-5)(8.3145\text{ J}\cdot\text{mol}^{-1}\cdot\text{K}^{-1})\left(\frac{1\text{ kJ}}{10^3\text{ J}}\right)(298\text{ K}) \\ &= -1166\text{ kJ}\cdot\text{mol}^{-1} + 12.4\text{ kJ}\cdot\text{mol}^{-1} = -1154\text{ kJ}\cdot\text{mol}^{-1}\end{aligned}$$

The values of $\Delta H^\circ_{\mathrm{rxn}}$ and $\Delta U^\circ_{\mathrm{rxn}}$ differ by 1.0%.

14-16. The number of moles of $O_2(g)$ that reacts when 0.1791 moles of Fe undergoes oxidation is

$$\text{moles of O}_2 = (0.1791\text{ mol Fe})\left(\frac{3\text{ mol O}_2}{4\text{ mol Fe}}\right) = 0.1343\text{ mol}$$

We have that $\Delta\nu_{\mathrm{gas}}$ is equal to the number of moles of $O_2(g)$ that reacted, or $\Delta\nu_{\mathrm{gas}} = -0.1343$ mol. Thus, the volume change is

$$\Delta V = \frac{\Delta\nu_{\mathrm{gas}} RT}{P} = \frac{(-0.1343\text{ mol})(0.083145\text{ bar}\cdot\text{L}\cdot\text{K}^{-1}\cdot\text{mol}^{-1})(298\text{ K})}{1.00\text{ bar}} = -3.33\text{ L}$$

The amount of work involved is

$$w = -P\Delta V = -(1.00\text{ bar})\left(\frac{10^5\text{ Pa}}{1\text{ bar}}\right)(-3.33\text{ L})\left(\frac{1\text{ m}^3}{10^3\text{ L}}\right) = 333\text{ J}$$

The value of ΔH for the oxidation of Fe(s) given in Problem 14-12 is −73.8 kilojoules. Thus, the energy change, ΔU_{rxn}, is given by

$$\begin{aligned}\Delta U_{\text{rxn}} &= \Delta H_{\text{rxn}} - P\Delta V \\ &= -73.8\text{ kJ} + 0.333\text{ kJ} = -73.5\text{ kJ}\end{aligned}$$

HESS'S LAW

14-18. The equation for the reaction is

$$\text{Ca(OH)}_2(s) \longrightarrow \text{CaO}(s) + \text{H}_2\text{O}(l)$$

For this reaction, $\Delta H^\circ_{\text{rxn}} = -(-64.50\text{ kJ}\cdot\text{mol}^{-1}) = 64.50\text{ kJ}\cdot\text{mol}^{-1}$. The energy as heat required to convert one gram of $\text{Ca(OH)}_2(s)$ to $\text{CaO}(s)$ is given by

$$q = (1.00\text{ g Ca(OH)}_2)\left(\frac{1\text{ mol Ca(OH)}_2}{74.10\text{ g Ca(OH)}_2}\right)\left(\frac{64.50\text{ kJ}}{1\text{ mol Ca(OH)}_2}\right) = 0.870\text{ kJ}\cdot\text{g}^{-1}$$

14-20. To obtain the third equation, reverse the first equation and add it to the second equation

$$\begin{array}{ll} \text{CuCl}_2(s) \longrightarrow \text{Cu}(s) + \text{Cl}_2(g) & \Delta H^\circ_{\text{rxn}} = +220.1\text{ kJ}\cdot\text{mol}^{-1} \\ 2\,\text{Cu}(s) + \text{Cl}_2(g) \longrightarrow 2\,\text{CuCl}(s) & \Delta H^\circ_{\text{rxn}} = -137.2\text{ kJ}\cdot\text{mol}^{-1} \\ \hline \text{CuCl}_2(s) + \text{Cu}(s) \longrightarrow 2\,\text{CuCl}(s) & \Delta H^\circ_{\text{rxn}} = 82.9\text{ kJ}\cdot\text{mol}^{-1} \end{array}$$

14-22. The equation that we want can be obtained from the two given equations by multiplying the first by 2, reversing the second equation, and then adding the two together.

$$\begin{array}{ll} 2\,\text{H}_2(g) + 2\,\text{F}_2(g) \longrightarrow 4\,\text{HF}(g) & \Delta H^\circ_{\text{rxn}} = (2)(-546.6\text{ kJ}\cdot\text{mol}^{-1}) = -1093\text{ kJ}\cdot\text{mol}^{-1} \\ 2\,\text{H}_2\text{O}(l) \longrightarrow 2\,\text{H}_2(g) + \text{O}_2(g) & \Delta H^\circ_{\text{rxn}} = 571.6\text{ kJ}\cdot\text{mol}^{-1} \\ \hline 2\,\text{F}_2(g) + 2\,\text{H}_2\text{O}(l) \longrightarrow 4\,\text{HF}(g) + \text{O}_2(g) & \Delta H^\circ_{\text{rxn}} = -521\text{ kJ}\cdot\text{mol}^{-1} \end{array}$$

14-24. The equations that correspond to the combustion reactions are

(1) $\underset{m\text{-xylene}}{(\text{CH}_3)_2\text{C}_6\text{H}_4(l)} + \frac{21}{2}\,\text{O}_2(g) \longrightarrow 8\,\text{CO}_2(g) + 5\,\text{H}_2\text{O}(l) \qquad \Delta H^\circ_{\text{rxn}} = -4550.1\text{ kJ}\cdot\text{mol}^{-1}$

(2) $\underset{p\text{-xylene}}{(\text{CH}_3)_2\text{C}_6\text{H}_4(l)} + \frac{21}{2}\,\text{O}_2(g) \longrightarrow 8\,\text{CO}_2(g) + 5\,\text{H}_2\text{O}(l) \qquad \Delta H^\circ_{\text{rxn}} = -4551.1\text{ kJ}\cdot\text{mol}^{-1}$

To obtain the desired equation, reverse equation (2) and add it to equation (1):

$$\underset{m\text{-xylene}}{(\text{CH}_3)_2\text{C}_6\text{H}_4(l)} \longrightarrow \underset{p\text{-xylene}}{(\text{CH}_3)_2\text{C}_6\text{H}_4(l)}$$

$$\Delta H^\circ_{\text{rxn}} = -4550.1\text{ kJ}\cdot\text{mol}^{-1} + 4551.1\text{ kJ}\cdot\text{mol}^{-1} = 1.0\text{ kJ}\cdot\text{mol}^{-1}$$

14-26. To obtain the desired equation, reverse the first equation and add it to the second equation

$$\begin{array}{ll} \text{XeF}_2(s) \longrightarrow \text{Xe}(g) + \text{F}_2(g) & \Delta H^\circ_{\text{rxn}} = 123\text{ kJ}\cdot\text{mol}^{-1} \\ \text{Xe}(g) + 2\,\text{F}_2(g) \longrightarrow \text{XeF}_4(s) & \Delta H^\circ_{\text{rxn}} = -262\text{ kJ}\cdot\text{mol}^{-1} \\ \hline \text{XeF}_2(g) + \text{F}_2(g) \longrightarrow \text{XeF}_4(s) & \Delta H^\circ_{\text{rxn}} = -139\text{ kJ}\cdot\text{mol}^{-1} \end{array}$$

MOLAR ENTHALPIES OF FORMATION

14-28. Using Equation 14.24 and the data given, we have

$$\begin{aligned}\Delta H^\circ_{\rm rxn} &= \Delta H^\circ_{\rm f}[\mathrm{Cu_2O}(s)] - \{\Delta H^\circ_{\rm f}[\mathrm{CuO}(s)] + \Delta H^\circ_{\rm f}[\mathrm{Cu}(s)]\}\\ &= (1)(-168.6\ \mathrm{kJ\cdot mol^{-1}}) - (1)(-157.3\ \mathrm{kJ\cdot mol^{-1}}) - (1)(0\ \mathrm{kJ\cdot mol^{-1}})\\ &= -11.3\ \mathrm{kJ\cdot mol^{-1}}\end{aligned}$$

14-30. (a) Using Equation 14.24 and the data in Table 14.3, we have

$$\begin{aligned}\Delta H^\circ_{\rm rxn} &= \{2\,\Delta H^\circ_{\rm f}[\mathrm{H_2O}(l)] + \Delta H^\circ_{\rm f}[\mathrm{O_2}(g)]\} - \{2\,\Delta H^\circ_{\rm f}[\mathrm{H_2O_2}(l)]\}\\ &= (2)(-285.8\ \mathrm{kJ\cdot mol^{-1}}) + (1)(0\ \mathrm{kJ\cdot mol^{-1}}) - (2)(-187.8\ \mathrm{kJ\cdot mol^{-1}})\\ &= -196.0\ \mathrm{kJ\cdot mol^{-1}} \qquad \text{exothermic}\end{aligned}$$

(b) Using Equation 14.24 and the data in Table 14.3, we have

$$\begin{aligned}\Delta H^\circ_{\rm rxn} &= \Delta H^\circ_{\rm f}[\mathrm{MgCO_3}(s)] - \{\Delta H^\circ_{\rm f}[\mathrm{MgO}(s)] + \Delta H^\circ_{\rm f}[\mathrm{CO_2}(g)]\}\\ &= (1)(-1095.8\ \mathrm{kJ\cdot mol^{-1}}) - (1)(-601.6\ \mathrm{kJ\cdot mol^{-1}}) - (1)(-393.5\ \mathrm{kJ\cdot mol^{-1}})\\ &= -100.7\ \mathrm{kJ\cdot mol^{-1}} \qquad \text{exothermic}\end{aligned}$$

(c) Using Equation 14.24 and the data in Table 14.3, we have

$$\begin{aligned}\Delta H^\circ_{\rm rxn} &= \{4\,\Delta H^\circ_{\rm f}[\mathrm{NO}(g)] + 6\,\Delta H^\circ_{\rm f}[\mathrm{H_2O}(g)]\} - \{4\,\Delta H^\circ_{\rm f}[\mathrm{NH_3}(g)] + 5\,\Delta H^\circ_{\rm f}[\mathrm{O_2}(g)]\}\\ &= (4)(91.3\ \mathrm{kJ\cdot mol^{-1}}) + (6)(-241.8\ \mathrm{kJ\cdot mol^{-1}})\\ &\quad - (4)(-45.9\ \mathrm{kJ\cdot mol^{-1}}) - (5)(0\ \mathrm{kJ\cdot mol^{-1}})\\ &= -902.0\ \mathrm{kJ\cdot mol^{-1}} \qquad \text{exothermic}\end{aligned}$$

14-32. (a) Using Equation 14.24 and the data in Table 14.3, we have

$$\begin{aligned}\Delta H^\circ_{\rm rxn} &= \{\Delta H^\circ_{\rm f}[\mathrm{CO_2}(g)] + 2\,\Delta H^\circ_{\rm f}[\mathrm{H_2O}(l)]\} - \left\{\Delta H^\circ_{\rm f}[\mathrm{CH_3OH}(l)] + \frac{3}{2}\,\Delta H^\circ_{\rm f}[\mathrm{O_2}(g)]\right\}\\ &= (1)(-393.5\ \mathrm{kJ\cdot mol^{-1}}) + (2)(-285.8\ \mathrm{kJ\cdot mol^{-1}})\\ &\quad - (1)(-239.2\ \mathrm{kJ\cdot mol^{-1}}) - \frac{3}{2}(0\ \mathrm{kJ\cdot mol^{-1}})\\ &= -725.9\ \mathrm{kJ\cdot mol^{-1}}\end{aligned}$$

The heat of combustion of $\mathrm{CH_3OH}(l)$ per gram is

$$\left(\frac{-725.9\ \mathrm{kJ}}{1\ \mathrm{mol}}\right)\left(\frac{1\ \mathrm{mol\ CH_3OH}}{32.04\ \mathrm{g\ CH_3OH}}\right) = -22.66\ \mathrm{kJ\cdot g^{-1}}$$

(b) Using Equation 14.24 and the data in Table 14.3, we have

$$\begin{aligned}\Delta H^\circ_{\rm rxn} &= \{\Delta H^\circ_{\rm f}[\mathrm{N_2}(g)] + 2\,\Delta H^\circ_{\rm f}[\mathrm{H_2O}(l)]\} - \{\Delta H^\circ_{\rm f}[\mathrm{N_2H_4}(l)] + \Delta H^\circ_{\rm f}[\mathrm{O_2}(g)]\}\\ &= (2)(0\ \mathrm{kJ\cdot mol^{-1}}) + (2)(-285.8\ \mathrm{kJ\cdot mol^{-1}})\\ &\quad - (1)(50.6\ \mathrm{kJ\cdot mol^{-1}}) - (1)(0\ \mathrm{kJ\cdot mol^{-1}})\\ &= -622.2\ \mathrm{kJ\cdot mol^{-1}}\end{aligned}$$

The heat of combustion of $N_2H_4(l)$ per gram is

$$\left(\frac{-622.2\text{ kJ}}{1\text{ mol}}\right)\left(\frac{1\text{ mol }N_2H_4}{32.05\text{ g }N_2H_4}\right) = -19.41\text{ kJ}\cdot\text{g}^{-1}$$

The combustion of CH_3OH produces 1.2 times as much energy as heat per gram as does the combustion of $N_2H_4(l)$.

14-34. For the reaction given, we have

$$\Delta H^\circ_{rxn} = \{12\,\Delta H^\circ_f[CO_2(g)] + 11\,\Delta H^\circ_f[H_2O(l)]\} - \{\Delta H^\circ_f[C_{12}H_{22}O_{11}(s)] + 12\,\Delta H^\circ_f[O_2(g)]\}$$

In this case, we are given ΔH°_{rxn} and must determine $\Delta H^\circ_f[C_{12}H_{22}O_{11}(s)]$. Using the data in Table 14.3, we have

$$\begin{aligned}-5639.7\text{ kJ}\cdot\text{mol}^{-1} &= (12)(-393.5\text{ kJ}\cdot\text{mol}^{-1}) + (11)(-285.8\text{ kJ}\cdot\text{mol}^{-1})\\ &\quad - (1)(\Delta H^\circ_f[C_{12}H_{22}O_{11}(s)]) - (12)(0\text{ kJ}\cdot\text{mol}^{-1})\end{aligned}$$

Solving for $\Delta H^\circ_f[C_{12}H_{22}O_{11}(s)]$, we have

$$\begin{aligned}\Delta H^\circ_f[C_{12}H_{22}O_{11}(s)] &= -4722.0\text{ kJ}\cdot\text{mol}^{-1} - 3143.8\text{ kJ}\cdot\text{mol}^{-1} + 5639.7\text{ kJ}\cdot\text{mol}^{-1}\\ &= -2226.1\text{ kJ}\cdot\text{mol}^{-1}\end{aligned}$$

14-36. (a) The equation for the reaction is

$$H_2O(l) \longrightarrow H_2O(g)$$

The energy as heat required to vaporize one mole of $H_2O(l)$ at 25°C is equal to ΔH°_{rxn}. Thus,

$$\text{energy as heat} = \Delta H^\circ_{rxn} = \Delta H^\circ_f[H_2O(g)] - \Delta H^\circ_f[H_2O(l)]$$

From Table 14.3, we have

$$\begin{aligned}\Delta H^\circ_{rxn} &= (1)(-241.8\text{ kJ}\cdot\text{mol}^{-1}) - (1)(-285.8\text{ kJ}\cdot\text{mol}^{-1})\\ &= 44.0\text{ kJ}\cdot\text{mol}^{-1}\end{aligned}$$

(b) The equation for the reaction is

$$CCl_4(l) \longrightarrow CCl_4(g)$$

The energy as heat required to vaporize one mole of $CCl_4(l)$ at 25°C is equal to ΔH°_{rxn}. Thus,

$$\text{energy as heat} = \Delta H^\circ_{rxn} = \Delta H^\circ_f[CCl_4(g)] - \Delta H^\circ_f[CCl_4(l)]$$

From Table 14.3, we have

$$\begin{aligned}\Delta H^\circ_{rxn} &= (1)(-95.7\text{ kJ}\cdot\text{mol}^{-1}) - (1)(-128.2\text{ kJ}\cdot\text{mol}^{-1})\\ &= 32.5\text{ kJ}\cdot\text{mol}^{-1}\end{aligned}$$

14-38. Using Equation 14.24 and the data in Table 14.3, we have

$$\begin{aligned}\Delta H^\circ_{rxn} &= \{2\,\Delta H^\circ_f[CaCO_3(s)] + \Delta H^\circ_f[CO_2(g)]\} \\ &\quad - \{2\,\Delta H^\circ_f[Ca(s)] + 4\,\Delta H^\circ_f[O_2(g)] + 3\,\Delta H^\circ_f[C(s)]\} \\ &= (2)(-1207.6\text{ kJ}\cdot\text{mol}^{-1}) + (1)(-393.5\text{ kJ}\cdot\text{mol}^{-1}) \\ &\quad - \{(2)(0\text{ kJ}\cdot\text{mol}^{-1}) + (4)(0\text{ kJ}\cdot\text{mol}^{-1}) + (3)(0\text{ kJ}\cdot\text{mol}^{-1})\} \\ &= -2808.7\text{ kJ}\cdot\text{mol}^{-1}\end{aligned}$$

The chemical equation as given represents the production of two moles of $CaCO_3(s)$, so the energy as heat evolved when 500.0 kg of $CaCO_3(s)$ are produced is given by

$$\Delta H^\circ = (500.0 \times 10^3\text{ g }CaCO_3)\left(\frac{1\text{ mol }CaCO_3}{100.09\text{ g }CaCO_3}\right)\left(\frac{-2808.7\text{ kJ}\cdot\text{mol}^{-1}}{2\text{ mol }CaCO_3}\right) = -7.015 \times 10^6\text{ kJ}$$

MOLAR BOND ENTHALPIES

14-40. The reaction involves breaking two moles of O—F bonds; thus

$$\Delta H^\circ_{rxn} \approx 2\,H_{bond}(\text{O—F})$$

Given the value of ΔH°_{rxn}, we can calculate $H_{bond}(\text{O—F})$

$$\begin{aligned}384\text{ kJ}\cdot\text{mol}^{-1} &= (2)\,H_{bond}(\text{O—F}) \\ H_{bond}(\text{O—F}) &= \frac{384\text{ kJ}\cdot\text{mol}^{-1}}{2} = 192\text{ kJ}\cdot\text{mol}^{-1}\end{aligned}$$

14-42. In the equation

$$CH_3OH(g) + F_2(g) \longrightarrow FCH_2OH(g) + HF(g)$$

we break one mole of C—H bonds and one mole of F—F bonds and we make one mole of C—F bonds and one mole of H—F bonds. The enthalpy required to break the bonds of the reactant molecules is given by

$$\begin{aligned}H_{bond}(\text{reactants}) &= H_{bond}(\text{C—H}) + H_{bond}(\text{F—F}) \\ &= (1)(414\text{ kJ}\cdot\text{mol}^{-1}) + (1)(155\text{ kJ}\cdot\text{mol}^{-1}) \\ &= 569\text{ kJ}\cdot\text{mol}^{-1}\end{aligned}$$

The enthalpy released upon the formation of the bonds in the product molecules is

$$\begin{aligned}H_{bond}(\text{products}) &= H_{bond}(\text{C—F}) + H_{bond}(\text{H—F}) \\ &= (1)(439\text{ kJ}\cdot\text{mol}^{-1}) + (1)(565\text{ kJ}\cdot\text{mol}^{-1}) \\ &= 1004\text{ kJ}\cdot\text{mol}^{-1}\end{aligned}$$

The enthalpy change of the reaction is given by

$$\begin{aligned}\Delta H^\circ_{rxn} &= H_{bond}(\text{reactants}) - H_{bond}(\text{products}) \\ &= 569\text{ kJ}\cdot\text{mol}^{-1} - 1004\text{ kJ}\cdot\text{mol}^{-1} = -435\text{ kJ}\cdot\text{mol}^{-1}\end{aligned}$$

14-44. From the reaction, we can write

$$\begin{aligned}\Delta H^\circ_{rxn} &= H_{bond}(\text{reactants}) - H_{bond}(\text{products}) \\ &= H_{bond}(\text{N}\equiv\text{N}) + 3\,H_{bond}(\text{H—H}) - 6\,H_{bond}(\text{N—H})\end{aligned}$$

The value of $\Delta H^\circ_{\text{rxn}}$ is given by

$$\begin{aligned}\Delta H^\circ_{\text{rxn}} &= 2\, H^\circ_{\text{f}}[\text{NH}_3(g)] \\ &= (2)(-45.9\ \text{kJ}\cdot\text{mol}^{-1}) = -91.8\ \text{kJ mol}^{-1}\end{aligned}$$

Thus, we can write

$$\begin{aligned}-91.8\ \text{kJ}\cdot\text{mol}^{-1} &= H_{\text{bond}}(\text{N}{\equiv}\text{N}) + 3\, H_{\text{bond}}(\text{H}{-}\text{H}) - 6\, H_{\text{bond}}(\text{N}{-}\text{H}) \\ &= (1)\, H_{\text{bond}}(\text{N}{\equiv}\text{N}) + (3)(435\ \text{kJ}\cdot\text{mol}^{-1}) - (6)(390\ \text{kJ}\cdot\text{mol}^{-1})\end{aligned}$$

Thus, $H(\text{N}{\equiv}\text{N}) = +943\ \text{kJ}\cdot\text{mol}^{-1}$, in agreement with the value listed in Table 14.5.

14-46. The relevant equation is

$$\text{CCl}_4(g) \longrightarrow \text{C}(g) + 4\,\text{Cl}(g)$$

We have that $\Delta H^\circ_{\text{rxn}} = 4\, H_{\text{bond}}(\text{C}{-}\text{Cl})$ and

$$\begin{aligned}\Delta H^\circ_{\text{rxn}} &= \Delta H^\circ_{\text{f}}[\text{C}(g)] + 4\,\Delta H^\circ_{\text{f}}[\text{Cl}(g)] - \Delta H^\circ_{\text{f}}[\text{CCl}_4(g)] \\ &= (1)(716.7\ \text{kJ}\cdot\text{mol}^{-1}) + (4)(121.3\ \text{kJ}\cdot\text{mol}^{-1}) - (1)(-95.7\ \text{kJ}\cdot\text{mol}^{-1}) \\ &= 1297.6\ \text{kJ}\cdot\text{mol}^{-1}\end{aligned}$$

Therefore, we have that $4\, H_{\text{bond}}(\text{C–Cl}) = 1297.6\ \text{kJ}\cdot\text{mol}^{-1}$, or

$$H_{\text{bond}}(\text{C–Cl}) = \frac{1297.6\ \text{kJ}\cdot\text{mol}^{-1}}{4} = 324.40\ \text{kJ}\cdot\text{mol}^{-1}$$

The value listed in Table 14.5 represents the molar bond energy of a carbon-chlorine bond averaged from many different molecules. The value that we obtained in this problem is specific for the carbon–chlorine bond in a CCl_4 molecule.

HEAT CAPACITY

14-48. Use Equation 14.27 with $q_{\text{P}} = 285\ \text{J}$ and $\Delta T = 3.74°\text{C} = 3.74\ \text{K}$. The total heat capacity is

$$c_{\text{P}} = \frac{q_{\text{P}}}{\Delta T} = \frac{285\ \text{J}}{3.74\ \text{K}} = 76.2\ \text{J}\cdot\text{K}^{-1}$$

for the 33.6-gram sample of $\text{C}_6\text{H}_{14}(l)$. The molar heat capacity is

$$C_{\text{P}} = \left(\frac{76.2\ \text{J}\cdot\text{K}^{-1}}{33.6\ \text{g C}_6\text{H}_{14}}\right)\left(\frac{86.17\ \text{g C}_6\text{H}_{14}}{1\ \text{mol C}_6\text{H}_{14}}\right) = 195\ \text{J}\cdot\text{K}^{-1}\cdot\text{mol}^{-1}$$

14-50. From Equation 14.30, we have

$$C_{\text{P}} = \frac{q_{\text{P}}}{n\Delta T}$$

Thus, the number of moles of $\text{Na}(l)$ required is given by

$$\begin{aligned}n &= \frac{q_{\text{P}}}{C_{\text{P}}\Delta T} = \frac{1.00 \times 10^6\ \text{J}}{(30.8\ \text{J}\cdot\text{K}^{-1}\cdot\text{mol}^{-1})(10\ \text{K})} \\ &= 3.247 \times 10^3\ \text{mol}\end{aligned}$$

The minimum mass of sodium that is needed is

$$\text{mass of Na} = (3.247 \times 10^3 \text{ mol})\left(\frac{22.99 \text{ g Na}}{1 \text{ mol Na}}\right) = 7.46 \times 10^4 \text{ g}$$

14-52. The sample of copper, being at a higher temperature than the water, will give up heat to the water. Using Equation 14.33, we have that

$$c_{P,\text{Cu}}\Delta T_{\text{Cu}} = -c_{P,\text{H}_2\text{O}}\Delta T_{\text{H}_2\text{O}}$$

For the sample of copper,

$$\Delta T_{\text{Cu}} = t_{\text{f}} - 100.0°\text{C}$$

and for the sample of water

$$\Delta T_{\text{H}_2\text{O}} = t_{\text{f}} - 1.5°\text{C}$$

where t_{f} is the final Celsius temperature. The value of $c_{P,\text{Cu}}$ is

$$c_{P,\text{Cu}} = (50.0 \text{ g})\left(\frac{1 \text{ mol Cu}}{63.55 \text{ g Cu}}\right)(24.4\,\text{J}\cdot\text{K}^{-1}\cdot\text{mol}^{-1}) = 18.61\,\text{J}\cdot\text{K}^{-1}$$

and the value of $c_{P,\text{H}_2\text{O}}$ is

$$c_{P,\text{H}_2\text{O}} = (250 \text{ mL})(1.00 \text{ g mL}^{-1})\left(\frac{1 \text{ mol H}_2\text{O}}{18.02 \text{ g H}_2\text{O}}\right)(75.3\,\text{J}\cdot\text{K}^{-1}\cdot\text{mol}^{-1}) = 1045\,\text{J}\cdot\text{K}^{-1}$$

Therefore,

$$(18.61\,\text{J}\cdot\text{K}^{-1})(t_{\text{f}} - 100.0°\text{C}) = -(1045\,\text{J}\cdot\text{K}^{-1})(t_{\text{f}} - 1.5°\text{C})$$
$$1064\, t_{\text{f}} = 3429°\text{C}$$

Solving for t_{f}

$$t_{\text{f}} = \frac{3429°\text{C}}{1064} = 3.2°\text{C}$$

14-54. Using Equation 14.33, we have

$$c_{P,\text{alloy}}\Delta T_{\text{alloy}} = -c_{P,\text{H}_2\text{O}}\Delta T_{\text{H}_2\text{O}}$$

For the sample of the alloy,

$$\Delta T_{\text{alloy}} = 48.1°\text{C} - 25.0°\text{C} = 23.1 \text{ K}$$

For the water,

$$\Delta T_{\text{H}_2\text{O}} = 48.1°\text{C} - 55.0°\text{C} = -6.9 \text{ K}$$

The value of $c_{P,\text{H}_2\text{O}}$ is

$$c_{P,\text{H}_2\text{O}} = (100.0 \text{ g})\left(\frac{1 \text{ mol H}_2\text{O}}{18.02 \text{ g H}_2\text{O}}\right)(75.3\,\text{J}\cdot\text{K}^{-1}\cdot\text{mol}^{-1}) = 417.9\,\text{J}\cdot\text{K}^{-1}$$

Substituting all this into the heat balance equation gives

$$(c_{\mathrm{P,alloy}})(23.1\ \mathrm{K}) = -(417.9\ \mathrm{J{\cdot}K^{-1}})(-6.9\ \mathrm{K})$$

$$c_{\mathrm{P,alloy}} = \frac{2900\ \mathrm{J}}{23.1\ \mathrm{K}} = 130\ \mathrm{J{\cdot}K^{-1}}$$

The specific heat of the alloy is

$$c_{\mathrm{P,alloy}} = \frac{130\ \mathrm{J{\cdot}K^{-1}}}{50.0\ \mathrm{g}} = 2.6\ \mathrm{J{\cdot}K^{-1}{\cdot}g^{-1}}$$

CALORIMETRY

14-56. The heat evolved by the reaction is given by Equation 14.36

$$\Delta H_{\mathrm{rxn}} = -c_{\mathrm{P,cal}}\Delta T = -(455\ \mathrm{J{\cdot}K^{-1}})(1.43\ \mathrm{K}) = -651\ \mathrm{J} = -0.651\ \mathrm{kJ}$$

This amount of heat is evolved when 0.0500 L of 0.500 M of each solution react. The number of moles that react is given by

$$n = MV = (0.500\ \mathrm{mol{\cdot}L^{-1}})(0.0500\ \mathrm{L}) = 0.0250\ \mathrm{mol}$$

The heat of reaction for one mole of reactants is

$$\Delta H^{\circ}_{\mathrm{rxn}} = \frac{-0.651\ \mathrm{kJ}}{0.0250\ \mathrm{mol}} = -26.0\ \mathrm{kJ{\cdot}mol^{-1}}$$

14-58. The value of ΔH_{rxn} is

$$\Delta H_{\mathrm{rxn}} = -(4.37\ \mathrm{kJ{\cdot}K^{-1}})(2.70\ \mathrm{K}) = -11.8\ \mathrm{kJ}$$

The 10.0 gram sample of CaO(s) is equivalent to

$$\text{moles of CaO} = (10.0\ \mathrm{g\ CaO})\left(\frac{1\ \mathrm{mol\ CaO}}{56.08\ \mathrm{g\ CaO}}\right) = 0.178\ \mathrm{mol}$$

and the 1.00 liter of $H_2O(l)$ is equivalent to

$$\text{moles of H}_2\text{O} = (1000\ \mathrm{mL\ H_2O})\left(\frac{1\ \mathrm{g\ H_2O}}{1\ \mathrm{mL\ H_2O}}\right)\left(\frac{1\ \mathrm{mol\ H_2O}}{18.02\ \mathrm{g\ H_2O}}\right) = 55.5\ \mathrm{mol}$$

Thus, the $H_2O(l)$ is in excess and the $CaO(s)$ is the limiting reactant. From the equation for the reaction, we see that 0.178 moles of $CaO(s)$ yields 0.178 moles of $Ca(OH)_2(s)$, and so the heat evolved in the formation of one mole of $Ca(OH)_2(s)$ is

$$\Delta H^{\circ}_{\mathrm{rxn}} = \frac{-11.8\ \mathrm{kJ}}{0.178\ \mathrm{mol}} = -66.3\ \mathrm{kJ{\cdot}mol^{-1}}$$

14-60. The energy as heat evolved is

$$\Delta H_{\mathrm{rxn}} = -(5.16\ \mathrm{kJ{\cdot}K^{-1}})(0.511\ \mathrm{K}) = -2.64\ \mathrm{kJ}$$

The number of moles of nitric acid that dissolved is

$$n = (5.00\ \mathrm{g})\left(\frac{1\ \mathrm{mol\ HNO_3}}{63.02\ \mathrm{g\ HNO_3}}\right) = 0.0793\ \mathrm{mol}$$

The molar hear of solution of nitric acid in water is

$$\Delta H^\circ_{\text{soln}} = \frac{-2.64\ \text{kJ}}{0.0793\ \text{mol}} = -33.3\ \text{kJ}\cdot\text{mol}^{-1}$$

BOMB CALORIMETERS

14-62. We have that the enthalpy of combustion for 5.00 grams of $C_6H_{12}O_6(s)$ is

$$\Delta H_{\text{rxn}} \approx \Delta U_{\text{rxn}} = -(29.7\ \text{kJ}\cdot\text{K}^{-1})(2.635\ \text{K}) = -78.26\ \text{kJ}$$

The standard enthalpy of combustion per gram of $C_6H_{12}O_6(s)$ is given by

$$\Delta H_{\text{comb}} = \frac{-78.26\ \text{kJ}}{5.00\ \text{g}} = -15.65\ \text{kJ}\cdot\text{g}^{-1}$$

The standard enthalpy of combustion per mole of $C_6H_{12}O_6(s)$ is given by

$$\Delta H^\circ_{\text{comb}} = (-15.65\ \text{kJ}\cdot\text{g}^{-1})\left(\frac{180.16\ \text{g}\ C_6H_{12}O_6}{1\ \text{mol}\ C_6H_{12}O_6}\right) = -2820\ \text{kJ}\cdot\text{mol}^{-1}$$

14-64. The energy released as heat is given by

$$\Delta H_{\text{released}} = (15.0\ \text{g}\ C_6H_{12}O_6)(-15.7\ \text{kJ}\cdot\text{g}^{-1}) = -236\ \text{kJ}$$

In terms of nutritional Calories, we have

$$\text{energy in Calories} = (-236\ \text{kJ})\left(\frac{1\ \text{Calorie}}{4.184\ \text{kJ}}\right) = -56.4\ \text{Calories}$$

14-66. We have that the enthalpy of combustion for 2.620 grams of $C_3H_6O_3$ is

$$\Delta H_{\text{rxn}} \approx \Delta U_{\text{rxn}} = -(21.70\ \text{kJ}\cdot\text{K}^{-1})(1.800\ \text{K}) = -39.06\ \text{kJ}$$

The standard molar enthalpy of combustion of $C_3H_6O_3(s)$ is given by

$$\Delta H^\circ_{\text{comb}} = \left(\frac{-39.06\ \text{kJ}}{2.620\ \text{g}\ C_3H_6O_3}\right)\left(\frac{90.08\ \text{g}\ C_3H_6O_3}{1\ \text{mol}\ C_3H_6O_3}\right) = -1343\ \text{kJ}\cdot\text{mol}^{-1}$$

The equation for the combustion of oxalic acid is

$$C_3H_6O_3(s) + 3\,O_2(g) \longrightarrow 3\,CO_2(g) + 3\,H_2O(l)$$

The value of $\Delta H^\circ_{\text{rxn}}$ for the combustion of lactic acid is given by

$$\Delta H^\circ_{\text{rxn}} = \{3\,\Delta H^\circ_f[CO_2(g)] + 3\,\Delta H^\circ_f[H_2O(l)]\} - \{\Delta H^\circ_f[C_3H_6O_3(s)] + 3\,\Delta H^\circ_f[O_2(g)]\}$$

Using the above value for $\Delta H^\circ_{\text{rxn}}$ and the data in Table 14.3, we have

$$\begin{aligned}-1343\ \text{kJ}\cdot\text{mol}^{-1} &= (3)(-393.5\ \text{kJ}\cdot\text{mol}^{-1}) + (3)(-285.8\ \text{kJ}\cdot\text{mol}^{-1})\\ &\quad - (1)\,\Delta H^\circ_f[C_3H_6O_3(s)] - 0\end{aligned}$$

Solving for $\Delta H^\circ_f[C_3H_6O_3(s)]$, we get

$$\Delta H^\circ_f[C_3H_6O_3(s)] = -695\ \text{kJ}\cdot\text{mol}^{-1}$$

ADDITIONAL PROBLEMS

14-68. In an endothermic reaction, $\Delta H_{rxn} < 0$. We have from Hess's law that

$$\Delta H_{rxn} = \Delta H_f(\text{products}) - \Delta H_f(\text{reactants})$$

Therefore, $\Delta H_f(\text{products}) < \Delta H_f(\text{reactants})$ for an endothermic reaction.

14-70. A thermometer measures temperature not heat. Consider, for example, the combustion of one gram of magnesium ribbon and that of one kilogram of magnesium ribbon. Both burn at about 1000°C, but the one-kilogram ribbon produces more heat than the one-gram ribbon. In order to collect and measure all the energy in the form of heat produced by a reaction, we must use a calorimeter.

14-72. Because the dissolution of ammonium chloride in water is endothermic, energy in the form of heat is required to dissolve it in water. Therefore, we expect that ammonium chloride is more soluble in warm water than cold water because more energy in the form of heat is available.

14-74. The total Calorie count is

$$\text{total Calories} = \left(\frac{10.5\ \text{ounce}}{1.125\ \text{ounce}\cdot\text{serving}^{-1}}\right)(130\ \text{Cal}\cdot\text{serving}^{-1}) = 1210\ \text{Cal}$$

The total energy content in kilojoules is

$$\text{total kilojoules} = (1210\ \text{Cal})\left(\frac{4.184\ \text{kJ}}{1\ \text{Cal}}\right) = 5060\ \text{kJ}$$

14-76. The energy as heat required to raise the temperature of 82 kilograms of water by 2°C is

$$\text{heat} = (82 \times 10^3\ \text{g H}_2\text{O})\left(\frac{1\ \text{mol H}_2\text{O}}{18.02\ \text{g H}_2\text{O}}\right)(75.3\ \text{J}\cdot\text{mol}^{-1}\cdot\text{K}^{-1})(2.2\ \text{K}) = 754\ \text{kJ}$$

This heat is provided by the burning of glucose in the body

$$(2802.5\ \text{kJ}\cdot\text{mol}^{-1})(\text{moles of glucose}) = 754\ \text{kJ}$$

$$\text{moles of glucose} = \frac{754\ \text{kJ}}{2802.5\ \text{kJ}\cdot\text{mol}^{-1}} = 0.269\ \text{mol}$$

The mass of glucose that must be burned is

$$\text{mass of glucose} = (0.269\ \text{mol C}_6\text{H}_{12}\text{O}_6)\left(\frac{180.16\ \text{g C}_6\text{H}_{12}\text{O}_6}{1\ \text{mol C}_6\text{H}_{12}\text{O}_6}\right) = 48\ \text{g}$$

14-78. We have the stoichiometric correspondence

$$2.0\ \text{g CO}_2 \approx 0.045\ \text{mol CO}_2$$

The change in the volume of the balloon is

$$\Delta V = \frac{\Delta n_{gas} RT}{P} = \frac{(0.045\ \text{mol})(0.083145\ \text{L}\cdot\text{bar}\cdot\text{K}^{-1}\cdot\text{mol}^{-1})(298\ \text{K})}{1.00\ \text{bar}} = 1.1\ \text{L}$$

The work done is

$$w = -P\Delta V = -(1.00\text{ bar})(1.1\text{ L}) = -1.1\text{ L·bar}$$
$$= -(1.1\text{ L·bar})\left(\frac{8.3145\text{ J·K}^{-1}\text{·mol}^{-1}}{0.083145\text{ L·bar·K}^{-1}\text{·mol}^{-1}}\right) = -110\text{ J}$$

14-80. The ionic charge of thallium can be +1 or +3. Thus the formula is either TlCl or $TlCl_3$. We set up the following table for the proposed formulas of the compounds of thallium and chlorine:

Proposed formula	Value of N in Dulong and Petit's rule	Predicted value of C_P/J·K^{-1}·mol^{-1}	Observed value of C_P from specific heat and proposed formula
TlCl	2	50	$(0.208\text{ J·K}^{-1}\text{·g}^{-1})\left(\frac{239.8\text{ g}}{1\text{ mol}}\right) = 49.9\text{ J·K}^{-1}\text{·mol}^{-1}$
$TlCl_3$	4	100	$(0.208\text{ J·K}^{-1}\text{·g}^{-1})\left(\frac{310.7\text{ g}}{1\text{ mol}}\right) = 64.6\text{ J·K}^{-1}\text{·mol}^{-1}$

Because of the agreement between the predicted and observed values of C_P, we conclude that the formula of the compound is TlCl(s).

14-82. In each case, we must write the equation for the combustion of the hydrocarbon, determine ΔH°_{rxn} from Table 14.3, and divide the result by the mass of the hydrocarbon burned.

(a) $2\,C_2H_2(g) + 5\,O_2(g) \longrightarrow 4\,CO_2 + 2\,H_2O(l)$

$$\Delta H^\circ_{rxn} = \{4\,\Delta H^\circ_f[CO_2(g)] + 2\,\Delta H^\circ_f[H_2O(l)]\} - \{2\,\Delta H^\circ_f[C_2H_2(g)] + 5\,\Delta H^\circ_f[O_2(g)]\}$$
$$= (4)(-393.5\text{ kJ·mol}^{-1}) + (2)(-285.8\text{ kJ·mol}^{-1})$$
$$- (2)(227.4\text{ kJ·mol}^{-1}) - (5)(0\text{ kJ·mol}^{-1})$$
$$= -2600.4\text{ kJ·mol}^{-1}$$

Two moles, or 52.07 grams, of $C_2H_2(g)$ are burned, and so the enthalpy of combustion per gram is

$$\text{enthalpy of combustion per gram} = \frac{-2600.4\text{ kJ}}{52.07\text{ g}} = -49.94\text{ kJ·g}^{-1}$$

(b) $C_2H_4(g) + 3\,O_2(g) \longrightarrow 2\,CO_2 + 2\,H_2O(l)$

$$\Delta H^\circ_{rxn} = \{2\,\Delta H^\circ_f[CO_2(g)] + 2\,\Delta H^\circ_f[H_2O(l)]\} - \{\Delta H^\circ_f[C_2H_4(g)] + 3\,\Delta H^\circ_f[O_2(g)]\}$$
$$= (2)(-393.5\text{ kJ·mol}^{-1}) + (2)(-285.8\text{ kJ·mol}^{-1})$$
$$- (1)(52.4\text{ kJ·mol}^{-1}) - (3)(0\text{ kJ·mol}^{-1})$$
$$= -1411.0\text{ kJ·mol}^{-1}$$

One mole, or 28.05 grams, of $C_2H_4(g)$ is burned, and so the enthalpy of combustion per gram is

$$\text{enthalpy of combustion per gram} = \frac{-1411\text{ kJ}}{28.05\text{ g}} = -50.30\text{ kJ·g}^{-1}$$

(c) $2\,C_2H_6(g) + 7\,O_2(g) \longrightarrow 4\,CO_2(g) + 6\,H_2O(l)$

$$\begin{aligned}\Delta H^\circ_{rxn} &= \{4\,\Delta H^\circ_f[CO_2(g)] + 6\,\Delta H^\circ_f[H_2O(l)]\} - \{2\,\Delta H^\circ_f[C_2H_6(g)] + 7\,\Delta H^\circ_f[O_2(g)]\}\\ &= (4)(-393.5\text{ kJ}\cdot\text{mol}^{-1}) + (6)(-285.8\text{ kJ}\cdot\text{mol}^{-1})\\ &\quad - (2)(-84.0\text{ kJ}\cdot\text{mol}^{-1}) - (7)(0\text{ kJ}\cdot\text{mol}^{-1})\\ &= -3120.8\text{ kJ}\cdot\text{mol}^{-1}\end{aligned}$$

Two moles, or 60.14 grams, of $C_2H_6(g)$ are burned, and so the enthalpy of combustion per gram is

$$\text{enthalpy of combustion per gram} = \frac{-3120.8\text{ kJ}}{60.14\text{ g}} = -51.89\text{ kJ}\cdot\text{g}^{-1}$$

The enthalpy of combustion per gram increases as the number of hydrogen atoms increases.

14-84. The equation for the net ionic reaction is

$$3\,Cu(s) + 8\,H^+(aq) + 2\,NO_3^-(aq) \longrightarrow 3\,Cu^{2+}(aq) + 2\,NO(g) + 4\,H_2O(l)$$

Thus, we have for ΔH°_{rxn}

$$\begin{aligned}\Delta H^\circ_{rxn} &= \{3\,\Delta H^\circ_f[Cu^{2+}(aq)] + 2\,\Delta H^\circ_f[NO(g)] + 4\,\Delta H^\circ_f[H_2O(l)]\}\\ &\quad - \{3\,\Delta H^\circ_f[Cu(s)] + 8\,\Delta H^\circ_f[H^+(aq)] + 2\,\Delta H^\circ_f[NO_3^-(aq)]\}\\ &= (3)(64.39\text{ kJ}\cdot\text{mol}^{-1}) + (2)(91.3\text{ kJ}\cdot\text{mol}^{-1}) + (4)(-285.8\text{ kJ}\cdot\text{mol}^{-1})\\ &\quad - (3)(0\text{ kJ}\cdot\text{mol}^{-1}) - (8)(0\text{ kJ}\cdot\text{mol}^{-1}) - (2)(-207.35\text{ kJ}\cdot\text{mol}^{-1})\\ &= -352.7\text{ kJ}\cdot\text{mol}^{-1}\end{aligned}$$

14-86. The equation for the reaction is

$$HCl(aq) + NaOH(aq) \longrightarrow NaCl(aq) + H_2O(l)$$

For 100 milliliters of 0.100 M $HCl(aq)$ or $NaOH(aq)$, we have

$$n = (0.100\text{ mol}\cdot\text{L}^{-1})(0.100\text{ L}) = 0.0100\text{ mol}$$

Therefore,

$$\Delta H = (0.0100\text{ mol})(-55.7\text{ kJ}\cdot\text{mol}^{-1}) = -0.557\text{ kJ}$$

The temperature change of the resulting solution (0.200 L) is given by Equation 14.36 as

$$\begin{aligned}\Delta H &= -c_{P,\text{cal}}\Delta T\\ -557\text{ J} &= -(200\text{ mL})\left(\frac{1.00\text{ g}}{1.00\text{ mL}}\right)\left(\frac{1\text{ mol}}{18.02\text{ g}}\right)(75.3\text{ J}\cdot\text{mol}^{-1})\Delta T\end{aligned}$$

or $\Delta T = 0.666°C$.

14-88. The standard enthalpy is given by

$$\Delta H_{rxn} \approx \Delta U_{rxn} = -c_P\Delta T = -(38.70\text{ kJ}\cdot\text{K}^{-1})(2.89\text{ K}) = -112\text{ kJ}$$

The number of Calories per gram is

$$\text{Calories per gram} = \left(\frac{112\text{ kJ}}{5.00\text{ g}}\right)\left(\frac{1\text{ Cal}}{4.184\text{ kJ}}\right) = 5.35\text{ Cal}\cdot\text{g}^{-1}$$

14-90. The number of moles of octane used per day is given by

$$\begin{aligned}\text{moles of octane per day} &= \left(\frac{32\text{ mi}}{25\text{ mi}\cdot\text{gal}^{-1}}\right)\left(\frac{3.7854\text{ L}}{1\text{ gal}}\right)\left(\frac{10^3\text{ mL}}{1\text{ L}}\right)\left(\frac{0.7\text{ g}}{1\text{ mL}}\right)\left(\frac{1\text{ mol C}_8\text{H}_{18}}{114.22\text{ g C}_8\text{H}_{18}}\right)\\ &= 29.7\text{ mol}\cdot\text{day}^{-1}\end{aligned}$$

The chemical equation for the combustion of $C_8H_{18}(l)$ at 25°C is

$$C_8H_{18}(l) + \frac{25}{2}O_2(g) \longrightarrow 8\,CO_2(g) + 9\,H_2O(l)$$

The value of ΔH°_{rxn} is

$$\begin{aligned}\Delta H^\circ_{rxn} &= \{8\,\Delta H^\circ_f[CO_2(g)] + 9\,\Delta H^\circ_f[H_2O(l)]\} - \left\{\Delta H^\circ_f[C_8H_{18}(l)] + \frac{25}{2}\Delta H^\circ_f[O_2(g)]\right\}\\ &= (8)(-393.5\text{ kJ}\cdot\text{mol}^{-1}) + (9)(-285.8\text{ kJ}\cdot\text{mol}^{-1}) - (1)(-250.1\text{ kJ}\cdot\text{mol}^{-1}) - 0\\ &= -5470.1\text{ kJ}\cdot\text{mol}^{-1}\end{aligned}$$

Thus we have

$$\text{energy required per day} = (29.7\text{ mol}\cdot\text{day}^{-1})(5470.1\text{ kJ}\cdot\text{mol}^{-1}) = 1.62\times10^5\text{ kJ}\cdot\text{day}^{-1}$$

$$\begin{aligned}\text{energy required per year} &= (1.62\times10^5\text{ kJ}\cdot\text{day}^{-1})(5\text{ day}\cdot\text{week}^{-1})(52\text{ week}\cdot\text{yr}^{-1})\\ &= 4.2\times10^7\text{ kJ}\cdot\text{yr}^{-1} = 42\,000\text{ MJ}\cdot\text{yr}^{-1}\\ &= \frac{42\,000\text{ MJ}\cdot\text{yr}^{-1}}{6000\text{ MJ}\cdot\text{barrel}^{-1}} = 7\text{ barrel}\cdot\text{yr}^{-1}\end{aligned}$$

14-92. The energy generated by the combustion of one molecule of glucose is given by (see Problem 14-76)

$$\Delta H_{rxn} = \left(\frac{-2802.5\text{ kJ}}{1\text{ mol}}\right)\left(\frac{1\text{ mol}}{6.02214\times10^{23}\text{ molecules}}\right) = -4.6537\times10^{-21}\text{ kJ}$$

The energy required to generate one molecule of ATP is given by

$$\Delta H_{rxn} = \left(\frac{30.5\text{ kJ}}{1\text{ mol}}\right)\left(\frac{1\text{ mol}}{6.02214\times10^{23}\text{ molecules}}\right) = 5.065\times10^{-23}\text{ kJ}\cdot\text{molecule}^{-1}$$

Thus, if all the energy from the combustion of one molecule of glucose is used to generate ATP, then the number of molecules of ATP generated is

$$\text{molecules of ATP} = \frac{4.6537\times10^{-21}\text{ kJ}}{5.065\times10^{-23}\text{ kJ}\cdot\text{molecule}^{-1}} = 92\text{ molecules}$$

14-94. The chemical equation for the reaction is

$$2\,NO(g) + 2\,CO(g) \longrightarrow N_2(g) + 2\,CO_2(g)$$

Thus,

$$\begin{aligned}\Delta H^\circ_{\text{rxn}} &= \{\Delta H^\circ_f[N_2(g)] + 2\,\Delta H^\circ_f[CO_2(g)]\} - \{2\,\Delta H^\circ_f[NO(g)] + 2\,\Delta H^\circ_f[CO(g)]\}\\ &= \{0\ \text{kJ}\cdot\text{mol}^{-1} + (2)(-393.5\ \text{kJ}\cdot\text{mol}^{-1})\}\\ &\quad - \{(2)(91.3\ \text{kJ}\cdot\text{mol}^{-1}) + (2)(-110.5\ \text{kJ}\cdot\text{mol}^{-1})\}\\ &= -748.6\ \text{kJ}\cdot\text{mol}^{-1}\end{aligned}$$

14-96. We have

$$\boxed{\begin{array}{c}V = 1\ \text{L}\\ P = 10\ \text{bar}\end{array}} \xrightarrow{1\ \text{bar}} \boxed{\begin{array}{c}V = 10\ \text{L}\\ P = 1\ \text{bar}\end{array}}$$

$$w = -P\Delta V = -9\ \text{L}\cdot\text{bar}$$

$$\boxed{\begin{array}{c}V = 1\ \text{L}\\ P = 10\ \text{bar}\end{array}} \xrightarrow{5\ \text{bar}} \boxed{\begin{array}{c}V = 2\ \text{L}\\ P = 5\ \text{bar}\end{array}} \xrightarrow{1\ \text{bar}} \boxed{\begin{array}{c}V = 10\ \text{L}\\ P = 1\ \text{bar}\end{array}}$$

$$\begin{aligned}w &= -P_1\Delta V_1 - P_2\Delta V_2\\ &= -(5\ \text{bar})(1\ \text{L}) - (1\ \text{bar})(8\ \text{L}) = -13.6\ \text{L}\cdot\text{bar}\end{aligned}$$

Note that the work is not the same in these two expansions although the initial states and final states are the same.

CHAPTER 15. Liquids and Solids

HEATS OF VAPORIZATION, FUSION, AND SUBLIMATION

15-2. The number of moles in 60.0 grams of benzene is

$$n = (60.0\ \text{g})\left(\frac{1\ \text{mol}}{78.11\ \text{g}}\right) = 0.768\ \text{mol}$$

The molar heat of vaporization is the energy as heat required to vaporize exactly one mole. If 23.6 kJ are required to vaporize 0.768 mol, then

$$\Delta H_{\text{vap}} = \frac{23.6\ \text{kJ}}{0.768\ \text{mol}} = 30.7\ \text{kJ}\cdot\text{mol}^{-1}$$

15-4. The number of moles of water in 16 one-ounce ice cubes is

$$n = (16\ \text{ice cubes})\left(\frac{1\ \text{oz}}{1\ \text{ice cube}}\right)\left(\frac{28\ \text{g}}{1\ \text{oz}}\right)\left(\frac{1\ \text{mol}}{18.02\ \text{g}}\right) = 24.86\ \text{mol}$$

The energy as heat released when the water is cooled from 18°C to 0°C is

$$q_{\text{P}} = nC_{\text{P}}\Delta T = (24.86\ \text{mol})(75.3\ \text{J}\cdot\text{mol}^{-1}\cdot\text{K}^{-1})(-18\ \text{K}) = -33700\ \text{J} = -33.7\ \text{kJ}$$

where we have used the fact that C_{P} for water is $75.3\ \text{J}\cdot\text{mol}^{-1}\cdot\text{K}^{-1}$ (Appendix D). The energy as heat released when the water is converted into ice is given by (see Table 15.3)

$$C_{\text{P}} = -n\Delta H_{\text{fus}} = -(24.86\ \text{mol})(6.01\ \text{kJ}\cdot\text{mol}^{-1}) = -149\ \text{kJ}$$

The total energy as heat released is

$$q_{\text{P}} = (-33.7\ \text{kJ}) + (-149\ \text{kJ}) = -183\ \text{kJ}$$

The mass of Freon-12 that is vaporized by absorbing 183 kJ is found from

$$183\ \text{kJ} = \Delta H_{\text{vap}}(\text{mass of Freon-12})$$

$$\text{mass of Freon-12} = \frac{(183\ \text{kJ})\left(\dfrac{1000\ \text{J}}{1\ \text{kJ}}\right)}{155\ \text{J}\cdot\text{g}^{-1}} = 1180\ \text{g} = 1.18\ \text{kg}$$

The mass of tetrafluoroethane required is given by

$$\text{mass of tetrafluoroethane} = \frac{(183\ \text{kJ})\left(\dfrac{1000\ \text{J}}{1\ \text{kJ}}\right)}{215.9\ \text{J}\cdot\text{g}^{-1}} = 848\ \text{g} = 0.848\ \text{kg}$$

15-6. The amount of energy as heat, q_P, required to vaporize 100 μg of einsteinium is

$$q_P = n\Delta H_{vap} = (100\ \mu g)\left(\frac{1\ g}{10^6\ \mu g}\right)\left(\frac{1\ mol}{252\ g}\right)(128\ kJ{\cdot}mol^{-1})$$
$$= 5.08 \times 10^{-5}\ kJ = 5.08 \times 10^{-2}\ J$$

15-8. The amount of energy as heat absorbed by the sublimation of 1.00 mole of $CO_2(s)$ is given by

$$q_P = (1.00\ mol)(25.2\ kJ{\cdot}mol^{-1}) = 25.2\ kJ$$

The amount of energy as heat absorbed by the sublimation of $CO_2(s)$ is the amount of energy evolved by the freezing of water. The energy evolved by the freezing of n moles of water is

$$q_P = -n(6.01\ kJ{\cdot}mol^{-1}) = -25.2\ kJ$$

Solving for n gives

$$n = \frac{25.2\ kJ}{6.01\ kJ{\cdot}mol^{-1}} = 4.19\ mol$$

HEATING CURVES

15-10. The number of moles in 10.0 grams of water is

$$n = (10.0\ g)\left(\frac{1\ mol}{18.02\ g}\right) = 0.555\ mol$$

The amount of energy required to raise the temperature of 10.0 grams of water from 50.0°C to 100.0°C is given by

$$q_P = nC_P\Delta T = (0.555\ mol)(75.3\ J{\cdot}mol^{-1}{\cdot}K^{-1})(50.0\ K) = 2090\ J$$

The amount of energy required to vaporize 10.0 grams of water is

$$q_P = n\Delta H_{vap} = (0.555\ mol)(40.65\ kJ{\cdot}mol^{-1}) = 22.6\ kJ = 22\ 600\ J$$

It would take longer to vaporize the water.

15-12. The quantity of energy absorbed by the propane is

$$q_P = (500.0\ J{\cdot}min^{-1})(38.9\ min) = 19450\ J = 19.45\ kJ$$

The number of moles of propane is

$$n = (45.0\ g)\left(\frac{1\ mol}{44.10\ g}\right) = 1.020\ mol$$

Thus, the molar enthalpy of vaporization of propane is

$$\Delta H_{vap} = \frac{19.45\ kJ}{1.020\ mol} = 19.1\ kJ{\cdot}mol^{-1}$$

VAN DER WAALS FORCES

15-14. H_2O is a polar molecule and exhibits mainly hydrogen bonding. He is nonpolar and so exhibits primarily London forces. Cl_2 is a nonpolar molecule and so exhibits primarily London forces. HCl is a polar molecule and so exhibits a dipolar force.

15-16. Only molecules in which a hydrogen atom is directly bonded to a highly electronegative atom, such as an oxygen, nitrogen, or fluorine atom, can form hydrogen bonds. The electronegativities of the hydrogen and iodine atoms are similar, and so HI does not form hydrogen bonds. The electronegativities of the hydrogen and sulfur atoms are similar, and so H_2S does not form hydrogen bonds. Although a CH_3OCH_3 molecule contains hydrogen atoms and a highly electronegative oxygen atom, the hydrogen atoms are not directly bonded to the oxygen atom, and so it does not form hydrogen bonds. The only polar molecule is CH_3CH_2OH and it exhibits hydrogen bonding. We predict that CH_3CH_2OH has an unusually high boiling point.

15-18. MgO is an ionic compound that contains a positive divalent ion and a negative divalent ion and so the two ions have the strongest attractions. KCl is also an ionic compound but contains a positively charged single ion and a negatively charged single ion. NH_3 is a polar molecule and exhibits hydrogen bonding. PH_3 is a polar molecule but does not form hydrogen bonds. Thus, we predict that the boiling points are ordered as

$$T_b[PH_3] < T_b[NH_3] < T_b[KCl] < T_b[MgO]$$

15-20. All four molecules are nonpolar molecules. The molar enthalpy of vaporization of nonpolar molecules depends upon the molecular mass. The value of ΔH_{vap} increases with increasing molecular mass. Therefore, the order of the molar enthalpies of vaporization is

$$\Delta H_{vap}[CH_4] < \Delta H_{vap}[CCl_4] < \Delta H_{vap}[SiCl_4] < \Delta H_{vap}[SiBr_4]$$

VAPOR PRESSURE

15-22. Using Table 15.7, we estimate that the equilibrium vapor pressure of water at 93°C is 0.76 atm. The normal atmospheric pressure in Mexico City is about 0.76 atm or 580 Torr.

15-24. The plot of atmospheric pressure versus altitude is

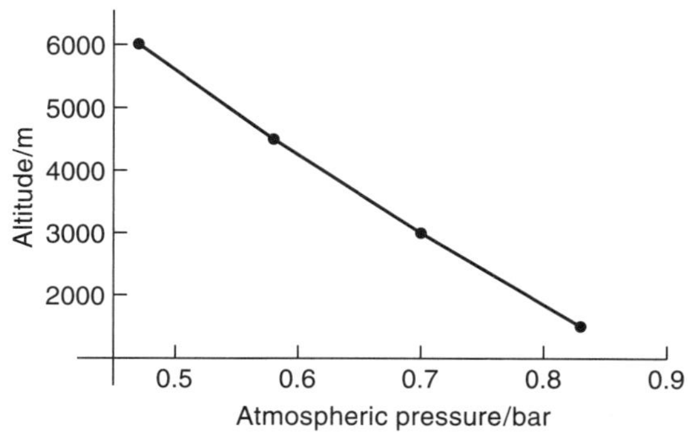

Using the above plot and Table 15.7 or Figure 15.23, we find the boiling point of water at the locations.

Location	Atmospheric pressure/bar	Boiling point/°C
Denver	0.8	95
Mt. Kilimanjaro	0.5	80
Mt. Washington	0.8	95
The Matterhorn	0.6	85

15-26. Relative humidity is given by

$$\text{relative humidity} = \frac{P_{H_2O}}{P^{\circ}_{H_2O}} \times 100$$

From Table 15.7, the equilibrium vapor pressure of water at 40 °C is 55.4 Torr. Thus, we have

$$92 = \frac{P_{H_2O}}{55.4\ \text{Torr}} \times 100$$

Solving for P_{H_2O}, we obtain

$$P_{H_2O} = \frac{(92)(55.4\ \text{Torr})}{100} = 51\ \text{Torr}$$

PROPERTIES OF LIQUIDS

15-28. The volume of the original drop is

$$V = \frac{4}{3}\pi r^3 = \frac{4}{3}\pi(3.0\ \text{mm})^3 = 113\ \text{mm}^3$$

The surface area of the original drop is

$$A = 4\pi r^2 = 4\pi(3\ \text{mm})^2 = 113\ \text{mm}^2$$

The energy of the surface is

$$\text{energy} = (72\ \text{mJ}\cdot\text{m}^{-2})(113\ \text{mm}^2)\left(\frac{1\ \text{m}}{10^3\ \text{mm}}\right)^2 = 0.00814\ \text{mJ} = 8.14\ \mu\text{J}$$

The volume of each new drop is

$$V = \frac{4}{3}\pi(3.0 \times 10^{-3}\ \text{mm})^3 = 113 \times 10^{-9}\ \text{mm}^3$$

Therefore, there are 10^9 new drops. Each new drop has a surface area of

$$A = 4\pi r^2 = 4\pi(3.0 \times 10^{-3}\ \text{mm})^2 = 1.13 \times 10^{-4}\ \text{mm}^2$$

The total new surface area is 10^9 times the area of one drop, or $1.13 \times 10^5\ \text{mm}^2$. The energy required to create this area is

$$\text{energy} = (72\ \text{mJ}\cdot\text{m}^{-2})(1.13 \times 10^5\ \text{mm}^2)\left(\frac{1\ \text{m}}{1000\ \text{mm}}\right)^2 = 8.1\ \text{mJ}$$

or must expend *one million* times the original surface energy to divide the drop. Compare with the result in Problem 15-27. (The amount of energy of 8.13 microjoules needed to break open the original drop is negligible here.)

15-30. Both water and ethanol can form hydrogen bonds. A water molecule can form hydrogen bonds with another water molecule or an ethanol molecule. Thus, ethanol will dissolve in water.

PHASE DIAGRAMS

15-32. (a) gas (b) solid (c) gas (d) solid

15-34. The phase diagram (not to scale) of nitrogen is shown below. For ease of plotting, the pressure is given in units of atmospheres.

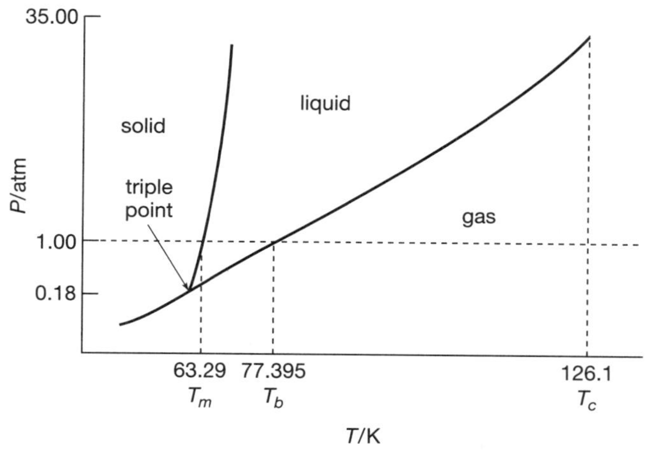

CRYSTAL STRUCTURES

15-36. There is a fluoride ion at each of the eight corners and one at each of the six faces of the unit cell. Each of those at the corners is shared by eight unit cells (for a total of one ion per unit cell) and each of those at the faces is shared by two unit cells (for a total of three ions per unit cell). Thus, there is a total of four fluoride ions per unit cell. There is one potassium ion at the center of the unit cell and one at each of the 12 edges. The one at the center belongs entirely to the unit cell and each of the 12 at the edges is shared by four unit cells (for a total of three ions per unit cell), giving a total of $1 + 3 = 4$ potassium ions per unit cell. Thus, there are four formula units of KF per unit cell.

15-38. We use Equation 15.8 to calculate the volume of the unit cell. There are two tantalum atoms per unit cell and so we have

$$\begin{aligned} V_{\text{unit cell}} &= \frac{nM}{dN_{\text{A}}} \\ &= \frac{(2\ \text{atom}\cdot\text{unit cell}^{-1})(180.95\ \text{g}\cdot\text{mol}^{-1})}{(16.653\ \text{g}\cdot\text{cm}^{-3})(6.022\times10^{23}\ \text{atom}\cdot\text{mol}^{-1})} \\ &= 3.6087\times10^{-23}\ \text{cm}^3\cdot\text{unit cell}^{-1} \end{aligned}$$

The length of an edge of a unit cell is

$$l = (3.6087\times10^{-23}\ \text{cm}^3\cdot\text{unit cell}^{-1})^{1/3} = 3.3051\times10^{-8}\ \text{cm} = 330.51\ \text{pm}$$

15-40. Solving Equation 15.8 for N_{A}, we have

$$N_{\text{A}} = \frac{nM}{d\,V_{\text{unit cell}}}$$

The volume of the unit cell is given by

$$\begin{aligned} V_{\text{unit cell}} &= l^3 = (288.4\times10^{-10}\ \text{m})^3 = (288.4\times10^{-10}\ \text{cm})^3 \\ &= 2.399\times10^{-23}\ \text{cm}^3 \end{aligned}$$

Therefore,

$$N_A = \frac{(2\ \text{atom}\cdot\text{unit cell}^{-1})(52.00\ \text{g}\cdot\text{mol}^{-1})}{(7.20\ \text{g}\cdot\text{cm}^{-3})(2.399\times10^{-23}\ \text{cm}^3)}$$

$$= 6.02\times10^{23}\ \text{atom}\cdot\text{mol}^{-1}$$

15-42. From Figure 15.30, we see that there is one CsBr formula unit per unit cell. Using Equation 15.8 gives

$$V_{\text{unit cell}} = \frac{nM}{d\,N_A}$$

$$= \frac{(1\ \text{formula unit}\cdot\text{unit cell}^{-1})(212.90\ \text{g}\cdot\text{mol}^{-1})}{(4.43\ \text{g}\cdot\text{cm}^{-3})(6.02214\times10^{23}\ \text{formula unit}\cdot\text{mol}^{-1})}$$

$$= 7.981\times10^{-23}\ \text{cm}^3\cdot\text{unit cell}^{-1}$$

The length of an edge of a unit cell is

$$l = (7.981\times10^{-23}\ \text{cm}^3)^{1/3} = 4.31\times10^{-8}\ \text{cm}$$

From Figure 15.29, we see that the nearest-neighborhood distance is one half of the main diagonal of the unit cell. The length of the main diagonal is $\sqrt{3}\,l$, and so the nearest-neighbor distance is $\sqrt{3}\,l/2$, or 3.73×10^{-8} cm, or 373 pm.

15-44. We solve Equation 15.8 for n:

$$n = \frac{V_{\text{unit cell}}\,d\,N_A}{M}$$

$$= \frac{(481.08\times10^{-10}\ \text{cm})^3(3.34\ \text{g}\cdot\text{cm}^{-3})(6.022\times10^{23}\ \text{formula unit}\cdot\text{mol}^{-1})}{56.08\ \text{g}\cdot\text{mol}^{-1}}$$

$$= 3.99\ \text{formula units} \approx 4\ \text{formula units}$$

The result tells us that there are four formula units of CaO in a unit cell, and so the unit cell must be the NaCl(s) type.

ADDITIONAL PROBLEMS

15-46. The temperature of the glass is close to 0°C, where the equilibrium vapor pressure of water is very low. The vapor pressure of water in the air near the glass is higher than the equilibrium vapor of water at 0°C and so water vapor will condense, forming moisture on the glass.

15-48. At equilibrium, molecules of the liquid are entering the gas phase at the same rate as the molecules of the gas are entering the liquid phase. Another example of a dynamic equilibrium is the constant exchange of sugar molecules between the aqueous phase and the solid phase in a cup of tea that has been saturated with sugar.

15-50. (a) A substance boils when its vapor pressure is greater than or equal to the atmospheric pressure. Evaporation occurs when the vapor pressure of a substance is less than the atmospheric pressure. The temperature of a liquid does not change during boiling, but can change during evaporation.

(b) A solid can evaporate, or turn directly into a gas. For example, ice cubes in a freezer "shrink" over time due to evaporation.

15-52. A substance boils when its vapor pressure is greater than or equal to the atmospheric pressure. Because the water used in a pressure cooker is under a pressure greater than one atmosphere, it boils at a temperature greater than 100 °C. Thus, the food in the closed pressure cooker is at a higher temperature than in boiling water in an open pan on the stove, and so cooks faster.

15-54. Your body uses energy to melt the snow. It requires

$$q_P = n\Delta H_{fus} = (1000\text{ g})\left(\frac{1\text{ mol}}{18.02\text{ g}}\right)(6.01\text{ kJ}\cdot\text{mol}^{-1}) = 334\text{ kJ}$$

$$= (334\text{ kJ})\left(\frac{1\text{ Cal}}{4.184\text{ kJ}}\right) = 79.7\text{ Cal}$$

to melt one liter of water from snow if the temperature of the snow is 0 °C. Your body must also expend energy to heat the water at 0 °C up to body temperature (37 °C). The energy required to heat one liter of water from 0 °C to 37 °C is given by

$$q_P = (1000\text{ g})\left(\frac{1\text{ mol}}{18.02\text{ g}}\right)(75.3\text{ J}\cdot\text{mol}^{-1}\cdot\text{K}^{-1})(37\text{ K})$$

$$= 154\ 000\text{ J} = 154\text{ kJ} = 37.2\text{ Cal}$$

Thus, it takes a total of 79.7 CaL + 37.2 CaL = 117 Cal to melt and to bring 1000 grams of snow to body temperature.

15-56. Using the values of the boiling point of the noble gases given in Table 15.3, we have

helium: $\Delta H_{vap} = (85\text{ J}\cdot\text{mol}^{-1}\cdot\text{K}^{-1})(4.22\text{ K}) = 360\text{ J}\cdot\text{mol}^{-1} = 0.36\text{ kJ}\cdot\text{mol}^{-1}$

neon: $\Delta H_{vap} = (85\text{ J}\cdot\text{mol}^{-1}\cdot\text{K}^{-1})(27.07\text{ K}) = 2300\text{ J}\cdot\text{mol}^{-1} = 2.3\text{ kJ}\cdot\text{mol}^{-1}$

argon: $\Delta H_{vap} = (85\text{ J}\cdot\text{mol}^{-1}\cdot\text{K}^{-1})(87.30\text{ K}) = 7400\text{ J}\cdot\text{mol}^{-1} = 7.4\text{ kJ}\cdot\text{mol}^{-1}$

krypton: $\Delta H_{vap} = (85\text{ J}\cdot\text{mol}^{-1}\cdot\text{K}^{-1})(119.93\text{ K}) = 10\ 000\text{ J}\cdot\text{mol}^{-1} = 10\text{ kJ}\cdot\text{mol}^{-1}$

xenon: $\Delta H_{vap} = (85\text{ J}\cdot\text{mol}^{-1}\cdot\text{K}^{-1})(165.04\text{ K}) = 14\ 000\text{ J}\cdot\text{mol}^{-1} = 14\text{ kJ}\cdot\text{mol}^{-1}$

15-58. Using Trouton's rule $\Delta H_{vap} = (85\text{ J}\cdot\text{mol}^{-1}\cdot\text{K}^{-1})\,T_b$, we have the following values of ΔH_{vap}:

Chloromethane:

$$\Delta H_{vap} = (85\text{ J}\cdot\text{mol}^{-1}\cdot\text{K}^{-1})(249.06\text{ K}) = 21\ 000\text{ J}\cdot\text{mol}^{-1} = 21\text{ kJ}\cdot\text{mol}^{-1}$$

Water:

$$\Delta H_{vap} = (85\text{ J}\cdot\text{mol}^{-1}\cdot\text{K}^{-1})(373.15\text{ K}) = 32\ 000\text{ J}\cdot\text{mol}^{-1} = 32\text{ kJ}\cdot\text{mol}^{-1}$$

Hydrogen sulfide:

$$\Delta H_{vap} = (85\text{ J}\cdot\text{mol}^{-1}\cdot\text{K}^{-1})(213.60\text{ K}) = 18\ 000\text{ J}\cdot\text{mol}^{-1} = 18\text{ kJ}\cdot\text{mol}^{-1}$$

The percentage error in each case is

Chloromethane:

$$\text{percentage error} = \frac{\text{experimental value} - \text{estimated value}}{\text{experimental value}} \times 100$$

$$= \frac{21.40\text{ kJ}\cdot\text{mol}^{-1} - 21\text{ kJ}\cdot\text{mol}^{-1}}{21.4\text{ kJ}\cdot\text{mol}^{-1}} \times 100 = 2\%$$

Water:

$$\text{percentage error} = \frac{40.65\text{ kJ}\cdot\text{mol}^{-1} - 32\text{ kJ}\cdot\text{mol}^{-1}}{40.65\text{ kJ}\cdot\text{mol}^{-1}} \times 100 = 20\%$$

Hydrogen sulfide:

$$\text{percentage error} = \frac{18.67\ \text{kJ}\cdot\text{mol}^{-1} - 18\ \text{kJ}\cdot\text{mol}^{-1}}{18.67\ \text{kJ}\cdot\text{mol}^{-1}} \times 100 = 4\%$$

Water is strongly hydrogen bonded, and so has strong specific intermolecular interactions. Trouton's rule does not apply to water. Trouton's rule applies to both chloromethane and hydrogen sulfide.

15-60. The extensive hydrogen bonding in H_2O holds the water molecules into a crystal lattice, but there is little hydrogen bonding in H_2S.

15-62. Both silicon carbide, $SiC(s)$, and boron nitride, $BN(s)$, form a diamondlike covalent crystal network.

15-64. The amount of energy as heat absorbed by an area 1.0 m^2 in one day is 8.1 MJ = 8100 kJ. The enthalpy of sublimation of ice is 50.9 $kJ\cdot mol^{-1}$. The amount of ice that sublimes is given by

$$q_P = n(50.9\ \text{kJ}\cdot\text{mol}^{-1}) = 8100\ \text{kJ}$$

or

$$n = \frac{8100\ \text{kJ}}{50.9\ \text{kJ}\cdot\text{mol}^{-1}} = 159\ \text{mol}$$

The mass of ice in 159 moles is

$$m = (159\ \text{mol})\left(\frac{18.02\ \text{g}}{1\ \text{mol}}\right) = 2870\ \text{g}$$

The volume occupied by 2870 grams of ice is given by the relation $d = m/V$. Solving for V, we have

$$V = \frac{m}{d} = \frac{2870\ \text{g}}{0.917\ \text{g}\cdot\text{cm}^{-1}} = 3130\ \text{cm}^3$$

The depth of ice removed in an area of 1 m^2 is

$$\text{depth} = \left(\frac{3130\ \text{cm}^3}{1\ \text{m}^2}\right)\left(\frac{1\ \text{m}^2}{10^4\ \text{cm}^2}\right) = 0.313\ \text{cm}$$

15-66. Water boils when $P_{\text{vap},H_2O} = P_{\text{atmosphere}}$. He could determine the atmospheric pressure at Lhasa from the boiling point of water. Plots of atmospheric pressure versus altitude (see Problem 15-24) were then used to find the altitude corresponding to the atmospheric pressure at Lhasa.

15-68. The equilibrium vapor pressure of water at 30°C is 31.8 Torr (Table 15.7). We can find the partial pressure of water from the definition of relative humidity

$$\text{relative humidity} = \frac{P_{H_2O}}{P^0_{H_2O}} \times 100$$

or

$$P_{H_2O} = \frac{(65)(31.8\ \text{Torr})}{100} = 21\ \text{Torr}$$

The temperature at which the equilibrium vapor pressure is 21 Torr is the dew point. From Table 15.7, we see that the dew point is about 23°C.

15-70. The unit cell of cesium chloride is body-centered cubic. The volume of the unit cell of CsCl(*s*) is

$$V_{\text{unit cell}} = (412.1\ \text{pm})^3 = (4.121 \times 10^{-8}\ \text{cm})^3 = 6.999 \times 10^{-23}\ \text{cm}^3$$

A unit cell consists of one chloride ion and one cesium ion (Figure 15.30). We solve Equation 15.8 for *d*

$$d = \frac{nM}{V_{\text{unit cell}} N_A}$$

$$= \frac{(1\ \text{formula unit}\cdot\text{unit cell}^{-1})(168.4\ \text{g}\cdot\text{mol}^{-1})}{(6.999 \times 10^{-23}\ \text{cm}^3)(6.022 \times 10^{23}\ \text{formula unit}\cdot\text{mol}^{-1})}$$

$$= 3.994\ \text{g}\cdot\text{cm}^{-3}$$

15-72. The equilibrium vapor pressure of water at 25°C is 23.8 Torr (Table 15.7). The concentration in moles per liter is given by the ideal-gas equation:

$$\frac{n}{V} = \frac{P}{RT} = \frac{23.8\ \text{Torr}}{(62.3637\ \text{L}\cdot\text{Torr}\cdot\text{mol}^{-1}\cdot\text{K}^{-1})(298\ \text{K})}$$

$$= 1.28 \times 10^{-3}\ \text{mol}\cdot\text{L}^{-1}$$

15-74. There are three triple points. When heated from 40°C to 200°C at 1 atm, sulfur goes from rhombic to monoclinic to liquid and then to the gaseous state. Sublimation will occur below 10^{-4} atm.

15-76. The σ bond framework is

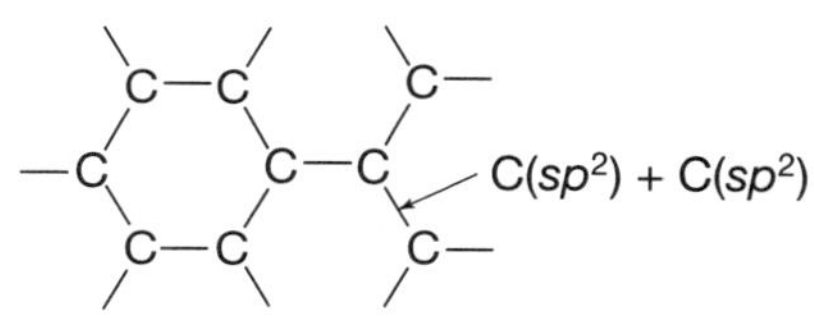

Because the bond geometry about each carbon atom within a layer is trigonal planar, it is appropriate to use sp^2 hybrid orbitals and to describe the carbon-carbon bonds as $\text{C}(sp^2)\text{–C}(sp^2)$. Each carbon atom accounts for one delocalized electron, which occupies delocalized π orbitals that spread over the entire network. Overlap of these π orbitals accounts for the weak bonding between the layers.

15-78. The vapor pressure of the solid is equal to the vapor pressure of the liquid at the triple point. Therefore, we write

$$24.513 - \frac{5892.5\ \text{K}}{T_t} = 17.357 - \frac{3479\ \text{K}}{T_t}$$

$$7.156 = \frac{2414\ \text{K}}{T_t}$$

Solving for T_t, we have

$$T_t = 337.3\ \text{K} = 64.1\,^\circ\text{C}$$

The logarithm of the pressure at the triple point is given by

$$\ln P_t = 17.357 - \frac{3479\ \text{K}}{337.3\ \text{K}} = 7.04$$

$$P_t = e^{7.04} = 1100\ \text{Torr}$$

15-80. Because the $H_2(g)$ is collected over water, the pressure within the collecting vessel is due to both $H_2(g)$ and water vapor, $P_{\text{total}} = P_{H_2} + P_{H_2O}$. The vapor pressure of water at 25 °C is 23.8 Torr, and so

$$P_{H_2} = P_{\text{total}} - P_{H_2O} = 756.0 \text{ Torr} - 23.8 \text{ Torr} = 732.2 \text{ Torr}$$

The number of moles of $H_2(g)$ collected is

$$n = \frac{PV}{RT} = \frac{(732.2 \text{ Torr})(0.263 \text{ L})}{(62.3637 \text{ L}\cdot\text{Torr}\cdot\text{mol}^{-1}\cdot\text{K}^{-1})(298 \text{ K})} = 0.01036 \text{ mol}$$

According to the chemical equation, this also corresponds to the number of moles of $Zn(s)$, and so we can write

$$0.01036 \text{ mol Zn} \approx 0.677 \text{ g Zn}$$

We divide through by 0.01036 to find that the atomic mass of zinc is $65.3 \text{ g}\cdot\text{mol}^{-1}$.

15-82. Substitute $P_2 = 760$ Torr, $T_2 = 373.2$ K, and $\Delta H_{\text{vap}} = 43\,990 \text{ J}\cdot\text{mol}^{-1}$ into the Clapeyron–Clausius equation given in the problem to write

$$\ln \frac{760 \text{ Torr}}{P} = \left(\frac{43\,990 \text{ J}\cdot\text{mol}^{-1}}{8.3145 \text{ J}\cdot\text{mol}^{-1}\cdot\text{K}^{-1}}\right)\left(\frac{1}{T} - \frac{1}{373.2 \text{ K}}\right)$$

Now use this equation to make the following table:

$T/°\text{C}$	T/K	$\ln \dfrac{760 \text{ Torr}}{P}$	P/Torr
95	368	0.2003	622
50	323	2.203	83.9
25	298	3.577	21.2
5	278	4.855	5.92

There is a small difference between these calculated values and the experimental values in Table 15.7 because ΔH_{vap} varies slightly with temperature.

CHAPTER 16. Colligative Properties of Solutions

MOLE FRACTION

16-2. The mass of methanal is 40 grams, the mass of methanol is 10 grams, and the mass of water is 50 grams in 100 grams of solution. The mole fractions are given by

$$x_i = \frac{n_i}{n_1 + n_2 + n_3} \qquad \text{where} \qquad i = 1, 2, \text{ or } 3$$

The value of $n_1 + n_2 + n_3$ is

$$n_1 + n_2 + n_3 = (40\text{ g})\left(\frac{1\text{ mol H}_2\text{CO}}{30.03\text{ g H}_2\text{CO}}\right) + (10\text{ g})\left(\frac{1\text{ mol CH}_3\text{OH}}{32.04\text{ g CH}_3\text{OH}}\right) + (50\text{ g})\left(\frac{1\text{ mol H}_2\text{O}}{18.02\text{ g H}_2\text{O}}\right)$$

$$= 1.332\text{ mol} + 0.312\text{ mol} + 2.775\text{ mol} = 4.419\text{ mol}$$

Therefore,

$$x_{\text{H}_2\text{CO}} = \frac{(40\text{ g})\left(\dfrac{1\text{ mol H}_2\text{CO}}{30.03\text{ g H}_2\text{CO}}\right)}{4.419\text{ mol}} = 0.30$$

$$x_{\text{CH}_3\text{OH}} = \frac{(10\text{ g})\left(\dfrac{1\text{ mol CH}_3\text{OH}}{32.04\text{ g CH}_3\text{OH}}\right)}{4.419\text{ mol}} = 0.071$$

$$x_{\text{H}_2\text{O}} = \frac{(50.0\text{ g})\left(\dfrac{1\text{ mol H}_2\text{O}}{18.02\text{ g H}_2\text{O}}\right)}{4.419\text{ mol}} = 0.63$$

Note that $x_{\text{H}_2\text{CO}} + x_{\text{CH}_3\text{OH}} + x_{\text{H}_2\text{O}} = 1$.

16-4. In 500.0 grams of solution, we have x grams of sucrose and $(500 - x)$ grams of water. The number of moles of sucrose and water in the solution are given by

$$n_{\text{sucrose}} = \frac{x\text{ g}}{342.30\text{ g}\cdot\text{mol}^{-1}} \qquad \text{and} \qquad n_{\text{H}_2\text{O}} = \frac{(500.0 - x)\text{ g}}{18.02\text{ g}\cdot\text{mol}^{-1}}$$

The mole fraction of sucrose is given by

$$x_{\text{sucrose}} = \frac{n_{\text{sucrose}}}{n_{\text{sucrose}} + n_{\text{H}_2\text{O}}}$$

$$0.125 = \frac{\dfrac{x\text{ g}}{342.30\text{ g}\cdot\text{mol}^{-1}}}{\dfrac{x\text{ g}}{342.30\text{ g}\cdot\text{mol}^{-1}} + \dfrac{(500.0 - x)\text{ g}}{18.02\text{ g}\cdot\text{mol}^{-1}}}$$

Solving for x yields $x = 365$ g. The solution is prepared by dissolving 365 grams of sucrose in $(500 - 365)$ g $= 135$ grams of water.

16-6. For convenience, we take a 100.0 gram sample of the alloy; thus, there are 95.0 grams of lead and 5.0 grams of calcium in the sample. The number of moles of lead and calcium in the sample are

$$n_{\text{Pb}} = \frac{95.0\text{ g}}{207.2\text{ g}\cdot\text{mol}^{-1}} = 0.458\text{ mol} \qquad \text{and} \qquad n_{\text{Ca}} = \frac{5.0\text{ g}}{40.08\text{ g}\cdot\text{mol}^{-1}} = 0.125\text{ mol}$$

The mole fraction of calcium is

$$x_{\text{Ca}} = \frac{n_{\text{Ca}}}{n_{\text{Ca}} + n_{\text{Pb}}} = \frac{0.125\text{ mol}}{0.125\text{ mol} + 0.458\text{ mol}} = 0.21$$

MOLALITY

16-8. A 1.75 m $Ba(NO_3)_2(aq)$ solution contains 1.75 moles of $Ba(NO_3)_2$ in 1000 grams of water. The mass of $Ba(NO_3)_2$ in the solution is

$$\text{mass} = [1.75\text{ mol Ba(NO}_3)_2]\left(\frac{261.35\text{ g Ba(NO}_3)_2}{1\text{ mol Ba(NO}_3)_2}\right) = 457\text{ g}$$

Thus, we dissolve 457 grams $Ba(NO_3)_2(s)$ in 1000 grams of water.

16-10. The number of moles in 18.0 grams of oxalic acid is

$$n = (18.0\text{ g})\left(\frac{1\text{ mol H}_2\text{C}_2\text{O}_4}{90.04\text{ g H}_2\text{C}_2\text{O}_4}\right) = 0.200\text{ mol}$$

We can find the mass of water from the definition of molality

$$\text{molality} = \frac{\text{moles of solute}}{\text{kilograms of solvent}}$$

$$0.050\text{ m} = \frac{0.200\text{ mol H}_2\text{C}_2\text{O}_4}{\text{kilograms of water}}$$

Solving for kilograms of water, we have

$$\text{kilograms of water} = \frac{0.200\text{ mol}}{0.050\text{ mol}\cdot\text{kg}^{-1}} = 4.0\text{ kg}$$

16-12. (a) There is one solute particle per CH_3OH formula unit because a CH_3OH molecule does not dissociate in water. Thus the colligative molality is 1.0 m_c.

(b) There are four ions per $Al(NO_3)_3$ formula unit because an $Al(NO_3)_3(aq)$ molecule dissociates into $Al^{3+}(aq)$ and $3\,NO_3^-(aq)$ ions in water. Thus, the colligative molality is 4.0 m_c.

(c) There are three ions per $Fe(NO_3)_2$ formula unit because a $Fe(NO_3)_2(aq)$ molecule dissociates into $Fe^{2+}(aq)$ and $2\,NO_3^-(aq)$ ions in water. Thus, the colligative molality is 3.0 m_c.

(d) There are three ions per $(NH_4)_2Cr_2O_7$ formula unit because an $(NH_4)_2Cr_2O_7(aq)$ molecule dissociates into 2 $NH_4^+(aq)$ and $Cr_2O_7^{2-}(aq)$ ions in water. Thus, the colligative molality is 3.0 m_c.

VAN'T HOFF FACTOR

16-14. (a) Na_2CO_3: a strong electrolyte; $i = 3$

(b) $(NH_4)_2SO_4$: a strong electrolyte; $i = 3$

(c) $Pb(NO_3)_2$: a strong electrolyte; $i = 3$

16-16. According to Section 12-3, only hydrochloric acid is a strong acid, and so it will completely dissociate in aqueous solution. Consequently, $HCl(aq)$ will have the largest van't Hoff i-factor.

RAOULT'S LAW AND VAPOR PRESSURE LOWERING

16-18. The mass of water is

$$\text{mass} = dV = (0.993\text{ g}\cdot\text{mL}^{-1})(100.0\text{ mL}) = 99.3\text{ g}$$

The mole fraction of water in the solution is

$$\begin{aligned} x_{H_2O} &= \frac{n_{H_2O}}{n_{H_2O} + n_{C_3H_8O_3}} \\ &= \frac{(99.3\text{ g})\left(\dfrac{1\text{ mol H}_2\text{O}}{18.02\text{ g H}_2\text{O}}\right)}{(99.3\text{ g})\left(\dfrac{1\text{ mol H}_2\text{O}}{18.02\text{ g H}_2\text{O}}\right) + (50.0\text{ g})\left(\dfrac{1\text{ mol C}_3\text{H}_8\text{O}_3}{92.09\text{ g C}_3\text{H}_8\text{O}_3}\right)} \\ &= 0.910 \end{aligned}$$

Raoult's law is $P_{H_2O} = x_{H_2O}P^\circ_{H_2O}$ and thus at 37°, we have

$$P_{H_2O} = (0.910)(6.27\text{ kPa}) = 5.71\text{ kPa}$$

The vapor pressure lowering is

$$P^\circ_{H_2O} - P_{H_2O} = 6.27\text{ kPa} - 5.71\text{ kPa} = 0.56\text{ kPa}$$

16-20. The mole fraction of water in the solution is

$$\begin{aligned} x_{H_2O} &= \frac{n_{H_2O}}{n_{H_2O} + n_{C_2H_6O_2}} \\ &= \frac{(100.0\text{ g})\left(\dfrac{1\text{ mol H}_2\text{O}}{18.02\text{ g H}_2\text{O}}\right)}{(100.0\text{ g})\left(\dfrac{1\text{ mol H}_2\text{O}}{18.02\text{ g H}_2\text{O}}\right) + (50.0\text{ g})\left(\dfrac{1\text{ mol C}_2\text{H}_6\text{O}_2}{62.07\text{ g C}_2\text{H}_6\text{O}_2}\right)} \\ &= 0.873 \end{aligned}$$

Raoult's law is $P_{H_2O} = x_{H_2O} P^\circ_{H_2O}$ and thus at 37°, we have

$$P_{H_2O} = (0.873)(1.00 \text{ atm}) = 0.873 \text{ atm}$$

The vapor pressure lowering is

$$P^\circ_{H_2O} - P_{H_2O} = 1.00 \text{ atm} - 0.873 \text{ atm} = 0.13 \text{ atm}$$

16-22. In a 100 mL sample, we have 40 mL of ethanol and 60 mL of water. The mass of ethanol and water are

$$\text{mass}_{\text{ethanol}} = dV = (0.79 \text{ g·mL}^{-1})(40 \text{ mL}) = 32 \text{ g}$$
$$\text{mass}_{\text{water}} = dV = (1.00 \text{ g·mL}^{-1})(60 \text{ mL}) = 60 \text{ g}$$

The mole fraction of ethanol is given by

$$\begin{aligned} x_{\text{ethanol}} &= \frac{n_{\text{ethanol}}}{n_{\text{ethanol}} + n_{\text{water}}} \\ &= \frac{(32 \text{ g})\left(\frac{1 \text{ mol ethanol}}{46.07 \text{ g ethanol}}\right)}{(32 \text{ g})\left(\frac{1 \text{ mol ethanol}}{46.07 \text{ g ethanol}}\right) + (60 \text{ g})\left(\frac{1 \text{ mol H}_2\text{O}}{18.02 \text{ g H}_2\text{O}}\right)} \\ &= 0.17 \end{aligned}$$

The vapor pressure of ethanol is calculated by using Raoult's law,

$$P_{\text{ethanol}} = (0.17)(40.0 \text{ Torr}) = 6.8 \text{ Torr}$$

16-24. The mole fraction of water in the solution is given by

$$x_{H_2O} = \frac{n_{H_2O}}{n_{H_2O} + n_{\text{solute}}} = \frac{(1000 \text{ g})\left(\frac{1 \text{ mol}}{18.02 \text{ g}}\right)}{(1000 \text{ g})\left(\frac{1 \text{ mol}}{18.02 \text{ g}}\right) + 0.30 \text{ mol}} = 0.9946$$

The vapor pressure of water is calculated by using Raoult's law,

$$P_{H_2O} = (0.9946)(62.8 \text{ mbar}) = 62.5 \text{ mbar}$$

16-26. We use Equation 16.10 $\Delta P_1 = x_2 P_1^\circ$ with $P_1^\circ = 31.7$ mbar in each case. We must convert the molality to mole fraction in each case. The number of moles of water in 1000 g is

$$\text{moles of water} = (1000 \text{ g})\left(\frac{1 \text{ mol}}{18.02 \text{ g}}\right) = 55.49 \text{ mol}$$

(a)

$$m_c = \frac{1.50 \text{ mol CH}_3\text{CH}_2\text{OH}}{1000 \text{ g H}_2\text{O}}$$

The mole fraction is

$$x_2 = \frac{1.50 \text{ mol CH}_3\text{CH}_2\text{OH}}{1.50 \text{ mol CH}_3\text{CH}_2\text{OH} + 55.49 \text{ mol H}_2\text{O}} = 0.0263$$

The vapor pressuring lowering is

$$\Delta P_1 = (0.0263)(31.7 \text{ mbar}) = 0.834 \text{ mbar}$$

(b)

$$m_c = \frac{\left(\frac{4\text{ mol ions}}{1\text{ mol TlCl}_3}\right)(0.50\text{ mol TlCl}_3)}{1000\text{ g H}_2\text{O}} = \frac{2.00\text{ mol ions}}{1000\text{ H}_2\text{O}}$$

$$x_2 = \frac{2.00\text{ mol ions}}{2.00\text{ mol ions} + 55.49\text{ mol H}_2\text{O}} = 0.0348$$

$$\Delta P_1 = (0.0348)(31.7\text{ mbar}) = 1.10\text{ mbar}$$

(c)

$$m_c = \frac{\left(\frac{3\text{ mol ions}}{1\text{ mol K}_2\text{SO}_4}\right)(0.25\text{ mol K}_2\text{SO}_4)}{1000\text{ g H}_2\text{O}} = \frac{0.75\text{ mol ions}}{1000\text{ H}_2\text{O}}$$

$$x_2 = \frac{0.75\text{ mol ions}}{0.75\text{ mol ions} + 55.49\text{ mol H}_2\text{O}} = 0.01334$$

$$\Delta P_1 = (0.0133)(31.7\text{ mbar}) = 0.423\text{ mbar}$$

16-28. Because a $CaCl_2$ molecule dissociates into one $Ca^{2+}(aq)$ ion and two $Cl^-(aq)$ ions in aqueous solution, we determine the mole fraction of solute particles (ions). The vapor pressure of water at 20 °C is 17.54 Torr (see Table 15.7).

$$x_2 = \frac{\Delta P_1}{P_1^\circ} = \frac{5.00\text{ Torr}}{17.54\text{ Torr}} = 0.285$$

We shall find the concentration of the solute particles in the units molality. Thus, we take 1000 grams of water. The number of moles in 1000 grams of water is given by

$$\text{moles of water} = (1000\text{ g H}_2\text{O})\left(\frac{1\text{ mol H}_2\text{O}}{18.02\text{ g H}_2\text{O}}\right) = 55.49\text{ mol}$$

We can calculate the number of moles of solute particles (ions) using the definition of mole fraction.

$$x_2 = \frac{\text{moles ions}}{\text{moles ions} + \text{moles H}_2\text{O}}$$

Let y = moles ions. We have that

$$x_2 = \frac{y}{y + 55.49} = 0.285$$

Solving for y yields $y = 22.1$ moles ions. The number of moles of $CaCl_2$ is one third the number of moles of ions or 7.37 moles. The concentration of $CaCl_2(aq)$ is

$$m = \frac{7.37\text{ mol}}{1000\text{ g H}_2\text{O}} = 7.37\text{ m}$$

16-30. We use Equation 16.10 $\Delta P_{H_2O} = x_{solute} P^\circ_{H_2O}$. At 20 °C, $P^\circ_{H_2O} = 17.5$ Torr (Table 15.7). The mole fraction is given by

$$x_{solute} = \frac{\text{moles of solute}}{\text{moles of solute} + \text{moles H}_2\text{O}}$$

If we multiply the denominator and numerator of the right side of the equation by 1000 g H_2O, we get

$$x_{\text{solute}} = \frac{\dfrac{\text{moles of solute}}{1000 \text{ g H}_2\text{O}}}{\dfrac{\text{moles of solute}}{1000 \text{ g H}_2\text{O}} + \dfrac{\text{moles H}_2\text{O}}{1000 \text{ g H}_2\text{O}}}$$

The molality is given by

$$m = \frac{\text{moles of solute}}{1000 \text{ g H}_2\text{O}}$$

and the number of moles in 1000 g of H_2O is 55.5 mol. So we have

$$x_{\text{solute}} = \frac{m_c}{m_c + 55.5 \text{ mol·kg}^{-1}}$$

(a) Potassium iodide, KI, yields two ions in solution per formula unit, and so $m_c = 2m$.

$$2.00 \text{ Torr} = \frac{2m}{2m + 55.5 \text{ mol·kg}^{-1}}(17.5 \text{ Torr})$$

$$\left(\frac{2.00}{17.5}\right)(2m) + \left(\frac{2.00}{17.5}\right)(55.5) = 2m$$

$$0.229m + 6.34 = 2m$$

$$m = 3.58 \text{ m}$$

(b) Strontium chloride, $SrCl_2$, yields three ions in solution per formula unit, and so $m_c = 3m$.

$$2.00 \text{ Torr} = \frac{3m}{3m + 55.5 \text{ mol·kg}^{-1}}(17.5 \text{ Torr})$$

$$\left(\frac{2.00}{17.5}\right)(3m) + \left(\frac{2.00}{17.5}\right)(55.5) = 3m$$

$$0.343m + 6.34 = 3m$$

$$m = 2.38 \text{ m}$$

(c) Ammonium sulfate yields three ions in solution per formula like strontium chloride; its concentration is also 2.38 m.

BOILING-POINT ELEVATION

16-32. The colligative molality of a solute that raises the boiling point of water by 1.0°C is

$$T_b - T_b^\circ = 1.0 \text{ K} = K_b m_c = (0.513 \text{ K·m}_c^{-1})(m_c)$$

$$m_c = \frac{1.0 \text{ K}}{0.513 \text{ K·m}_c^{-1}} = 1.95 \text{ m}_c$$

The molality of the NaCl(aq) solution is 1.95 $m_c/2 = 0.975$ m. Thus, there are 0.975 moles of NaCl dissolved in 1000 grams of water. The mass of NaCl(s) in 1000 grams of water is

$$\text{mass of NaCl} = (0.975 \text{ mol})\left(\frac{58.44 \text{ g NaCl}}{1 \text{ mol NaCl}}\right) = 57 \text{ g}$$

16-34. The boiling point elevation is

$$T_b - T_b^\circ = K_b m_c = (0.513\ \text{K}\cdot\text{m}_c^{-1})(1.10\ \text{m}_c) = 0.564\ \text{K} = 0.564\,^\circ\text{C}$$

The boiling point is

$$T_b = 100.0\,^\circ\text{C} + 0.564\,^\circ\text{C} = 100.6\,^\circ\text{C}$$

The mole fraction of water in a 1.10 m solution is

$$x_{\text{H}_2\text{O}} = \frac{n_{\text{H}_2\text{O}}}{n_{\text{H}_2\text{O}} + n_{\text{solute}}} = \frac{(1000\ \text{g})\left(\dfrac{1\ \text{mol H}_2\text{O}}{18.02\ \text{g H}_2\text{O}}\right)}{(1000\ \text{g})\left(\dfrac{1\ \text{mol H}_2\text{O}}{18.02\ \text{g H}_2\text{O}}\right) + 1.10\ \text{mol}} = 0.981$$

The vapor pressure is given by Raoult's law

$$P_{\text{H}_2\text{O}} = (0.981)(12.79\ \text{Torr}) = 12.5\ \text{Torr}$$

16-36. Urea and thiourea do not dissociate in trichloromethane; thus the colligative molality of the solution is

$$m_c = \frac{(25.0\ \text{g})\left(\dfrac{1\ \text{mol urea}}{60.06\ \text{g urea}}\right) + (25.0\ \text{g})\left(\dfrac{1\ \text{mol thiourea}}{76.13\ \text{g thiourea}}\right)}{0.500\ \text{kg}} = 1.49\ \text{m}_c$$

The boiling-point elevation (Table 16.2) is

$$T_b - T_b^\circ = K_b m_c = (3.80\ \text{K}\cdot\text{m}_c^{-1})(1.49\ \text{m}_c) = 5.66\ \text{K} = 5.66\,^\circ\text{C}$$

The boiling point of the solution (Table 16.2) is

$$T_b = 61.17\,^\circ\text{C} + 5.66\,^\circ\text{C} = 66.83\,^\circ\text{C}$$

FREEZING-POINT DEPRESSION

16-38. The molality of the $\text{Ca(NO}_3)_2(aq)$ solution is

$$m = \frac{(20.0\ \text{g})\left(\dfrac{1\ \text{mol Ca(NO}_3)_2}{164.1\ \text{g Ca(NO}_3)_2}\right)}{0.500\ \text{kg water}} = 0.244\ \text{m}$$

Because a $\text{Ca(NO}_3)_2(aq)$ molecule dissociates into $\text{Ca}^{2+}(aq)$ and $2\,\text{NO}_3^-(aq)$ ions, the colligative molality of the solution is

$$m_c = (3)(0.244) = 0.732\ \text{m}_c$$

The freezing-point depression (Table 16.2) is

$$T_f^\circ - T_f = K_f m_c = (1.86\ \text{K}\cdot\text{m}_c^{-1})(0.732\ \text{m}_c) = 1.36\ \text{K} = 1.36\,^\circ\text{C}$$

The freezing-point (Table 16.2) is

$$T_f = 0.00\,^\circ\text{C} - 1.36\,^\circ\text{C} = -1.36\,^\circ\text{C}$$

16-40. We can find the molality of the solution from the freezing-point depression:

$$T_f^\circ - T_f = K_f m_c$$

$$6.42\ \text{K} = (20.8\ \text{K}\cdot\text{m}_c^{-1})\, m_c$$

$$m_c = \frac{6.42\ \text{K}}{20.8\ \text{K}\cdot\text{m}_c^{-1}} = 0.309\ \text{m}_c$$

Because the mass is given, we have the correspondence

$$0.309\ \text{mol}\cdot\text{kg}^{-1} \approx \frac{1.00\ \text{g quinine}}{0.0100\ \text{kg cyclohexane}} = 100.0\ \text{g}\cdot\text{kg}^{-1}$$

and therefore,

$$0.309\ \text{mol} \approx 100.0\ \text{g}$$

Dividing both sides by 0.309, we have

$$1\ \text{mol} \approx 324\ \text{g}$$

The molecular mass of quinine is 324.

16-42. We can find the colligative molality of the solution from the freezing-point depression:

$$T_f^\circ - T_f = K_f m_c$$

$$0.00\,^\circ\text{C} - (-57\,^\circ\text{C}) = 57\ \text{K} = (1.86\ \text{K}\cdot\text{m}_c^{-1})\, m_c$$

$$m_c = \frac{57\ \text{K}}{1.86\ \text{K}\cdot\text{m}_c^{-1}} = 30.6\ \text{m}_c$$

The concentration of $CaCl_2$ in the pond is

$$\text{molality} = \frac{m_c}{3} = \frac{30.6}{3} = 10\ \text{m}$$

16-44. The empirical formula of the compound is computed by taking a 100 g sample, as follows:

$$39.12\ \text{g C} \approx 8.76\ \text{g H} \approx 52.12\ \text{O}$$

Converting to the number of moles of each element, we have

$$3.257\ \text{mol C} \approx 8.690\ \text{mol H} \approx 3.258\ \text{mol O}$$

Dividing through by the smallest value of the number of moles, 3.257, we have

$$1.00\ \text{mol C} \approx 2.668\ \text{mol H} \approx 1.00\ \text{mol O}$$

Multiplying through by 3 to convert all values to whole numbers, we have

$$3.00\ \text{mol C} \approx 8.00\ \text{mol H} \approx 3.00\ \text{mol O}$$

Thus, the empirical formula is $C_3H_8O_3$, and its formula mass is 92.09.
The colligative molality of the solute in the camphor solution is given by

$$m_c = \frac{\dfrac{2.67\ \text{g}}{M}}{0.0653\ \text{kg}} = \frac{40.9\ \text{g}\cdot\text{kg}^{-1}}{M}$$

where M is the molar mass. We can use the freezing-point depression to find a second expression for the colligative molality.

$$T_f^\circ - T_f = K_f m_c$$

$$176\,^\circ\text{C} - 159.2\,^\circ\text{C} = 16.8\,^\circ\text{C} = 16.8\text{ K} = (37.8\,^\circ\text{K}\cdot\text{m}_c^{-1})\,m_c$$

$$m_c = \frac{16.8\text{ K}}{37.8\text{ K}\cdot\text{m}_c^{-1}} = 0.444\text{ m}_c$$

Setting the two expressions for m_c equals, we have

$$0.444\text{ m}_c = \frac{40.9\text{ g}\cdot\text{kg}^{-1}}{M}$$

Solving for the molar mass M, we have

$$M = \frac{40.9\text{ g}\cdot\text{kg}^{-1}}{0.444\text{ m}_c} = 92.1\text{ g}\cdot\text{mol}^{-1}$$

Thus, the molecular formula and the empirical formula are the same.

16-46. We can find the colligative molality of the solution from the freezing-point depression:

$$T_f^\circ - T_f = K_f m_c$$

$$0.00\,^\circ\text{C} - (-1.41\,^\circ\text{C}) = 1.41\text{ K} = (1.86\text{ K}\cdot\text{m}_c^{-1})\,m_c$$

$$m_c = \frac{1.41\text{ K}}{1.86\text{ K}\cdot\text{m}_c^{-1}} = 0.758\text{ m}_c$$

The colligative molality is three times the molality. Thus, $K_2HgI_4(s)$ in water yields three ions per formula unit

$$\text{K}_2\text{HgI}_4(s) \xrightarrow[\text{H}_2\text{O}(l)]{} 2\,\text{K}^+(aq) + \text{HgI}_4^{2-}(aq)$$

OSMOTIC PRESSURE

16-48. The osmotic pressure is given by

$$\Pi = RTM_c$$

Thus

$$\Pi = (0.08206\text{ L}\cdot\text{atm}\cdot\text{K}^{-1}\cdot\text{mol}^{-1})(310\text{ K})(1.10\text{ mol}\cdot\text{L}^{-1}) = 28.0\text{ atm}$$

16-50. The concentration of pepsin in the aqueous solution is

$$M_c = \frac{\Pi}{RT} = \frac{0.162\text{ Torr}}{(62.3637\text{ L}\cdot\text{Torr}\cdot\text{K}^{-1}\cdot\text{mol}^{-1})(298\text{ K})}$$
$$= 8.72 \times 10^{-6}\text{ mol}\cdot\text{L}^{-1}$$

The molecular mass can be determined from the concentration. We have the correspondence

$$8.72 \times 10^{-6}\text{ mol}\cdot\text{L}^{-1} \approx \frac{3.00 \times 10^{-3}\text{ g}}{0.0100\text{ L}} = 0.300\text{ g}\cdot\text{L}^{-1}$$

and therefore,

$$8.72 \times 10^{-6}\ \text{mol} \approx 0.300\ \text{g}$$

Dividing both sides by 8.72×10^{-6}, we have

$$1.00\ \text{mol} \approx 3.44 \times 10^{4}\ \text{g}$$

The molecular mass of pepsin is about 34 400.

16-52. The concentration of a solution for which an applied pressure of 75 bar is just sufficient to cause reverse osmosis is

$$M_c = \frac{75\ \text{bar}}{(0.08314\ \text{L}\cdot\text{bar}\cdot\text{K}^{-1}\cdot\text{mol}^{-1})(293\ \text{K})} = 3.08\ \text{mol}\cdot\text{L}^{-1} = 3.08\text{M}_c$$

The concentration of seawater is 1.1 M_c. The number of moles of ions in the seawater will remain the same after reverse osmosis:

$$\text{moles of ions before osmosis} = \text{moles of ions after reverse osmosis}$$

$$M_b V_b = M_a V_a$$

$$(1.1\ \text{M}_c)\,V_b = (3.08\ \text{M}_c)\,V_a$$

We want 15 liters of fresh water, that is, $\Delta V = 15$ L:

$$V_b = 15\ \text{L} + V_a \qquad \text{or} \qquad V_a = V_b - 15\ \text{L}$$

$$(1.1\ \text{M}_c)\,V_b = (3.08\ \text{M}_c)(V_b - 15\ \text{L})$$

Solving for V_b, we have

$$46.2\ \text{L} = 1.98\,V_b$$

$$V_b = \frac{46.2\ \text{L}}{1.98} = 23\ \text{L}$$

RAOULT'S LAW FOR TWO COMPONENTS

16-54. The partial pressures are calculated using Raoult's law

$$P_{\text{prop}} = x_{\text{prop}} P^{\circ}_{\text{prop}} \qquad \text{and} \qquad P_{\text{iso}} = x_{\text{iso}} P^{\circ}_{\text{iso}}$$

At $x_{\text{prop}} = 0.25$ and $x_{\text{iso}} = 0.75$, we have

$$P_{\text{prop}} = (0.25)(20.9\ \text{Torr}) = 5.2\ \text{Torr} \qquad \text{and} \qquad P_{\text{iso}} = (0.75)(45.2\ \text{Torr}) = 34\ \text{Torr}$$

At $x_{\text{prop}} = 0.50$ and $x_{\text{iso}} = 0.50$, we have

$$P_{\text{prop}} = (0.50)(20.9\ \text{Torr}) = 10\ \text{Torr} \qquad \text{and} \qquad P_{\text{iso}} = (0.50)(45.2\ \text{Torr}) = 23\ \text{Torr}$$

At $x_{\text{prop}} = 0.75$ and $x_{\text{iso}} = 0.25$, we have

$$P_{\text{prop}} = (0.75)(20.9\ \text{Torr}) = 16\ \text{Torr} \qquad \text{and} \qquad P_{\text{iso}} = (0.25)(45.2\ \text{Torr}) = 11\ \text{Torr}$$

The mole fractions in the vapor phase are computed from the partial pressures

$$y_{\text{prop}} = \frac{P_{\text{prop}}}{P_{\text{prop}} + P_{\text{iso}}} \qquad \text{and} \qquad y_{\text{iso}} = 1 - y_{\text{prop}}$$

Thus, for $x_{\text{prop}} = 0.25$, we have

$$y_{\text{prop}} = \frac{5.2\ \text{Torr}}{5.2\ \text{Torr} + 34\ \text{Torr}} = 0.13 \qquad \text{and} \qquad y_{\text{iso}} = 1 - 0.13 = 0.87$$

Thus, for $x_{\text{prop}} = 0.50$, we have

$$y_{\text{prop}} = \frac{10\ \text{Torr}}{10\ \text{Torr} + 23\ \text{Torr}} = 0.30 \qquad \text{and} \qquad y_{\text{iso}} = 1 - 0.30 = 0.70$$

Thus, for $x_{\text{prop}} = 0.75$, we have

$$y_{\text{prop}} = \frac{16\ \text{Torr}}{16\ \text{Torr} + 11\ \text{Torr}} = 0.59 \qquad \text{and} \qquad y_{\text{iso}} = 1 - 0.13 = 0.41$$

HENRY'S LAW

16-56. The Henry's law constants are 1.6×10^3 bar·M^{-1} for $N_2(g)$, 7.9×10^2 bar·M^{-1} for $O_2(g)$, and 29 bar·M^{-1} for $CO_2(g)$ (Table 16.3). The concentration of $N_2(g)$ in water is given by

$$M_{N_2} = \frac{P_{N_2}}{k_h} = \frac{1.0\ \text{bar}}{1.6 \times 10^3\ \text{bar·M}^{-1}} = 6.3 \times 10^{-4}\ \text{M}$$

The concentration of $O_2(g)$ in water is given by

$$M_{O_2} = \frac{P_{O_2}}{k_h} = \frac{1.0\ \text{bar}}{7.9 \times 10^2\ \text{bar·M}^{-1}} = 1.3 \times 10^{-3}\ \text{M}$$

The concentration of $CO_2(g)$ in water is given by

$$M_{CO_2} = \frac{P_{CO_2}}{k_h} = \frac{1.0\ \text{bar}}{29\ \text{bar·M}^{-1}} = 3.4 \times 10^{-2}\ \text{M}$$

Thus, $CO_2(g)$ at 1.0 bar pressure has the highest concentration in water. Note that because the concentration of dissolved gas is inversely proportional to its Henry's law constant, under the same pressure conditions, the gas with the smallest Henry's law constant will have the highest solubility.

16-58. The molarity of the dissolved gas is given by Henry's law $M_{\text{gas}} = P_{\text{gas}}/k_h$. The values of the Henry's law constants for oxygen and nitrogen are given in Table 16.3. We first must convert 760 Torr to the unit bar. One bar is equivalent to 750.1 Torr, so 760 Torr is equivalent to 1.013 bar. The partial pressure of oxygen in the air is (0.21)(1.013 bar) = 0.213 bar. The partial pressure of nitrogen in the air is (0.78)(1.013 bar) = 0.790 bar. Thus, we have

$$M_{O_2} = \frac{0.213\ \text{bar}}{7.9 \times 10^2\ \text{bar·M}^{-1}} = 2.7 \times 10^{-4}\ \text{M}$$

$$M_{N_2} = \frac{0.790\ \text{bar}}{1.6 \times 10^3\ \text{bar·M}^{-1}} = 4.9 \times 10^{-4}\ \text{M}$$

The mass of the dissolved gas (m) is equal to the number of moles of dissolved gas times its molar mass. The number of moles of dissolved gas is equal to the molarity times the volume of the solution. Thus, we have

$$m_{O_2} = (2.7 \times 10^{-4}\ \text{mol·L}^{-1})(1.00\ \text{L})(32.00\ \text{g·mol}^{-1}) = 8.6 \times 10^{-3}\ \text{g} = 8.6\ \text{mg}$$

$$m_{N_2} = (4.9 \times 10^{-4}\ \text{mol·L}^{-1})(1.00\ \text{L})(28.02\ \text{g·mol}^{-1}) = 1.4 \times 10^{-2}\ \text{g} = 14\ \text{mg}$$

ADDITIONAL PROBLEMS

16-60. The formula mass of sucrose, $C_{12}H_{22}O_{11}$, is 342.30 $g \cdot mol^{-1}$. To prepare a 0.100 m solution, you would dissolve 34.23 grams of sucrose in 1.00 kilogram of water. The resulting solution would not be 0.100 M in sucrose because the volume of the resulting solution is not one liter.

16-62. The addition of salt to the boiling water causes an elevation in the boiling point of the resulting solution. The solution will cease to boil until its temperature rises to the new (elevated) boiling point.

16-64. A mixture of milk, cream, eggs, sugar, and other ingredients has a freezing point that is lower than that of pure water, and so ice cannot freeze the mixture. The addition of salt to the ice bath lowers the freezing point of the bath below that of the freezing point of the ice-cream mixture, and so can freeze the ice cream.

16-66. The concentration of salts in seawater is greater than that of the cells in our body. Drinking seawater dehydrates the cells in order to equalize the osmotic pressure between the cells of our body and the seawater in our bodily fluids. Thus, drinking seawater will make a person more thirsty rather than less.

16-68. Set up the following table:

	formula	molality/m	colligative molality/m_c
(a)	$C_6H_{12}O_6$	0.1	0.1
(b)	$MgCl_2$	0.1	0.3
(c)	$NaHCO_3$	0.1	0.2
(d)	NH_4NO_3	0.1	0.2

The magnesium chloride solution will have the highest boiling point because it has the largest colligative molality.

16-70. A 100 mL sample of wine contains 12 mL of ethanol and 88 mL of water. The mass of each is given by

$$\text{mass of ethanol} = dV = (0.79\ g\cdot mL^{-1})(12\ mL) = 9.48\ g$$
$$\text{mass of water} = dV = (1.00\ g\cdot mL^{-1})(88\ mL) = 88.0\ g = 0.0880\ kg$$

The molality of ethanol in wine is

$$m = \frac{(9.48\ g)\left(\dfrac{1\ \text{mol ethanol}}{46.07\ \text{g ethanol}}\right)}{0.088\ kg} = 2.34\ m$$

The freezing-point depression is

$$T_f^\circ - T_f = (1.86\ K\cdot m_c^{-1})(2.34\ m_c) = 4.4\ K = 4.4^\circ C$$

The freezing point of wine is $T_f = 0.00^\circ C - 4.4^\circ C = -4.4^\circ C$.
A 100 mL sample of vodka contains 40 mL of ethanol and 60 mL of water. The mass of each is given by

$$\text{mass of ethanol} = dV = (0.79\ g\cdot mL^{-1})(40\ mL) = 31.6\ g$$
$$\text{mass of water} = dV = (1.00\ g\cdot mL^{-1})(60\ mL) = 60.0\ g = 0.0600\ kg$$

The molality of ethanol in vodka is

$$m = \frac{(31.6\text{ g})\left(\dfrac{1\text{ mol ethanol}}{46.07\text{ g ethanol}}\right)}{0.0600\text{ kg}} = 11.4\text{ m}$$

The freezing-point depression is

$$T_f^\circ - T_f = (1.86\text{ K}\cdot\text{m}_c^{-1})(11.4\text{ m}_c) = 21\text{ K} = 21\,^\circ\text{C}$$

The freezing point of vodka is $T_f = 0.00\,^\circ\text{C} - 21\,^\circ\text{C} = -21\,^\circ\text{C}$.

16-72. According to Wikipedia, the freezing point of pure ethylene glycol is $-12.9\,^\circ\text{C}$, whereas a 50% by mass ethylene glycol and water solution has a freezing point of about $-33\,^\circ\text{C}$ (Figure 16.9). The freezing point of a solution is lower than the freezing point of the pure species and so the 50-50 mixture will provide your friend's car with greater protection against freezing than pure ethylene glycol would. Pure ethylene glycol may cause damage to the engine because of its greater viscosity.

16-74. There will be a net flow of water from the solution with the lower osmotic pressure to the solution with the higher osmotic pressure. Because $\Pi = RTM_c$, the net flow of water will be from the solution with the lower value of M_c to the solution with the higher value of M_c.

(a) Because M_c ($= 0.20\text{ M}_c$) is the same in both cases, there is no net flow.

(b) The values of M_c are

$$M_c[\text{for } 0.10\text{ M Al(NO}_3)_3] = (4)(0.10\text{ M}) = 0.40\text{ M}_c$$
$$M_c[\text{for } 0.20\text{ M NaNO}_3] = (2)(0.20\text{ M}) = 0.40\text{ M}_c$$

Because M_c is the same in both cases, there is no net flow.

(c) Because both are solutions of $CaCl_2(aq)$, the net flow will be from the 0.10 M solution to the 0.50 M solutions.

16-76. (a) The molality of the $K_2SO_4(aq)$ solution is

$$m = \frac{(5.00\text{ g})\left(\dfrac{1\text{ mol K}_2\text{SO}_4}{174.27\text{ g K}_2\text{SO}_4}\right)}{0.250\text{ kg H}_2\text{O}} = 0.115\text{ m}$$

In water a K_2SO_4 molecule dissociates into $2\,K^+(aq) + SO_4^{2-}(aq)$ ions; thus its colligative molality is $m_c = 3m = 0.345\text{ m}_c$. The freezing point of the solution is given by

$$\begin{aligned} T_f &= T_f^\circ - K_f m_c \\ &= 0.00\,^\circ\text{C} - (1.86\text{ K}\cdot\text{m}_c^{-1})(0.345\text{ m}_c) = -0.64\,^\circ\text{C} \end{aligned}$$

The boiling point of the solution is

$$\begin{aligned} T_b &= T_b^\circ + K_b m_c \\ &= 100.00\,^\circ\text{C} + (0.513\text{ K}\cdot\text{m}_c^{-1})(0.345\text{ m}_c) = 100.18\,^\circ\text{C} \end{aligned}$$

(b) An ethanol molecule does not dissociate in aqueous solution, so $m = m_c$ and

$$\mathrm{m_c} = \frac{(5.00\ \mathrm{g})\left(\dfrac{1\ \mathrm{mol\ CH_3CH_2OH}}{46.07\ \mathrm{g\ CH_3CH_2OH}}\right)}{0.250\ \mathrm{kg\ H_2O}} = 0.434\ \mathrm{m}$$

The freezing point of the solution is

$$\begin{aligned} T_f &= T_f^\circ - K_f m_c \\ &= 0.00\,^\circ\mathrm{C} - (1.86\ \mathrm{K\cdot m_c^{-1}})(0.434\ \mathrm{m_c}) = -0.81\,^\circ\mathrm{C} \end{aligned}$$

The boiling point of the solution is

$$\begin{aligned} T_b &= T_b^\circ + K_b m_c \\ &= 100.00\,^\circ\mathrm{C} + (0.513\ \mathrm{K\cdot m_c^{-1}})(0.434\ \mathrm{m_c}) = 100.22\,^\circ\mathrm{C} \end{aligned}$$

16-78. We can find the colligative molality of the solution from the boiling point elevation of carbon disulfide.

$$m_c = \frac{T_b - T_b^\circ}{K_b} = \frac{46.71\,^\circ\mathrm{C} - 46.30\,^\circ\mathrm{C}}{2.34\ \mathrm{K\cdot m_c^{-1}}} = 0.175\ \mathrm{m_c}$$

We can find the molar mass of phosphorus from the definition of molality. The mass of carbon disulfide is given by

$$m = dV = (1.261\ \mathrm{g\cdot mL^{-1}})(100.0\ \mathrm{mL}) = 126.1\ \mathrm{g} = 0.1261\ \mathrm{kg}$$

Let M be the molecular mass of phosphorus and P_x be the molecular formula of phosphorus. The molality of the solution is given by

$$m_c = \frac{(2.74\ \mathrm{g})\left(\dfrac{1\ \mathrm{mol\ P_x}}{M\ \mathrm{P_x}}\right)}{0.1261\ \mathrm{kg}} = \frac{21.73}{M}\ \mathrm{g\cdot kg^{-1}}$$

Assuming that $m = m_c$, we have the equality

$$\frac{21.73\ \mathrm{g\cdot kg^{-1}}}{M} = 0.175\ \mathrm{mol\cdot kg^{-1}}$$

$$M = \frac{21.73\ \mathrm{g\cdot kg^{-1}}}{0.175\ \mathrm{mol\cdot kg^{-1}}} = 124\ \mathrm{g\cdot mol^{-1}}$$

The atomic mass of phosphorus is 30.97 and so

$$x = \frac{124}{30.97} = 4.01$$

Thus, there are four phosphorus atoms per formula unit and the molecular formula of phosphorus is P_4.

16-80. The mass of the solution prepared by dissolving 2.00 moles of NaOH(s) in 1000 grams of H_2O is

$$\mathrm{mass_{sol}} = \mathrm{mass_{NaOH}} + \mathrm{mass_{H_2O}}$$

Thus

$$\mathrm{mass_{sol}} = (2.00\ \mathrm{mol})\left(\frac{40.00\ \mathrm{g\ NaOH}}{1\ \mathrm{mol\ NaOH}}\right) + 1000\ \mathrm{g} = 1080\ \mathrm{g}$$

The volume of the solution can be found from the density

$$V = \frac{m}{d} = \frac{1080\ \text{g}}{1.22\ \text{g}\cdot\text{mL}^{-1}} = 885\ \text{mL} = 0.885\ \text{L}$$

The molarity is

$$\text{molarity} = \frac{2.00\ \text{mol NaOH}}{0.885\ \text{L solution}} = 2.26\ \text{M}$$

16-82. The molality of the solution is

$$\text{molality} = \frac{(50.0\ \text{g})\left(\dfrac{1\ \text{mol ethylene glycol}}{62.07\ \text{g ethylene glycol}}\right)}{0.0500\ \text{kg water}} = 16.1\ \text{m}$$

The colligative molality is 16.1 m_c. The boiling point is

$$T_\text{b} = T_\text{b}^\circ + K_\text{b} m_\text{c} = 100.00^\circ\text{C} + (0.513\ \text{K}\cdot\text{m}_\text{c}^{-1})(16.1\ \text{m}_\text{c}) = 108.26^\circ\text{C}$$

16-84. First we must convert mass percent to molality and then to colligative molality. Mass percent of NaCl is defined by

$$\text{mass \% NaCl} = \frac{g_\text{NaCl}}{g_\text{NaCl} + g_{\text{H}_2\text{O}}} \times 100$$

Let the mass of H_2O be 1000 grams and write

$$\text{mass \% NaCl} = \frac{g_\text{NaCl}}{g_\text{NaCl} + 1000\ \text{g}} \times 100$$

Solving for the mass of NaCl, g_NaCl, we have

$$\text{mass of NaCl} = \frac{(1000\ \text{g})(\text{mass \% NaCl})}{100 - \text{mass \% NaCl}}$$

Now set up the following table:

mass % NaCl	mass of NaCl/g	moles of NaCl/mol	colligative molality/mol·kg^{-1}	calculated freezing point/°C	actual freezing point/°C	% error
0.50	5.025	0.0860	0.172	−0.32	−0.30	7%
1.0	10.10	0.173	0.346	−0.64	−0.59	8%
5.0	52.63	0.901	1.80	−3.35	−3.05	9.8%
10.0	111.1	1.90	3.80	−7.07	−6.56	7.7%

where the molar mass of NaCl is 58.44 g·mol^{-1}. The colligative molality must be less than 0.50 to produce an error of 2%.

16-86. We have from the definition of mole fraction that

$$x_2 = \frac{\text{moles of solute}}{\text{moles of solute} + \text{moles of water}}$$

From the definition of colligative molality, we have

$$m_c = \frac{\text{moles of solute}}{1000 \text{ grams of water}}$$

If we take 1000 g of water, then

$$\text{moles of solute} = m_c$$

$$\text{moles of } H_2O = (1000 \text{ g } H_2O)\left(\frac{1 \text{ mol } H_2O}{18.02 \text{ g } H_2O}\right) = 55.5 \text{ mol}$$

Thus we have that

$$x_2 = \frac{m_c}{m_c + 55.5 \text{ mol}\cdot\text{kg}^{-1}}$$

When $m_c \ll 55.5 \text{ mol}\cdot\text{kg}^{-1}$, we can ignore m_c in the denominator to get

$$x_2 = \frac{m_c}{55.5 \text{ mol}\cdot\text{kg}^{-1}}$$

16-88. We use Equation 16.15 and solve for im.

$$im = \frac{T_f^\circ - T_f}{K_f} = \frac{3.74 \text{ K}}{1.86 \text{ K}\cdot\text{m}_c^{-1}} = 2.01 \text{ m}_c$$

where we obtained K_f from Table 16.2. Solving for i, the number of solute particle per formula unit, we have

$$i = \frac{m_c}{m} = \frac{2.01}{1.00} = 2.01$$

Note that 1% of the second protons dissociated.

16-90.* The empirical formula of the compound is given by

$$35.00 \text{ g N} \Leftrightarrow 59.96 \text{ g O} \Leftrightarrow 5.04 \text{ g H}$$
$$2.498 \text{ mol N} \Leftrightarrow 3.748 \text{ mol O} \Leftrightarrow 5.00 \text{ mol H}$$
$$1.00 \text{ mol N} \Leftrightarrow 1.50 \text{ mol O} \Leftrightarrow 2.00 \text{ mol H}$$
$$2 \text{ mol N} \Leftrightarrow 3 \text{ mol O} \Leftrightarrow 4 \text{ mol H}$$

Therefore, the empirical formula is $H_4N_2O_3$.
The colligative molality of the solution is

$$m_c = \frac{0.372 \text{ K}}{1.86 \text{ K}\cdot\text{m}_c^{-1}} = 0.200 \text{ m}_c$$

The solution consists of 2.00 grams dissolved in 250 mL of water or 8.00 g in 1000 mL of water. Taking the density of water to be $1.00 \text{ g}\cdot\text{mL}^{-1}$, we have 8.00 grams of the compound dissolved in one kilogram of water. Thus, we have the stoichiometric correspondence

$$\frac{8.00 \text{ g}}{1 \text{ kg water}} \Leftrightarrow \frac{1}{i}\frac{0.200 \text{ mol}}{1 \text{ kg water}}$$

$$8.00 \text{ g} \Leftrightarrow \frac{1}{i}(0.200 \text{ mol})$$

$$40.0 \text{ g} \Leftrightarrow \frac{1}{i}(0.200 \text{ mol})$$

Thus, the molecular mass is 40 or a multiple of 40 if it's an ionic compound. The molecular mass of $H_4N_2O_3$ is 80, which corresponds to a van't Hoff factor of 2. The compound produces two ions in solution. The compound is ammonium nitrate, NH_4NO_3.

16-92.* The two beakers will eventually come to the same vapor pressure. Because the vapor pressure over each solution depends upon its concentration, the two solutions will also reach the same concentration eventually. The solutions contain

$$\begin{aligned}\text{moles of NaCl in solution 1} &= (200.0 \text{ mL})(0.100 \text{ M}) = (200.0 \text{ mL})(0.100 \text{ mol}\cdot\text{L}^{-1})\\ &= 0.0200 \text{ mol NaCl}\end{aligned}$$

$$\begin{aligned}\text{moles of NaCl in solution 2} &= (300.0 \text{ mL})(0.200 \text{ M}) = (300.0 \text{ mL})(0.200 \text{ mol}\cdot\text{L}^{-1})\\ &= 0.0600 \text{ mol NaCl}\end{aligned}$$

Let x be the final volume of solution 1. Then 500 mL $- x$ is the final volume of solution 2, where 500 mL is the total volume of the two solutions. When the two solutions are at the same concentration,

$$\frac{0.0200 \text{ mol}}{x} = \frac{0.0600 \text{ mol}}{500.0 \text{ mL} - x}$$

Solving for x yields $x = 500.0 \text{ mL}/4 = 125.0$ mL. The other solution will contain 375.0 mL of water. The concentrations in the final solutions are

$$M_1 = \frac{0.0200 \text{ mol}}{0.125 \text{ L}} = 0.16 \text{ M}$$

$$M_2 = \frac{0.0600 \text{ mol}}{0.375 \text{ L}} = 0.16 \text{ M}$$

CHAPTER 17. Chemical Kinetics: Rate Laws

REACTION RATES

17-2. There are two moles of SO_2 in the balanced equation as written; therefore, the rate law in terms of SO_2 is given by

$$\text{rate of concentration change} = -\frac{1}{2}\frac{\Delta[SO_2]}{\Delta t}$$

There is one mole of O_2 produced in the balanced equation; thus, we have

$$\text{rate of concentration change} = -\frac{\Delta[O_2]}{\Delta t}$$

Note that a minus sign is placed in front of this expression to ensure that the rate of reaction is a positive quantity. There are two moles of SO_3 produced in the balanced equation; thus,

$$\text{rate of concentration change} = \frac{1}{2}\frac{\Delta[SO_3]}{\Delta t}$$

Note that the rate of reaction will be positive because $[SO_3]$ is increasing.

17-4. (a) As the reaction proceeds, oxygen gas is produced; therefore, the pressure of oxygen increases as the reaction proceeds. We can measure the rate of the reaction by measuring the rate of increase of the pressure of oxygen.

(b) Bromine gas is reddish-brown in color, whereas both HBr and H_2 are colorless. We can measure the rate of the reaction by measuring the rate of the increase in the intensity of the color of the gaseous mixture using a spectrophotometer.

17-6. We are given that

$$\text{rate of reaction} = (1.3 \times 10^{-3}\ \text{M}^{-2}\cdot\text{min}^{-1})[\text{NO}]^2[\text{Br}_2]$$

Substituting the initial concentrations into the reaction rate law, the initial reaction rate is

$$\begin{aligned}(\text{rate of reaction})_0 &= (1.3 \times 10^{-3}\ \text{M}^{-2}\cdot\text{min}^{-1})(3.0 \times 10^{-4}\ \text{M})^2(3.0 \times 10^{-4}\ \text{M})\\ &= 3.5 \times 10^{-14}\ \text{M}\cdot\text{min}^{-1}\end{aligned}$$

The overall order is third order.

17-8. (a) The average rate of reaction from 0 minutes to 100 minutes is

$$\text{average rate of formation of NO}_2 = \frac{2.36 \times 10^{-2}\ \text{M} - 0\ \text{M}}{100\ \text{min} - 0\ \text{min}} = 2.36 \times 10^{-4}\ \text{M}\cdot\text{min}^{-1}$$

(b) Draw a line tangent to the curve at $t = 50$ min to obtain or estimate the rate at 50 min from

$$\text{instantaneous rate of formation at } t = 50\ \text{min} \approx \frac{2.08 \times 10^{-2}\ \text{M} - 1.74 \times 10^{-2}\ \text{M}}{60\ \text{min} - 40\ \text{min}}$$
$$= 1.7 \times 10^{-4}\ \text{M}\cdot\text{min}^{-1}$$

(c) Similarly,

$$\text{instantaneous rate of formation at } t = 0\ \text{min} \approx \frac{0.64 \times 10^{-2}\ \text{M} - 0\ \text{M}}{10\ \text{min} - 0\ \text{min}}$$
$$= 6.4 \times 10^{-4}\ \text{M}\cdot\text{min}^{-1}$$

17-10. The average rate of disappearance of $H^+(aq)$ is

$$\text{rate} = -\frac{\Delta[\text{H}^+]}{\Delta t}$$

For the interval 0 min to 31 min, we have

$$\text{average rate} = -\frac{1.90\ \text{M} - 2.12\ \text{M}}{31\ \text{min} - 0\ \text{min}} = 7.1 \times 10^{-3}\ \text{M}\cdot\text{min}^{-1}$$

For the interval 31 min to 61 min, we have

$$\text{average rate} = -\frac{1.78\ \text{M} - 1.90\ \text{M}}{61\ \text{min} - 31\ \text{min}} = 4.0 \times 10^{-3}\ \text{M}\cdot\text{min}^{-1}$$

For the interval 61 min to 121 min, we have

$$\text{average rate} = -\frac{1.61\ \text{M} - 1.78\ \text{M}}{121\ \text{min} - 61\ \text{min}} = 2.8 \times 10^{-3}\ \text{M}\cdot\text{min}^{-1}$$

From the reaction stoichiometry, we conclude that the average rate of disappearance of $CH_3OH(aq)$ is equal to the average rate of disappearance of $H^+(aq)$.

$$\text{average rate} = -\frac{\Delta[\text{CH}_3\text{OH}]}{\Delta t} = -\frac{\Delta[\text{H}^+]}{\Delta t}$$

Similarly, we find that the average rate of appearance of $CH_3Cl(aq)$ is

$$\text{average rate} = \frac{\Delta[\text{CH}_3\text{Cl}]}{\Delta t} = -\frac{\Delta[\text{H}^+]}{\Delta t}$$

INITIAL RATES

17-12. The rate of the reaction increases by a factor of four when [NOBr] is doubled, by a factor of nine when [NOBr] is tripled, and by a factor of 16 when [NOBr] is quadrupled. Assuming that the rate law does not vary with time, we deduce that the order of the reaction is second order. The second-order reaction rate law is

$$\text{rate of reaction} = k[\text{NOBr}]^2$$

We can calculate the value of k using any of the data points.

$$k = \frac{\text{rate of reaction}}{[\text{NOBr}]^2} = \frac{0.80\ \text{mol·L}^{-1}\text{·s}^{-1}}{(0.20\ \text{mol·L}^{-1})^2} = 20\ \text{mol}^{-1}\text{·L·s}^{-1}$$

$$= \frac{3.20\ \text{mol·L}^{-1}\text{·s}^{-1}}{(0.40\ \text{mol·L}^{-1})^2} = 20\ \text{mol}^{-1}\text{·L·s}^{-1}$$

$$= \frac{7.20\ \text{mol·L}^{-1}\text{·s}^{-1}}{(0.60\ \text{mol·L}^{-1})^2} = 20\ \text{mol}^{-1}\text{·L·s}^{-1}$$

$$= \frac{12.80\ \text{mol·L}^{-1}\text{·s}^{-1}}{(0.80\ \text{mol·L}^{-1})^2} = 20\ \text{mol}^{-1}\text{·L·s}^{-1}$$

17-14. Using Runs 1 and 2 (arbitrarily), we write

$$\frac{(\text{rate of reaction})_2}{(\text{rate of reaction})_1} = \frac{14.2\ \text{Torr·s}^{-1}}{5.76\ \text{Torr·s}^{-1}} = \frac{(314\ \text{Torr})^x}{(200\ \text{Torr})^x} = \left(\frac{314}{200}\right)^x = (1.57)^x$$

or

$$2.465 = (1.57)^x$$

Therefore, $x = 2$ and the reaction is second order or

$$\text{rate of reaction} = kP^2_{\text{C}_5\text{H}_6}$$

Using Run 2 (arbitrarily) to evaluate the rate constant:

$$k = \frac{(\text{rate of reaction})_0}{P^2_{\text{C}_5\text{H}_6}} = \frac{14.2\ \text{Torr·s}^{-1}}{(314\ \text{Torr})^2} = 1.44 \times 10^{-4}\ \text{Torr}^{-1}\text{·s}^{-1}$$

The rate law is

$$\text{rate of reaction} = (1.44 \times 10^{-4}\ \text{Torr}^{-1}\text{·s}^{-1})\,P^2_{\text{C}_5\text{H}_6}$$

17-16. When $P_{\text{N}_2\text{O}_3}$ is increased by a factor of $1.4/0.91 = 1.5$, the initial rate of reaction is increased by a factor of $8.4/5.5 = 1.5$. Thus, the rate of reaction is first order in $P_{\text{N}_2\text{O}_3}$. The rate law for the reaction is

$$\text{rate of reaction} = kP_{\text{N}_2\text{O}_3}$$

We can calculate the rate constant from the data of any of the three runs. The value of the rate constant is given by

$$k = \frac{\text{rate of reaction}}{P_{\text{N}_2\text{O}_3}} = \frac{5.5\ \text{Torr·s}^{-1}}{0.91\ \text{Torr}} = 6.0\ \text{s}^{-1}$$

$$= \frac{8.4\ \text{Torr·s}^{-1}}{1.4\ \text{Torr}} = 6.0\ \text{s}^{-1}$$

$$= \frac{13\ \text{Torr·s}^{-1}}{2.1\ \text{Torr}} = 6.2\ \text{s}^{-1}$$

17-18. When $[\text{CoBr(NH}_3)_5^{2+}]_0$ is doubled and $[\text{OH}^-]_0$ remains constant, the initial rate of reaction doubles. Thus, the rate law is first order in $[\text{CoBr(NH}_3)_5^{2+}]$. When $[\text{OH}^-]_0$ is tripled and

$[CoBr(NH_3)_5^{2+}]_0$ remains constant, the initial rate of reaction triples. Thus, the rate law is first order in $[OH^-]$. The rate law is

$$\text{rate of reaction} = k\left[CoBr(NH_3)_5^{2+}\right][OH^-]$$

The rate law is second order overall. We can calculate the rate constant using data from any of the four runs. The value of the rate constant is given by

$$\begin{aligned} k = \frac{\text{rate of reaction}}{\left[CoBr(NH_3)_5^{2+}\right][OH^-]} &= \frac{1.37 \times 10^{-3}\ M\cdot s^{-1}}{(0.030\ M)(0.030\ M)} = 1.5\ M^{-1}\cdot s^{-1} \\ &= \frac{2.74 \times 10^{-3}\ M\cdot s^{-1}}{(0.060\ M)(0.030\ M)} = 1.5\ M^{-1}\cdot s^{-1} \\ &= \frac{4.11 \times 10^{-3}\ M\cdot s^{-1}}{(0.030\ M)(0.090\ M)} = 1.5\ M^{-1}\cdot s^{-1} \\ &= \frac{1.23 \times 10^{-2}\ M\cdot s^{-1}}{(0.090\ M)(0.090\ M)} = 1.5\ M^{-1}\cdot s^{-1} \end{aligned}$$

17-20. When $[CH_3COCH_3]_0$ is increased by a factor of 1.75 with $[Br_2]_0$ and $[H^+]_0$ constant, the initial rate of reaction increases by a factor of $7.0 \times 10^{-3}/4.0 \times 10^{-3} = 1.75$. Thus, the rate law is first order in $[CH_3COCH_3]$. When $[Br_2]_0$ is increased by a factor of 1.40 with $[CH_3COCH_3]_0$ and $[H^+]_0$ constant, the initial rate increases by a factor of $(9.8 \times 10^{-3})/(7.0 \times 10^{-3}) = 1.4$. Therefore, the reaction is first order in $[Br_2]$. When $[H^+]_0$ and $[Br_2]_0$ are increased and $[CH_3COCH_3]_0$ remains constant, the initial rate of reaction increases by $(11.3 \times 10^{-3})/(4.0 \times 10^{-3}) = 2.83$. To determine the order of the reaction in $[H^+]$, we write

$$2.83 = \left(\frac{1.40}{1.00}\right)\left(\frac{2.00}{1.00}\right)^x$$

which gives us $x = 1$. Thus, the rate law is first order in $[H^+]$. The rate law is

$$\text{rate of reaction} = k[CH_3COCH_3][H^+][B_2]$$

The rate law is third order overall. We can calculate the rate constant using data from any of the four runs. The value of the rate constant is given by

$$\begin{aligned} k = \frac{\text{rate of reaction}}{[CH_3COCH_3][H^+][B_2]} &= \frac{4.0 \times 10^{-3}\ M\cdot s^{-1}}{(1.00\ M)(1.00\ M)(1.00\ M)} = 4.0 \times 10^{-3}\ M^{-1}\cdot s^{-1} \\ &= \frac{7.0 \times 10^{-3}\ M\cdot s^{-1}}{(1.75\ M)(1.00\ M)(1.00\ M)} = 4.0 \times 10^{-3}\ M^{-1}\cdot s^{-1} \\ &= \frac{9.8 \times 10^{-3}\ M\cdot s^{-1}}{(1.75\ M)(1.00\ M)(1.40\ M)} = 4.0 \times 10^{-3}\ M^{-1}\cdot s^{-1} \\ &= \frac{11.3 \times 10^{-3}\ M\cdot s^{-1}}{(1.00\ M)(2.00\ M)(1.40\ M)} = 4.0 \times 10^{-3}\ M^{-1}\cdot s^{-1} \end{aligned}$$

FIRST-ORDER REACTIONS

17-22. The value of the half life is given by

$$t_{1/2} = \frac{0.693}{k}$$

Thus,

$$t_{1/2} = \frac{0.693}{5.5 \times 10^{-4}\ \text{s}^{-1}} = 1.26 \times 10^3\ \text{s}$$
$$= (1.26 \times 10^3\ \text{s})\left(\frac{1\ \text{min}}{60\ \text{s}}\right)\left(\frac{1\ \text{h}}{60\ \text{min}}\right) = 0.35\ \text{h}$$

We use Equation 17.17 to calculate the concentration of cyclopropane after 2.0 hours.

$$\ln[\text{A}] = \ln[\text{A}]_0 - kt$$

Thus, we have

$$\ln[\text{cyclopropane}]/\text{M} = \ln(1.00 \times 10^{-3}) - (5.5 \times 10^{-4}\ \text{s}^{-1})(2.0\ \text{h})\left(\frac{60\ \text{min}}{1\ \text{h}}\right)\left(\frac{60\ \text{s}}{1\ \text{min}}\right)$$
$$= -10.9$$

Taking antilogarithms of both sides, we have

$$[\text{cyclopropane}] = e^{-10.9} = 1.9 \times 10^{-5}\ \text{M}$$

17-24. The fraction of a reactant remaining after time t is given by

$$\ln \frac{[\text{A}]}{[\text{A}]_0} = -kt$$

Thus,

$$\ln \frac{[\text{CH}_3\text{I}]}{[\text{CH}_3\text{I}]_0} = -(1.5 \times 10^{-4}\ \text{s}^{-1})(60\ \text{s}) = -9.0 \times 10^{-3}$$

Taking antilogarithms, we have

$$\frac{[\text{CH}_3\text{I}]}{[\text{CH}_3\text{I}]_0} = \text{fraction remaining} = e^{-9.0\times10^{-3}} = 0.99$$

In other words, very little CH_3I decomposes in one minute.

17-26. We first set up the following table:

t/min	$[\text{SO}_2\text{Cl}_2]$/M	$\ln([\text{SO}_2\text{Cl}_2]/\text{M})$
0.0	0.0345	−3.367
3.8	0.0245	−3.710
5.6	0.0212	−3.854
9.3	0.0154	−4.173
14.0	0.0103	−4.576

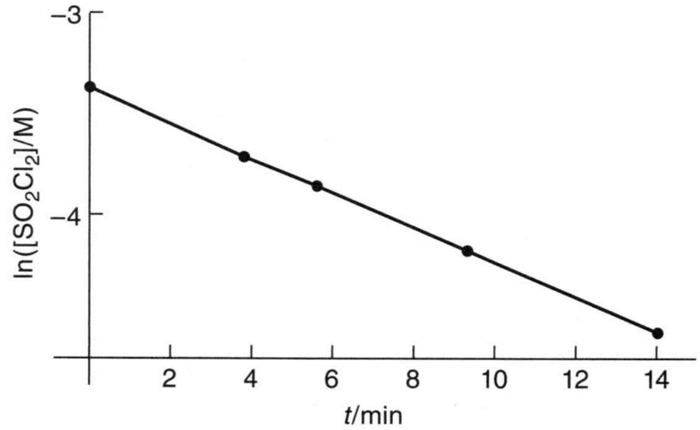

A plot of $\ln([SO_2Cl_2]/M)$ versus time shown in the figure above is a straight line, confirming that the reaction is first order. The rate constant can be calculated from the slope of the line. Using the first and last data points, for instance, we have

$$k = \frac{-3.367 - (-4.576)}{14.0\text{ min} - 0\text{ min}} = 0.0864\text{ min}^{-1}$$

An alternative method to calculate the rate constant is to use Equation 17.18.

17-28. The total pressure is given by

$$P_{\text{tot}} = P_{H_2C_2O_4} + P_{CO_2} + P_{HCHO_2}$$

From the reaction stoichiometry, we have

$$P_{H_2C_2O_4} = (P_{H_2C_2O_4})_0 - P_{CO_2} \qquad \text{and} \qquad P_{CO_2} = P_{HCHO_2}$$

Thus,

$$P_{\text{tot}} = P_{H_2C_2O_4} + 2\,P_{CO_2}$$

To find the relation between P_{CO_2} and $P_{H_2C_2O_4}$, we use the fact that $P_{CO_2} = (P_{H_2C_2O_4})_0 - P_{H_2C_2O_4}$. We now write

$$P_{\text{tot}} = P_{H_2C_2O_4} + 2\,[(P_{H_2C_2O_4})_0 - P_{H_2C_2O_4}]$$

Thus, we have that

$$P_{H_2C_2O_4} = 2(P_{H_2C_2O_4})_0 - P_{\text{tot}}$$

Application of the above equation to the data yields the following table:

Run	$(P_{H_2C_2O_4})_0$/Torr	$P_{H_2C_2O_4}$ at 2×10^4 s/Torr
1	5.0	2.8
2	7.0	4.0
3	8.4	4.8

A first order decomposition reaction obeys the equation $[A] = [A]_0 e^{-kt}$ or $[A]/[A]_0 = e^{-kt}$. For a fixed time, 2.00×10^4 s in this case, the ratio of the concentrations for all runs should be a constant.

$$\frac{[A]}{[A]_0} = \frac{P_{H_2C_2O_4}}{(P_{H_2C_2O_4})_0} = \frac{2.8\text{ Torr}}{5.0\text{ Torr}} = 0.56$$

$$= \frac{4.0\text{ Torr}}{7.0\text{ Torr}} = 0.57$$

$$= \frac{4.8\text{ Torr}}{8.4\text{ Torr}} = 0.57$$

Thus, the rate law is first order in $P_{H_2C_2O_4}$ where

$$\text{rate of reaction} = kP_{H_2C_2O_4}$$

We can calculate the rate constant k using Equation 17.17

$$\ln[A] = \ln[A]_0 - kt$$

Choosing the data from, say, Run 2 yields

$$\ln 4.0 = \ln 7.0 - k(2.00 \times 10^4 \text{ s})$$

Thus,

$$k = 2.8 \times 10^{-5} \text{ s}^{-1}$$

RATES OF NUCLEAR DECAY

17-30. Using Equation 17.23, we have

$$\ln \frac{N}{N_0} = -\frac{0.693\, t}{t_{1/2}} = -\frac{(0.693)(1 \text{ d})\left(\dfrac{24 \text{ h}}{1 \text{ d}}\right)}{35.3 \text{ h}} = -0.4712$$

Taking antilogarithms of both sides, we have

$$\frac{N}{N_0} = e^{-0.4712} = 0.624$$

The fraction of the sample of bromine-82 that remains after one day is 0.624.

17-32. We use Equation 17.23,

$$\ln \frac{N}{N_0} = -\frac{0.693\, t}{t_{1/2}}$$

When an isotope decays to 0.10% of its original value, then $N/N_0 = 0.0010$. Solving Equation 17.23 for t, we have

$$t = -\frac{(\ln 0.0010)(30.2 \text{ y})}{0.693} = 300 \text{ y}$$

17-34. We use Equation 17.23,

$$\ln \frac{N}{N_0} = -\frac{0.693\, t}{t_{1/2}} = -\frac{(0.693)(74 \text{ y})}{29.1 \text{ y}} = -1.76$$

Taking antilogarithms,

$$\frac{N}{N_0} = 0.17$$

The fraction of strontium-90 remaining after 74 years is 0.17.

DATING

17-36. We use Equation 17.27 with $R = 13.19$ disintegrations$\cdot$min$^{-1}\cdot$g^{-1} of carbon.

$$t = (8.27 \times 10^3 \text{ years}) \ln\left(\frac{15.3 \text{ disintegrations}\cdot\text{min}^{-1}\cdot\text{g}^{-1}}{R}\right)$$

$$= (8.27 \times 10^3 \text{ years}) \ln\left(\frac{15.3 \text{ disintegrations}\cdot\text{min}^{-1}\cdot\text{g}^{-1}}{13.19 \text{ disintegrations}\cdot\text{min}^{-1}\cdot\text{g}^{-1}}\right) = 1230 \text{ years}$$

The log is about 1230 years old.

17-38. We use Equation 17.27 with $R = (0.65)(15.3\text{ disintegrations}\cdot\text{min}^{-1}\cdot\text{g}^{-1})$. Thus,

$$t = (8.27 \times 10^3\text{ years}) \ln \frac{15.3\text{ disintegrations}\cdot\text{min}^{-1}\cdot\text{g}^{-1}}{R} = (8.27 \times 10^3\text{ years}) \ln \frac{1}{0.65}$$
$$= 3.56 \times 10^3\text{ years}$$

The oak is about 3600 years old.

17-40. The age of the uranite is given by solving Equation 17.23 for t.

$$t = \frac{t_{1/2}}{0.693} \ln \frac{N_0}{N} = \left(\frac{4.51 \times 10^9\text{ y}}{0.693}\right) \ln \frac{N_0}{N}$$

where

$$\frac{N_0}{N} = \frac{\text{initial mass of } {}^{238}_{92}\text{U}}{\text{present mass of } {}^{238}_{92}\text{U}}$$

The initial mass of ${}^{238}_{92}\text{U}$ is the sum of the present mass plus the mass that has decayed.

$$\frac{N_0}{N} = \frac{\text{mass of } {}^{238}_{92}\text{U that has decayed} + \text{present mass of } {}^{238}_{92}\text{U}}{\text{present mass of } {}^{238}_{92}\text{U}}$$
$$= \frac{\text{mass of } {}^{238}_{92}\text{U that has decayed}}{\text{present mass of } {}^{238}_{92}\text{U}} + 1$$

The mass of ${}^{206}_{82}\text{Pb}$ resulting from the decay of ${}^{238}_{92}\text{U}$ is

$$\text{mass of } {}^{206}_{82}\text{Pb} = (\text{mass of } {}^{238}_{92}\text{U that has decayed}) \left(\frac{1\text{ mol } {}^{238}_{92}\text{U}}{238\text{ g } {}^{238}_{92}\text{U}}\right)$$
$$\times \left(\frac{1\text{ mol } {}^{206}_{82}\text{Pb}}{1\text{ mol } {}^{238}_{92}\text{U}}\right) \left(\frac{206\text{ g } {}^{206}_{82}\text{Pb}}{1\text{ mol } {}^{206}_{82}\text{Pb}}\right)$$
$$= \left(\frac{206}{238}\right) (\text{mass of } {}^{238}_{92}\text{U that has decayed})$$

The mass ratio ${}^{206}_{82}\text{Pb}/{}^{238}_{92}\text{U}$ is given as

$$\text{mass ratio} = \frac{\text{present mass of } {}^{206}_{82}\text{Pb}}{\text{present mass of } {}^{238}_{92}\text{U}} = 0.395$$
$$= \frac{\left(\frac{206}{238}\right)(\text{mass of } {}^{238}_{92}\text{U that has decayed})}{\text{present mass of } {}^{238}_{92}\text{U}}$$

Therefore,

$$\frac{\text{mass of } {}^{238}_{92}\text{U that has decayed}}{\text{present mass of } {}^{238}_{92}\text{U}} = \left(\frac{238}{206}\right)(0.395)$$

and

$$\frac{N_0}{N} = \left(\frac{238}{206}\right)(0.395) + 1 = 1.456$$

and the age of the uranite is

$$t = \frac{4.51 \times 10^9 \text{ y}}{0.693} \ln 1.456 = 2.44 \times 10^9 \text{ y}$$

SECOND-ORDER REACTIONS

17-42. To find the concentration of $BrO^-(aq)$ 1.00 minute later, we use the second-order rate equation, Equation 17.29.

$$\frac{1}{[BrO^-]} = \frac{1}{[BrO^-]_0} + kt$$

Substituting in the values of $[BrO^-]_0$, k, and t yields

$$\frac{1}{[BrO^-]} = \frac{1}{0.212 \text{ M}} + (0.056 \text{ M}^{-1}\cdot\text{s}^{-1})(60.0 \text{ s}) = 8.08 \text{ M}^{-1}$$

Solving for $[BrO^-]$ yields

$$[BrO^-] = \frac{1}{8.08 \text{ M}^{-1}} = 0.12 \text{ M}$$

17-44. We set up the following table:

t/s	[HBr]/M	1/[HBr]/M^{-1}
0.00	0.0714	14.01
1.02	0.0520	19.23
1.81	0.0430	23.26
2.53	0.0371	26.95
3.16	0.0332	30.12

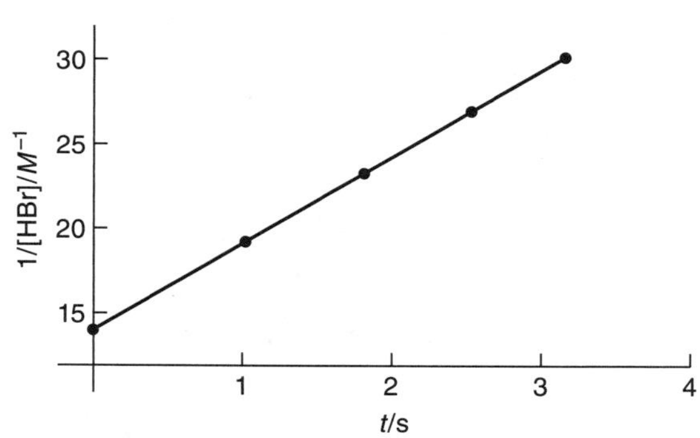

The plot of 1/[HBr] versus time, shown above, confirms that the reaction is second order. The rate constant can be calculated from the slope of the line in the above figure. Using the second and fourth data points, we have

$$k = \frac{26.95 \text{ M}^{-1} - 19.23 \text{ M}^{-1}}{2.53 \text{ s} - 1.02 \text{ s}} = 5.11 \text{ M}^{-1}\cdot\text{s}^{-1}$$

17-46. Using Equation 17.29 and the values of $[NO_2]_0$ and k given, we have that

$$\begin{aligned}\frac{1}{[NO_2]} &= \frac{1}{[NO_2]_0} + kt \\ &= \frac{1}{1.25\text{ M}} + (0.54\text{ M}^{-1}\cdot\text{s}^{-1})(120\text{ s}) = 65.6\text{ M}^{-1} \\ [NO_2] &= 0.015\text{ M}\end{aligned}$$

ADDITIONAL PROBLEMS

17-48. Under initial conditions, the concentrations of the reactants do not vary appreciably from their initial values, and so the initial rate of reaction is essentially constant. To locate the period of time over which a reaction is proceeding under initial conditions, we locate the initial region of the plot of concentration versus time where the curve describing the rate of reaction is essentially a straight line. Looking at Figure 17.3, we see that the three curves are all essentially straight lines over the time interval from $t = 0$ to about $t = 20$ minutes. Past this time interval, the three curves start to deviate from straight lines. This time interval is not the same for all reactions because not all reactions proceed at the same rate.

17-50. The units of the rate constant, k, are given by

$$-\frac{\Delta[\text{A}]}{\Delta t} = k[\text{A}]^n \qquad \text{or} \qquad k = -\frac{1}{[\text{A}]^n}\frac{\Delta[\text{A}]}{\Delta t}$$

where n is the order of the reaction. Therefore, k has units of

$$\text{units of } k = \frac{\text{M}^{1-n}}{\text{time}} = \text{M}^{1-\text{n}}\cdot\text{time}^{-1}$$

Therefore, we have

$$\begin{aligned} n &= 0 \qquad \text{M}\cdot\text{time}^{-1} \\ n &= 1 \qquad \text{time}^{-1} \\ n &= 2 \qquad \text{M}^{-1}\cdot\text{time}^{-1} \\ n &= 3 \qquad \text{M}^{-2}\cdot\text{time}^{-1}\end{aligned}$$

17-52. The total pressure in the reaction vessel is

$$P_{\text{total}} = P_{\text{CO}} + P_{\text{CO}_2}$$

If P^0_{CO} is the initial pressure of CO(g), then the partial pressure of CO(g) is equal to

$$P_{\text{CO}} = P^0_{\text{CO}} - 2\,P_{\text{CO}_2}$$

because each mole of $CO_2(g)$ produced consumes two moles of CO(g). Thus, substituting $P_{\text{CO}_2} = P_{\text{total}} - P_{\text{CO}}$ into this equation yields

$$P_{\text{CO}} = P^0_{\text{CO}} - 2(P_{\text{total}} - P_{\text{CO}})$$

Solving for P_{CO} yields

$$P_{\text{CO}} = 2\,P_{\text{total}} - P^0_{\text{CO}} = 2\,P_{\text{total}} - 250\text{ Torr}$$

Using the data given in the problem and the preceding equation, we obtain

t/s	P_{CO}/Torr	ln P_{CO}/Torr	P_{CO}^{-1}/Torr^{-1}
0	250	5.521	0.00400
398	226	5.421	0.00442
1002	198	5.288	0.00505
1801	170	5.136	0.00588
3000	142	4.956	0.00704

We must first test these data to see if the reaction is first order or second order in P_{CO}. To do so, we calculated both ln P_{CO} and $1/P_{CO}$ as given in the table above. The plots of both versus time are shown below.

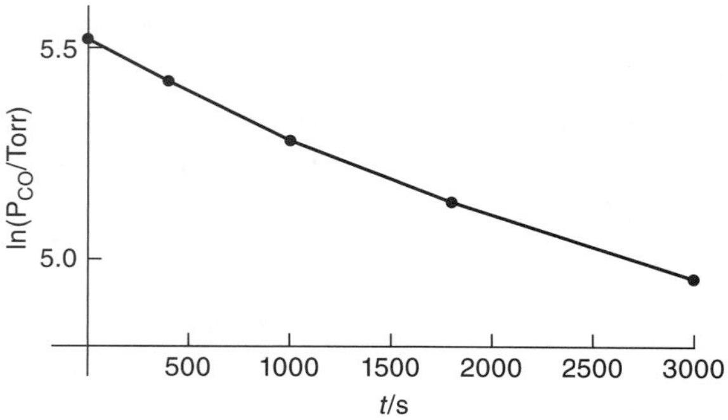

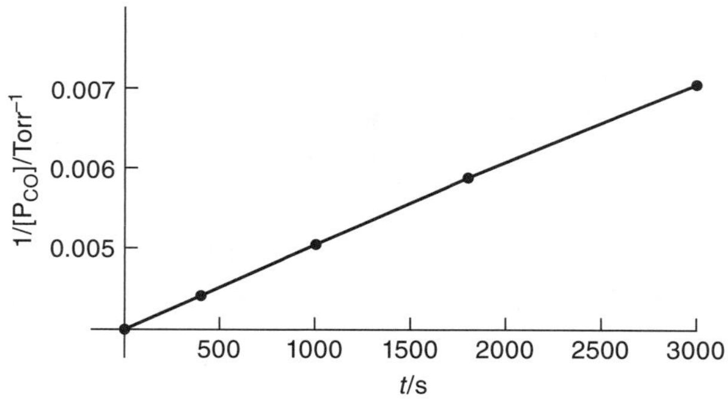

We see from the plots that the reaction is second order and the rate law is

$$\text{rate of reaction} = kP_{CO}^2$$

The value of the rate constant k can be found from the slope of the second plot taking, say, the second and fifth points, Thus, we have

$$k = \frac{(0.00704 - 0.00442)\ \text{Torr}^{-1}}{(3000 - 398)\ \text{s}} = 1.01 \times 10^{-6}\ \text{Torr}^{-1}\cdot\text{s}^{-1}$$

and find that

$$\text{rate of reaction} = (1.01 \times 10^{-6}\ \text{Torr}^{-1}\cdot\text{s}^{-1})\,P_{CO}^2$$

17-54. The initial rate of reaction is

$$\begin{aligned}\text{rate of reaction} &= (2.99 \times 10^6\ \text{M}^{-1}\cdot\text{s}^{-1})(2.0 \times 10^{-6}\ \text{M})(6.0 \times 10^{-5}\ \text{M})\\ &= 3.6 \times 10^{-4}\ \text{M}\cdot\text{s}^{-1}\end{aligned}$$

The rate of production of $NO_2(g)$ is given by

$$\text{rate of production} = \frac{\Delta[NO_2]}{\Delta t} = 3.6 \times 10^{-4}\ \text{M}\cdot\text{s}^{-1}$$

The amount of $NO_2(g)$ produced in one hour under these conditions is

$$\begin{aligned}[NO_2] &= \text{rate} \times \text{time}\\ &= (3.6 \times 10^{-4}\ \text{M}\cdot\text{s}^{-1})\left(\frac{60\ \text{s}}{1\ \text{min}}\right)\left(\frac{60\ \text{min}}{1\ \text{h}}\right)(1\ \text{h})\\ &= 1.3\ \text{mol}\cdot\text{L}^{-1}\end{aligned}$$

17-56. The time for 99.9% of the reaction to take place is given by

$$\ln \frac{[A]}{[A]_0} = -kt_1$$

where

$$\frac{[A]}{[A]_0} = 1 - 0.999 = 0.001$$

Thus, we have $\ln(0.001) = -kt_1$. Solving for kt_1, we have that $kt_1 = 6.91$.
The time for 50% of the reaction to take place is given by

$$\ln \frac{[A]}{[A]_0} = -kt_2$$

where

$$\frac{[A]}{[A]_0} = 1 - 0.50 = 0.50$$

Thus, we have $\ln(0.50) = -kt_2$. Solving for kt_2, we have that $kt_2 = 0.693$. The ratio of t_1 to t_2 is given by

$$\frac{kt_1}{kt_2} = \frac{t_1}{t_2} = \frac{6.91}{0.693} \approx 10$$

Thus, it takes 10 times as long for 99.9% of the reaction to take place than for 50%.

17-58. Use Equation 17.23,

$$\ln \frac{N}{N_0} = -\frac{0.693t}{t_{1/2}}$$

The time required for 10% to decay, $N = 0.90\ N_0$, is given by

$$t = \frac{(2.41 \times 10^9\ \text{y})(\ln 0.90)}{0.693} = 3.7 \times 10^8\ \text{years}$$

When 99% decays, $N = 0.010\ N_0$

$$t = -\frac{(2.41 \times 10^9\ \text{y})(\ln 0.010)}{0.693} = 1.6 \times 10^{10}\ \text{years}$$

It takes a very long time for the radioactive material to decay to a safe level.

17-60. We have (Equation 17.23)

$$\ln \frac{N}{N_0} = -\frac{0.693\,t}{t^{1/2}} = -\frac{(0.693)(6\ \text{h})\left(\frac{60\ \text{min}}{1\ \text{h}}\right)}{25.00\ \text{min}} = -9.98$$

$$\frac{N}{N_0} = e^{-9.98} = 4.64 \times 10^{-5}$$

The rate of decay is proportional to the number of nuclei present; thus,

$$\frac{N}{N_0} = \frac{\text{rate of decay}}{(\text{rate of decay})_0} = \frac{\text{activity}}{(\text{activity})_0}$$

and

$$\begin{aligned}\text{activity at 2 P.M.} &= (4.64 \times 10^{-5})(10\,000 \text{ disintegrations}\cdot\text{min}^{-1})\\ &= 0.464 \text{ disintegrations}\cdot\text{min}^{-1}\end{aligned}$$

or about one disintegration every two minutes.

17-62. Assuming that the activity due to phosphorus-32 remains constant during the time of the experiment we write

$$(\text{activity})_{\text{start}} = (\text{activity})_{\text{later}}$$

or

$$50\,000 \text{ distintegration}\cdot\text{min}^{-1} = (10.0 \text{ distintegration}\cdot\text{min}^{-1}\cdot\text{mL}^{-1})(\text{volume of blood})$$

Solving for the volume of blood, we find that

$$\begin{aligned}\text{volume of blood} &= \frac{50\,000 \text{ distintegration}\cdot\text{min}^{-1}}{10.0 \text{ distintegration}\cdot\text{min}^{-1}\cdot\text{mL}^{-1}}\\ &= 5000 \text{ mL} = 5.00 \text{ L}\end{aligned}$$

17-64. When $[BrO_3^-]_0$ is increased by a factor of 1.8 with $[I^-]_0$ and $[H^+]_0$ constant, the initial rate increases by a factor of $5.40 \times 10^{-4}/3.00 \times 10^{-4} = 1.8$. Thus, the rate is first order in $[BrO_3^-]$. When $[I^-]_0$ decreases by a factor of 1.4 with $[BrO_3^-]_0$ and $[H^+]_0$ constant, the initial rate decreases by a factor of $7.56 \times 10^{-4}/5.40 \times 10^{-4} = 1.4$. Thus, the rate is first order in $[I^-]$. When $[I^-]_0$ increases by a factor of $0.31/0.14 = 2.2$ and $[H^+]_0$ is doubled with $[BrO_3^-]_0$ constant, the initial rate increases by a factor of $1.67 \times 10^{-3}/7.56 \times 10^{-4} = 2.2$. Because we know that the rate law is first order in $[I^-]_0$, we conclude that the rate law does not depend upon $[H^+]_0$, or is zero-order in $[H^+]$. Thus, the rate law is

$$\text{rate} = k[BrO_3^-][I^-]$$

The value of k calculated from any run is $0.030 \text{ M}^{-1}\cdot\text{s}^{-1}$.

17-66. The production of bacteria is a first-order reaction with a certain doubling time. Thus, the number of bacteria present after n doubling times is given by

$$\text{number of bacteria} = (\text{number of bacteria})_0(2)^n$$

The number of doubling times in 10 days is

$$n = \frac{(10 \text{ d})\left(\frac{24 \text{ h}}{1 \text{ d}}\right)}{39 \text{ h}} = 6.15$$

Thus, if there are 20 000 bacteria per milliliter present initially, after 10 days at 40°F, there are

$$\text{number of bacteria} = (20\,000)(2)^{6.15} = 1.4 \times 10^6 \text{ per milliliter}$$

17-68. The time for 99.99% of the reaction to take place is given by

$$\ln \frac{[\mathrm{A}]}{[\mathrm{A}]_0} = -kt_1$$

where

$$\frac{[\mathrm{A}]}{[\mathrm{A}]_0} = 1 - 0.9999 = 0.0001$$

Thus, we have $\ln(0.0001) = -kt_1$. Solving for kt_1, we have that $kt_1 = 9.21$.
The time for 99.0% of the reaction to take place is given by

$$\ln \frac{[\mathrm{A}]}{[\mathrm{A}]_0} = -kt_2$$

where

$$\frac{[\mathrm{A}]}{[\mathrm{A}]_0} = 1 - 0.99 = 0.010$$

Thus, we have $\ln(0.010) = -kt_2$. Solving for kt_2, we have that $kt_2 = 4.61$. The ratio of t_1 to t_2 is given by

$$\frac{kt_1}{kt_2} = \frac{t_1}{t_2} = \frac{9.21}{4.61} = 2.00$$

Thus, it takes twice as long for the reaction to be 99.99% reacted as 99%.

17-70. We use Equation 17.29

$$\frac{1}{[\mathrm{NO_2}]} = \frac{1}{[\mathrm{NO_2}]_0} + kt$$

Substituting the values given yields

$$\frac{1}{0.200\ \mathrm{M}} = \frac{1}{2.00\ \mathrm{M}} + (0.54\ \mathrm{M^{-1}\cdot s^{-1}})\,t$$

Solving for t yields

$$t = 8.3\ \mathrm{s}$$

17-72. We will first calculate the number of moles of $O_2(g)$ generated at each time using

$$n = \frac{PV}{RT}$$

with $P_{O_2} = 730.0$ Torr $-$ 17.4 Torr $= 712.6$ Torr, $V = V_{O_2}$, and $T = 293.2$ K, where 17.4 Torr is due to the vapor pressure of H_2O at 20°C. Next we calculate the number of moles of $H_2O_2(aq)$ remaining using

$$\begin{pmatrix}\text{mol } \mathrm{H_2O_2}\\ \text{remaining}\end{pmatrix} = \begin{pmatrix}\text{mol } \mathrm{H_2O_2}\\ \text{initially}\end{pmatrix} - \begin{pmatrix}\text{mol } \mathrm{O_2}\\ \text{generated}\end{pmatrix}\left(\frac{2\text{ mol } \mathrm{H_2O_2}}{1\text{ mol } \mathrm{O_2}}\right)$$

with

$$\text{mol } \mathrm{H_2O_2} \text{ initially} = (0.250\ \mathrm{M})(0.0500\ \mathrm{L}) = 12.5 \times 10^{-3}\ \text{mol}$$

We then divide by the volume of the $H_2O_2(aq)$ solution (0.0500 L) to calculate $[H_2O_2]$ at each time. The results are given in the following table:

t/min	V_{O_2}/mL	$O_2/\times 10^{-3}$ mol	$H_2O_2/\times 10^{-3}$ mol	$[H_2O_2]$/M
0	0	0	12.5	0.250
10	16.0	0.624	11.3	0.226
20	29.5	1.15	10.2	0.204
35	47.8	1.86	8.8	0.18
50	63.9	2.49	7.5	0.15
65	77.4	3.02	6.5	0.13

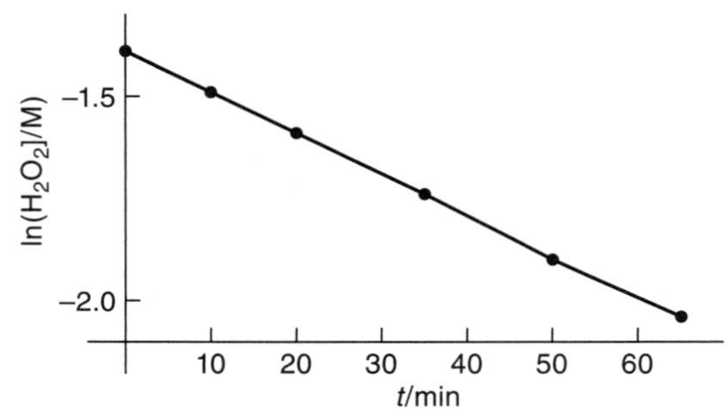

A plot of $\ln[H_2O_2]$ against t, shown above, shows that the reaction is first order in $[H_2O_2]$. The slope of the line is equal to the rate constant, and so

$$k = 0.0102\ \text{min}^{-1}$$

17-74. We do this problem in the same way as the initial rate problems in Section 17-2. The rate law has the form

$$(\text{rate of reaction})_0 = kP_{H_2}^x P_{NO}^y$$

When P_{NO} is increased by a factor of $300/159 = 1.89$ with P_{H_2} constant, the initial rate increases by a factor of $125/34 = 3.68 = (1.89)^2$. Therefore, $y = 2$ and the reaction is second order in $NO(g)$.

When P_{H_2} is decreased by a factor of $205/289 = 0.71$ with P_{NO} constant, the initial rate decreases by a factor of $110/160 = 0.69 \approx 0.71$. Therefore, $x = 1$ and the reaction is first order in $H_2(g)$.

The rate law for the reaction is

$$(\text{rate of reaction})_0 = kP_{H_2} P_{NO}^2$$

We can substitute each set of data given in the table into the rate law to calculate the average rate constant. Below is a table of the rate constant found for each initial rate in the table.

initial rate/Torr·s^{-1}	34	125	160	110	79
k/Torr^{-2}·s^{-1}	3.36×10^{-6}	3.47×10^{-6}	3.46×10^{-6}	3.35×10^{-6}	3.36×10^{-6}

The average value of k is 3.40×10^{-6} Torr^{-2}·s^{-1}.

17-76. We use Equation 17.23

$$\ln \frac{N}{N_0} = -\frac{0.693t}{t_{1/2}}$$

where $t_{1/2}$ is equal to 1.248×10^9 years, N_0 is the original number of potassium-40 nuclei, and N is the present number of potassium-40 nuclei. The original number of potassium-40 nuclei is equal to the present number of potassium-40 nuclei plus the number of potassium-40 nuclei that decayed. We are given that 10.7% of the potassium-40 nuclei that decay produce argon-40 nuclei. Thus, the number of potassium-40 nuclei that decayed is given by

$$\text{number of potassium-40 that decayed} = \frac{\text{present number of argon-40}}{0.107}$$

The original number of potassium-40 nuclei is given by

$$\begin{aligned} N_0 &= \text{present number of potassium-40} + \frac{\text{present number of argon-40}}{0.107} \\ &= N + \frac{\text{present number of argon-40}}{0.107} \end{aligned}$$

However, we are given that

$$\frac{\text{present number of argon-40}}{N} = 0.0102$$

Thus

$$\text{present number of argon-40} = (0.0102)\,N$$

and so

$$N_0 = N + \frac{(0.0102)\,N}{0.107}$$

Therefore,

$$\frac{N}{N_0} = \frac{N}{N + \dfrac{(0.0102)\,N}{0.107}} = \frac{N}{(1.095)\,N} = 0.913$$

Thus,

$$\ln 0.913 = -0.0910 = -\frac{0.693\,t}{1.248 \times 10^9 \text{ years}}$$

Solving for t, we have that

$$t = 1.7 \times 10^8 \text{ years}$$

The sedimentary rocks are 1.7×10^8 or 170 million years old.

17-78.* The rate law is given as

$$\text{rate of reaction} = -\frac{d[\text{A}]}{dt} = k[\text{A}]$$

Rearranging, we have

$$\frac{d[\mathrm{A}]}{[\mathrm{A}]} = -kdt$$

If we integrate this equation from $[\mathrm{A}]_0$ to [A] and from 0 to t, respectively,

$$\int_{[\mathrm{A}]_0}^{[\mathrm{A}]} \frac{d[\mathrm{A}]}{[\mathrm{A}]} = -k\int_0^t dt$$

then we obtain

$$\ln[\mathrm{A}] - \ln[\mathrm{A}]_0 = -kt$$

17-80.* The rate equation for a zero-order reaction is

$$-\frac{d[\mathrm{A}]}{dt} = k$$

or

$$d[\mathrm{A}] = -kdt$$

If we integrate this equation from $[\mathrm{A}]_0$ to [A] and from 0 to t, respectively,

$$\int_{[\mathrm{A}]_0}^{[\mathrm{A}]^2} d[\mathrm{A}] = -k\int_0^k dt$$

then we obtain

$$[\mathrm{A}] - [\mathrm{A}]_0 = -kt$$

A plot of [A] versus time is linear if the reaction is zero order.
Substitute $[\mathrm{A}] = [\mathrm{A}]_0/2$ to obtain the half life

$$t_{1/2} = \frac{[\mathrm{A}]_0}{2k}$$

CHAPTER 18. Chemical Kinetics: Mechanisms

ARRHENIUS EQUATION

18-2. The Arrhenius equation is

$$\ln \frac{k_2}{k_1} = \frac{E_a}{R}\left(\frac{T_2 - T_1}{T_1 T_2}\right)$$

Thus, substituting in the values of k_1, k_2, T_1, T_2, and R, we have

$$\ln \frac{0.750\ \mathrm{M^{-1}\cdot s^{-1}}}{0.0234\ \mathrm{M^{-1}\cdot s^{-1}}} = \frac{E_a}{8.3145\ \mathrm{J\cdot mol^{-1}\cdot K^{-1}}}\left[\frac{773\ \mathrm{K} - 673\ \mathrm{K}}{(673\ \mathrm{K})(773\ \mathrm{K})}\right]$$

$$3.467 = E_a(2.312 \times 10^{-5}\ \mathrm{J^{-1}\cdot mol})$$

Solving for E_a, we have

$$E_a = \frac{3.467}{2.312 \times 10^{-5}\ \mathrm{J^{-1}\cdot mol}} = 1.50 \times 10^5\ \mathrm{J\cdot mol^{-1}} = 1.50 \times 10^2\ \mathrm{kJ\cdot mol^{-1}}$$

18-4. The Arrhenius equation is

$$\ln \frac{k_2}{k_1} = \frac{E_a}{R}\left(\frac{T_2 - T_1}{T_1 T_2}\right)$$

Thus, substituting in the values of k_1, T_1, T_2, E_a, and R, we have

$$\ln \frac{k_2}{5.0 \times 10^{-4}\ \mathrm{s^{-1}}} = \left(\frac{102 \times 10^3\ \mathrm{J\cdot mol^{-1}}}{8.3145\ \mathrm{J\cdot mol^{-1}\cdot K^{-1}}}\right)\left[\frac{338.2\ \mathrm{K} - 318.2\ \mathrm{K}}{(318.2\ \mathrm{K})(338.2\ \mathrm{K})}\right]$$
$$= 2.28$$

Taking the antilogarithm of both sides, we have

$$\frac{k_2}{5.0 \times 10^{-4}\ \mathrm{s^{-1}}} = e^{2.28} = 9.8$$

or solving for k_2,

$$k_2 = (9.8)(5.0 \times 10^{-4}\ \mathrm{s^{-1}}) = 4.9 \times 10^{-3}\ \mathrm{s^{-1}}$$

18-6. The Arrhenius equation is

$$\ln \frac{k_2}{k_1} = \frac{E_a}{R}\left(\frac{T_2 - T_1}{T_1 T_2}\right)$$

We shall consider the pulse rate to be the rate constant k, thus, substituting in the values of k_1, T_1, T_2, E_a, and R, we have

$$\ln \frac{k_2}{75\text{ beats}\cdot\text{min}^{-1}} = \left(\frac{30 \times 10^3\text{ J}\cdot\text{mol}^{-1}}{8.314\text{ J}\cdot\text{mol}^{-1}\cdot\text{K}^{-1}}\right)\left[\frac{295\text{ K} - 310\text{ K}}{(310\text{ K})(295\text{ K})}\right]$$
$$= -0.59$$

Taking the antilogarithm of both sides, we have

$$\frac{k_2}{75\text{ beats}\cdot\text{min}^{-1}} = e^{-0.59} = 0.55$$

or solving for k_2,

$$k_2 = (0.55)(75\text{ beats}\cdot\text{min}^{-1}) = 41\text{ beats}\cdot\text{min}^{-1}$$

REACTION MECHANISMS

18-8. (a) rate of reaction $= k[\text{K}][\text{HCl}]$

(b) rate of reaction $= k[\text{H}_2\text{O}_2]$

(c) rate of reaction $= k[\text{O}_2]^2[\text{Cl}]$

(d) rate of reaction $= k[\text{NO}_3][\text{CO}]$

18-10. (a) Second order; first order in [K], first order in [HCl]; bimolecular reaction

(b) First order; first order in $[\text{H}_2\text{O}_2]$; unimolecular reaction

(c) Third order; second order in $[\text{O}_2]$, first order in [Cl]; termolecular reaction

(d) Second order; first order in $[\text{NO}_3]$, first order in [CO]; bimolecular reaction

18-12. The rate law is determined by the slow elementary step. The rate law for the slow step is

$$\text{rate of reaction} = k[\text{NO}_2]^2$$

The experimental rate law is

$$\text{rate of reaction} = k[\text{NO}_2]^2$$

Thus, the mechanism is consistent with the rate equation because the mechanism leads to the same rate law as found experimentally.

18-14. Step 1 is a bimolecular reaction and step 2 is a unimolecular reaction. The reaction intermediate is $\text{NO}_3(g)$.

18-16. The second step is the rate-determining step, and so

$$\text{rate of reaction} = k_2[\text{NO}_2\text{Br}_2]$$

We can eliminate the intermediate $[NO_2Br_2]$ by using step 1 to write

$$k_1[NO_2][Br_2] = k_{-1}[NO_2Br_2]$$

and so the rate law

$$\text{rate of reaction} = \frac{k_2 k_1}{k_{-1}}[NO_2][Br_2] = k[NO_2][Br_2]$$

This mechanism is consistent with the experimentally determined rate law.

18-18. (a) For mechanism a, the rate law is

$$\text{rate of reaction} = k_1[H_2][I_2] = k[H_2][I_2]$$

which is consistent with the experimentally determined rate law.
For mechanism b, the rate law is given by the rate-determining step

$$\text{rate of reaction} = k_2[H_2][I]^2$$

We can use step 1 to eliminate the intermediate [I] from the rate law

$$k_1[I_2] = k_{-1}[I]^2$$

and write the rate law as

$$\text{rate of reaction} = \frac{k_1 k_2}{k_{-1}}[H_2][I_2] = k[H_2][I_2]$$

which is also consistent with the experimentally determined rate law.
For mechanism c, the rate law is given by step 3, the rate-determining step

$$\text{rate of reaction} = k_3[H_2I]$$

We can eliminate the intermediate $[H_2I]$ from this expression by using step 2

$$k_2[H_2][I] = k_{-2}[H_2I]$$

to write

$$\text{rate of reaction} = \frac{k_2 k_3}{k_{-2}}[H_2][I]$$

Now eliminate the intermediate [I] by using step 1

$$k_1[I_2] = k_{-1}[I]^2$$

and write

$$\text{rate of reaction} = \frac{k_2 k_3}{k_{-2}}\left(\frac{k_1}{k_{-1}}\right)^{1/2}[H_2][I_2]^{1/2} = k[H_2][I_2]^{1/2}$$

Only the first two mechanisms are consistent with the observed rate law.

(b) The second and third mechanisms involve an $I(g)$ intermediate, and so both are consistent with the additional observation. But only the second mechanism is consistent with the rate law, and so we can consider mechanism b as consistent with the experimental observations.

18-20. From step 1 of the reaction mechanism, the fast equilibrium allows us to write

$$k_1[O_2NNH_2] = k_{-1}[O_2NNH^-][H^+]$$

and so the rate equation for the second step (rate determining) of the reaction mechanism becomes

$$\text{rate of reaction} = k_2[O_2NNH^-] = k_2\left(\frac{k_1}{k_{-1}}\right)\frac{[O_2NNH_2]}{[H^+]} = k\frac{[O_2NNH_2]}{[H^+]}$$

This is consistent with the observed rate law with

$$k = \frac{k_2k_1}{k_{-1}}$$

18-22. The observed rate law is (from Problem 18-19)

$$\text{rate of reaction} = k[Cl_2]^{3/2}[CO]$$

From step 3 of the reaction mechanism, we write

$$\text{rate of reaction} = k_3[Cl_3][CO] \qquad (1)$$

Because steps 2 and 1 establish a fast equilibrium, we write

$$k_2[Cl][Cl_2] = k_{-2}[Cl_3] \qquad (2)$$

Also,

$$k_1[Cl_2] = k_{-1}[Cl]^2$$

and so

$$[Cl] = \left(\frac{k_1}{k_{-1}}\right)^{1/2}[Cl_2]^{1/2}$$

Substituting this expression for [Cl] into equation 2 gives

$$[Cl_3] = \frac{k_2}{k_{-2}}\left(\frac{k_1}{k_{-1}}\right)^{1/2}[Cl_2]^{3/2}$$

and substituting this expression into equation 1 gives

$$\text{rate of reaction} = \frac{k_3k_2k_1^{1/2}}{k_{-2}k_{-1}^{1/2}}[Cl_2]^{3/2}[CO] = k[Cl_2]^{3/2}[CO]$$

which corresponds to the observed rate law. To determine whether this mechanism or the one given in Problem 18-19 occurs, we might determine whether a Cl_3 intermediate is produced during the reaction.

CATALYSIS

18-24. When $[H_2O_2]_0$ is doubled with $[I^-]_0$ and $[H^+]_0$ constant, the initial rate doubles. Therefore, the reaction is first order in $[H_2O_2]$. When $[I^-]_0$ is doubled with $[H^+]_0$ and $[H_2O_2]_0$ constant, the initial rate quadruples. The reaction is second order in $[I^-]$. When $[H^+]_0$ is doubled with

$[I^-]$ and $[H_2O_2]_0$ constant, the initial rate doubles. Thus, the reaction is first order in $[H^+]_0$. The rate law is

$$\text{rate of reaction} = k[H_2O_2][I^-]^2[H^+]$$

The catalysts are $H^+(aq)$ and $I^-(aq)$ because they do not appear in the equation for the overall reaction as reactants.

18-26. The observed rate law is consistent with the mechanism

$$NH_3(g) \longrightarrow NH_3(\text{surface}) \qquad (\text{fast})$$
$$2\,NH_3(\text{surface}) \longrightarrow N_2(g) + 3\,H_2(g) \qquad (\text{slow})$$

If the number of surface sites for adsorption of NH_3 is small compared to the number of $NH_3(g)$ molecules, then all surface sites will be occupied for a wide range of P_{NH_3} values and the reaction rate will be independent of P_{NH_3}.

ENZYME KINETICS

18-28. The following table shows the data as reciprocals

$[S]^{-1}/10^{-3}$ L·μmol^{-1}	333	200	100	33.3	11.1
$R^{-1}/10^{-3}$ L·min·μmol^{-1}	96.2	69.0	44.4	29.6	24.7

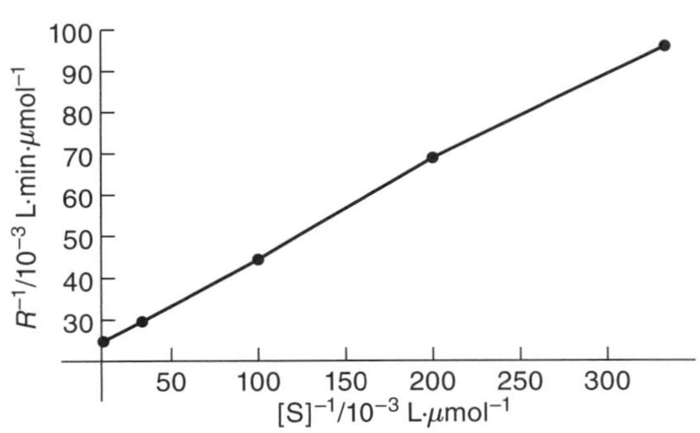

The above figure shows R^{-1} plotted against $[S]^{-1}$. Using the third and fourth data points arbitrarily, the slope comes out to be

$$\text{slope} = \frac{29.6\times10^{-3}\ \text{L·min·}\mu\text{mol}^{-1} - 44.4\times10^{-3}\ \text{L·min·}\mu\text{mol}^{-1}}{33.3\times10^{-3}\ \text{L·}\mu\text{mol}^{-1} - 100\times10^{-3}\ \text{L·}\mu\text{mol}^{-1}}$$
$$= 0.222\ \text{min}$$

We estimate the intercept visually as 20, or 0.020 L·min·μmol^{-1}. Therefore,

$$R_{\max} = \frac{1}{0.020\ \text{L·min·}\mu\text{mol}^{-1}} = 50\ \mu\text{mol·L}^{-1}\text{·min}^{-1}$$

and

$$K_M = (0.222\ \text{min})\,R_{\max} = 11\ \mu\text{mol·L}^{-1}$$

You should realize that these results depend upon the fact that we arbitrarily chose the third and fourth data points to calculate the slope. Furthermore, we estimated the intercept visually. There is a standard numerical procedure called the method of least-squares (or linear regression) that uses all the data points and provides an optimum estimate of the slope and the intercept. This method is provided in many computer packages such as Excel. Using this method, we obtain $R_{max} = 44.6\ \mu\text{mol·L}^{-1}\text{·min}^{-1}$ and $K_M = 10.0\ \mu\text{mol·L}^{-1}$.

18-30. The following table shows the data as reciprocals

$[S]^{-1}/\text{L·mol}^{-1}$	28.0	17.0	12.0	6.49	4.50
$R^{-1}/\text{L·min·mmol}^{-1}$	0.142	0.112	0.0952	0.0813	0.0769

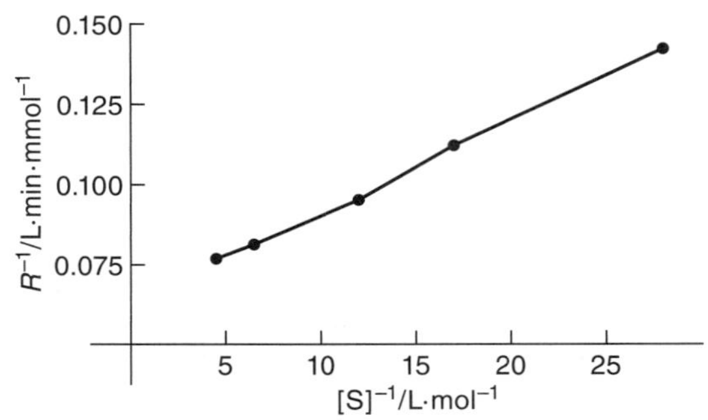

The above figure shows R^{-1} plotted against $[S]^{-1}$. Using the third and fourth data points arbitrarily, the slope comes out to be

$$\text{slope} = \frac{0.0813\ \text{L·min·mmol}^{-1} - 0.0952\ \text{L·min·mmol}^{-1}}{6.49\ \text{L·mol}^{-1} - 12.0\ \text{L·mol}^{-1}}$$
$$= 2.52 \times 10^{-3}\ \text{min·mol·mmol}^{-1} = 2.52\ \text{min}$$

We estimate the intercept visually as 0.065, or 0.065 L·min·mmol^{-1}. Therefore,

$$R_{max} = \frac{1}{0.065\ \text{L·min·mmol}^{-1}} = 15\ \text{mmol·L}^{-1}\text{·min}^{-1}$$

and

$$K_M = (2.52\ \text{min})(15\ \text{mmol·L}^{-1}\text{·min}^{-1}) = 38\ \text{mmol·L}^{-1}$$

You should realize that these results depend upon the fact that we arbiatrarily chose the third and fourth data points to calculate the slope. Furthermore, we estimated the intercept visually. There is a standard numerical procedure called the method of least-squares that uses all the data points and provides an optimum estimate of the slope and the intercept. This method is provided in many computer packages such as Excel. Using this method, we obtain $R_{max} = 15.8\ \text{mmol·L}^{-1}\text{·min}^{-1}$ and $K_M = 44.4\ \text{mmol·L}^{-1}$.

18-32. The following table shows the data as reciprocals

$[H_2O_2]^{-1}/L{\cdot}mmol^{-1}$	1.0	0.50	0.20
$R^{-1}/L{\cdot}s{\cdot}mmol^{-1}$	0.725	0.375	0.167

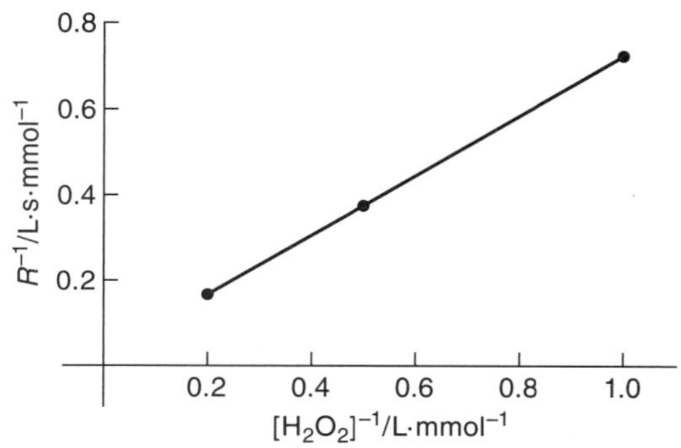

The above figure shows R^{-1} plotted against $[H_2O_2]^{-1}$. Using the second and third data points arbitrarily, the slope comes out to be

$$\begin{aligned}\text{slope} &= \frac{0.167\ \text{L}\cdot\text{s}\cdot\text{mmol}^{-1} - 0.375\ \text{L}\cdot\text{s}\cdot\text{mmol}^{-1}}{0.20\ \text{L}\cdot\text{mmol}^{-1} - 0.50\ \text{L}\cdot\text{mmol}^{-1}}\\ &= 0.69\ \text{s}\end{aligned}$$

We estimate the intercept visually as 0.025, or 0.025 $\text{L}\cdot\text{s}\cdot\text{mmol}^{-1}$. Therefore,

$$R_{\max} = \frac{1}{0.025\ \text{L}\cdot\text{s}\cdot\text{mmol}^{-1}} = 40\ \text{mmol}\cdot\text{L}^{-1}\cdot\text{s}^{-1}$$

and

$$K_M = (0.69\ \text{s})(40\ \text{mmol}\cdot\text{L}^{-1}\cdot\text{s}^{-1}) = 28\ \text{mmol}\cdot\text{L}^{-1}\cdot\text{s}^{-1}$$

You should realize that these results depend upon the fact that we arbiatrarily chose the third and fourth data points to calculate the slope. Furthermore, we estimated the intercept visually. There is a standard numerical procedure called the method of least-squares that uses all the data points and provides an optimum estimate of the slope and the intercept. This method is provided in many computer packages such as Excel. Using this method, we obtain $R_{\max} = 37.7\ \text{mmol}\cdot\text{L}^{-1}\cdot\text{s}^{-1}$ and $K_M = 26.3\ \text{mmol}\cdot\text{L}^{-1}$.

18-34. Substituting in the values for $R_{\max}$ and $[E]_t$ into the equation for the turnover number yields

$$k_2 = \frac{R_{\max}}{[E]_t} = \frac{15\ \text{mmol}\cdot\text{L}^{-1}\cdot\text{min}^{-1}}{1.1\ \mu\text{mol}\cdot\text{L}^{-1}} = 1.4\times10^{4}\ \text{min}^{-1}$$

18-36. Substituting in the values for $R_{\max}$ and $[E]_t$ into the equation for the turnover number yields

$$k_2 = \frac{R_{\max}}{[E]_t} = \frac{40\ \text{mmol}\cdot\text{L}^{-1}\cdot\text{s}^{-1}}{4.0\ \text{nmol}\cdot\text{L}^{-1}} = 1.0\times10^{7}\ \text{s}^{-1}$$

ADDITIONAL PROBLEMS

18-38. No. Often the molecules must collide in some preferred relative orientation for a reaction to occur.

18-40. An intermediate is usually a short-lived species that is generated in one step and then consumed in a later step of a reaction mechanism. It may have a lifetime millions of times longer than an activated complex and can often be detected experimentally, or occasionally even isolated. An activated complex is an intrinsically unstable species that sits at the maximum in the activation energy diagram (Figure 18.5). It cannot be isolated or detected by ordinary experimental means, but can be inferred from fast spectroscopic data. Neither appear in the overall equation for a reaction; however, intermediates are necessarily included in the equations that describe a reaction mechanism, whereas activated complexes generally are not.

18-42. The student made two mistakes: a catalyst increases the rate of a reaction, but it does not affect the yield of the reaction; and a catalyst is not consumed during the reaction, but it does participate in the reaction by changing its mechanism. A better sentence might read "a catalyst increases the rate of a reaction by lowering the activation energy, but is not consumed in the reaction."

18-44. A zero-order reaction is a reaction that has a *constant rate*. The rate law of a zero-order reaction is given by rate of reaction $= k[A]^0 = k$. The rate of reaction depends on only the rate constant and not upon the concentration of the reactants or products of the reaction.

18-46. Conduct the study at two or more different temperatures.

18-48. The formation of a covalent bond from two radicals does not involve any bond-breaking process, only the movement of electrons. Electrons move very rapidly. For two $CH_3\cdot$ radicals

$$2\,CH_3\cdot \longrightarrow H_3C\text{–}CH_3$$

Thus, $E_a \approx 0$. The only limitation to the reaction would be the collision frequency.

18-50. Filling in the table, we have

Change	Effect on the value of k	Effect on the rate of reaction
Increasing [NO]:	none	increase
Adding a catalyst:	increase	increase
Decreasing [O_2]:	none	decrease
Decreasing T:	decrease	decrease

18-52. Start with $P_A = P_A^0 - kt$, and then write

$$t = \frac{275\text{ Torr} - \frac{1}{2}(275\text{ Torr})}{7.4 \times 10^{-1}\text{ Torr}\cdot\text{s}^{-1}} = 186\text{ s} = 3.1\text{ min}$$

18-54. We first must calculate $\ln k$ and $1/T$ as given in the following table:

$\ln(k/\text{M}^{-1}\cdot\text{s}^{-1})$	$(1/T)/10^{-3}\text{ K}^{-1}$
−0.357	1.67
0.604	1.60
1.495	1.54
3.082	1.43

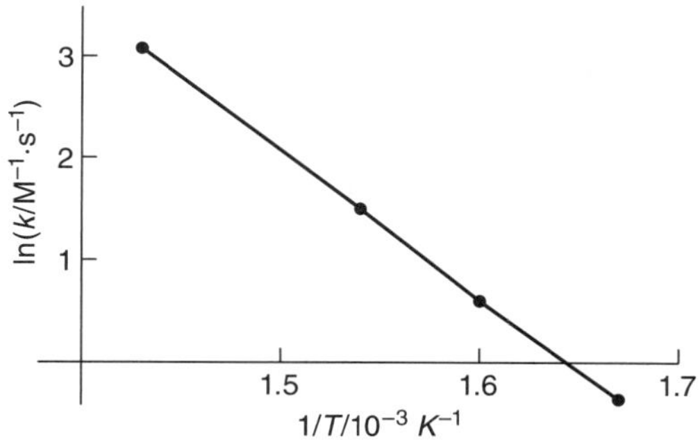

A plot of $\ln(k/\mathrm{M}^{-1}\cdot\mathrm{s}^{-1})$ versus $1/T/10^{-3}\ \mathrm{K}^{-1}$ is a straight line as shown in the above figure. To estimate k at 500 K, we read the value of $\ln(k/\mathrm{M}^{-1}\cdot\mathrm{s}^{-1})$ at $1/(500\ \mathrm{K}) = 2.00 \times 10^{-3}\ \mathrm{K}^{-1}$ from the plot. We see that

$$\ln(k/\mathrm{M}^{-1}\cdot\mathrm{s}^{-1}) = -5.30$$

Taking antilogarithms of both sides and solving for k yields

$$k = 5.0 \times 10^{-3}\ \mathrm{M}^{-1}\cdot\mathrm{s}^{-1}$$

From Equation 18.8, we have

$$\ln k = -\frac{E_\mathrm{a}}{RT} + \ln A$$

and so the slope of the line in our problem is equal to $-E_\mathrm{a}/R$. Because the plot is a straight line, we can choose any two points to find the slope, say the first and the third, as

$$\text{slope} = -\frac{E_\mathrm{a}}{R} = \frac{-0.3567 - 1.495}{(1.67 - 1.54) \times 10^{-3}\ \mathrm{K}^{-1}} = -1.4 \times 10^{4}\ \mathrm{K}$$

From which we find

$$\begin{aligned} E_\mathrm{a} &= (1.4 \times 10^{4}\ \mathrm{K})(8.3145\ \mathrm{J}\cdot\mathrm{mol}^{-1}\cdot\mathrm{K}^{-1}) \\ &= 1.2 \times 10^{5}\ \mathrm{J}\cdot\mathrm{mol}^{-1} = 120\ \mathrm{kJ}\cdot\mathrm{mol}^{-1} \end{aligned}$$

We could also have calculated E_a by using the Arrhenius equation and any pair of data.

$$\ln\frac{k_2}{k_1} = \frac{E_\mathrm{a}}{R}\left(\frac{T_2 - T_1}{T_1 T_2}\right)$$

$$\ln\frac{4.46\ \mathrm{M}^{-1}\cdot\mathrm{s}^{-1}}{1.83\ \mathrm{M}^{-1}\cdot\mathrm{s}^{-1}} = \frac{E_\mathrm{a}(650\ \mathrm{K} - 625\ \mathrm{K})}{(8.3145\ \mathrm{J}\cdot\mathrm{mol}^{-1}\cdot\mathrm{K}^{-1})(650\ \mathrm{K})(625\ \mathrm{K})}$$

Solving for E_a, we have

$$E_\mathrm{a} = 1.2 \times 10^{5}\ \mathrm{J}\cdot\mathrm{mol}^{-1} = 120\ \mathrm{kJ}\cdot\mathrm{mol}^{-1}$$

18-56. The units indicate that the reaction is second order. The activation energy is $46.6\ \mathrm{kJ}\cdot\mathrm{mol}^{-1}$. The value of the rate constant at 250 K is

$$\begin{aligned} k &= (3.3 \times 10^{-7}\ \mathrm{cm}^3\cdot\mathrm{molecule}^{-1}\cdot\mathrm{s}^{-1})e^{-46.6\times10^{3}\ \mathrm{J}\cdot\mathrm{mol}^{-1}/(8.3145\ \mathrm{J}\cdot\mathrm{mol}^{-1}\cdot\mathrm{K}^{-1})(250\ \mathrm{K})} \\ &= (3.3 \times 10^{-7}\ \mathrm{cm}^3\cdot\mathrm{molecule}^{-1}\cdot\mathrm{s}^{-1})e^{-22.4} \\ &= 6.16 \times 10^{-17}\ \mathrm{cm}^3\cdot\mathrm{molecule}^{-1}\cdot\mathrm{s}^{-1} = 3.6 \times 10^{7}\ \mathrm{cm}^3\cdot\mathrm{mol}^{-1}\cdot\mathrm{s}^{-1} = 3.6 \times 10^{4}\ \mathrm{L}\cdot\mathrm{mol}^{-1}\cdot\mathrm{s}^{-1} \end{aligned}$$

Because k is given in terms of concentration, we cannot use pressure directly. A pressure at 50.0 mbar at 250 K corresponds to a density of

$$\rho = \frac{P}{RT} = \frac{(50.0 \times 10^{-3}\ \text{bar})(10^5\ \text{Pa}\cdot\text{bar}^{-1})}{(8.3145\ \text{J}\cdot\text{K}^{-1}\cdot\text{mol}^{-1})(250\ \text{K})}$$
$$= 2.41\ \text{mol}\cdot\text{m}^{-3} = 0.00241\ \text{mol}\cdot\text{L}^{-1}$$

The concentration of $N_2O_4(g)$ remaining after 10.0 milliseconds is given by

$$\frac{1}{[N_2O_4]} = \frac{1}{[N_2O_4]_0} + kt = \frac{1}{0.00241\ \text{mol}\cdot\text{L}^{-1}} + (3.6 \times 10^4\ \text{L}\cdot\text{mol}^{-1}\cdot\text{s}^{-1})(10.0 \times 10^{-3}\ \text{s})$$

or $[N_2O_4] = 0.0013\ \text{mol}\cdot\text{L}^{-1}$. The corresponding pressure is given by

$$P = [N_2O_4]\,RT = (0.0013\ \text{mol}\cdot\text{L}^{-1})(0.083145\ \text{L}\cdot\text{bar}\cdot\text{K}^{-1}\cdot\text{mol}^{-1})(250\ \text{K})$$
$$= 0.027\ \text{bar} = 27\ \text{mbar}$$

18-58. (a) Step 2 is rate determining, so

$$\text{rate of reaction} = k_2[\text{Br}]\,[\text{H}_2]$$

We can eliminate the concentration of the intermediate Br(g) from this rate expression by using Step 1 to write

$$k_1[\text{Br}_2] = k_{-1}[\text{Br}]^2$$

Solving this for [Br] and substituting into the rate equation above gives

$$\text{rate of reaction} = \frac{k_1 k_2}{k_{-1}}[\text{H}_2]\,[\text{Br}_2]^{1/2} = k[\text{H}_2]\,[\text{Br}_2]^{1/2}$$

in agreement with the observed rate law.

(b) The units of k are given by

$$\text{rate of reaction} = -\frac{d[\text{Br}_2]}{dt} = k[\text{H}_2]\,[\text{Br}_2]^{1/2}$$

or $\text{L}^{1/2}\cdot\text{mol}^{-1/2}\cdot\text{s}^{-1}$.

(c) No. a single-step mechanism would have a rate law $k[\text{H}_2]\,[\text{Br}_2]$.

18-60. The Arrhenius equation is

$$\ln\frac{k_2}{k_1} = \frac{E_a}{R}\left(\frac{T_2 - T_1}{T_1 T_2}\right)$$

We are given that $k_2 = 2k_1$. Thus,

$$\ln\frac{2k_1}{k_1} = \frac{E_a(303\ \text{K} - 293\ \text{K})}{(8.3145\ \text{J}\cdot\text{mol}^{-1}\cdot\text{K}^{-1})(303\ \text{K})(293\ \text{K})}$$
$$0.693 = E_a(1.35 \times 10^{-5}\ \text{J}^{-1}\cdot\text{mol})$$
$$E_a = 51\ 000\ \text{J}\cdot\text{mol}^{-1} = 51\ \text{kJ}\cdot\text{mol}^{-1}$$

18-62. We first make a table of ln k versus $1/T$ for the data given and then plot the data.

$\ln(k/\mathrm{s}^{-1})$	$(1/T)/10^{-3}\ \mathrm{K}^{-1}$
−10.269	3.36
−8.910	3.25
−7.782	3.14
−6.496	3.05
−5.319	2.96

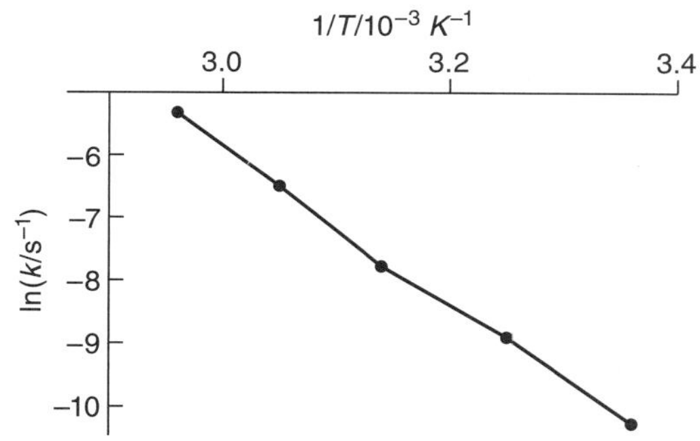

The above figure shows that the plot of ln k versus $1/T$ is a straight line. We can determine the slope of the line using any two data points.

$$\text{slope} = \frac{-6.496 - (-8.910)}{(3.05 - 3.25) \times 10^{-3}\ \mathrm{K}^{-1}} = -12\ 100\ \mathrm{K}$$

The slope of the line is related to the activation energy by (Equation 18.8 and Problem 18-53)

$$\text{slope} = -\frac{E_a}{R}$$

Thus, we have that

$$\begin{aligned} E_a &= -(\text{slope})(R) = -(-12\ 100\ \mathrm{K})(8.3145\ \mathrm{J\cdot mol^{-1}\cdot K^{-1}}) \\ &= 101\ 000\ \mathrm{J\cdot mol^{-1}} = 101\ \mathrm{kJ\cdot mol^{-1}} \end{aligned}$$

You should realize that these results depend upon the fact that we arbitrarily chose the third and fourth data points to calculate the slope. Furthermore, we estimated the intercept visually. There is a standard numerical procedure called the method of least-squares (or linear regression) that uses all the data points and provides an optimum estimate of the slope and the intercept. This method is provided in many computer packages such as *Excel.* Using this method, we obtain slope $= -E_a/R = -1.23 \times 10^4$ K and so $E_a = 102\ \mathrm{kJ\cdot mol^{-1}}$.

18-64.* Start with

$$k_1 P_A (1 - \theta_A) = k_2 \theta_A$$

Solving for θ_A gives

$$\theta_A = \frac{k_1 P_A}{k_1 P_A + k_2}$$

Substituting this result into rate of reaction $= k\theta_A$ gives

$$\text{rate of reaction} = \frac{kk_1 P_A}{k_1 P_A + k_2} = \frac{k(k_1/k_2) P_A}{(k_1/k_2) P_A + 1} = \frac{kb P_A}{b P_A + 1}$$

For small pressures, we can neglect $b P_A$ compared to 1 in the denominator and write

$$\text{rate of reaction} = kb P_A$$

which is first order in P_A. For large pressures, we can neglect 1 compared to $b P_A$ in the denominator and write

$$\text{rate of reaction} = k$$

which is zero order in P_A. The rate is zero order at high pressures because the surface is fully covered and a reactant molecule must wait around until a surface site becomes available.

The reaction has the same mechanism as an enzyme catalyzed reaction, and so the rate equation has the same form as the Michaelis–Menten equation

$$\text{rate of reaction} = \frac{R_{max}[\text{S}]}{K_M + [\text{S}]}$$

18-66. The following table shows the data as reciprocals

$[\text{ATP}]^{-1}/10^{-3}\ \text{L}\cdot\mu\text{mol}^{-1}$	133	80.0	50.0	23.0	16.0
$R^{-1}/10^{-3}\ \text{L}\cdot\text{s}\cdot\text{pmol}^{-1}$	14.9	10.5	8.40	6.45	6.02

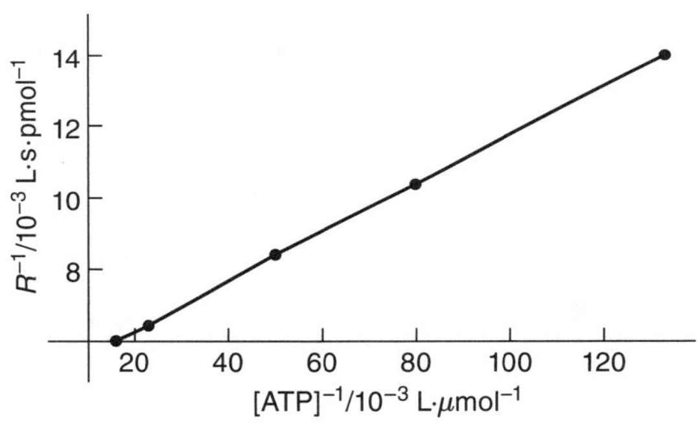

The above figure shows R^{-1} plotted against $[\text{ATP}]^{-1}$. Using the third and fourth data points arbitrarily, the slope comes out to be

$$\text{slope} = \frac{(6.45 - 8.40) \times 10^{-3}\ \text{L}\cdot\text{s}\cdot\text{pmol}^{-1}}{(23.0 - 50.0) \times 10^{-3}\ \text{L}\cdot\mu\text{mol}^{-1}}$$
$$= 0.0722\ \mu\text{mol}\cdot\text{pmol}^{-1}\cdot\text{s} = 7.22 \times 10^4\ \text{s}$$

We estimate the intercept visually as 4.8, or $4.8 \times 10^{-3}\ \text{L}\cdot\text{s}\cdot\text{pmol}^{-1}$. Therefore,

$$R_{max} = \frac{1}{4.8 \times 10^{-3}\ \text{L}\cdot\text{s}\cdot\text{pmol}^{-1}} = 208\ \text{pmol}\cdot\text{L}^{-1}\cdot\text{s}^{-1}$$

and

$$K_M = (7.22 \times 10^4\ \text{s})(208\ \text{pmol}\cdot\text{L}^{-1}\cdot\text{s}^{-1}) = 1.50 \times 10^7\ \text{pmol}\cdot\text{L}^{-1} = 15.0\ \mu\text{mol}\cdot\text{L}^{-1}$$

You should realize that these results depend upon the fact that we arbitrarily chose the third and fourth data points to calculate the slope. Furthermore, we estimated the intercept visually. There is a standard numerical procedure called the method of least-squares that uses all the data points and provides an optimum estimate of the slope and the intercept. This method is provided in many computer packages such as Excel. Using this method, we obtain $R_{max} = 213$ pmol·L^{-1}·s^{-1} and $K_M = 16.2$ μmol·L^{-1}.

18-68. The following table shows the data as reciprocals

$[CO_2]^{-1}/10^{-2}$ L·mmol^{-1}	80.0	40.0	20.0	5.0
$R^{-1}/10^{-2}$ L·s·μmol^{-1}	3.57	2.08	1.25	0.645

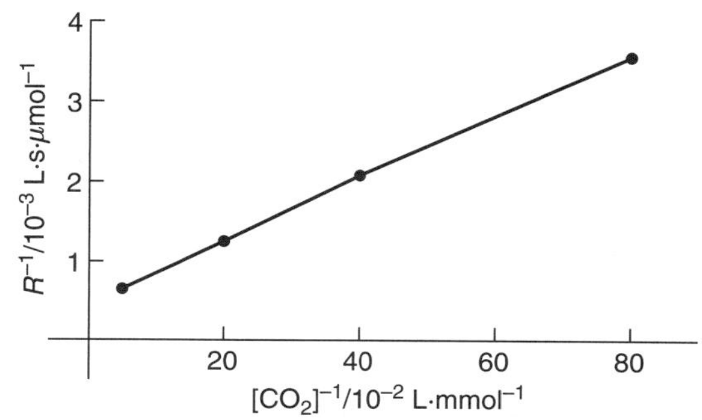

The above figure shows R^{-1} plotted against $[CO_2]^{-1}$. Using the second and third data points arbitrarily, the slope comes out to be

$$\text{slope} = \frac{(1.25 - 2.08) \times 10^{-2}\ \text{L·min·}\mu\text{mol}^{-1}}{(20.0 - 40.0) \times 10^{-2}\ \text{L·mmol}^{-1}}$$
$$= 0.0415\ \text{mmol·}\mu\text{mol}^{-1}\text{·s} = 41.5\ \text{s}$$

We estimate the intercept visually as 0.45, or 0.45×10^{-2} L·s·μmol^{-1}.

$$R_{max} = \frac{1}{0.45 \times 10^{-2}\ \text{L·s·}\mu\text{mol}^{-1}} = 222\ \mu\text{mol·L}^{-1}\text{·s}^{-1}$$

and

$$K_M = (41.5\ \text{s})(222\ \mu\text{mol·L}^{-1}\text{·s}^{-1}) = 9200\ \mu\text{mol·L}^{-1} = 9.20\ \text{mmol·L}^{-1}$$

You should realize that these results depend upon the fact that we arbitrarily chose the third and fourth data points to calculate the slope. Furthermore, we estimated the intercept visually. There is a standard numerical procedure called the method of least-squares that uses all the data points and provides an optimum estimate of the slope and the intercept. This method is provided in many computer packages such as Excel. Using this method, we obtain $R_{max} = 211$ μmol·L^{-1}·s^{-1} and $K_M = 8.23$ mmol·L^{-1}.

18-70. The Michaelis–Menten equation is

$$R = \frac{k_2[S]_t[E]}{K_M[S]} = \frac{R_{max}[S]}{K_M + [S]}$$

The turnover number for a single active site catalyst is equal to $R_{\max}/[\mathrm{E}]_{\mathrm{t}}$, so

$$R_{\max} = (4.0 \times 10^{7}\ \mathrm{s}^{-1})(0.016\ \mu\mathrm{M}) = 0.64\ \mathrm{M{\cdot}s}^{-1}$$

and

$$R = \frac{(0.64\ \mathrm{M{\cdot}s}^{-1})(4.32\ \mu\mathrm{M})}{25\ \mathrm{mM} + 4.32\ \mu\mathrm{M}} = 110\ \mu\mathrm{M{\cdot}s}^{-1}$$

18-72.* The rate of production of $O_2(g)$ is given by step 2 as

$$\text{rate of reaction} = \frac{\Delta[\mathrm{O_2}]}{\Delta t} = k_2[\mathrm{NO_2}][\mathrm{NO_3}]$$

We can eliminate $[NO_3]$, the concentration of the intermediate species, by assuming that $[NO_3]$ is at a steady state. From step 1, we have that the rate of formation of the $NO_3(g)$ intermediate is given by

$$\text{rate of formation of NO}_3 = k_1[\mathrm{N_2O_5}]$$

and from steps 1 and 2 that the rate of consumption of the $NO_3(g)$ intermediate is given by

$$\text{rate of consumption of NO}_3 = k_{-1}[\mathrm{NO_2}][\mathrm{NO_3}] + k_2[\mathrm{NO_2}][\mathrm{NO_3}]$$

In order for $[NO_3]$ to reach a steady state, its rate of formation must equal its rate of consumption, and so

$$k_1[\mathrm{N_2O_5}] = k_{-1}[\mathrm{NO_2}][\mathrm{NO_3}] + k_2[\mathrm{NO_2}][\mathrm{NO_3}]$$

Solving for $[NO_3]$ gives

$$[\mathrm{NO_3}] = \frac{k_1[\mathrm{N_2O_5}]}{(k_2 + k_{-1})[\mathrm{NO_2}]}$$

and substituting this result into the “rate expression” gives

$$\text{rate of reaction} = \frac{\Delta[\mathrm{O_2}]}{\Delta t} = \frac{k_1 k_2[\mathrm{N_2O_5}]}{k_2 + k_{-1}}$$

18-74.* At high concentration, $k = k_1 k_2/k_{-1}$. From Equation 18.7, we have

$$k = Ae^{-E_{\mathrm{a}}/RT}$$

If each step of the reaction mechanism shows Arrhenius behavior, then each of the rate constants, k_1, k_{-1}, and k_2 can be written in terms of an Arrhenius equation,

$$k_1 = A_1 e^{-E_{\mathrm{a},1}/RT}$$
$$k_{-1} = A_{-1} e^{-E_{\mathrm{a},-1}/RT}$$
$$k_2 = A_2 e^{-E_{\mathrm{a},2}/RT}$$

Substituting these equations into $k = k_1 k_2/k_{-1}$ gives

$$k = \frac{(A_1 e^{-E_{\mathrm{a},1}/RT})(A_2 e^{-E_{\mathrm{a},2}/RT})}{A_{-1} e^{-E_{\mathrm{a},-1}/RT}}$$

or

$$k = \frac{A_2 A_1}{A_{-1}} e^{(-E_{a,1}-E_{a,2}+E_{a,,-1})/RT}$$

Thus, we have that

$$E_a = E_{a,1} - E_{a,-1} + E_{a,2}$$

and

$$A = \frac{A_2 A_1}{A_{-1}}$$

We now see that measured values for A and E_a do not correspond to a single step of the reaction but are influenced by each step of the reaction mechanism.

CHAPTER 19. Chemical Equilibrium

CALCULATION OF EQUILIBRIUM CONCENTRATION

19-2. The number of moles per liter of $O_2(g)$ that react are (notice that this reaction is shifting from right to left as written to reach equilibrium) is

$$\left(\begin{array}{c}\text{moles per liter}\\ \text{of } O_2 \text{ consumed}\end{array}\right) = 0.436\ \text{M} - 0.387\ \text{M} = 0.049\ \text{M}$$

From the reaction stoichiometry, we have that the number of moles per liter of $SO_2(g)$ consumed is twice the number of moles per liter of $O_2(g)$ consumed. Thus, at equilibrium

$$\begin{aligned}[SO_2] &= [SO_2]_0 - (2)(0.049\ \text{M})\\ &= 0.625\ \text{M} - 0.098\ \text{M} = 0.527\ \text{M}\end{aligned}$$

Similarly, the number of moles per liter of $SO_3(g)$ produced is twice the number of moles per liter of $O_2(g)$ consumed, and so

$$\begin{aligned}[SO_3] &= [SO_3]_0 + (2)(0.049\ \text{M})\\ &= 0.176\ \text{M} + 0.098\ \text{M} = 0.274\ \text{M}\end{aligned}$$

EQUILIBRIUM-CONSTANT EXPRESSION

19-4. Each product concentration factor, raised to a power equal to its balancing coefficient, appears in the numerator of the K_c expression, and each reactant concentration factor, raised to a power equal to its balancing coefficient, appears in the denominator of the K_c expression. Pure solids and liquids do not appear in the K_c expression.

(a) $K_c = \dfrac{[SO_3]^2}{[SO_2]^2[O_2]}$ units $= \dfrac{\text{M}^2}{\text{M}^3} = \text{M}^{-1}$

(b) $K_c = [CO_2][H_2O]$ units $= \text{M}^2$

(c) $K_c = \dfrac{[CH_4]}{[H_2]^2}$ units $= \dfrac{\text{M}}{\text{M}^2} = \text{M}^{-1}$

19-6. Each product concentration factor, raised to a power equal to its balancing coefficient, appears in the numerator of the K_c expression, and each reactant concentration factor, raised to a power equal to its balancing coefficient, appears in the denominator of the K_c expression. Pure solids and liquids do not appear in the K_c expression.

(a) $K_c = [CO_2]\,[NH_3]^2$ units $= M^3$

(b) $K_c = [O_2]$ units $= M$

(c) $K_c = \dfrac{[N_2O_4]}{[N_2]\,[O_2]^2}$ units $= \dfrac{M}{M^3} = M^{-2}$

19-8. Each product pressure factor, raised to a power equal to its balancing coefficient, appears in the numerator of the K_p expression, and each reactant pressure factor, raised to a power equal to its balancing coefficient, appears in the denominator of the K_p expression. Pure solids and liquids do not appear in the K_p expression.

(a) $K_p = P^2_{NH_3} P_{CO_2}$ units $= \text{bar}^3$

(b) $K_p = P_{O_2}$ units $= \text{bar}$

(c) $K_p = \dfrac{P_{N_2O_4}}{P_{N_2} P^2_{O_2}}$ units $= \text{bar}^{-2}$

CALCULATION OF EQUILIBRIUM CONSTANTS

19-10. The K_p expression for the reaction is

$$K_p = \frac{P_{CO} P^3_{H_2}}{P_{CH_4} P_{H_2O}}$$

Substituting the values of the equilibrium pressures into the K_p expression, we find that

$$K_p = \frac{(0.58\text{ bar})(2.29\text{ bar})^3}{(0.31\text{ bar})(0.84\text{ bar})} = 27\text{ bar}^2 \qquad \text{at } 1000^\circ\text{C}$$

19-12. The K_c expression for the equation is

$$K_c = \frac{[PCl_3]\,[Cl_2]}{[PCl_5]}$$

We now set up a concentration table.

Concentration	$PCl_5(g)$	$\rightleftharpoons$	$PCl_3(g)$	+	$Cl_2(g)$
Initial	1.10 M		0		0
Equilibrium	0.33 M				

From the table, we see that the change in concentration of $PCl_5(g)$ is 0.77 M. From the reaction stoichiometry, we can complete the table as follows:

Concentration	$PCl_5(g)$	$\rightleftharpoons$	$PCl_3(g)$	+	$Cl_2(g)$
Initial	1.10 M		0		0
Equilibrium	0.33 M		0.77 M		0.77 M

Note that at equilibrium, $[PCl_3] = [Cl_2]$ because $PCl_3(g)$ and $Cl_2(g)$ are formed in a 1:1 ratio on decomposition of $PCl_5(g)$. Substitution of the values of the equilibrium concentrations into the K_c expression yields

$$K_c = \frac{(0.77\text{ M})(0.77\text{ M})}{0.33\text{ M}} = 1.8\text{ M} \qquad \text{at } 250^\circ\text{C}$$

19-14. The K_p expression for the equation is

$$K_p = \frac{P_{NO}^2 P_{O_2}}{P_{NO_2}^2}$$

We set up a pressure table. Let x be the pressure of $O_2(g)$ that is produced by the reaction.

Pressure	$2\,NO_2(g)$	$\rightleftharpoons$	$2\,NO(g)$	+	$O_2(g)$
Initial	0.500 bar		0		0
Change	$-2x$		$+2x$		$+x$
Equilibrium	0.500 bar $- 2x$		$2x$		x

From Dalton's law of partial pressures, we have that

$$P_{tot} = P_{NO_2} + P_{NO} + P_{O_2}$$

The total pressure at equilibrium is 0.732 bar, and so

$$0.732\text{ bar} = (0.500\text{ bar} - 2x) + (2x) + (x)$$

Solving for x, we find that $x = 0.232$ bar. Thus, at equilibrium, we have that

$$P_{NO_2} = 0.500\text{ bar} - 2x = 0.036\text{ bar}$$
$$P_{NO} = 2x = 0.464\text{ bar}$$
$$P_{O_2} = x = 0.232\text{ bar}$$

Substituting the values of the equilibrium pressures into the K_p expression, we find that

$$K_p = \frac{(0.464\text{ bar})^2(0.232\text{ bar})}{(0.036\text{ bar})^2} = 39\text{ bar} \qquad \text{at 1000 K}$$

19-16. We set up a pressure table for the reaction. Let x be the pressure of $S_2(g)$ that is produced by the reaction.

Pressure	$2\,H_2S(g)$	$\rightleftharpoons$	$2\,H_2(g)$	+	$S_2(g)$
Initial	0.956 bar		0		0
Change	$-2x$		$+2x$		$+x$
Equilibrium	0.956 bar $- 2x$		$2x$		x

From Dalton's law of partial pressures, we have that

$$P_{tot} = P_{H_2S} + P_{H_2} + P_{S_2} = 1.26\text{ bar}$$

Substituting in the expressions from the pressure table, we obtain

$$(0.956\text{ bar} - 2x) + 2x + x = 1.26\text{ bar}$$

Solving for x, we have that $x = 0.30$ bar. At equilibrium,

$$P_{H_2S} = 0.956\text{ bar} - 2x = 0.36\text{ bar}$$
$$P_{H_2} = 2x = 0.60\text{ bar}$$
$$P_{S_2} = x = 0.30\text{ bar}$$

Substituting the values of the equilibrium pressures into the K_p expression, we find that

$$K_p = \frac{P_{H_2}^2 P_{S_2}}{P_{H_2S}^2} = \frac{(0.60\text{ bar})^2(0.30\text{ bar})}{(0.36\text{ bar})^2} = 0.83\text{ bar}$$

EQUILIBRIUM CALCULATIONS

19-18. The K_p expression for the equation is

$$K_p = \frac{P_{CO}^2}{P_{CO_2}} = 1.90\text{ bar}$$

Substituting the values of the equilibrium pressure of CO(g) into the K_p expression, we have that

$$1.90\text{ bar} = \frac{(1.50\text{ bar})^2}{P_{CO_2}}$$

Solving for P_{CO_2} yields

$$P_{CO_2} = \frac{(1.50\text{ bar})^2}{1.90\text{ bar}} = 1.18\text{ bar}$$

19-20. The equilibrium-constant expression is

$$K_c = \frac{[CS_2]}{[S_2]} = 9.40$$

We set up a concentration table. The initial concentration of $S_2(g)$ is 10.0 mol/5.00 L = 2.00 M. Let x be the number of moles per liter of $CS_2(g)$ produced by the reaction of $S_2(g)$ with C(s).

Concentration	$S_2(g)$	$+\ C(s)$	$\rightleftharpoons$	$CS_2(g)$
Initial	2.00 M	–		0
Change	$-x$	–		$+x$
Equilibrium	$2.00\text{ M} - x$	–		x

Substituting the equilibrium concentration expressions into the K_c expression, we have

$$K_c = \frac{x}{2.00\text{ M} - x} = 9.40$$

Thus,

$$x = (9.40)(2.00\text{ M} - x) = 18.80\text{ M} - 9.40x$$
$$10.40x = 18.8\text{ M}$$
$$x = \frac{18.8\text{ M}}{10.40} = 1.808\text{ M}$$

The number of moles of $CS_2(g)$ prepared is

$$\text{moles of CS}_2 = (1.808\text{ mol}\cdot\text{L}^{-1})(5.00\text{ L}) = 9.04\text{ mol}$$

The mass of $CS_2(g)$ prepared is

$$\text{mass of CS}_2 = (9.04\text{ mol})\left(\frac{76.15\text{ g}}{1\text{ mol}}\right) = 688\text{ g}$$

19-22. The equilibrium-constant expression is

$$K_p = \frac{P_{CH_4}}{P_{H_2}^2} = 0.263\text{ bar}^{-1}$$

The concentration of $CH_4(g)$ is 0.250 mol/4.00 L = 0.0625 M. Let x be the number of moles per liter of $CH_4(g)$ that react. From the reaction stoichiometry, the number of moles of $H_2(g)$ produced is $2x$. We set up a concentration table as

Concentration	$C(s)$	+	$2\,H_2(g)$	$\rightleftharpoons$	$CH_4(g)$
Initial	—		0		0.0625 M
Change	—		$+2x$		$-x$
Equilibrium	—		$2x$		$0.0625\text{ M} - x$

Substituting the equilibrium concentration expressions into the K_p expression, and using the ideal-gas equation, $P = nRT/V$, we have

$$K_p = \frac{P_{CH_4}}{P_{H_2}^2} = \frac{[CH_4](RT)}{[H_2]^2(RT)^2} = \frac{(0.0625\text{ M} - x)(RT)}{(2x)^2(RT)^2} = 0.263\text{ bar}^{-1}$$

and so

$$0.0625\text{ M} - x = (4x^2)(0.263\text{ bar}^{-1})(RT)$$

Thus, at 1000°C

$$0.0625\text{ M} - x = (4)(0.263\text{ bar}^{-1})(0.083145\text{ L}\cdot\text{bar}\cdot\text{K}^{-1}\cdot\text{mol}^{-1})(1273\text{ K})x^2$$
$$0.0625\text{ M} - x = (111\text{ M}^{-1})x^2$$
$$111x^2 + (1\text{ M})x - 0.0625\text{ M}^2 = 0$$

Using the quadratic formula, we obtain

$$x = \frac{-1\text{ M} \pm \sqrt{1\text{ M}^2 - (4)(111)(-0.0625\text{ M}^2)}}{(2)(111)}$$

Taking the positive root so that the pressures are all positive values, we have

$$x = 0.0196 \text{ M}$$

Thus $[CH_4] = 0.0625 \text{ M} - x = 0.0625 \text{ M} - 0.0196 \text{ M} = 0.0429 \text{ M}$. The value of P_{CH_4} at equilibrium is

$$\begin{aligned} P_{CH_4} &= [CH_4](RT) = (0.0429 \text{ M})(0.083145 \text{ L·bar·K}^{-1}\text{·mol}^{-1})(1273 \text{ K}) \\ &= 4.54 \text{ bar} \end{aligned}$$

19-24. The equilibrium-constant expression is

$$K_p = P_{CO_2} P_{H_2O}$$

Let x be the equilibrium pressure of $CO_2(g)$. From the reaction stoichiometry and because we started with only $NaHCO_3(s)$, we have $P_{CO_2} = P_{H_2O}$ at equilibrium. Thus,

$$K_p = (x)(x) = x^2 = 0.26 \text{ bar}^2$$

Taking square roots yields

$$x = 0.51 \text{ bar}$$

At equilibrium $P_{CO_2} = P_{H_2O} = 0.51$ bar.

19-26. The equilibrium-constant expression is

$$K_c = \frac{[COCl_2]}{[CO][Cl_2]} = 4.00 \text{ M}^{-1}$$

The initial concentration of $CO(g)$ is $5.00 \text{ mol}/10.0 \text{ L} = 0.500$ M, and the initial concentration of $Cl_2(g)$ is $2.50 \text{ mol}/10.0 \text{ L} = 0.250$ M. Let x be the number of moles per liter of CO(g) and $Cl_2(g)$ that react. From the reaction stoichiometry, the number of moles per liter of $COCl_2(g)$ produced is x. Setting up a concentration table, we have that

Concentration	$CO(g)$	+	$Cl_2(g)$	$\rightleftharpoons$	$COCl_2(g)$
Initial	0.500 M		0.250 M		0
Change	$-x$		$-x$		$+x$
Equilibrium	$0.500 \text{ M} - x$		$0.250 \text{ M} - x$		x

Substituting the equilibrium concentration expressions into the K_c expression, we have

$$\frac{x}{(0.500 \text{ M} - x)(0.250 \text{ M} - x)} = \frac{x}{0.125 \text{ M}^2 - 0.750 \text{ M}x + x^2} = 4.00 \text{ M}^{-1}$$

Multiplying out and collecting like terms yields

$$(4.00 \text{ M}^{-1})x^2 - 4.00x + 0.500 \text{ M} = 0$$

The solutions of this quadratic equation are given by the quadratic formula

$$x = \frac{4.00 \pm \sqrt{16.0 - (4)(4.00 \text{ M}^{-1})(0.500 \text{ M})}}{(2)(4.00 \text{ M}^{-1})}$$

The two roots are $x = 0.854$ M and $x = 0.146$ M. We reject the 0.854 M root because this value would yield a negative concentration of $CO(g)$ and $Cl_2(g)$. Thus, we have at equilibrium

$$[COCl_2] = 0.146 \text{ M}$$
$$[Cl_2] = 0.250 \text{ M} - 0.146 \text{ M} = 0.104 \text{ M}$$
$$[CO] = 0.500 \text{ M} - 0.146 \text{ M} = 0.354 \text{ M}$$

As a check, we note that $[COCl_2]/[Cl_2][CO] = 3.97 \text{ M}^{-1} \approx 4.00 \text{ M}^{-1}$.

19-28. The equilibrium-constant expression is

$$K_p = \frac{P_{HI}^2}{P_{H_2}} = 8.6 \text{ bar}$$

We know from Dalton's law of partial pressure that

$$P_{tot} = P_{HI} + P_{H_2} = 4.5 \text{ bar}$$

at equilibrium. Solving for P_{HI}, we have

$$P_{HI} = 4.5 \text{ bar} - P_{H_2}$$

Substituting this expression for P_{HI} into the K_p expression, we have

$$\frac{(4.5 \text{ bar} - P_{H_2})^2}{P_{H_2}} = \frac{20.25 \text{ bar}^2 - (9.0 \text{ bar})P_{H_2} + P_{H_2}^2}{P_{H_2}} = 8.6 \text{ bar}$$

or

$$20.25 \text{ bar}^2 - (9.0 \text{ bar})P_{H_2} + P_{H_2}^2 = (8.6 \text{ bar})P_{H_2}$$

Rearranging to the standard form of a quadratic equation, we have

$$P_{H_2}^2 - (17.6 \text{ bar})P_{H_2} + 20.25 \text{ bar}^2 = 0$$

The solutions of this quadratic equation are given by the quadratic formula

$$P_{H_2} = \frac{17.6 \text{ bar} \pm \sqrt{309.76 \text{ bar}^2 - (4)(1)(20.25 \text{ bar}^2)}}{2}$$

The two roots are 16.4 bar and 1.25 bar. We can rule out the value 16.4 bar because it is larger than the total pressure. Thus, at equilibrium $P_{H_2} = 1.3$ bar and $P_{HI} = 4.5 \text{ bar} - 1.3 \text{ bar} = 3.2$ bar to two significant figures. As a check, we note that $P_{HI}^2/P_{H_2} = 8.8$ atm (using $P_{H_2} = 1.25$ bar).

19-30. The equilibrium-constant expression is

$$K_p = \frac{P_{CH_4}}{P_{H_2}^2} = 0.263 \text{ bar}^{-1}$$

We know from Dalton's law of partial pressure that

$$P_{tot} = P_{CH_4} + P_{H_2} = 2.11 \text{ bar}$$

at equilibrium. Solving for P_{CH_4}, we have

$$P_{CH_4} = 2.11 \text{ bar} - P_{H_2}$$

Substituting this expression for P_{CH_4} into the K_p expression, we have

$$\frac{2.11 \text{ bar} - P_{H_2}}{P_{H_2}^2} = 0.263 \text{ bar}^{-1}$$

or

$$2.11 \text{ bar} - P_{H_2} = 0.263 \text{ bar}^{-1} P_{H_2}^2$$

Rearranging to the standard form of a quadratic equation, we have

$$0.263 \text{ bar}^{-1} P_{H_2}^2 + P_{H_2} - 2.11 \text{ bar} = 0$$

The solutions of this quadratic equation are given by the quadratic formula

$$P_{H_2} = \frac{-1 \pm \sqrt{(-1)^2 - (4)(0.263 \text{ bar}^{-1})(-2.11 \text{ bar})}}{(2)(0.263 \text{ bar}^{-1})}$$

The two roots are 1.51 bar and −5.30 bar. We can rule out the value −5.30 bar as physically impossible. At equilibrium $P_{H_2} = 1.51$ bar and $P_{CH_4} = 2.11 \text{ bar} - 1.51 \text{ bar} = 0.60$ bar. As a check, we note that $P_{CH_4}/P_{H_2}^2 = 0.263 \text{ bar}^{-1}$.

EQUILIBRIUM CONSTANTS OF EQUATIONS FOR COMBINATIONS OF REACTIONS

19-32. The equation

$$C(s) + CO_2(g) \rightleftharpoons 2\,CO(g)$$

is obtained by adding the first equation and two times the second equation

$$C(s) + 2\,H_2O(g) \rightleftharpoons CO_2(g) + 2\,H_2(g)$$
$$2\,H_2(g) + 2\,CO_2(g) \rightleftharpoons 2\,H_2O(g) + 2\,CO(g)$$

Thus, the equilibrium constant is given by

$$K_p = K_{p_1} K_{p_2}^2 = (3.85 \text{ bar})(0.71)^2 = 1.9 \text{ bar}$$

19-34. The equation

$$CaCO_3(s) + C(s) \rightleftharpoons CaO(s) + 2\,CO(g)$$

is the sum of the two given equations. Thus, the equilibrium constant is given by

$$K_p = K_{p_1} K_{p_2} = (0.040 \text{ bar})(1.9 \text{ bar}) = 0.076 \text{ bar}^2$$

QUALITATIVE APPLICATION OF LE CHÂTELIER'S PRINCIPLE

19-36. (a) An increase in the concentration of $Br_2(g)$ will cause the reaction equilibrium to shift to the right, thereby increasing the equilibrium concentration of $NOBr(g)$ and decreasing the equilibrium concentration of $NO(g)$.

19-38. (a) $\rightarrow$ The reaction is endothermic. An increase in temperature shifts the equilibrium to the right.

(b) $\rightarrow$ There are more moles of gas on the right, and thus a shift to the right will partially offset the decrease in the number of moles per unit volume.

(c) $\leftarrow$ Addition of $H_2O(l)$ decreases the concentration of $N_2(aq)$, and so the reaction shifts to the left to increase the concentration of $N_2(aq)$.

(d) $\leftarrow$ A shift in equilibrium to the left will decrease the pressure of $N_2(g)$.

19-40. (a) No change.

(b) $\leftarrow$ to the left; the $NaCl(s)$ supplies additional $Cl^-(aq)$ ions.

(c) $\rightarrow$ to the right; the addition of $H_2O(l)$ decreases $[Pb^{2+}]$ and $[Cl^-]$.

(d) $\rightarrow$ to the right; the addition of $AgNO_3(s)$ precipitates $Cl^-(aq)$ as $AgCl(s)$, and so decreases $[Cl^-]$.

19-42. Because the reaction involving the formation of $CH_3OH(g)$ from $CO(g)$ and $H_2(g)$ is exothermic and because there are more moles of gas on the left, the use of Le Châtelier's principle suggests that low temperature and high pressure (small reaction volume) favors the formation of methanol.

QUANTITATIVE APPLICATION OF LE CHÂTELIER'S PRINCIPLE

19-44. We first must calculate the value of the equilibrium constant K_p.

$$K_p = \frac{P_{PCl_5}}{P_{PCl_3}P_{Cl_2}} = \frac{217\text{ Torr}}{(13.2\text{ Torr})^2} = 1.25\text{ Torr}^{-1}$$

The total pressure is given by

$$P_{tot} = P_{PCl_5} + P_{PCl_3} + P_{Cl_2}$$

Substituting the given values into this relation gives the original initial total pressure

$$P_{tot} = 217\text{ Torr} + 13.2\text{ Torr} + 13.2\text{ Torr} = 243\text{ Torr}$$

Therefore, 263 Torr − 243 Torr = 20 Torr of $Cl_2(g)$ are injected into the reaction mixture. To determine the new equilibrium conditions, we set up a pressure table.

Pressure	$PCl_3(g)$	+	$Cl_2(g)$	$\rightleftharpoons$	$PCl_5(g)$
Initial	13.2 Torr		33.2 Torr		217 Torr
Change	$-x$		$-x$		$+x$
Equilibrium	13.2 Torr $- x$		33.2 Torr $- x$		217 Torr $+ x$

We know from Le Châtelier's principle that x will be a positive quantity. We substitute these equilibrium pressures into the equilibrium-constant expression to get

$$\frac{217\text{ Torr} + x}{(13.2\text{ Torr} - x)(33.2\text{ Torr} - x)} = 1.25\text{ Torr}^{-1}$$

This expression can be written out as

$$\begin{aligned}217\text{ Torr} + x &= (1.25\text{ Torr}^{-1})[438.2\text{ Torr}^2 - (46.4\text{ Torr})x + x^2] \\ &= 548\text{ Torr} - 58.0\,x + (1.25\text{ Torr}^{-1})x^2\end{aligned}$$

or

$$(1.25 \text{ Torr}^{-1})\,x^2 - 59.0x + 331 \text{ Torr} = 0$$

We use the quadratic formula for the roots to this equation

$$x = \frac{59.0 \pm \sqrt{3481 - (4)(1.25 \text{ Torr}^{-1})(331 \text{ Torr})}}{(2)(1.25 \text{ Torr}^{-1})}$$

The two roots are 6.52 Torr and 101.7 Torr. We reject the second root because P_{Cl_2} must be a positive quantity. Thus, the equilibrium partial pressures are

$$P_{PCl_3} = 13.2 \text{ Torr} - 6.52 \text{ Torr} = 6.7 \text{ Torr}$$
$$P_{PCl_5} = 217 \text{ Torr} + 6.52 \text{ Torr} = 224 \text{ Torr}$$
$$P_{Cl_2} = 33.2 \text{ Torr} - 6.52 \text{ Torr} = 26.7 \text{ Torr}$$

As a final check, note that

$$\frac{P_{PCl_5}}{P_{PCl_3} P_{Cl_2}} = \frac{(224 \text{ Torr})}{(6.7 \text{ Torr})(26.7 \text{ Torr})} = 1.25 \text{ Torr}^{-1}$$

19-46. We first must calculate the value of the equilibrium constant K_p.

$$K_p = \frac{P^2_{NO_2}}{P_{N_2O_4}}$$

The original total pressure is given by

$$P_{tot} = P_{N_2O_4} + P_{NO_2} = 0.770 \text{ bar}$$

To determine $P_{N_2O_4}$ and P_{NO_2} at equilibrium, we set up a pressure table.

Pressure	$N_2O_4(g)$	$\rightleftharpoons$	$2\,NO_2(g)$
Initial	0.554 bar		0
Change	$-x$		$+2x$
Equilibrium	$0.554 \text{ bar} - x$		$2x$

Substituting these expressions into the equation for the total pressure gives

$$P_{tot} = 0.554 \text{ bar} - x + 2x = 0.770 \text{ bar}$$
$$x = 0.216 \text{ bar}$$

Therefore, $P_{N_2O_4} = 0.338$ bar and $P_{NO_2} = 0.432$ bar. The value of the equilibrium constant K_p is

$$K_p = \frac{P^2_{NO_2}}{P_{N_2O_4}} = \frac{(0.432 \text{ bar})^2}{0.338 \text{ bar}} = 0.552 \text{ bar}$$

The quantity of $NO_2(g)$ injected is 0.906 bar − 0.770 bar = 0.136 bar, and so the new pressure of $NO_2(g)$ is 0.433 bar + 0.136 bar = 0.568 bar. To find the new equilibrium pressures, we set up a pressure table.

Pressure	$N_2O_4(g)$	$\rightleftharpoons$	$2\,NO_2(g)$
Initial	0.338 bar		0.568 bar
Change	$+x$		$-2x$
Equilibrium	$0.338\text{ bar} + x$		$0.568\text{ bar} - 2x$

Substituting these pressures into the equilibrium constant K_p yields

$$\frac{(0.568\text{ bar} - 2x)^2}{0.338\text{ bar} + x} = 0.552\text{ bar}$$

$$0.3226\text{ bar}^2 - (2.272\text{ bar})\,x + 4x^2 = 0.1866\text{ bar}^2 + (0.552\text{ bar})\,x$$

$$4x^2 - (2.824\text{ bar})\,x + 0.1360\text{ bar}^2 = 0$$

We use the quadratic formula to find the solutions to the above quadratic equation

$$x = \frac{-(-2.824) \pm \sqrt{(2.824\text{ bar})^2 - (4)(4)(0.1360\text{ bar}^2)}}{(2)(4)}$$

The two roots are 0.0520 bar and 0.654 bar. We reject the second root because P_{NO_2} must be a positive quantity. The two equilibrium pressures are

$$P_{N_2O_4} = 0.338\text{ bar} + 0.0520\text{ bar} = 0.390\text{ bar}$$
$$P_{NO_2} = 0.568\text{ bar} - 0.104\text{ bar} = 0.464\text{ bar}$$

As a check, we note that $P^2_{NO_2}/P_{N_2O_4} = 0.522$ bar. Finally, the total pressure after equilibrium is reestablished is

$$P_{tot} = P_{N_2O_4} + P_{NO_2} = 0.390\text{ bar} + 0.464\text{ bar} = 0.854\text{ bar}$$

19-48. We first must calculate the value of the equilibrium constant K_p.

$$K_p = P_{NH_3}P_{HBr}$$

The total pressure is given by

$$P_{tot} = P_{NH_3} + P_{HBr} = 26.4\text{ Torr}$$

Because only $NH_4Br(s)$ is introduced into the reaction container, we have

$$P_{NH_3} = P_{HBr} = \frac{1}{2}P_{tot} = 13.2\text{ Torr}$$

at equilibrium. Thus,

$$K_p = (13.2\text{ Torr})^2 = 174\text{ Torr}^2$$

If the volume of the reaction container is halved, then the initial pressure of each gas will double from 13.2 Torr to 26.4 Torr. We set up the following pressure table:

Pressure	$NH_4Br(s)$	$\rightleftharpoons$	$NH_3(g)$	+	$HBr(g)$
Initial	–		26.4 Torr		26.4 Torr
Change	–		$-x$		$-x$
Equilibrium	–		26.4 Torr $- x$		26.4 Torr $- x$

Substituting the new equilibrium pressures into the K_p expression yields

$$(26.4\text{ Torr} - x)^2 = 174\text{ Torr}^2$$

Taking the square root of both sides yields

$$26.4\text{ Torr} - x = \pm 13.2\text{ Torr}$$

The two values of x are 13.2 Torr and 39.6 Torr. We reject the second value because $P_{NH_3} = P_{HBr}$ must be a positive quantity. Thus, $P_{NH_3} = P_{HBr} = 13.2$ Torr, and the total pressure is

$$P_{tot} = 13.2\text{ Torr} + 13.2\text{ Torr} = 26.4\text{ Torr}$$

Notice that because $K_p = P_{NH_3} P_{HBr}$, it does not matter what the volume of the container is. As long as the temperature is constant, the partial pressure of each gas will always be 13.2 Torr and the total pressure 26.4 Torr. Only the total number of moles of gas in the container will change.

REACTION QUOTIENT CALCULATIONS

19-50. The Q_p expression for the reaction is

$$Q_p = \frac{(P_{CH_4})_0}{(P_{H_2})_0^2}$$

From the data given, we compute

$$Q_p = \frac{(3.0\text{ bar})}{(0.20\text{ bar})^2} = 75\text{ bar}^{-1}$$

Because Q_p does not equal K_p, the reaction is not at equilibrium.

$$\frac{Q_p}{K_p} = \frac{75\text{ bar}^{-1}}{2.69 \times 10^3\text{ bar}^{-1}} = 2.8 \times 10^{-2} < 1$$

Because the value of Q_p/K_p is less than 1, the reaction proceeds from left to right toward equilibrium.

19-52. The Q_p expression for the reaction is

$$Q_p = \frac{(P_{CH_3OH})_0}{(P_{H_2})_0^2 (P_{CO})_0}$$

Thus,

$$Q_p = \frac{10.0 \text{ atm}}{(0.010 \text{ atm})^2(0.0050 \text{ atm})} = 2.0 \times 10^7 \text{ atm}^{-2}$$

$$\frac{Q_p}{K_p} = \frac{2.0 \times 10^7 \text{ atm}^{-2}}{2.25 \times 10^4 \text{ atm}^{-2}} = 8.9 \times 10^2 > 1$$

Because the value of Q_p/K_p is greater than 1, the reaction proceeds from right to left toward equilibrium.

ADDITIONAL PROBLEMS

19-54. An equilibrium constant is given by the ratio of the concentrations or pressures of the products over those of the reactants, each raised to a power equal to its respective stoichiometric coefficient in the balanced chemical equation. An equilibrium state is given by the values of the actual concentrations of the reactants and products at equilibrium. While there is only one possible value of the equilibrium constant for a stated chemical equation at a given temperature, there is often more than one possible equilibrium state that will satisfy the equilibrium constant expression.

19-56. They are the same for this equation because the total number of moles of gaseous reactants is equal to the total number of gaseous products.

19-58. The value of K_c is fixed by the equilibrium state. The value of Q_c is arbitrary, and any concentrations may be used.

19-60. (a) The vapor pressure will increase and the color will become more intense.

(b) The vapor pressure will decrease and the color will become less intense.

(c) No change.

19-62. A better answer might be, "As a reaction approaches equilibrium from the left, the rate of the forward reaction decreases and the rate of the reverse reaction increases until the rates of the two are equal. At this point, a dynamic balance is reached and the reaction is at equilibrium."

19-64. (a) The equilibrium shifts toward the side with the smaller total number of moles of gaseous species.

(b) When the number of moles of gaseous species on both sides of the reaction equation are equal, increasing the volume has no effect on the equilibrium state.

(c) Increasing the amount of solvent is equivalent to increasing the volume of a gas-phase reaction.

19-66. The K_p expression for the equation is

$$K_p = \frac{P^2_{NH_3}}{P_{N_2} P^3_{H_2}}$$

The relation between K_c and K_p is obtained from Equation 19.18 as

$$K_c = \frac{K_p}{(RT)^{\Delta\nu_{gas}}} = K_p(RT)^{-\Delta\nu_{gas}}$$

For the given chemical equation, $\Delta\nu_{gas} = 2 - 4 = -2$, and so

$$K_c = K_p(RT)^2 = (0.099\ \text{bar}^{-2})[(0.083145\ \text{L·bar·K}^{-1}\text{·mol}^{-1})(500\ \text{K})]^2 = 170\ \text{M}^{-2}$$

at 227°C.

19-68. The equilibrium-constant expression is

$$K_p = \frac{P_{CH_3OH}}{P_{H_2}^2 P_{CO}} = 2.19 \times 10^4\ \text{bar}^{-2}$$

(a) Substituting the given values of P_{H_2} and P_{CO} into the K_p expression, we have

$$K_p = \frac{P_{CH_3OH}}{(0.020\ \text{bar})^2(0.010\ \text{bar})} = 2.19 \times 10^4\ \text{bar}^{-2}$$

$$P_{CH_3OH} = (2.19 \times 10^4\ \text{bar}^{-2})(4.0 \times 10^{-6}\ \text{bar}^3) = 0.088\ \text{bar}$$

(b) The total pressure is equal to the sum of the partial pressures

$$P_{tot} = P_{H_2} + P_{CO} + P_{CH_3OH} = 10.0\ \text{bar}$$

Substituting in the value of P_{H_2} yields

$$0.020\ \text{bar} + P_{CO} + P_{CH_3OH} = 10.0\ \text{bar}$$

Solving for P_{CO}, we have

$$P_{CO} = 9.98\ \text{bar} - P_{CH_3OH}$$

Substituting the expression for P_{CO} and the value of P_{H_2} into the K_p expression, we have

$$\frac{P_{CH_3OH}}{(0.020\ \text{bar})^2(9.98\ \text{bar} - P_{CH_3OH})} = 2.19 \times 10^4\ \text{bar}^{-2}$$

or

$$\begin{aligned} P_{CH_3OH} &= (2.19 \times 10^4\ \text{bar}^{-2})[3.99 \times 10^{-3}\ \text{bar}^3 - (4.00 \times 10^{-4}\ \text{bar}^2)P_{CH_3OH}] \\ &= 87.4\ \text{bar} - 8.76P_{CH_3OH} \end{aligned}$$

Collect like terms

$$9.76P_{CH_3OH} = 87.4\ \text{bar}$$

and solve for P_{CH_3OH} to get

$$P_{CH_3OH} = \frac{87.4\ \text{bar}}{9.76} = 8.95\ \text{bar}$$

and

$$P_{CO} = 9.98\ \text{bar} - 8.95\ \text{bar} = 1.03\ \text{bar}$$

As a check, we note that $P_{CH_3OH}/P_{H_2}^2 P_{CO} = 2.17 \times 10^4\ \text{bar}^{-1}$.

19-70. The K_c expression for the equation is

$$K_c = \frac{[NO_2]^2}{[N_2O_4]} = 0.20\,M$$

The relation between K_c and K_p is obtained from Equation 19.18 as

$$K_p = K_c(RT)^{\Delta\nu_{gas}}$$

For the given chemical equation, $\Delta\nu_{gas} = 2 - 1 = 1$, and so

$$K_p = K_c(RT) = (0.20\text{ mol}\cdot\text{L}^{-1})(0.083145\text{ L}\cdot\text{bar}\cdot\text{K}^{-1}\cdot\text{mol}^{-1})(373\text{ K}) = 6.2\text{ bar}$$

at 100°C.

19-72. The equilibrium-constant expression is

$$K_c = \frac{[SO_3][NO]}{[SO_2][NO_2]} = 3.0$$

We set up a concentration table.

Concentration	$SO_2(g)$	+	$NO_2(g)$	$\rightleftharpoons$	$SO_3(g)$	+	$NO(g)$
Initial	2.4 mol		x		0		0
Change	2.4 mol − 1.2 mol		x − 1.2 mol		+1.2 mol		+1.2 mol
Equilibrium	1.2 mol		x − 1.2 mol		1.2 mol		1.2 mol

because the number of moles of gaseous reactants is equal to the number of moles of gaseous products, we can work directly with moles rather than concentration. Substituting the equilibrium concentration expressions into the K_c expression, we have

$$K_c = \frac{(1.2\text{ mol})^2}{(1.2\text{ mol})(x - 1.2\text{ mol})} = 3.0$$

and

$$1.2\text{ mol} = 3.0x - 3.6\text{ mol}$$

Solving for x, we see that 1.60 moles of $NO_2(g)$ must be added. As a check, we note that $[SO_3][NO]/[SO_2][NO_2] = 3.0$.

19-74. Because the number of moles of gaseous products equals the number of moles of gaseous reactants in the chemical equation, $K_p = K_c$, and so

$$K_c = \frac{[HI]^2}{[H_2][I_2]} = K_p = \frac{P_{HI}^2}{P_{H_2}P_{I_2}} = 85$$

(a) With $P_{HI} = P_{I_2} = P_{H_2}$, we have

$$Q_p = \frac{(P_{HI}^2)_0}{(P_{H_2})_0(P_{I_2})_0} = \frac{(P_{HI}^2)_0}{(P_{HI})_0(P_{HI})_0} = 1 \neq K_p$$

and thus, an equilibrium reaction mixture is not possible with these concentrations.

(b) The initial value of [HI] is

$$[\mathrm{HI}] = \frac{(5.0\ \mathrm{g\ HI})\left(\dfrac{1\ \mathrm{mol\ HI}}{127.91\ \mathrm{g\ HI}}\right)}{2.00\ \mathrm{L}} = 0.0195\ \mathrm{M}$$

We now set up a concentration table.

Concentration	$H_2(g)$	+	$I_2(g)$	⇌	$2\,HI(g)$
Initial	0		0		0.0195 M
Change	$+x$		$+x$		$-2x$
Equilibrium	x		x		$0.0195\ \mathrm{M} - 2x$

Thus,

$$\frac{(0.0195\ \mathrm{M} - 2x)^2}{x^2} = 85$$

Taking the square root of both sides, we obtain

$$\frac{(0.0195\ \mathrm{M} - 2x)}{x} = \pm 9.22$$

Solving for x, we get

$$x = \frac{0.0195\ \mathrm{M}}{11.22} = 0.00174\ \mathrm{M}$$

We discard the negative value because x must be a positive quantity. Thus, in the new equilibrium state, we have

$$[\mathrm{H_2}] = [\mathrm{I_2}] = 1.7 \times 10^{-3}\ \mathrm{M}$$

$$[\mathrm{HI}] = 0.0195\ \mathrm{M} - 2(0.00174\ \mathrm{M}) = 0.016\ \mathrm{M}$$

As a check, we note that $[\mathrm{HI}]^2/[\mathrm{H_2}]\,[\mathrm{I_2}] = 89$.

19-76. The initial number of moles of $O_2(g)$ is computed by using the ideal-gas equation at 25°C

$$n = \frac{PV}{RT} = \frac{(1.00\ \mathrm{atm})\,(2.00\ \mathrm{L})}{(0.08206\ \mathrm{L{\cdot}atm{\cdot}K^{-1}{\cdot}mol^{-1}})\,(298\ \mathrm{K})} = 0.0818\ \mathrm{mol}$$

The number of moles of $MgCl_2(s)$ is

$$n = \frac{50.0\ \mathrm{g}}{95.21\ \mathrm{g{\cdot}mol^{-1}}} = 0.525\ \mathrm{mol}$$

Thus, $MgCl_2$ is in excess, and $O_2(g)$ is the limiting reactant. The K_p expression is

$$K_p = \frac{P_{\mathrm{Cl_2}}}{P_{\mathrm{O_2}}^{1/2}} = 1.75\ \mathrm{atm}^{1/2}$$

The initial pressure of $O_2(g)$ at 823 K is given by

$$P_{\mathrm{O_2}} = (1.00\ \mathrm{atm})\left(\frac{823\ \mathrm{K}}{298\ \mathrm{K}}\right) = 2.76\ \mathrm{atm}$$

If we let $x = P_{\mathrm{Cl_2}}$ produced, then we have the pressure table

Pressure	$MgCl_2(s)$ +	$\frac{1}{2}O_2(g)$	⇌ $MgO(s)$	+ $Cl_2(g)$
Initial	—	2.76 atm	—	0
Change	—	$-\frac{1}{2}x$	—	$+x$
Equilibrium	—	2.76 atm $-\frac{1}{2}x$	—	x

Substituting into the equilibrium-constant expression, we have

$$\frac{x}{\left(2.76\text{ atm} - \dfrac{x}{2}\right)^{1/2}} = 1.75\text{ atm}^{1/2}$$

Squaring both sides of the equation yields

$$\frac{x^2}{2.76\text{ atm} - \dfrac{x}{2}} = 3.06\text{ atm}$$

or

$$x^2 + (1.53\text{ atm})\,x - 8.45\text{ atm}^2 = 0$$

The solutions to the quadratic equation are

$$x = \frac{-1.53\text{ atm} \pm \sqrt{(1.53\text{ atm})^2 + 4(8.45\text{ atm})}}{2}$$
$$= 2.24\text{ atm} \quad \text{and} \quad -3.77\text{ atm}$$

We reject the negative root as being physically impossible, and so we have at equilibrium

$$P_{Cl_2} = 2.24\text{ atm}$$
$$P_{O_2} = 2.76\text{ atm} - \frac{2.24\text{ atm}}{2} = 1.64\text{ atm}$$

As a check, we note that $P_{Cl_2}/P_{O_2}^{1/2} = 1.75\text{ atm}^{1/2}$.

19-78. High pressure favors the more dense form. Thus, the brown, crystalline form is more stable at high pressure.

19-80. We set up a concentration table.

Concentration	$PCl_5(g)$	⇌ $PCl_3(g)$	+ $Cl_2(g)$
Initial	x	0	0
Change	−1.00 M	+1.00 M	+1.00 M
Equilibrium	x − 1.00 M	1.00 M	1.00 M

We can find the value of K_c from the value of K_p.

$$K_c = (RT)^{-1}K_p = \frac{1.78\text{ atm}}{(0.08206\text{ L·atm·K}^{-1}\text{·mol}^{-1})(523\text{ K})} = 0.0415\text{ M}$$

Substituting in the equilibrium expression into the K_c expression gives

$$K_c = \frac{[PCl_3]\,[Cl_2]}{[PCl_5]} = \frac{(1.00\text{ M})^2}{x - 1.00\text{ M}} = 0.0415\text{ M}$$

Solving for x, we find that $x = 25.1$ M or 25.1 moles of $PCl_5(g)$ must be added to the 1.0-liter container. As a check, we note that $[PCl_3]\,[Cl_2]/[PCl_5] = 0.0415$ M.

19-82. We set up a concentration table.

Concentration	$H_2(g)$	+	$I_2(g)$	$\rightleftharpoons$	$2\,HI(g)$
Initial	0.200 M		0.200 M		0
Change	$-x$		$-x$		$+2x$
Equilibrium	0.200 M $- x$		0.200 M $- x$		$2x$

In this case, $K_c = K_p$, Substituting in the equilibrium expressions, we have

$$K_c = \frac{[HI]^2}{[H_2]\,[I_2]} = \frac{(2x)^2}{(0.200\text{ M} - x)^2} = 54.4$$

Taking the square root of both sides gives $x = 0.157$ M and 0.274 M. We reject 0.274 M because it gives a percent conversion greater than 100%. Thus,

$$\%\text{ conversion} = \frac{0.157\text{ M}}{0.200\text{ M}} \times 100 = 78.5\%$$

As a check, note that $[HI]^2/[H_2]\,[I_2] = 54.3$.

19-84. We set up a pressure table.

Pressure	$CO_2(g)$	+	$H_2(g)$	$\rightleftharpoons$	$CO(g)$	+	$H_2O(g)$
Initial	26.1 Torr		26.1 Torr		0		0
Change	$-x$		$-x$		$+x$		$+x$
Equilibrium	26.1 Torr $- x$		26.1 Torr $- x$		x		x

We know from Le Châtelier's principle that x will be a positive quantity. The equilibrium-constant expression is

$$K_p = \frac{P_{CO}\,P_{H_2O}}{P_{CO_2}\,P_{H_2}} = \frac{x^2}{(26.1\text{ Torr} - x)^2} = 0.719$$

Taking the square root of both sides yields

$$\frac{x}{26.1\text{ Torr} - x} = \pm 0.848$$

The only physically acceptable solution is $x = 12.0$ Torr. The partial pressures are

$$P_{CO} = P_{H_2O} = 12.0\text{ Torr}$$

$$P_{CO_2} = P_{H_2} = 26.1\text{ Torr} - 12.0\text{ Torr} = 14.1\text{ Torr}$$

Note that

$$\frac{P_{CO}P_{H_2O}}{P_{CO_2}P_{H_2}} = \frac{(12.0\text{ Torr})(12.0\text{ Torr})}{(14.1\text{ Torr})(14.1\text{ Torr})} = 0.72$$

19-86. Because equal molar quantities of both $NH_3(g)$ and $HCl(g)$ are involved, we have initially that $P_{total} = P_{NH_3} + P_{HCl} = 8.87$ bar, and so

$$P_{NH_3} = P_{HCl} = \frac{8.87\text{ bar}}{2} = 4.435\text{ bar}$$

We set up a pressure table.

Pressure	$NH_4Cl(s)$	$\rightleftharpoons$	$NH_3(g)$	+	$HCl(g)$
Initial	–		4.435 bar		4.435 bar
Change	–		$-x$		$-x$
Equilibrium	–		4.435 bar – x		4.435 bar – x

The equilibrium-constant expression is

$$K_p = P_{NH_3}P_{HCl} = (4.435\text{ bar} - x)^2 = 5.82\times 10^{-2}\text{ bar}^2$$

Taking the square root of both sides yields

$$4.435\text{ bar} - x = \pm 0.241\text{ bar}$$

The only physically acceptable solution is $x = 4.19$ bar. The partial pressures of the two gases at equilibrium are

$$P_{NH_3} = P_{HCl} = 4.435\text{ bar} - 4.19\text{ bar} = 0.24\text{ bar}$$

As a check, note that $P_{NH_3}P_{HCl} = 0.058\text{ bar}^2$.
The number of moles of $NH_3(g)$ and HCl(g) that react can be found using the ideal gas equation.

$$n = \frac{PV}{RT} = \frac{(4.19\text{ bar})(2.00\text{ L})}{(0.083145\text{ L·bar·K}^{-1}\text{·mol}^{-1})(573\text{ K})} = 0.176\text{ mol}$$

The number of grams of $NH_4C(s)$ produced is

$$\text{mass of NH}_4\text{Cl} = (0.176\text{ mol HCl})\left(\frac{1\text{ mol NH}_4\text{Cl}}{1\text{ mol HCl}}\right)\left(\frac{53.49\text{ g NH}_4\text{Cl}}{1\text{ mol NH}_4\text{Cl}}\right) = 9.41\text{ g}$$

19-88. We first calculate the value of the equilibrium constant from the initial equilibrium concentrations.

$$K_c = \frac{[CO_2][H_2]}{[CO][H_2O]} = \frac{\left(\frac{0.80\text{ mol}}{V}\right)\left(\frac{0.20\text{ mol}}{V}\right)}{\left(\frac{0.10\text{ mol}}{V}\right)\left(\frac{0.40\text{ mol}}{V}\right)} = 4.0$$

where we have let V be the volume of the container. We see that the volume cancels in the equilibrium-constant expression, and so we can use just the number of moles of each constituent. Let x be the number of moles of $CO_2(g)$ to be added. Note that the increase in the number of moles of CO(g) is 0.20 mol − 0.10 mol = 0.10 mol. The concentration table is

Concentration	$CO(g)$	+	$H_2O(g)$	$\rightleftharpoons$	$CO_2(g)$	+	$H_2(g)$
Initial	0.10 mol		0.40 mol		0.80 mol		0.20 mol
Change	+0.10 mol		+0.10 mol		$+x-0.10$ mol		−0.10 mol
Equilibrium	0.20 mol		0.50 mol		0.70 mol + x		0.10 mol

Substituting the above expressions into the equilibrium-constant expression yields

$$K_c = \frac{(0.70\text{ mol} + x)(0.10\text{ mol})}{(0.20\text{ mol})(0.50\text{ mol})} = 4.0$$

Solving for x yields $x = 3.3$ mol. Therefore, we must add 3.3 moles of $CO_2(g)$. As a check, note that $[CO_2][H_2]/[CO][H_2O] = 4.0$.

19-90. Consider a gaseous reaction described by the general equation

$$\nu_A A + \nu_B B \rightleftharpoons \nu_C C + \nu_D D$$

Then

$$K_p = \frac{P_C^{\nu_c} P_D^{\nu_D}}{P_A^{\nu_A} P_B^{\nu_B}}$$

If the pressures are expressed in atmospheres, then the value of $K_{p,atm}$ is $z(\text{atm})^{\nu_C+\nu_D-\nu_B-\nu_A} = z(\text{atm})^{\Delta\nu_{gas}}$, where z is just the numerical value of K_p when it is expressed in units of atmospheres. Now, 1 atm = 1.01325 bar, and so

$$K_{p,bar} = z(\text{atm})^{\Delta\nu_{gas}}\left(\frac{1.01325\text{ bar}}{1\text{ atm}}\right)^{\Delta\nu_{gas}}$$
$$= K_{p,atm}\left(\frac{1.01325\text{ bar}}{1\text{ atm}}\right)^{\Delta\nu_{gas}}$$

19-92. From Dalton's law of partial pressures,

$$P_{tot} = P_{HC_2H_3O_2} + P_{(HC_2H_3O_2)_2} = 1.50\text{ bar}$$

Thus,

$$P_{HC_2H_3O_2} = 1.50\text{ bar} - P_{(HC_2H_3O_2)_2}$$

The equilibrium-constant expression for the reaction is

$$K_p = \frac{P_{(HC_2H_3O_2)_2}}{P^2_{HC_2H_3O_2}} = 3.67\text{ bar}^{-1}$$

Substituting in the expression for $P_{HC_2H_3O_2}$, we have

$$K_p = \frac{P_{(HC_2H_3O_2)_2}}{(1.50\text{ bar} - P_{(HC_2H_3O_2)_2})^2} = 3.67\text{ bar}^{-1}$$

Rearranging, we obtain

$$(3.67\text{ bar}^{-1})\,P^2_{(HC_2H_3O_2)_2} - (12.01)\,P_{(HC_2H_3O_2)_2} + 8.2575\text{ bar} = 0$$

The two roots are 2.29 bar and 0.982 bar. We reject the root 2.29 bar because it is greater than the original total pressure.

$$P_{(HC_2H_3O_2)_2} = 0.982 \text{ bar}$$

As a check, note that $P_{(HC_2H_3O_2)_2}/P^2_{HC_2H_3O_2} = 3.68 \text{ bar}^{-1}$.

19-94. The K_p expression for the reaction is

$$K_p = P^2_{NH_3} P_{CO_2}$$

The total pressure is given by

$$P_{tot} = P_{NH_3} + P_{CO_2}$$

but from the reaction stoichiometry

$$P_{NH_3} = 2P_{CO_2}$$

Thus,

$$P_{tot} = 3P_{CO_2}$$

or

$$P_{CO_2} = \frac{1}{3}P_{tot}$$

Therefore,

$$K_p = (2P_{CO_2})^2 P_{CO_2} = 4P^3_{CO_2} = 4\left(\frac{1}{3}P_{tot}\right)^3$$
$$= \frac{4}{27}P^3_{tot}$$

19-96.* We can calculate the value of K_c by substituting the values of the equilibrium concentrations.

$$K_c = \frac{[CH_4][H_2O]}{[CO][H_2]^3} = \frac{(0.387\text{ M})(0.387\text{ M})}{(0.613\text{ M})(0.387\text{ M})^3} = 4.22\text{ M}^{-2}$$

Let x be the number of moles per liter of $H_2O(g)$ that is produced by the reaction after the removal of the water vapor. We set up a concentration table.

Concentration	$CO(g)$	+	$3\,H_2(g)$	$\rightleftharpoons$	$CH_4(g)$	+	$H_2O(g)$
Initial	0.613 M		0.387 M		0.387 M		0
Change	$-x$		$-3x$		$+x$		$+x$
Equilibrium	$0.613\text{ M} - x$		$0.387\text{ M} - 3x$		$0.387\text{ M} + x$		x

Substituting the equilibrium concentration expressions into the K_c expression, we have

$$K_c = \frac{(0.387\text{ M} + x)(x)}{(0.613\text{ M} - x)(0.387\text{ M} - 3x)^3} = 4.22\text{ M}^{-2}$$

By trial and error, we obtain

$$x = [H_2O] = 0.0559 \text{ M}$$
$$[CH_4] = 0.443 \text{ M}$$
$$[CO] = 0.557 \text{ M}$$
$$[H_2] = 0.219 \text{ M}$$

As a check, note that $K_c = 4.23 \text{ M}^{-2}$.

CHAPTER 20. Acids and Bases

STRONG ACIDS AND BASES

20-2. Because $KOH(aq)$ is a strong base (see Table 20.1), it is completely dissociated, and thus,

$$[OH^-] = 0.25\ \text{M} \qquad \text{and} \qquad [K^+] = 0.25\ \text{M}$$

We can calculate $[H_3O^+]$ from the K_w expression:

$$K_w = [H_3O^+]\,[OH^-] = 1.00 \times 10^{-14}\ \text{M}^2$$

Solving for $[H_3O^+]$, we get

$$[H_3O^+] = \frac{1.00 \times 10^{-14}\ \text{M}^2}{[OH^-]} = \frac{1.00 \times 10^{-14}\ \text{M}^2}{0.25\ \text{M}} = 4.0 \times 10^{-14}\ \text{M}$$

Because $[OH^-] > [H_3O^+]$, the solution is basic.

20-4. We first must find the number of moles in 0.600 grams of $Ca(OH)_2(aq)$.

$$n = (0.600\ \text{g})\left(\frac{1\ \text{mol Ca(OH)}_2}{74.10\ \text{g Ca(OH)}_2}\right) = 8.10 \times 10^{-3}\ \text{mol}$$

The molarity of the solution is

$$\text{molarity} = \frac{\text{moles of solute}}{\text{volume of solution}} = \frac{8.10 \times 10^{-3}\ \text{mol}}{1.00\ \text{L}} = 8.10 \times 10^{-3}\ \text{M}$$

Because $Ca(OH)_2$ is a strong base in water (see Table 20.1), it is completely dissociated, and yields two moles of $OH^-(aq)$ per mole of $Ca(OH)_2$. Thus,

$$[Ca^{2+}] = 8.10 \times 10^{-3}\ \text{M}$$
$$[OH^-] = (2)(8.10 \times 10^{-3}\ \text{M}) = 0.0162\ \text{M}$$

We can calculate $[H_3O^+]$ from the K_w epxression:

$$K_w = [H_3O^+]\,[OH^-] = 1.00 \times 10^{-14}\ \text{M}^2$$

to get

$$[\mathrm{H_3O^+}] = \frac{1.00 \times 10^{-14}\ \mathrm{M^2}}{0.0162\ \mathrm{M}} = 6.17 \times 10^{-13}\ \mathrm{M}$$

pH CALCULATIONS

20-6. Because $\mathrm{CsOH}(aq)$ is a strong base, it is completely dissociated, and thus,

$$[\mathrm{OH^-}] = 0.20\ \mathrm{M}$$

We can calculate $[\mathrm{H_3O^+}]$ from the K_w epxression:

$$[\mathrm{H_3O^+}] = \frac{K_w}{[\mathrm{OH^-}]} = \frac{1.00 \times 10^{-14}\ \mathrm{M^2}}{0.20\ \mathrm{M}} = 5.0 \times 10^{-14}\ \mathrm{M}$$

The pH is given by

$$\mathrm{pH} = -\log([\mathrm{H_3O^+}]/\mathrm{M}) = -\log(5.0 \times 10^{-14}) = 13.30$$

Because pH > 7, the solution is basic.

20-8. Because $\mathrm{Ba(OH)_2}(aq)$ is a strong base, it is completely dissociated to yield two moles of $\mathrm{OH^-}(aq)$ per one mole of $\mathrm{Ba(OH)_2}(aq)$. Thus

$$[\mathrm{OH^-}] = (2)(0.020\ \mathrm{M}) = 0.040\ \mathrm{M}$$

The pOH is given by

$$\mathrm{pOH} = -\log([\mathrm{OH^-}]/\mathrm{M}) = -\log(0.040) = 1.40$$

The pH of the solution is

$$\mathrm{pH} = 14.00 - \mathrm{pOH} = 14.00 - 1.40 = 12.60$$

20-10. We first must find the number of moles in 6.3 grams of $\mathrm{NaOH}(s)$.

$$n = (6.3\ \mathrm{g})\left(\frac{1\ \mathrm{mol\ NaOH}}{40.00\ \mathrm{g\ NaOH}}\right) = 0.16\ \mathrm{mol}$$

The molarity of the solution is

$$\text{molarity} = \frac{\text{moles of solute}}{\text{volume of solution}} = \frac{0.16\ \mathrm{mol}}{0.100\ \mathrm{L}} = 1.6\ \mathrm{M}$$

Because $\mathrm{NaOH}(aq)$ is a strong base, it is completely dissociated, and thus,

$$[\mathrm{OH^-}] = 1.6\ \mathrm{M}$$

The pOH is given by

$$\mathrm{pOH} = -\log([\mathrm{OH^-}]/\mathrm{M}) = -\log(1.6) = -0.20$$

The pH of the solution is

$$\mathrm{pH} = 14.00 - \mathrm{pOH} = 14.00 - (-0.20) = 14.20$$

CALCULATION OF $[H_3O^+]$ AND $[OH^-]$ FROM pH

20-12. We calculate the value of $[H_3O^+]$ using Equation 20.7,

$$[H_3O^+] = 10^{-pH}\text{ M}$$

The pH of the soap solution is given as 11.0, and so

$$[H_3O^+] = 10^{-11.0}\text{ M} = 1 \times 10^{-11}\text{ M}$$

We can calculate $[OH^-]$ from the K_w expression:

$$[OH^-] = \frac{1.00 \times 10^{-14}\text{ M}^2}{[H_3O^+]} = \frac{1.00 \times 10^{-14}\text{ M}^2}{1 \times 10^{-11}\text{ M}} = 1 \times 10^{-3}\text{ M}$$

20-14. We use Equation 20.7. The pH of normal rainwater is 5.6, and so

$$[H_3O^+] = 10^{-pH}\text{ M} = 10^{-5.6}\text{ M} = 3 \times 10^{-6}\text{ M}$$

The pH of acid rain is 3.0, and so

$$[H_3O^+] = 10^{-3.0}\text{ M} = 1 \times 10^{-3}\text{ M}$$

The ratio of the concentration of $H_3O^+(aq)$ in acid rain to that in normal rain is given by

$$\text{ratio} = \frac{[H_3O^+]_{\text{acid rain}}}{[H_3O^+]_{\text{normal rain}}} = \frac{1 \times 10^{-3}\text{ M}}{3 \times 10^{-6}\text{ M}} = 300$$

20-16. We use Equation 20.7. The pH of the world's oceans is 8.15, and so

$$[H_3O^+] = 10^{-pH}\text{ M} = 10^{-8.15}\text{ M} = 7.1 \times 10^{-9}\text{ M}$$

We can calculate $[OH^-]$ from the K_w expression:

$$[OH^-] = \frac{1.00 \times 10^{-14}\text{ M}^2}{[H_3O^+]} = \frac{1.00 \times 10^{-14}\text{ M}^2}{7.1 \times 10^{-9}\text{ M}} = 1.4 \times 10^{-6}\text{ M}$$

CALCULATION OF K_a FROM pH

20-18. The acid-dissociation reaction is given by the equation

$$CH_3(CH_2)_4COOH(aq) + H_2O(l) \rightleftharpoons H_3O^+(aq) + CH_3(CH_2)_4COO^-(aq)$$

The acid-dissociation-constant expression is

$$K_a = \frac{[H_3O^+][CH_3(CH_2)_4COO^-]}{[CH_3(CH_2)_4COOH]}$$

We use Equation 20.7 to find the value of $[H_3O^+]$ from the pH of the solution

$$[H_3O^+] = 10^{-pH}\text{ M} = 10^{-2.78}\text{ M} = 1.7 \times 10^{-3}\text{ M}$$

We set up a table of initial and equilibrium concentrations

Concentration	$CH_3(CH_2)_4COOH(aq)$	+	$H_2O(l)$	$\rightleftharpoons$	$H_3O^+(aq)$	+	$CH_3(CH_2)_4COO^-(aq)$
Initial	0.20 M		–		≈ 0 M		0 M
Change	$-x$		–		$+x$		$+x$
Equilibrium	$0.20\text{ M} - x$		–		x		x

At equilibrium, $x = [H_3O^+]$, and so we have

$$[H_3O^+] = [CH_3(CH_2)_4COO^-] = 1.7 \times 10^{-3}\text{ M}$$

and

$$[CH_3(CH_2)_4COOH] = 0.20\text{ M} - 1.7 \times 10^{-3}\text{ M} = 0.20\text{ M}$$

Substituting the values of the equilibrium concentrations into the K_a expression yields

$$K_a = \frac{(1.7 \times 10^{-3}\text{ M})(1.7 \times 10^{-3}\text{ M})}{0.20\text{ M}} = 1.4 \times 10^{-5}\text{ M}$$

20-20. The acid-dissociation reaction is given by the equation

$$HCOOH(aq) + H_2O(l) \rightleftharpoons H_3O^+(aq) + HCOO^-(aq)$$

The acid-dissociation-constant expression is

$$K_a = \frac{[H_3O^+][HCOO^-]}{[HCOOH]}$$

We use Equation 20.7 to find the value of $[H_3O^+]$ from the pH of the solution

$$[H_3O^+] = 10^{-\text{pH}}\text{ M} = 10^{-2.38}\text{ M} = 4.2 \times 10^{-3}\text{ M}$$

We set up a table of initial and equilibrium concentrations

Concentration	$HCOOH(aq)$	+	$H_2O(l)$	$\rightleftharpoons$	$H_3O^+(aq)$	+	$HCOO^-(aq)$
Initial	0.10 M		–		≈ 0 M		0 M
Change	$-x$		–		$+x$		x
Equilibrium	$0.10\text{ M} - x$		–		x		x

At equilibrium, $x = [H_3O^+]$, and so we have

$$[H_3O^+] = [HCOO^-] = 4.2 \times 10^{-3}\text{ M}$$

and

$$[HCOOH] = 0.10\text{ M} - 4.2 \times 10^{-3}\text{ M} = 0.10\text{ M}$$

Substituting the values of the equilibrium concentrations into the K_a expression yields

$$K_a = \frac{(4.2 \times 10^{-3}\text{ M})(4.2 \times 10^{-3}\text{ M})}{0.10\text{ M}} = 1.8 \times 10^{-4}\text{ M}$$

CALCULATION OF pH FROM K_a

20-22. The equation is

$$HClO(aq) + H_2O(l) \rightleftharpoons H_3O^+(aq) + ClO^-(aq)$$

We can set up a table of initial and equilibrium concentrations

Concentration	$HClO(aq)$	+	$H_2O(l)$	$\rightleftharpoons$	$H_3O^+(aq)$	+	$ClO^-(aq)$
Initial	0.15 M		–		≈ 0 M		0 M
Change	$-x$		–		$+x$		x
Equilibrium	$0.15\text{ M} - x$		–		x		x

Substituting the equilibrium concentration expressions into the K_a expression, we have

$$K_a = \frac{[H_3O^+][ClO^-]}{[HClO]} = \frac{x^2}{0.15\text{ M} - x} = 4.0 \times 10^{-8}\text{ M}$$

We shall use the method of successive approximations to find x. We ignore x in the denominator to obtain

$$\frac{x^2}{0.15\text{ M}} = 4.0 \times 10^{-8}\text{ M}$$

Solving for x, we have

$$x = 7.7 \times 10^{-5}\text{ M}$$

We see that x is very small compared to 0.15 M. Because $[H_3O^+] = x$, the pH of the solution is

$$\text{pH} = -\log(7.7 \times 10^{-5}) = 4.11$$

The concentrations of the other species are

$$[ClO^-] = [H_3O^+] = 7.7 \times 10^{-5}\text{ M}$$
$$[HClO] = 0.15\text{ M} - 7.8 \times 10^{-5}\text{ M} \approx 0.15\text{ M}$$
$$[OH^-] = \frac{1.00 \times 10^{-14}\text{ M}^2}{7.7 \times 10^{-5}\text{ M}} = 1.3 \times 10^{-10}\text{ M}$$

20-24. The equation is

$$ClCH_2COOH(aq) + H_2O(l) \rightleftharpoons H_3O^+(aq) + ClCH_2COO^-(aq)$$

We can set up a table of initial and equilibrium concentrations

Concentration	$ClCH_2COOH(aq)$	+	$H_2O(l)$	$\rightleftharpoons$	$H_3O^+(aq)$	+	$ClCH_2COO^-(aq)$
Initial	0.10 M		–		≈ 0 M		0 M
Change	$-x$		–		$+x$		$+x$
Equilibrium	$0.10\text{ M} - x$		–		x		x

Substituting the equilibrium concentration expressions into the K_a expression, we have

$$K_a = \frac{[H_3O^+][ClCH_2COO^-]}{[ClCH_2COOH]} = \frac{x^2}{0.10\text{ M} - x} = 1.4 \times 10^{-3}\text{ M}$$

Using the method of successive approximations three times, we obtain $x = 0.012$ M, 0.011 M, and 0.011 M. The pH of the solution is

$$pH = -\log(0.011) = 1.96$$

The concentrations of the other species are

$$[ClCH_2COO^-] = [H_3O^+] = 0.011\text{ M}$$
$$[ClCH_2COOH] = 0.10\text{ M} - 0.011\text{ M} = 0.09\text{ M}$$
$$[OH^-] = \frac{1.00 \times 10^{-14}\text{ M}^2}{1.1 \times 10^{-2}\text{ M}} = 9.1 \times 10^{-13}\text{ M}$$

CALCULATIONS INVOLVING K_b

20-26. The base-protonation equation is

$$CH_3CH_2NH_2(aq) + H_2O(l) \rightleftharpoons CH_3CH_2NH_3^+(aq) + OH^-(aq)$$

The pOH of the solution is given by

$$pOH = 14.00 - pH = 14.00 - 12.17 = 1.83$$

and the concentration of $OH^-(aq)$ is

$$[OH^-] = 10^{-pOH}\text{ M} = 10^{-1.83}\text{ M} = 0.015\text{ M}$$

We can set up a concentration table:

Concentration	$CH_3CH_2NH_2(aq)$	+	$H_2O(l)$	$\rightleftharpoons$	$CH_3CH_2NH_3^+(aq)$	+	$OH^-(aq)$
Initial	0.50 M		—		0 M		≈ 0 M
Change	$-x$		—		$+x$		$+x$
Equilibrium	$0.50\text{ M} - x$		—		x		x

At equilibrium, $x = [OH^-] = 0.015$ M, and so we have

$$[CH_3CH_2NH_3^+] = [OH^-] = 0.015\text{ M}$$

and

$$[CH_3CH_2NH_2] = 0.50\text{ M} - 0.015\text{ M} = 0.49\text{ M}$$

The base-protonation constant expression is

$$K_b = \frac{[CH_3CH_2NH_3^+][OH^-]}{[CH_3CH_2NH_2]} = \frac{(0.015\text{ M})(0.015\text{ M})}{0.49\text{ M}}$$
$$= 4.6 \times 10^{-4}\text{ M}$$

20-28. The equation is

$$HONH_2(aq) + H_2O(l) \rightleftharpoons HONH_3^+(aq) + OH^-(aq)$$

We can set up a table of initial and equilibrium concentrations

Concentration	$HONH_2(aq)$	+	$H_2O(l)$	$\rightleftharpoons$	$HONH_3^+(aq)$	+	$OH^-(aq)$
Initial	0.125 M		–		0 M		≈ 0 M
Change	$-x$		–		$+x$		$+x$
Equilibrium	0.125 M − x		–		x		x

Substituting the equilibrium concentration expressions into the K_b expression, we have

$$K_b = \frac{[HONH_3^+][OH^-]}{[HONH_2]} = \frac{x^2}{0.125\text{ M} - x} = 8.7 \times 10^{-9}\text{ M}$$

where we have used the value of K_b in Table 20.5. Because K_b is so small, we expect that x will be small and thus negligible compared with 0.125 M. The expression for K_b becomes

$$\frac{x^2}{0.125\text{ M}} = 8.7 \times 10^{-9}\text{ M}$$

Solving for x,

$$x = 3.3 \times 10^{-5}\text{ M}$$

We see that x is much smaller than 0.125 M. The full method of successive approximations also confirms this result. The pOH of the solution is given by

$$\text{pOH} = -\log([OH^-]/\text{M}) = -\log(3.3 \times 10^{-5}) = 4.48$$

and the pH is

$$\text{pH} = 14.00 - \text{pOH} = 14.00 - 4.48 = 9.52$$

20-30. We can set up a concentration table.

Concentration	$NH_3(aq)$	+	$H_2O(l)$	$\rightleftharpoons$	$NH_4^+(aq)$	+	$OH^-(aq)$
Initial	0.20 M		–		0 M		≈ 0 M
Change	$-x$		–		$+x$		$+x$
Equilibrium	0.20 M − x		–		x		x

Substituting the equilibrium concentration expressions into the K_b expression, we have

$$K_b = \frac{[NH_4^+][OH^-]}{[NH_3]} = \frac{x^2}{0.20\text{ M} - x} = 1.8 \times 10^{-5}\text{ M}$$

The method of successive approximations gives $x = [OH^-] = 1.9 \times 10^{-3}$ M. The pOH of the solution is given by

$$\text{pOH} = -\log([OH^-]/\text{M}) = -\log(1.9 \times 10^{-3}) = 2.72$$

and the pH is

$$\text{pH} = 14.00 - \text{pOH} = 14.00 - 2.72 = 11.28$$

LE CHÂTELIER'S PRINCIPLE

20-32. (a) The equilibrium is shifted from left to right. The addition of $OH^-(aq)$ to the solution removes $H_3O^+(aq)$ by the reaction between $H_3O^+(aq)$ and $OH^-(aq)$ to produce $H_2O(l)$.

(b) The equilibrium is shifted from right to left.

(c) Dilution shifts the equilibrium to the right because there are more moles of solute particles on the right than on the left.

(d) Addition of HCl shifts the equilibrium from right to left because of the increase in $[H_3O^+]$.

20-34. (a) The equilibrium is shifted from right to left because $\Delta H^0_{rxn} < 0$.

(b) The equilibrium is shifted from right to left because $NO_2^-(aq)$ has been added.

(c) The equilibrium is shifted from left to right. The $OH^-(aq)$ added reacts with $H_3O^+(aq)$ so that the concentration of $H_3O^+(aq)$ is decreased.

(d) The equilibrium is shifted from left to right.

CONJUGATE ACIDS AND BASES

20-36. (a) $NO_3^-(aq)$ (b) $HCOO^-(aq)$ (c) $C_6H_5COO^-(aq)$ (d) $CH_3NH_2(aq)$

20-38. (a) Acid; the conjugate base is $ClCH_2COO^-(aq)$

(b) Base; the conjugate acid is $NH_4^+(aq)$

(c) Base; the conjugate acid is $HClO(aq)$

(d) Base; the conjugate acid is $HCOOH(aq)$

(e) Acid; the conjugate base is $N_3^-(aq)$

(f) Base; the conjugate acid is $HNO_2(aq)$

20-40. We have that $K_a = \dfrac{K_w}{K_b}$,

(a) $K_a = \dfrac{1.00 \times 10^{-14}\ \text{M}^2}{1.5 \times 10^{-9}\ \text{M}} = 6.7 \times 10^{-6}\ \text{M}$ for $C_6H_5NH^+$

(b) $K_a = \dfrac{1.00 \times 10^{-14}\ \text{M}^2}{2.1 \times 10^{-5}\ \text{M}} = 4.8 \times 10^{-10}\ \text{M}$ for HCN

(c) $K_a = \dfrac{1.00 \times 10^{-14}\ \text{M}^2}{8.7 \times 10^{-9}\ \text{M}} = 1.1 \times 10^{-6}\ \text{M}$ for NH_3OH^+

(d) $K_a = \dfrac{1.00 \times 10^{-14}\ \text{M}^2}{5.4 \times 10^{-4}\ \text{M}} = 1.9 \times 10^{-11}\ \text{M}$ for $(CH_3)_2NH_2^+$

ACID-BASE PROPERTIES OF SALTS

20-42. (a) acidic (acidic cation, neutral anion)

$$[Fe(H_2O)_6]^{3+}(aq) + H_2O(l) \rightleftharpoons [Fe(OH)(H_2O)_5]^{2+}(aq) + H_3O^+(aq)$$

(b) neutral (neutral cation, neutral anion)

(c) acidic (neutral cation, acidic anion)

$$HSO_4^-(aq) + H_2O(l) \rightleftharpoons SO_4^{2-}(aq) + H_3O^+(aq)$$

(d) basic (neutral cation, basic anion)

$$F^-(aq) + H_2O(l) \rightleftharpoons HF(aq) + OH^-(aq)$$

20-44. (a) neutral cation, basic anion; basic solution

(b) acidic cation, neutral anion; acidic solution

(c) neutral cation, acidic anion; acidic solution

(d) neutral cation, neutral anion; neutral solution

20-46. The aluminum salts produce acidic solutions because $Al^{3+}(aq)$ is an acidic cation and $SO_4^{2-}(aq)$ is a weak basic anion. See Tables 20.7 and 20.8.

pH CALCULATIONS OF SALT SOLUTIONS

20-48. The salt $NaCH_3CH_2COO(s)$ dissociates completely in water to yield $Na^+(aq)$ and $CH_3CH_2COO^-(aq)$. The $Na^+(aq)$ is a neutral cation and the equation of the reaction of the (basic) anion with water is

$$CH_3CH_2COO^-(aq) + H_2O(l) \rightleftharpoons CH_3CH_2COOH(aq) + OH^-(aq)$$

The value of the equilibrium constant is

$$K_b = \frac{K_w}{K_a} = \frac{1.00 \times 10^{-14}\ M^2}{1.4 \times 10^{-5}\ M} = 7.1 \times 10^{-10}\ M$$

We can set up a concentration table.

Concentration	$CH_3CH_2COO^-(aq)$	+	$H_2O(l)$	$\rightleftharpoons$	$CH_3CH_2COOH(aq)$	+	$OH^-(aq)$
Initial	0.20 M		–		0 M		≈ 0 M
Change	$-x$		–		$+x$		$+x$
Equilibrium	0.20 M $- x$		–		x		x

Substituting the equilibrium concentration expressions into the K_b expression, we have

$$K_b = \frac{[CH_3CH_2COOH]\,[OH^-]}{[CH_3CH_2COO^-]} = \frac{x^2}{0.20\ M - x} = 7.1 \times 10^{-10}\ M$$

The method of successive approximations gives $x = [OH^-] = 1.2 \times 10^{-5}$ M. Using the ion-product constant of water, we obtain

$$[H_3O^+] = \frac{1.00 \times 10^{-14}\ M^2}{[OH^-]} = \frac{1.00 \times 10^{-14}\ M^2}{1.2 \times 10^{-5}\ M} = 8.3 \times 10^{-10}\ M$$

The pH of the solution is

$$pH = -\log([H_3O^+]/M) = -\log(8.3 \times 10^{-10}) = 9.08$$

20-50. The $Na^+(aq)$ is a neutral cation and the equation of the reaction of the (basic) anion with water is

$$NO_2^-(aq) + H_2O(l) \rightleftharpoons HNO_2(aq) + OH^-(aq)$$

The value of the equilibrium constant is $K_b = 1.8 \times 10^{-11}$ M (Table 20.6). We can set up a concentration table.

Concentration	$NO_2^-(aq)$	+	$H_2O(l)$	$\rightleftharpoons$	$HNO_2(aq)$	+	$OH^-(aq)$
Initial	0.25 M		–		0 M		≈ 0 M
Change	$-x$		–		$+x$		x
Equilibrium	0.25 M $- x$		–		x		x

Substituting the equilibrium concentration expressions into the K_b expression, we have

$$K_b = \frac{[HNO_2]\,[OH^-]}{[OH^-]} = \frac{x^2}{0.25\text{ M} - x} = 1.8 \times 10^{-11}\text{ M}$$

The method of successive approximations gives $x = [OH^-] = 2.1 \times 10^{-6}$ M, and so

$$[HNO_2] = 2.1 \times 10^{-6}\text{ M}$$

$$[NO_2^-] = 0.25\text{ M} - x = 0.25\text{ M} - 2.1 \times 10^{-6}\text{ M} \approx 0.25\text{ M}$$

Using the ion-product constant of water, we obtain

$$[H_3O^+] = \frac{1.00 \times 10^{-14}\text{ M}^2}{[OH^-]} = \frac{1.00 \times 10^{-14}\text{ M}^2}{2.1 \times 10^{-6}\text{ M}} = 4.8 \times 10^{-9}\text{ M}$$

The pH of the solution is

$$\text{pH} = -\log([H_3O^+]/\text{M}) = -\log(4.8 \times 10^{-9}) = 8.32$$

20-52. The number of moles of in 25.0 grams of $Ba(CH_3COO)_2(aq)$ is

$$n = (25.0\text{ g})\left(\frac{1\text{ mol }Ba(CH_3COO)_2}{255.42\text{ g }Ba(CH_3COO)_2}\right) = 0.0979\text{ mol}$$

The molarity of the $Ba(CH_3COO)_2(aq)$ solution is

$$\text{molarity} = \frac{0.0979\text{ mol}}{1.00\text{ L}} = 0.0979\text{ M}$$

Because each $Ba(CH_3COO)_2$ molecule yields two $CH_3COO^-(aq)$ ions in solution, the concentration of $CH_3COO^-(aq)$ is $(2)(0.0979\text{ M}) = 0.196$ M. The $Ba^{2+}(aq)$ is a neutral cation and the equation of the reaction of the (basic) anion with water is

$$CH_3COO^-(aq) + H_2O(l) \rightleftharpoons CH_3COOH(aq) + OH^-(aq)$$

We can set up a concentration table.

Concentration	$CH_3COO^-(aq)$	+	$H_2O(l)$	$\rightleftharpoons$	$CH_3COOH(aq)$	+	$OH^-(aq)$
Initial	0.196 M		—		0 M		≈ 0 M
Change	$-x$		—		$+x$		$+x$
Equilibrium	$0.196\text{ M} - x$		—		x		x

Substituting the equilibrium concentration expressions into the K_b expression (Table 20.6), we have

$$K_b = \frac{[OH^-]\,[CH_3COOH]}{[CH_3COO^-]} = \frac{x^2}{0.196\text{ M} - x} = 5.6 \times 10^{-10}\text{ M}$$

The method of successive approximations gives $x = [OH^-] = 1.0 \times 10^{-5}$ M. Using the ion-product constant of water, we obtain

$$[H_3O^+] = \frac{1.00 \times 10^{-14}\text{ M}^2}{[OH^-]} = \frac{1.00 \times 10^{-14}\text{ M}^2}{1.0 \times 10^{-5}\text{ M}} = 1.0 \times 10^{-9}\text{ M}$$

The pH of the solution is

$$\text{pH} = -\log([\text{H}_3\text{O}^+]/\text{M}) = -\log(1.0 \times 10^{-9}) = 9.00$$

20-54. In aqueous solution $Tl^{3+}(aq)$ exists as $[Tl(H_2O)_6]^{3+}(aq)$ and $Br^-(aq)$ is a neutral anion. We can set up a concentration table.

Concentration	$[Tl(H_2O)_6]^{3+}(aq)$	+	$H_2O(l)$	$\rightleftharpoons$	$[Tl(OH)(H_2O)_5]^{2+}(aq)$	+	$H_3O^+(aq)$
Initial	0.10 M		–		0 M		≈ 0 M
Change	$-x$		–		$+x$		$+x$
Equilibrium	0.10 M $- x$		–		x		x

Substituting the equilibrium concentration expressions into the K_a expression, we have

$$K_a = \frac{[H_3O^+][Tl(OH)(H_2O)_5^{2+}]}{[Tl(H_2O)_6^{3+}]} = \frac{x^2}{0.10\text{ M} - x} = 7.0 \times 10^{-2}\text{ M}$$

Because the value of K_a is large, we must use the quadratic formula to solve for $[H_3O^+]$, which gives $x = [H_3O^+] = 0.056$ M. The pH of the solution is

$$\text{pH} = -\log([\text{H}_3\text{O}^+]/\text{M}) = -\log(0.056) = 1.25$$

POLYPROTIC ACIDS

20-56. The two pK_a values are 1.85 and 7.2, and so $K_{a_1} = 0.014$ M and $K_{a_2} = 6 \times 10^{-8}$ M. Using the fact that $K_{a_1} \gg K_{a_2}$, we can neglect the second dissociation as a source of $H_3O^+(aq)$. We set up a concentration table.

Concentration	$H_2SO_3(aq)$	+	$H_2O(l)$	$\rightleftharpoons$	$H_3O^+(aq)$	+	$HSO_3^-(aq)$
Initial	0.50 M		–		≈ 0 M		0 M
Change	$-x$		–		$+x$		$+x$
Equilibrium	0.50 M $- x$		–		x		x

The acid-dissociation-constant expression is

$$K_{a_1} = \frac{[H_3O^+]^2}{0.50\text{ M} - [H_3O^+]} = 0.014\text{ M}$$

Because the value of K_a is large, we must use the quadratic formula to solve for $[H_3O^+]$, which gives $[H_3O^+] = 0.077$ M, or pH = 1.11.

20-58. The three pK_a values are 3.13, 4.76, and 6.40 and so $K_{a_1} = 7.4 \times 10^{-4}$ M, $K_{a_2} = 1.7 \times 10^{-5}$ M, and $K_{a_3} = 4.0 \times 10^{-7}$ M. Using the fact that $K_{a_1} > K_{a_2}$ and K_{a_3}, we can neglect the second and third dissociations as a source of $H_3O^+(aq)$. We set up a concentration table.

Concentration	$HO(CH_2COOH)_2COOH(aq)$ +	$H_2O(l)$ $\rightleftharpoons$	$H_3O^+(aq)$ +	$HO(CH_2COOH)_2COO^-(aq)$
Initial	0.050 M	–	≈ 0 M	0 M
Change	$-x$	–	$+x$	$+x$
Equilibrium	0.050 M $- x$	–	x	x

The first acid-dissociation-constant expression is

$$K_{a_1} = \frac{x^2}{0.050\ \text{M} - x} = 7.4 \times 10^{-4}\ \text{M}$$

which yields $x = [\text{H}_3\text{O}^+]_1 = 5.7 \times 10^{-3}$ M, or pH = 2.24. Assuming that each dissociation reaction can be treated independently, we can calculate contribution to the concentration of $\text{H}_3\text{O}^+(aq)$ by the subsequent acid-dissociation reactions by setting up the relevant concentration tables and acid-dissociation reaction equations.

Concentration	$\text{HO(CH}_2\text{COOH)}_2\text{COO}^-(aq)$	$+\ \text{H}_2\text{O}(l)\ \rightleftharpoons$	$\text{H}_3\text{O}^+(aq)$	$+\ \text{HO(CH}_2\text{COO}^-)_2\text{COOH}(aq)$
Initial	5.7×10^{-3} M	–	5.7×10^{-3} M	0 M
Change	$-y$	–	$+y$	$+y$
Equilibrium	$5.7 \times 10^{-3}\ \text{M} - y$	–	$5.7 \times 10^{-3}\ \text{M} - y$	y

for which

$$K_{a_2} = 1.7 \times 10^{-5}\ \text{M} = \frac{[\text{H}_3\text{O}^+][\text{HO(CH}_2\text{COO}^-)_2\text{COOH}]}{[\text{HO(CH}_2\text{COOH)}_2\text{COO}^-]} = \frac{(5.7 \times 10^{-3}\ \text{M} + y)(y)}{5.7 \times 10^{-3}\ \text{M} - y} \approx y$$

Therefore, the contribution of y to the total concentration of $\text{H}_3\text{O}^+(aq)$ is $y = [\text{H}_3\text{O}^+]_2 = 1.7 \times 10^{-5}$ M, and $[\text{H}_3\text{O}^+] = 5.7 \times 10^{-3}\ \text{M} + 1.7 \times 10^{-5}\ \text{M} = 5.7 \times 10^{-3}$ M. For the third acid-dissociation reaction, we have

Concentration	$\text{HO(CH}_2\text{COO}^-)_2\text{COOH}(aq)$	$+\ \text{H}_2\text{O}(l)\ \rightleftharpoons$	$\text{H}_3\text{O}^+(aq)$	$+\ \text{HO(CH}_2\text{COO}^-)_2\text{COO}^-(aq)$
Initial	1.7×10^{-5} M	–	5.7×10^{-3} M	0 M
Change	$-z$	–	$+z$	$+z$
Equilibrium	$1.7 \times 10^{-5}\ \text{M} - z$	–	$5.7 \times 10^{-3}\ \text{M} - z$	z

for which

$$K_{a_3} = 4.0 \times 10^{-7}\ \text{M} = \frac{[\text{H}_3\text{O}^+][\text{HO(CH}_2\text{COO}^-)_2\text{COO}-]}{[\text{HO(CH}_2\text{COO}^-)_2\text{COOH}]} = \frac{(5.7 \times 10^{-3}\ \text{M} + z)(z)}{5.7 \times 10^{-3}\ \text{M} - z} \approx z$$

Therefore, the contribution of z to the total concentration of $\text{H}_3\text{O}^+(aq)$ is $z = [\text{H}_3\text{O}^+]_3 = 4.0 \times 10^{-7}$ M, and $[\text{H}_3\text{O}^+] = 5.7 \times 10^{-3}\ \text{M} + 1.7 \times 10^{-5}\ \text{M} + 4.0 \times 10^{-7}\ \text{M} = 5.7 \times 10^{-3}$ M. The fraction of total $[\text{H}_3\text{O}^+]$ for each dissociation is given by

$$f_1 = \frac{[\text{H}_3\text{O}^+]_1}{[\text{H}_3\text{O}^+]_1 + [\text{H}_3\text{O}^+]_2 + [\text{H}_3\text{O}^+]_3} = \frac{5.7 \times 10^{-3}\ \text{M}}{5.7 \times 10^{-3}\ \text{M}} = 1.0$$

$$f_2 = \frac{[\text{H}_3\text{O}^+]_2}{5.7 \times 10^{-3}\ \text{M}} = \frac{1.7 \times 10^{5}\ \text{M}}{5.7 \times 10^{-3}\ \text{M}} \approx 3.0 \times 10^{-3}$$

$$f_3 = \frac{[\text{H}_3\text{O}^+]_3}{5.7 \times 10^{-3}\ \text{M}} = \frac{4.0 \times 10^{-7}\ \text{M}}{5.7 \times 10^{-3}\ \text{M}} \approx 7.0 \times 10^{-5}$$

ADDITIONAL PROBLEMS

20-60. (a) The greater the value of K_a or the lower the value of pK_a, the greater the strength of the acid.

(b) The greater the value of K_b or the lower the value of pK_b, the greater the strength of the base.

20-62. Yes. For example, the pH of a 6.0 M HCl(*aq*) solution is −0.78.

20-64. (a) NH_3 yields $OH^-(aq)$ in aqueous solution and thus is an Arrhenius base. It is also a proton acceptor and thus is a Brønsted–Lowry base. It can act as an electron-pair acceptor and thus is a Lewis base.

(b) Br^- is neither an Arrhenius acid nor a Brønsted–Lowry acid. It can act as an electron-pair donor in aqueous solution and hence is a Lewis base.

(c) NaOH yields $OH^-(aq)$ in aqueous solution and thus is an Arrhenius base; it can act as a proton acceptor and thus is a Brønsted–Lowry base; and it can act as an electron-pair donor and thus is a Lewis base.

20-66. The hydrogen halides, HCl(*g*) amd HBr(*g*), dissociate in water to form protons, $H^+(aq)$ and neutral anions. The protons react with water to form hydronium ions, making aqueous solutions acidic. The metal hydrides, NaH(*s*) and KH(*s*), dissociate in water to form neutral cations and hydride anions, $H^-(aq)$. The hydride ions react with water to form hydrogen gas and hydroxide ions, making aqueous solutions basic.

20-68. Lithium is an active metal (see Chapter 10), and so reacts with water according to the equation

$$2\,\text{Li}(s) + 2\,\text{H}_2\text{O}(l) \rightleftharpoons 2\,\text{LiOH}(aq) + \text{H}_2(g)$$

Because LiOH(*aq*) is a strong base, the resulting solution is basic. In contrast, LiCl(*s*) is a salt composed of a neutral cation and a neutral anion, and so a LiCl(*aq*) solution is neutral.

20-70. The acid dissociation constant expression is

$$K_a = \frac{[\text{H}_3\text{O}^+]\,[\text{NO}_2^-]}{[\text{HNO}_2]} = 5.6 \times 10^{-4}\ \text{M}$$

The ratio of the concentrations of $HNO_2(aq)$ to $NO_2^-(aq)$ is

$$\text{ratio} = \frac{[\text{HNO}_2]}{[\text{NO}_2^-]} = \frac{[\text{H}_3\text{O}^+]}{K_a} = \frac{[\text{H}_3\text{O}^+]}{5.6 \times 10^{-4}\ \text{M}} = \frac{0.10\ \text{M}}{5.6 \times 10^{-4}\ \text{M}} = 180$$

Most of the $HNO_2(aq)$ is undissociated at this pH.

20-72. The ion-product constant for water at 37°C is

$$K_w = [\text{H}_3\text{O}^+]\,[\text{OH}^-] = 2.40 \times 10^{-14}\ \text{M}^2$$

In a neutral solutions, $[H_3O^+] = [OH^-]$. Therefore

$$K_w = [\text{H}_3\text{O}^+]\,[\text{OH}^-] = [\text{H}_3\text{O}^+]^2 = 2.40 \times 10^{-14}\ \text{M}^2$$

Taking the square root of both sides yields

$$[\text{H}_3\text{O}^+] = 1.55 \times 10^{-7}\ \text{M}$$

The pH of the neutral solution is

$$\text{pH} = -\log([\text{H}_3\text{O}^+]/\text{M}) = -\log(1.55 \times 10^{-7}) = 6.81$$

A solution with a pH of 7.00 is basic at 37°C because the pH of the solution is greater than 6.81, the pH of a neutral solution at 37°C.

20-74. The pOH of the solution is given by

$$\text{pOH} = 14.00 - \text{pH} = 14.00 - 10.52 = 3.48$$

and the value of $[\text{OH}^-]$ is given by

$$[\text{OH}^-] = 10^{-\text{pOH}}\ \text{M} = 10^{-3.48}\ \text{M} = 3.3 \times 10^{-4}\ \text{M}$$

Because one mole of $\text{Mg(OH)}_2(aq)$ yields two moles of $\text{OH}^-(aq)$ in water, the concentration of $\text{Mg(OH)}_2(aq)$ is

$$[\text{Mg(OH)}_2] = \frac{1}{2}[\text{OH}^-] = 1.65 \times 10^{-4}\ \text{M}$$

Solubility often is expressed as the number of grams per 100 mL of solution. The number of moles in 100 mL of solution is

$$n = \text{molarity} \times \text{volume} = (1.65 \times 10^{-4}\ \text{M})(0.100\ \text{L}) = 1.65 \times 10^{-5}\ \text{mol}$$

The mass in 1.65×10^{-5} moles of Mg(OH)_2 is

$$\text{mass} = (1.65 \times 10^{-5}\ \text{mol})\left(\frac{58.32\ \text{g Mg(OH)}_2}{1\ \text{mol Mg(OH)}_2}\right) = 9.62 \times 10^{-4}\ \text{g}$$

The solubility of $\text{Mg(OH)}_2(s)$ is 9.62×10^{-4} grams per 100 mL of solution.

20-76. Take 1000 grams of solution (one liter). The composition is 55 grams of acetic acid, or 0.916 moles, or a concentration of 0.916 mol·L^{-1}. We can set up a concentration table.

Concentration	$\text{CH}_3\text{COOH}(aq)$	+	$\text{H}_2\text{O}(l)$	$\rightleftharpoons$	$\text{H}_3\text{O}^+(aq)$	+	$\text{CH}_3\text{COO}^-(aq)$
Initial	0.916 M		–		≈ 0 M		0 M
Change	$-x$		–		$+x$		x
Equilibrium	$0.916\ \text{M} - x$		–		x		x

Substituting the equilibrium concentration expressions into the K_a expression (Table 20.6), we have

$$K_a = \frac{[\text{H}_3\text{O}^+][\text{CH}_3\text{COO}^-]}{[\text{CH}_3\text{COOH}]} = \frac{x^2}{0.916\ \text{M} - x} = 1.8 \times 10^{-5}\ \text{M}$$

The method of successive approximations yields $[\text{H}_3\text{O}^+] = 4.1 \times 10^{-3}$ M and the pH of the solution is 2.39.

20-78. The equation of the reaction is

$$\text{C}_6\text{H}_5\text{COOH}(aq) + \text{H}_2\text{O}(l) \rightleftharpoons \text{C}_6\text{H}_5\text{COO}^-(aq) + \text{H}_3\text{O}^+(aq)$$

The acid dissociation constant expression is

$$K_a = \frac{[H_3O^+]\,[C_6H_5COO^-]}{[C_6H_5COOH]} = 6.3 \times 10^{-5}\ M$$

The ratio of the concentrations of the benzoic acid to benzoate is

$$\text{ratio} = \frac{[C_6H_5COOH]}{[C_6H_5COO^-]} = \frac{[H_3O^+]}{K_a} = \frac{[H_3O^+]}{6.3 \times 10^{-5}\ M}$$

We can find the value of $[H_3O^+]$ from the pH of the solution

$$[H_3O^+] = 10^{-pH}\ M = 10^{-3.00}\ M = 1.0 \times 10^{-3}\ M$$

The value of the ratio is

$$\text{ratio} = \frac{1.0 \times 10^{-3}\ M}{6.3 \times 10^{-5}\ M} = 16$$

20-80. Because pK_{a_2} is so much greater than $pK_{a_1} = 4.04$, we neglect the second dissociation constant of ascorbic acid. The stoichiometric concentration of ascorbic acid is

$$[\text{ascorbic acid}] = \frac{(0.500\text{ g ascorbic acid})\left(\dfrac{1\text{ mol ascorbic acid}}{176.12\text{ g ascorbic acid}}\right)}{1.00\text{ L}} = 2.84 \times 10^{-3}\ M$$

The equation for the reaction is

$$\text{ascorbic acid}(aq) + H_2O(l) \rightleftharpoons \text{ascorbate}(aq) + H_3O^+(aq)$$

and the corresponding equilibrium constant expression is

$$K_{a_1} = \frac{[H_3O^+]\,[\text{ascorbate}]}{[\text{ascorbic acid}]} = \frac{[H_3O^+]^2}{2.84 \times 10^{-3}\ M - [H_3O^+]} = 9.1 \times 10^{-5}\ M$$

where

$$K_{a_1} = 10^{-pK_{a_1}} = 10^{-4.04} = 9.1 \times 10^{-5}\ M$$

The method of successive approximations gives $[H_3O^+] = 4.6 \times 10^{-4}$ M. The pH of the solution is

$$pH = -\log([H_3O^+]/M) = -\log(4.7 \times 10^{-4}) = 3.34$$

20-82. (a) At equilibrium, we have

$$[CH_3CH_2OH_2^+]\,[CH_3CH_2O^-] = 8 \times 10^{-20}\ M^2$$

In a neutral ethanol solution, we have that $[CH_3CH_2OH_2^+] = [CH_3CH_2O^-]$. Thus,

$$[CH_3CH_2OH_2^+] = \sqrt{8 \times 10^{-20}\ M^2} = 3 \times 10^{-10}\ M$$

The pH is

$$pH = -\log([CH_3CH_2OH_2^+]/M) = -\log(3 \times 10^{-10}) = 9.5$$

(b) We have that $[CH_3CH_2O^-] = 0.010$ M. From the ion-product constant expression, we obtain

$$[CH_3CH_2OH_2^+] = \frac{8 \times 10^{-20}\ M^2}{0.010\ M} = 8 \times 10^{-18}\ M$$

The pH is

$$pH = -\log(8 \times 10^{-18}) = 17.1$$

20-84. The ion-product constant for water at 0°C is

$$K_w = [H_3O^+][OH^-] = 0.12 \times 10^{-14}\ M^2$$

In a neutral solution $[H_3O^+] = [OH^-]$. Therefore,

$$K_w = [H_3O^+]^2 = 0.12 \times 10^{-14}\ M^2$$

Taking the square root of both sides yields

$$[H_3O^+] = 3.5 \times 10^{-8}\ M$$

The pH of the solution is

$$pH = -\log(3.5 \times 10^{-8}) = 7.46$$

At 0°C, an aqueous solution with a pH of 7.25 is acidic. The pH of the solution is less than 7.46, the pH of a neutral solution.

20-86. The equation of the reaction is

$$[Ga(H_2O)_6]^{3+}(aq) + H_2O(l) \rightleftharpoons [Ga(OH)(H_2O)_5]^{2+}(aq) + H_3O^+(aq)$$

We can set up a concentration table.

Concentration	$[Ga(H_2O)_6]^{3+}(aq)$	+	$H_2O(l)$	$\rightleftharpoons$	$H_3O^+(aq)$	+	$[Ga(OH)(H_2O)_5]^{2+}(aq)$
Initial	0.200 M		—		≈ 0 M		0 M
Change	$-x$		—		$+x$		$+x$
Equilibrium	0.200 M $- x$		—		x		x

The acid dissociation constant expression is (Table 20.8)

$$K_a = \frac{[H_3O^+][[Ga(OH)(H_2O)_5]^{2+}]}{[[Ga(H_2O)_6]^{3+}]} = \frac{x^2}{0.200\ M - x} = 1.7 \times 10^{-3}\ M$$

The method of successive approximations yields $x = [H_3O^+] = 0.018$ and the pH of the solution is

$$pH = -\log(0.018) = 1.74$$

20-88. We can set up a concentration table.

Concentration	$(CH_3)_2NH(aq)$	+	$H_2O(l)$	$\rightleftharpoons$	$(CH_3)_2NH_2^+(aq)$	+	$OH^-(aq)$
Initial	0.150 M		–		0 M		≈ 0 M
Change	$-x$		–		$+x$		$+x$
Equilibrium	$0.150\text{ M} - x$		–		x		x

Substituting the equilibrium concentration expressions into the K_b expression (Table 20.5), we have

$$K_b = \frac{[(CH_3)_2NH_2^+][OH^-]}{[(CH_3)_2NH]} = \frac{x^2}{0.150\text{ M} - x} = 5.4 \times 10^{-4}\text{ M}$$

The method of successive approximations yields

$$x = [OH^-] = 8.7 \times 10^{-3}\text{ M}$$

The concentrations of the species in the solution are

$$[(CH_3)_2NH_2^+] = [OH^-] = 8.7 \times 10^{-3}\text{ M}$$

$$[(CH_3)_2NH] = 0.150\text{ M} - 8.7 \times 10^{-3}\text{ M} = 0.141\text{ M}$$

$$[H_3O^+] = \frac{1.00 \times 10^{-14}\text{ M}^2}{8.7 \times 10^{-3}\text{ M}} = 1.15 \times 10^{-12}\text{ M}$$

$$\text{pH} = -\log(1.15 \times 10^{-12}) = 11.94$$

20-90. We can set up a concentration table.

Concentration	$HCNO(aq)$	+	$H_2O(l)$	$\rightleftharpoons$	$H_3O^+(aq)$	+	$CNO^-(aq)$
Initial	0.100 M		–		≈ 0 M		0 M
Change	$-x$		–		$+x$		$+x$
Equilibrium	$0.100\text{ M} - x$		–		x		x

Substituting the equilibrium concentration expressions into the K_a expression (Table 20.4), we have

$$K_a = \frac{[H_3O^+][CNO^-]}{[HCNO]} = \frac{x^2}{0.100\text{ M} - x} = 3.5 \times 10^{-4}\text{ M}$$

The method of successive approximations yields

$$x = [H_3O^+] = [CNO^-] = 5.7 \times 10^{-3}\text{ M}$$

The percent of the acid that is dissociated is

$$\%\text{ dissociated} = \frac{[CNO^-]}{[HCNO]_0} \times 100 = \frac{5.7 \times 10^{-3}\text{ M}}{0.100\text{ M}} \times 100 = 5.7\%$$

20-92.* The pH is less than 7.0 because of dissolved $CO_2(g)$ from the atmosphere. Carbon dioxide reacts with water according to

$$CO_2(g) + 2\,H_2O(l) \longrightarrow HCO_3^-(aq) + H_3O^+(aq)$$

Thus, the water is acidic.

20-94.* If we take $[H_3O^+] = 2.60 \times 10^{-8}$ M, then we calculate for the pH

$$\text{pH} = -\log(2.60 \times 10^{-8}) = 7.585$$

But this answer is wrong because an acidic solution at 25°C has a pH less than 7.00. The given concentration of $HCl(aq)$ is so low that we cannot neglect the dissociation of water as an important source of $H_3O^+(aq)$. Thus, we must consider the dissociation

$$2\,H_2O(l) \rightleftharpoons H_3O^+(aq) + OH^-(aq)$$

At equilibrium, $[H_3O^+] = 2.60 \times 10^{-8}\ \text{M} + [OH^-]$. Using the K_w expression, we have

$$K_w = 1.00 \times 10^{-14}\ \text{M}^2 = [H_3O^+]\,[OH^-] = (2.60 \times 10^{-8}\ \text{M} + [OH^-])[OH^-]$$

This gives a quadratic equation in $[OH^-]$:

$$[OH^-]^2 + (2.60 \times 10^{-8}\ \text{M})[OH^-] - 1.00 \times 10^{-14}\ \text{M}^2 = 0$$

Using the quadratic formula, we have

$$[OH^-] = \frac{-2.60 \times 10^{-8}\ \text{M} \pm \sqrt{6.76 \times 10^{-16}\ \text{M}^2 + 4.00 \times 10^{-14}\ \text{M}^2}}{2}$$

$$8.78 \times 10^{-8}\ \text{M} \qquad (\text{and } -1.13 \times 10^{-7}\ \text{M})$$

Thus, we have

$$[H_3O^+] = 2.60 \times 10^{-8}\ \text{M} + 8.78 \times 10^{-8}\ \text{M} = 1.14 \times 10^{-7}\ \text{M}$$

and

$$\text{pH} = -\log(1.14 \times 10^{-7}) = 6.943$$

Note that $\text{pH} < 7.00$; the solution is acidic.

CHAPTER 21. Buffers and the Titration of Acids and Bases

BUFFER CALCULATIONS

21-2. We first identify the acid and the base:

$$\text{acid}: \mathrm{CH_3COOH}(aq) \qquad \text{base}: \mathrm{CH_3COO^-}(aq) \quad \text{from } \mathrm{NaCH_3COO}(aq)$$

The stoichiometric concentrations of the acid and base forms are

$$[\text{acid}]_0 = 0.10\ \mathrm{M} \qquad \text{and} \qquad [\text{base}]_0 = 0.20\ \mathrm{M}$$

The value of the pK_a of acetic acid is 4.74 (Table 20.6). From the Henderson–Hasselbalch equation we have

$$\begin{aligned} \mathrm{pH} &= \mathrm{p}K_a + \log\frac{[\text{base}]_0}{[\text{acid}]_0} = 4.74 + \log\frac{0.20\ \mathrm{M}}{0.10\ \mathrm{M}} \\ &= 4.74 + 0.30 = 5.04 \end{aligned}$$

21-4. The stoichiometric concentrations of the acid and base forms are

$$\begin{aligned} [\text{acid}]_0 &= [\mathrm{HNO_2}]_0 = 0.20\ \mathrm{M} \\ [\text{base}]_0 &= [\mathrm{NO_2^-}]_0 = 0.15\ \mathrm{M} \quad \text{from } \mathrm{NaNO_2}(aq) \end{aligned}$$

The value of the pK_a of nitrous acid is 3.25 (Table 20.4). From the Henderson–Hasselbalch equation we have

$$\mathrm{pH} = \mathrm{p}K_a + \log\frac{[\text{base}]_0}{[\text{acid}]_0} = 3.25 + \log\frac{0.15\ \mathrm{M}}{0.20\ \mathrm{M}} = 3.25 - 0.125 = 3.13$$

21-6. The stoichiometric concentrations of the acid and base forms are

$$\begin{aligned} [\text{acid}]_0 &= [\mathrm{NH_4^+}]_0 = 0.40\ \mathrm{M} \quad \text{from } \mathrm{NH_4Cl}(aq) \\ [\text{base}]_0 &= [\mathrm{NH_3}]_0 = 0.20\ \mathrm{M} \end{aligned}$$

The value of the pK_a of conjugate acid-base pair NH_4^+/NH_3 is $14.00 - pK_b = 14.00 - 4.75 = 9.25$ (Table 20.5). From the Henderson–Hasselbalch equation we have

$$pH = pK_a + \log\frac{[\text{base}]_0}{[\text{acid}]_0} = 9.25 + \log\frac{0.20\text{ M}}{0.40\text{ M}} = 9.25 - 0.30 = 8.95$$

ADDITION OF ACIDS AND BASES TO BUFFERS

21-8. (a) The number of millimoles of $H_3O^+(aq)$ in 5.00 milliliters of 0.100 M $HCl(aq)$ is

$$\text{millimoles of } H_3O^+ = MV = (0.100\text{ M})(5.00\text{ mL}) = 0.500\text{ mmol}$$

The $H_3O^+(aq)$ reacts with $NH_3(aq)$ in the buffer via the reaction

$$NH_3(aq) + H_3O^+(aq) \rightleftharpoons NH_4^+(aq) + H_2O(l)$$

The number of millimoles of $NH_3(aq)$ in the buffer solution before the addition of $HCl(aq)$ is

$$\text{millimoles of } NH_3 = MV = (0.100\text{ M})(100.0\text{ mL}) = 10.0\text{ mmol}$$

The number of millimoles of $NH_3(aq)$ in the buffer solution after the addition of $HCl(aq)$ is

$$\begin{aligned}\text{millimoles of } NH_3 &= \text{millimoles of } NH_3 \text{ before} - \text{millimoles of } H_3O^+ \text{ added}\\ &= 10.0\text{ mmol} - 0.500\text{ mmol} = 9.5\text{ mmol}\end{aligned}$$

The concentration of $NH_3(aq)$ after the addition of $HCl(aq)$ is

$$[NH_3] = \frac{9.5\text{ mmol}}{105.0\text{ mL}} = 0.0905\text{ M}$$

The number of millimoles of $NH_4^+(aq)$ in the buffer solution before the addition of $HCl(aq)$ is

$$\text{millimoles of } NH_4^+ = MV = (0.100\text{ M})(100.0\text{ mL}) = 10.0\text{ mmol}$$

The number of millimoles of $NH_4^+(aq)$ in the buffer solution after the addition of $HCl(aq)$ is

$$\begin{aligned}\text{millimoles of } NH_4^+ &= \text{millimoles of } NH_4^+ \text{ before} + \text{millimoles of } H_3O^+ \text{ added}\\ &= 10.0\text{ mmol} + 0.500\text{ mmol} = 10.5\text{ mmol}\end{aligned}$$

The concentration of $NH_4^+(aq)$ after the addition of $HCl(aq)$ is

$$[NH_4^+] = \frac{10.5\text{ mmol}}{105.0\text{ mL}} = 0.100\text{ M}$$

The initial pH of the buffer is (Table 20.6)

$$pH = pK_a + \log\frac{[\text{base}]_0}{[\text{acid}]_0} = 9.25 + \log\frac{0.100\text{ M}}{0.100\text{ M}} = 9.25$$

The final pH of the buffer is

$$pH = pK_a + \log\frac{[\text{base}]_0}{[\text{acid}]_0} = 9.25 + \log\frac{0.0905\text{ M}}{0.100\text{ M}} = 9.21$$

The change in pH is $9.21 - 9.25 = -0.04$.

(b) The $OH^-(aq)$ reacts with $NH_4^+(aq)$ in the buffer via the reaction

$$NH_4^+(aq) + OH^-(aq) \longrightarrow NH_3(aq) + H_2O(l)$$

The number of millimoles of $OH^-(aq)$ in 5.00 milliliters of 0.100 M $NaOH(aq)$ is

$$\text{millimoles of OH}^- = MV = (0.100\text{ M})(5.00\text{ mL}) = 0.500\text{ mmol}$$

The number of millimoles of $NH_3(aq)$ ion in the buffer solution after the addition of $NaOH(aq)$ is

$$\begin{aligned}\text{millimoles of NH}_3 &= \text{millimoles of NH}_3\text{ before} + \text{millimoles of OH}^-\text{ added}\\ &= 10.0\text{ mmol} + 0.500\text{ mmol} = 10.5\text{ mmol}\end{aligned}$$

The concentration of $NH_3(aq)$ in the buffer solution after the addition of $NaOH(aq)$ is

$$[NH_3] = \frac{10.5\text{ mmol}}{105.0\text{ mL}} = 0.100\text{ M}$$

The number of millimoles of $NH_4^+(aq)$ in the buffer solution after the addition of $NaOH(aq)$ is

$$\begin{aligned}\text{millimoles of NH}_4^+ &= \text{millimoles of NH}_4^+\text{ before} - \text{millimoles of OH}^-\text{ added}\\ &= 10.0\text{ mmol} - 0.500\text{ mmol} = 9.5\text{ mmol}\end{aligned}$$

The concentration of $NH_4^+(aq)$ in the buffer solution after the addition of $NaOH(aq)$ is

$$[NH_4^+] = \frac{9.5\text{ mmol}}{105.0\text{ mL}} = 0.0905\text{ M}$$

The final pH of the buffer is

$$\text{pH} = \text{p}K_a + \log\frac{[\text{base}]_0}{[\text{acid}]_0} = 9.25 + \log\frac{0.100\text{ M}}{0.0905\text{ M}} = 9.29$$

The change in pH is $9.29 - 9.25 = 0.04$.

21-10. The number of moles of pyridine available is

$$\text{moles of pyridine available} = (6.52\text{ g})\left(\frac{1\text{ mol C}_5\text{H}_5\text{N}}{79.10\text{ g C}_5\text{H}_5\text{N}}\right) = 0.0824\text{ mol}$$

The number of moles of $HCl(aq)$ available is

$$\text{moles of acid available} = MV = (0.950\text{ M})(0.0300\text{ L}) = 0.0285\text{ mol}$$

The equation for the reaction upon addition of acid is

$$C_5H_5N(aq) + H_3O^+(aq) \longrightarrow C_5H_5NH^+(aq) + H_2O(l)$$

Because $HCl(aq)$ is a strong acid, this reaction goes essentially to completion. Thus the stoichiometric concentrations of pyridine and the pyridinium ion are

$$[C_5H_5N]_0 = \frac{0.0824\text{ mol} - 0.0285\text{ mol}}{0.0360\text{ L}} = 1.50\text{ M}$$

$$[C_5H_5NH^+]_0 = \frac{0.0285\text{ mol}}{0.0360\text{ L}} = 0.792\text{ M}$$

The pH of the resulting buffer is calculated by using the Henderson–Hasselbalch equation (Table 20.6),

$$\text{pH} = \text{p}K_a + \log\frac{[\text{base}]_0}{[\text{acid}]_0} = 5.23 + \log\frac{1.50\text{ M}}{0.792\text{ M}} = 5.23 + 0.28 = 5.51$$

21-12. The equation for the reaction is

$$H_2PO_4^-(aq) + H_3O^+(aq) \longrightarrow H_3PO_4(aq) + H_2O(l)$$

Because $HCl(aq)$ is a strong acid, this reaction goes essentially to completion. The initial number of millimoles of $H_3PO_4(aq)$ and $H_2PO_4^-(aq)$ available are

$$n = MV = (0.20\text{ M})(200\text{ mL}) = 40\text{ mmol}$$

Because the concentration of acid increases and that of base decreases, we take the buffer to fail when

$$\frac{[H_2PO_4^-]}{[H_3PO_4]} = \frac{40\text{ mmol} - x}{40\text{ mmol} + x} = 0.10$$

where x is the number of millimoles of $HCl(aq)$ added. Solving for x, we have that $x =$ 33 mmol. The volume of 0.10 M $HCl(aq)$ that contains 33 mmoles is

$$V = \frac{33\text{ mmol}}{0.10\text{ M}} = 330\text{ mL}$$

PREPARING BUFFER SOLUTIONS

21-14. If we use equal concentrations of conjugate acid and base, so that $[\text{base}]_0 = [\text{acid}]_0$, then

$$\text{pH} = \text{p}K_a + \log\frac{[\text{base}]_0}{[\text{acid}]_0} = \text{p}K_a$$

To obtain a pH buffered at 5.20, we want $\text{p}K_a \approx 5.20$. From Table 20.6, we find that $\text{p}K_a = 5.23$ for pyridium ion, and so a solution of almost equal concentrations of pyridinium chloride and pyridine would act as a buffer at pH = 5.20.

21-16. (a) The solution is a stoichiometric mixture of a strong acid and a strong base, and is not a buffer solution.

(b) The solution is a mixture of a weak base, $NH_3(aq)$, and its conjugate acid, $NH_4^+(aq)$, and so is a buffer solution. Its pH is given by (see Table 20.6)

$$\text{pH} = \text{p}K_a + \log\frac{[NH_3]_0}{[NH_4^+]_0} = 9.25 + \log\frac{\left(\frac{15.0\text{ mL}}{40.0\text{ mL}}\right)(0.100\text{ M})}{\left(\frac{25.0\text{ mL}}{40.0\text{ mL}}\right)(0.100\text{ M})} = 9.03$$

(c) The reaction is described by

$$\begin{array}{ccccccccc} NH_4Cl(aq) & + & NaOH(aq) & \longrightarrow & NaCl(aq) & + & NH_3(aq) & + & H_2O(l) \\ (2.50\text{ mmol}) & & (2.50\text{ mmol}) & & (2.50\text{ mmol}) & & (2.50\text{ mmol}) & & \end{array}$$

The resulting solution does not contain $NH_4^+(aq)$, the conjugate acid to $NH_3(aq)$, and so is not a buffer solution.

(d) The reaction is described by

$$\begin{array}{ccccccc} NH_4Cl(aq) & + & NaOH(aq) & \longrightarrow & NaCl(aq) & + & NH_3(aq) \quad + H_2O(l) \\ (2.50\text{ mmol}) & & (1.50\text{ mmol}) & & (1.50\text{ mmol}) & & (1.50\text{ mmol}) \end{array}$$

with 1.00 mmol of $NH_4^+(aq)$ in excess. The pH of the resulting buffer solution is given by

$$pH = pK_a + \log\frac{[NH_3]_0}{[NH_4^+]_0} = 9.25 + \log\frac{1.50\text{ mmol}}{1.00\text{ mmol}} = 9.43$$

21-18. Start with the Henderson–Hasselbalch equation,

$$pH = pK_a + \log\frac{[SO_3^{2-}]_0}{[HSO_3^-]_0} = 7.17 + \log\frac{[SO_3^{2-}]_0}{[HSO_3^-]_0} = 7.00$$

Thus, we want

$$\log\frac{[SO_3^{2-}]_0}{[HSO_3^-]_0} = 7.00 - 7.17 = -0.17$$

or

$$\frac{[SO_3^{2-}]_0}{[HSO_3^-]_0} = 10^{-0.17} = 0.631$$

Let x be the volume of the 0.100 M $K_2SO_3(aq)$ solution and (50.0 mL $- x$) be the volume of the 0.200 M $KHSO_3(aq)$ solution to be used. Then, we must have

$$\frac{x(0.100\text{ M})}{(50.0\text{ mL} - x)(0.200\text{ M})} = 0.631$$

or $x = 28.7$ mL. Thus, we mix 28.7 mL of 0.100 M $K_2SO_3(aq)$ with 50.0 mL $-$ 28.7 mL $=$ 21.3 mL of 0.200 M $KHSO_3(aq)$ to form the desired buffer solution.

21-20. Start with the Henderson–Hasselbalch equation,

$$pH = pK_a + \log\frac{[\text{lactate}]_0}{[\text{lactic acid}]_0} = 3.86 + \log\frac{[\text{lactate}]_0}{[\text{lactic acid}]_0} = 4.00$$

Thus, we want

$$\log\frac{[\text{lactate}]_0}{[\text{lactic acid}]_0} = 4.00 - 3.86 = 0.14$$

or

$$\frac{[\text{lactate}]_0}{[\text{lactic acid}]_0} = 10^{0.14} = 1.38$$

The lactic acid reacts with the $Ba(OH)_2(aq)$ according to

$$2\,CH_3CHOHCOOH(aq) + Ba(OH)_2(aq) \longrightarrow Ba(CH_3CHOHCOO)_2(aq) + 2\,H_2O(l)$$

Let x be the volume of the 0.150 M lactic acid solution and (200.0 mL − x) be the volume of the 0.100 M $Ba(OH)_2(aq)$ solution to be used. Then, we want x to satisfy the condition

$$\frac{[\text{lactate}]_0}{[\text{lactic acid}]_0} = \frac{2(200.0\text{ mL} - x)(0.100\text{ M})}{(0.150\text{ M})x - 2(200.0\text{ mL} - x)(0.100\text{ M})} = 1.38$$

where we have not included the total volume in the numerator and denominator as they are the same in each case. Solving for x gives 139 mL. Thus, we react 139 mL of the lactic acid solution with 61.0 mL of the $Ba(OH)_2(aq)$ solution.

INDICATORS

21-22. Using Figure 21.7, we see that the pH at which methyl orange is yellow is greater than 4.5 and the pH at which bromcresol purple is yellow is less than 5. The pH of the solution is between 4.5 and 5.

21-24. From Figure 21.7, we see that the pH where Nile blue is red and Alizarin yellow R is a deep yellow lies in the range 11.0 to 12.0.

21-26. We see from Figure 21.7 that the transition color range of neutral red is between pH 7 and 8. When neutral red is added to the nutrient broth, an orange color indicates that the pH of the broth is between 7 and 8. When the broth is red, the pH is below 7 and when the broth is yellow, the pH is above 8.

21-28. At the color transition point, pH $\approx$ pK_{ai}. In the case of Nile blue, the middle of the color change occurs at pH $\approx$ 10.5, Thus, pK_{ai} = 10.5, and so

$$K_{ai} = 10^{-10.5}\text{ M} = 3\times 10^{-11}\text{ M}$$

TITRATIONS INVOLVING STRONG ACIDS AND STRONG BASES

21-30. The net ionic equation for the reaction is

$$H_3O^+(aq) + OH^-(aq) \longrightarrow 2\,H_2O(l)$$

(a) The number of millimoles of base available is

$$(40.00\text{ mL})(0.10\text{ M}) = 4.0\text{ mmol}$$

The number of millimoles of acid added is

$$(20.0\text{ mL})(0.20\text{ M}) = 4.0\text{ mmol}$$

The number of millimoles of acid added is equal to the number of millimoles of base available; thus, the solution is neutralized and the pH is equal to 7.00.

(b) Proceeding as in part (a), we have

$$\text{millimoles of OH}^-(\text{aq})\text{ available} = (20.0\text{ mL})(0.15\text{ M}) = 3.0\text{ mmol}$$
$$\text{millimoles of H}_3\text{O}^+(\text{aq})\text{ added} = (20.0\text{ mL})(0.20\text{ M}) = 4.0\text{ mmol}$$

Thus, the acid is in excess. The value of $[H_3O^+]$ in the resulting solution is

$$[H_3O^+] = \frac{4.0\text{ mmol} - 3.0\text{ mmol}}{20.0\text{ mL} + 20.0\text{ mL}} = 0.025\text{ M}$$

The pH of the solution is

$$\text{pH} = -\log(0.025) = 1.60$$

21-32. (a) The equivalence point is the point at which the number of moles of base added is equal to the number of moles of acid initially present.

$$\text{millimoles of } H_3O^+(aq) \text{ available} = (30.0\text{ mL})(0.600\text{ M}) = 18.0\text{ mmol}$$

The volume in milliliters of 0.100 M $NaOH(aq)$ that contains 18.0 millimoles of $OH^-(aq)$ is

$$(0.100\text{ M})\,V = 18.0\text{ mmol}$$

$$V = 180\text{ mL}$$

The total volume at the equivalence point, then, is 210 mL.

(b) Proceeding as in part (a), we have

$$\text{millimoles of } H_3O^+(aq) \text{ available} = (50.0\text{ mL})(0.400\text{ M}) = 20.0\text{ mmol}$$

The volume in milliliters of 0.100 M $NaOH(aq)$ that contains 20.0 millimoles of $OH^-(aq)$ is

$$(0.100\text{ M})\,V = 20.0\text{ mmol}$$

$$V = 200\text{ mL}$$

The total volume at the equivalence point, then, is 250 mL.

21-34. The concentration of $H_3O^+(aq)$ in a 0.25 M $HNO_3(aq)$ solution is 0.25 M. The pH of the solution initially is

$$\text{pH} = -\log[H_3O^+] = -\log 0.25 = 0.60$$

Because $HNO_3(aq)$ is a monoprotic acid, the volume of $KOH(aq)$ to reach the equivalence point is given by

$$V_{\text{base}} = \frac{M_{\text{acid}}V_{\text{acid}}}{M_{\text{base}}} = \frac{(0.25\text{ mol}\cdot\text{L}^{-1})(50.0\text{ mL})}{0.50\text{ mol}\cdot\text{L}^{-1}} = 25\text{ mL}$$

Because $HNO_3(aq)$ is a strong acid and $KOH(aq)$ is a strong base, the pH at the equivalence point is 7.00. The pH after adding 25 mL of $KOH(aq)$ can be obtained by realizing that once the equivalence point is reached, any additional $KOH(aq)$ remains as $OH^-(aq)$. Thus, when another 25 mL of $KOH(aq)$ is added, there are

$$\text{moles of } OH^-(aq) = MV_{\text{excess}} = (0.50\text{ M})(0.025\text{ L}) = 0.013\text{ mol}$$

The total volume of the solution is 50 mL + 50 mL = 100 mL, and so $[OH^-]$ is given by

$$[OH^-] = \frac{0.013\text{ mol}}{0.100\text{ L}} = 0.13\text{ M}$$

Therefore,

$$\text{pOH} = -\log([\text{OH}^-]/\text{M}) = \log(0.13) = 0.89$$

and the pH of the solution is

$$\text{pH} = -\log(8.00 \times 10^{-14}) = 13.10$$

The titration curve is shown below

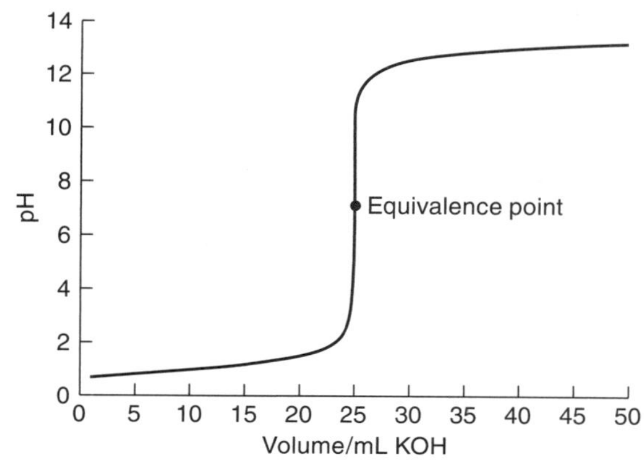

21-36. The equation that describes the neutralization reaction is

$$\text{HCl}(aq) + \text{NaOH}(aq) \longrightarrow \text{NaCl}(aq) + \text{H}_2\text{O}(l)$$

The number of millimoles of sodium hydroxide neutralized is given by

$$\text{millimoles of NaOH} = (34.7\ \text{mL HCl})(0.125\ \text{M HCl})\left(\frac{1\ \text{mmol NaOH}}{1\ \text{mmol HCl}}\right) = 4.34\ \text{mmol}$$

The volume of sodium hydroxide titrated is 15.0 mL, and so the concentration of sodium hydroxide is

$$M = \frac{n}{V} = \frac{4.34\ \text{mmol}}{15.0\ \text{mL}} = 0.289\ \text{M}$$

TITRATION INVOLVING WEAK ACIDS OR WEAK BASES

21-38. The equation for the reaction is

$$\text{HNO}_2(aq) + \text{OH}^-(aq) \rightleftharpoons \text{NO}_2^-(aq) + \text{H}_2\text{O}(l)$$

The number of millimoles of NaOH(aq) added to reach the equivalence point is given by

$$\text{millimoles of NaOH} = (50.0\ \text{mL HNO}_2)(0.100\ \text{M})\left(\frac{1\ \text{mmol NaOH}}{1\ \text{mmol HNO}_2}\right) = 5.00\ \text{mmol}$$

and so the volume required is

$$V = \frac{n}{M} = \frac{5.00\ \text{mmol}}{0.150\ \text{M}} = 33.3\ \text{mL}$$

At the equivalence point,

$$\text{millimoles of } NO_2^- = \text{millimoles of } HNO_2 = 5.00 \text{ mmol}$$

The volume of the solution at the equivalence point is 50.0 mL + 33.3 mL = 83.3 mL. The concentration of $NO_2^-(aq)$ is

$$[NO_2^-] = \frac{5.00 \text{ mmol}}{83.3 \text{ mL}} = 0.0600 \text{ M}$$

The nitrite ion is a weak base because it is the conjugate base of a weak acid. The equation of the reaction of the nitrate ion with water is

$$NO_2^-(aq) + H_2O(l) \rightleftharpoons HNO_2(aq) + OH^-(aq)$$

The value of K_b is

$$K_b = \frac{K_w}{K_a} = \frac{1.00 \times 10^{-14} \text{ M}^2}{5.6 \times 10^{-4} \text{ M}} = 1.8 \times 10^{-11} \text{ M}$$

Setting up a concentration table, we have

Concentration	$NO_2^-(aq)$	+	$H_2O(l)$	$\rightleftharpoons$	$HNO_2(aq)$	+	$OH^-(aq)$
Initial	0.0600 M		—		0 M		≈ 0 M
Change	$-x$		—		$+x$		$+x$
Equilibrium	$0.0600 \text{ M} - x$		—		x		x

Substituting the equilibrium concentrations into the expression for K_b yields

$$K_b = \frac{[HNO_2][OH^-]}{[NO_2^-]} = \frac{x^2}{0.0600 \text{ M} - x} = 1.8 \times 10^{-11} \text{ M}$$

Using the method of successive approximations yields

$$x = [OH^-] = 1.04 \times 10^{-6} \text{ M}$$

The pOH of the solution is

$$\text{pOH} = -\log([OH^-]/\text{M}) = -\log(1.04 \times 10^{-6}) = 5.99$$

The pH of the solution is

$$\text{pH} = 14.00 - \text{pOH} = 14.00 - 5.99 = 8.01$$

Referring to Figure 21.7, we see that thymol blue or phenolphthalein is a suitable indicator.

21-40. (a) The equation for the initial reaction equilibrium is

$$CH_3COOH(aq) + H_2O(l) \rightleftharpoons CH_3COO^-(aq) + H_3O^+(aq)$$

The K_a expression is (Table 20.6)

$$K_a = \frac{[CH_3COO^-][H_3O^+]}{[CH_3COOH]} = 1.8 \times 10^{-5} \text{ M}$$

Before any base is added, we have the following concentration table

Concentration	$CH_3COOH(aq)$	+	$H_2O(l)$	$\rightleftharpoons$	$CH_3COO^-(aq)$	+	$H_3O^+(aq)$
Initial	0.200 M		—		0 M		≈ 0 M
Change	$-x$		—		$+x$		$+x$
Equilibrium	0.200 M $-x$		—		x		x

Substituting these expressions into the K_a expression yields

$$\frac{x^2}{0.200\text{ M} - x} = 1.8 \times 10^{-5}\text{ M}$$

Using the method of successive approximations yields

$$x = [H_3O^+] = 1.9 \times 10^{-3}\text{ M}$$

The pH of the solution is

$$\text{pH} = -\log(1.9 \times 10^{-3}) = 2.72$$

(b) The equation for the reaction is

$$CH_3COOH(aq) + OH^-(aq) \rightleftharpoons CH_3COO^-(aq) + H_2O(l)$$

The number of millimoles of base added is

$$\text{millimoles of base added} = (5.00\text{ mL})(0.200\text{ M}) = 1.00\text{ mmol}$$

The number of millimoles of acid available is

$$\text{millimoles of acid available} = (25.00\text{ mL})(0.200\text{ M}) = 5.00\text{ mmol}$$

The number of millimoles of $CH_3COO^-(aq)$ is equal to the number of millimoles of $OH^-(aq)$ added because the acid is in excess. Thus, we have that the number of millimoles of $CH_3COO^-(aq)$ is 1.00 mmol and the number of millimoles of $CH_3COOH(aq)$ is 5.00 mmol − 1.00 mmol = 4.00 mmol, and so the solution is a buffer. Applying the Henderson–Hasselbalch equation and using the value of pK_a from Table 20.6, we have

$$\text{pH} = \text{p}K_a + \frac{[\text{base}]_0}{[\text{acid}]_0} = 4.74 + \log\frac{1.00\text{ mmol}}{4.00\text{ mmol}} = 4.14$$

(c) Proceeding as in part (b), we start with 5.00 millimoles of acid. We now have

$$\text{millimoles of base added} = (12.50\text{ mL})(0.200\text{ M}) = 2.50\text{ mmol}$$

and so we have that the number of millimoles of $CH_3COO^-(aq)$ is 2.50 mmol and the number of millimoles of $CH_3COOH(aq)$ is 5.00 mmol − 2.50 mmol = 2.50 mmol. Therefore, we are at the midpoint of the titration, and so

$$\text{pH} = \text{p}K_a = 4.74$$

(d) The number of millimoles of base added is (25.00 mL) (0.200 M) = 5.00 mmol, which is equal to the number of millimoles of acid originally present. Thus, the solution is at the equivalence point.

$$[\mathrm{CH_3COO^-}]_0 = \frac{5.00\ \mathrm{mmol}}{50.00\ \mathrm{mL}} = 0.100\ \mathrm{M}$$

The acetate ion is a weak base because it is the conjugate base of a weak acid. The equation for the reaction of the acetate ion with water is

$$\mathrm{CH_3COO^-}(aq) + \mathrm{H_2O}(l) \rightleftharpoons \mathrm{CH_3COOH}(aq) + \mathrm{OH^-}(aq)$$

The value of K_b is (Table 20.6)

$$K_b = \frac{[\mathrm{CH_3COOH}]\,[\mathrm{OH^-}]}{[\mathrm{CH_3COO^-}]} = 5.6 \times 10^{-10}\ \mathrm{M}$$

Setting up a concentration table:

Concentration	$CH_3COO^-(aq)$	+	$H_2O(l)$	$\rightleftharpoons$	$CH_3COOH(aq)$	+	$OH^-(aq)$
Initial	0.100 M		–		0 M		≈ 0 M
Change	$-x$		–		$+x$		$+x$
Equilibrium	0.1000 M $-x$		–		x		x

Substituting these values into the K_b expression yields

$$\frac{x^2}{0.100\ \mathrm{M} - x} = 5.6 \times 10^{-10}\ \mathrm{M}$$

The method of successive approximations yields

$$x = [\mathrm{OH^-}] = 7.5 \times 10^{-6}\ \mathrm{M}$$

The pOH of the solution is

$$\mathrm{pOH} = -\log([\mathrm{OH^-}]/\mathrm{M}) = -\log(7.5 \times 10^{-6}) = 5.12$$

The pH of the solution is

$$\mathrm{pH} = 14.00 - \mathrm{pOH} = 14.00 - 5.12 = 8.88$$

(e) In this case, the solution is beyond the equivalence point. The stoichiometric concentration of $CH_3COO^-(aq)$ is

$$[\mathrm{CH_3COO^-}]_0 = \frac{5.00\ \mathrm{mmol}}{25.00\ \mathrm{mL} + 26.00\ \mathrm{ml}} = 0.0980\ \mathrm{M}$$

The stoichiometric concentration of $OH^-(aq)$ is

$$[\mathrm{OH^-}]_0 = \frac{(\text{total mmoles of base added}) - (\text{total mmoles of acid initially})}{\text{total volume}}$$

$$= \frac{(26.0\ \mathrm{mL})\,(0.200\ \mathrm{M}) - 5.00\ \mathrm{mmol}}{51.00\ \mathrm{mL}} = 0.00392\ \mathrm{M}$$

Setting up a concentration table:

Concentration	$CH_3COO^-(aq)$	+	$H_2O(l)$	$\rightleftharpoons$	$CH_3COOH(aq)$	+	$OH^-(aq)$
Initial	0.0980 M		–		≈ 0 M		0.00392 M
Change	$-x$		–		$+x$		$+x$
Equilibrium	0.0980 M $-x$		–		x		$0.00392 + x$

Substituting these values into the K_b expression yields

$$\frac{[CH_3COOH][OH^-]}{[CH_3COO^-]} = \frac{(x)(0.00392\text{ M} + x)}{0.0980\text{ M} - x} = 5.6 \times 10^{-10}\text{ M}$$

The method of successive approximations yields

$$x = [CH_3COOH] = 1.4 \times 10^{-8}\text{ M}$$

Thus, we have

$$[OH^-] = 0.00392\text{ M} - 1.4 \times 10^{-8}\text{ M} = 0.00392\text{ M}$$

The pOH of the solution is

$$\text{pOH} = -\log([OH^-]/\text{M}) = -\log(0.00392) = 2.41$$

The pH of the solution is

$$\text{pH} = 14.00 - \text{pOH} = 14.00 - 2.41 = 11.59$$

21-42. (a) The equation for the inital reaction equilibrium is

$$CH_3NH_2(aq) + H_2O(l) \rightleftharpoons CH_3NH_3^+(aq) + OH^-(aq)$$

The K_b expression is (Table 20.5)

$$K_b = \frac{[CH_3NH_3^+][OH^-]}{[CH_3NH_2]} = 4.6 \times 10^{-4}\text{ M}$$

Setting up a concentration table:

Concentration	$CH_3NH_2(aq)$	+	$H_2O(l)$	$\rightleftharpoons$	$CH_3NH_3^+(aq)$	+	$OH^-(aq)$
Initial	0.200 M		–		0 M		≈ 0 M
Change	$-x$		–		$+x$		$+x$
Equilibrium	0.200 M $-x$		–		x		x

Substituting these expressions into the K_b expression yields

$$\frac{x^2}{0.200\text{ M} - x} = 4.6 \times 10^{-4}\text{ M}$$

The method of successive approximations yields

$$x = [OH^-] = 9.36 \times 10^{-3}\text{ M}$$

The pOH of the solution is

$$pOH = -\log([OH^-]/M) = -\log(9.36 \times 10^{-3}) = 2.03$$

The pH of the solution is

$$pH = 14.00 - pOH = 14.00 - 2.03 = 11.97$$

(b) The number of millimoles of acid added is

$$\text{millimoles of acid} = MV = (17.5\text{ mL})(0.200\text{ M}) = 3.50\text{ mmol}$$

which is equal to the number of millimoles of $CH_3NH_3^+(aq)$ formed. The number of millimoles of $CH_3NH_2(aq)$ remaining is

$$\text{millimoles of } CH_3NH_2 = (35.0\text{ mL})(0.200\text{ M}) = 3.25\text{ mmol}$$

Thus, we see that we are at the midpoint of the titration, and so (Table 20.5)

$$pH = pK_a = 14.00 - pK_b = 14.00 - 3.34 = 10.66$$

(c) The number of millimoles of $HCl(aq)$ added is given by

$$n = MV = (34.9\text{ mL})(0.200\text{ M}) = 6.98\text{ mmol}$$

which is equal to the number of millimoles of $C_5H_5NH^+(aq)$ formed. The number of millimoles of $C_5H_5N(aq)$ remaining is

$$\text{millimoles of } C_5H_5N = (35.0\text{ mL})(0.200\text{ M}) - 6.98\text{ mmol}- = 0.02\text{ mmol}$$

Because $[\text{base}]_0/[\text{acid}]_0 < 0.1$, we cannot use the Henderson–Hasselbalch equation (although it turns out to give the same result in this case). Instead, we find that the stoichiometric concentrations are

$$[CH_3NH_3^+]_0 = \frac{6.98\text{ mmol}}{35.00\text{ mL} + 34.9\text{ ml}} = 0.0999\text{ M}$$

$$[CH_3NH_2]_0 = \frac{0.02\text{ mmol}}{35.00\text{ mL} + 34.9\text{ mL}} = 2.9 \times 10^{-4}\text{ M}$$

Substituting these values into the K_b expression yields

$$\frac{([OH^-])(0.0999\text{ M} + [OH^-])}{2.9 \times 10^{-4}\text{ M} - [OH^-]} = 4.6 \times 10^{-4}\text{ M}$$

The method of successive approximations yields

$$[OH^-] = 1.3 \times 10^{-6}\text{ M}$$

The pOH of the solution is

$$pOH = -\log([OH^-]/M) = -\log(1.3 \times 10^{-6}) = 5.89$$

The pH of the solution is

$$pH = 14.00 - pOH = 14.00 - 5.89 = 8.11$$

(d) The number of millimoles of acid added is

$$\text{millimoles of acid} = MV = (35.0\text{ mL})(0.200\text{ M}) = 7.00\text{ mmol}$$

The number of millimoles of base available is

$$\text{millimoles of base} = MV = (35.0\text{ mL})(0.200\text{ M}) = 7.00\text{ mmol}$$

Thus, the solution is at the equivalence point. The stoichiometric concentration of $CH_3NH_3^+(aq)$ is

$$[CH_3NH_3^+]_0 = \frac{7.00\text{ mmol}}{70.0\text{ mL}} = 0.100\text{ M}$$

The equation for the reaction that occurs is

$$CH_3NH_3^+(aq) + H_2O(l) \rightleftharpoons CH_3NH_2(aq) + H_3O^+(aq)$$

The K_a expression is

$$K_a = \frac{[CH_3NH_2][H_3O^+]}{[CH_3NH_3^+]} = \frac{1.00 \times 10^{-14}\text{ M}^2}{4.6 \times 10^{-4}\text{ M}} = 2.17 \times 10^{-11}\text{ M}$$

Setting up a concentration table:

Concentration	$CH_3NH_3^+(aq)$	+	$H_2O(l)$	$\rightleftharpoons$	$CH_3NH_2(aq)$	+	$H_3O^+(aq)$
Initial	0.100 M		–		0 M		≈ 0 M
Change	$-x$		–		$+x$		$+x$
Equilibrium	0.100 M $-x$		–		x		x

Substituting these values into the K_a expression yields

$$\frac{x^2}{0.100\text{ M} - x} = 2.17 \times 10^{-11}\text{ M}$$

The method of successive approximations yields

$$x = [H_3O^+] = 1.5 \times 10^{-6}\text{ M}$$

The pH of the solution is

$$\text{pH} = -\log([H_3O^+]/\text{M}) = -\log(1.5 \times 10^{-6}) = 5.82$$

(e) In this case, the solution is beyond the equivalence point. The stoichiometric concentration of $CH_3NH_3^+(aq)$ is

$$[CH_3NH_3^+]_0 = \frac{(35.0\text{ mL})(0.200\text{ M})}{35.00\text{ mL} + 35.1\text{ ml}} = 0.0999\text{ M}$$

The stoichiometric concentration of $H_3O^+(aq)$ is

$$[H_3O^+]_0 = \frac{(35.1\text{ mL})(0.200\text{ M}) - (35.0\text{ mL})(0.200\text{ M})}{70.1\text{ mL}} = 2.85 \times 10^{-4}\text{ M}$$

The equilibrium is described by

$$CH_3NH_3^+(aq) + H_2O(l) \rightleftharpoons C_5H_5N(aq) + H_3O^+(aq)$$

Setting up a concentration table:

Concentration	$CH_3NH_3^+(aq)$	+	$H_2O(l)$	$\rightleftharpoons$	$CH_3NH_2(aq)$	+	$H_3O^+(aq)$
Initial	0.0999 M		—		0 M		2.85×10^{-4} M
Change	$-x$		—		$+x$		$+x$
Equilibrium	0.0999 M $-x$		—		x		2.85×10^{-4} M $+x$

Substituting these values into the K_a expression yields

$$\frac{[CH_3NH_2][H_3O^+]}{[CH_3NH_3^+]} = \frac{(x)(2.85 \times 10^{-4}\ \text{M})}{0.0999\ \text{M} - x} = 2.17 \times 10^{-11}\ \text{M}$$

The method of successive approximations yields

$$x = [CH_3NH_2] = 7.6 \times 10^{-9}\ \text{M}$$

Thus, we have

$$[H_3O^+] = 2.85 \times 10^{-4}\ \text{M} + 7.6 \times 10^{-9}\ \text{M} = 2.85 \times 10^{-4}\ \text{M}$$

The pH of the solution is

$$\text{pH} = -\log(2.85 \times 10^{-4}) = 3.55$$

21-44. The titration curve is shown below.

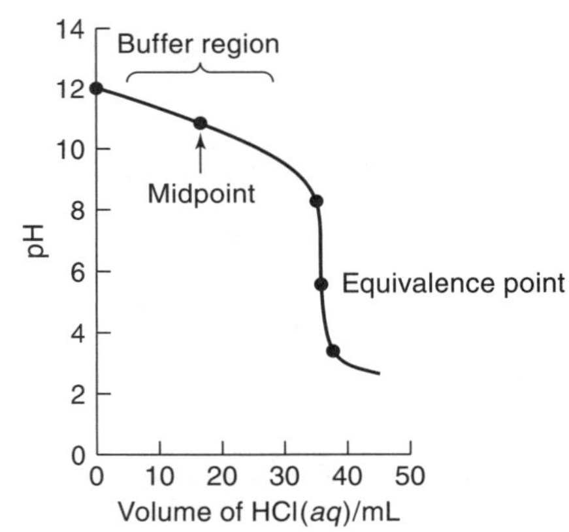

21-46. At the equivalence point, the number of moles of acid present is equal to the number of moles of base added. The number of moles of acid is given by

$$\text{moles of acid} = (0.250\ \text{mol}\cdot\text{L}^{-1})(0.0253\ \text{L}) = 0.00633\ \text{mol}$$

We have the correspondence

$$0.772\ \text{g benzoic acid} \approx 0.00633\ \text{mol benzoic acid}$$

Dividing by 0.00633, we have

$$122 \text{ g benzoic acid} \Leftrightarrow 1.00 \text{ mol benzoic acid}$$

Thus, the molecular mass of benzoic acid is 122.

ADDITIONAL PROBLEMS

21-48. No. An equal molar mixture of a strong acid and a strong base forms a neutral salt solution, not a buffer solution.

21-50. At the equivalence point, stoichiometrically equivalent amounts of acid and base have reacted, and the reaction is complete. At the end point, an indicator color change or other signal occurs showing that the titration is complete. Although an indicator should be chosen so that the end point corresponds as closely as possible with the equivalence point, these two points may or may not coincide exactly.

21-52. There are several methods that can be used: (1) Combine approximately equal volumes of the acetic acid and sodium acetate solutions; (2) combine one volume of the acidic acid solution with one half volume of the sodium hydroxide solution, or (3) combine one volume of the sodium acetate solution with one-half volume of the hydrochloric acid solution.

21-54. When a strong acid is titrated to its equivalence point with a strong base, the resulting solution consists of an aqueous neutral salt, and so the pH of the solution is 7.00. When a weak acid is titrated to its equivalence point with a strong base, the resulting solution consists of an aqueous solution of the conjugate base of the acid, and so the pH of the solution is not 7.00.

21-56. We can write the equation for the dissociation of a weak acid as

$$\text{HA}(aq) + \text{H}_2\text{O}(l) \rightleftharpoons \text{A}^-(aq) + \text{H}_3\text{O}^+(aq)$$

The addition of a strong acid to this equilibrium system increases the amount of $H_3O^+(aq)$ present, causing a shift in the equilibrium position to the left in accordance with Le Châtelier's principle. Thus, the percentage of the weak acid, HA(aq), that is dissociated decreases.

21-58. As discussed at the end of Section 21-12, $HCO_3^-(aq)$ is an amphoteric species that can act as either an acid or a base according to

$$\text{HCO}_3^-(aq) + \text{H}_2\text{O}(l) \rightleftharpoons \text{H}_3\text{O}^+(aq) + \text{CO}_3^{2-}(aq)$$

and

$$\text{HCO}_3^-(aq) + \text{H}_2\text{O}(l) \rightleftharpoons \text{H}_2\text{CO}_3(aq) + \text{OH}^-(aq)$$

Thus, a solution containing $HCO_3^-(aq)$ ions can neutralize either an added acid or an added base, and so acts as a buffer solution.

21-60. The 0.500 M HCl(aq) solution would be best because it will require about 25 mL to reach the end point. The 0.250 M HCl(aq) solution may exceed the capacity of the burette and the 2.00 M solution would require too small a volume to achieve good accuracy. The 6 M concentrated HCl(aq) solution is not known precisely enough to be useful and once again would require too small a volume.

21-62. The number of moles in 1.00 gram of $Mg(OH)_2$ is

$$\text{moles of Mg(OH)}_2 = (1.00\text{ g})\left(\frac{1\text{ mol Mg(OH)}_2}{58.32\text{ g Mg(OH)}_2}\right) = 0.0171\text{ mol}$$

At neutralization (the equivalence point)

$$\text{moles of OH}^-(aq) = \text{moles of H}_3\text{O}^+(aq)$$

There are two moles of $OH^-(aq)$ per mole of $Mg(OH)_2(s)$, and so we have

$$\text{moles of OH}^- = (0.0171\text{ mol Mg(OH)}_2)\left(\frac{2\text{ mol OH}^-}{1\text{ mol Mg(OH)}_2}\right) = 0.0342\text{ mol}$$

The volume of $HCl(aq)$ that is neutralized by 0.0342 moles of $Mg(OH)_2$ is

$$V_a = \frac{0.0342\text{ mol}}{M_a} = \frac{0.0342\text{ mol}}{0.10\text{ mol}\cdot\text{L}^{-1}} = 0.34\text{ L} = 340\text{ mL}$$

21-64. We must first calculate the concentrations of $KH_2PO_4(aq)$ and $Na_2HPO_4(aq)$ in the buffer solution.

$$[\text{KH}_2\text{PO}_4]_0 = \frac{(3.40\text{ g})\left(\dfrac{1\text{ mol KH}_2\text{PO}_4}{136.09\text{ g KH}_2\text{PO}_4}\right)}{1.00\text{ L}} = 0.0250\text{ M}$$

$$[\text{Na}_2\text{HPO}_4]_0 = \frac{(3.55\text{ g})\left(\dfrac{1\text{ mol Na}_2\text{HPO}_4}{141.96\text{ g Na}_2\text{HPO}_4}\right)}{1.00\text{ L}} = 0.0250\text{ M}$$

The stoichiometric concentrations of the acid and base forms are

$$[\text{acid}]_0 = [\text{H}_2\text{PO}_4^-]_0 = 0.0250\text{ M}$$

$$[\text{base}]_0 = [\text{HPO}_4^{2-}]_0 = 0.0250\text{ M}$$

The value of the pK_a of the conjugate acid/base pair is is 7.21 (Table 20.6). From the Henderson–Hasselbalch equation we have

$$\text{pH} = \text{p}K_a + \log\frac{[\text{base}]_0}{[\text{acid}]_0} = 7.21 + \log\frac{0.0250\text{ M}}{0.0250\text{ M}} = 7.21$$

21-66. One equilibrium that is established in the solution is

$$\text{CH}_3\text{COOH}(aq) + \text{H}_2\text{O}(l) \rightleftharpoons \text{H}_3\text{O}^+(aq) + \text{CH}_3\text{COO}^-(aq)$$

The acid-dissociation-constant expression is (Table 20.6)

$$K_a = \frac{[\text{CH}_3\text{COO}^-][\text{H}_3\text{O}^+]}{[\text{CH}_3\text{COOH}]} = 1.8 \times 10^{-5}\text{ M}$$

We set up a table of initial and equilibrium concentrations

Concentration	$CH_3COOH(aq)$	+	$H_2O(l)$	$\rightleftharpoons$	$H_3O^+(aq)$	+	$CH_3COO^-(aq)$
Initial	0.100 M		–		≈ 0 M		0.100 M
Change	$-x$		–		$+x$		$+x$
Equilibrium	0.100 M $-x$		–		x		0.100 M $+x$

Note that $[CH_3COO^-]$ is not equal to $[H_3O^+]$ in this case. Substituting the values of the equilibrium concentrations into the K_a expression yields

$$K_a = \frac{[H_3O^+][CH_3COO^-]}{[CH_3COOH]} = \frac{(x)(0.100\text{ M} + x)}{0.100\text{ M} - x} = 1.8 \times 10^{-5}\text{ M}$$

Using the method of successive approximations yields

$$x = [H_3O^+] = 1.8 \times 10^{-5}\text{ M}$$

The equilibrium concentrations are

$$[CH_3COOH] = 0.100\text{ M} - 1.8 \times 10^{-5}\text{ M} \approx 0.100\text{ M}$$
$$[CH_3COO^-] = 0.100\text{ M} + 1.8 \times 10^{-5}\text{ M} \approx 0.100\text{ M}$$

and so we see that

$$[CH_3COOH] \approx [CH_3COOH]_0$$
$$[CH_3COO^-] \approx [CH_3COO^-]_0$$

These relations arise because $[H_3O^+]$ is negligible compared to either $[CH_3COOH]$ or $[CH_3COO^-]$.

21-68. The schematic equation for the reactions is

$$\begin{array}{l} 3\,HCO_3^-(aq) + H_3C_6H_5O_7(aq) \rightleftharpoons 3\,H_2CO_3(aq) + C_6H_5O_7^{3-}(aq) \\ \qquad\qquad\qquad\qquad\qquad\qquad\quad \uparrow\downarrow \\ \qquad\qquad\qquad\qquad\qquad\quad 3\,CO_2(aq) \ + \ 3\,H_2O(l) \\ \qquad\qquad\qquad\qquad\qquad\qquad\quad \uparrow\downarrow \\ \qquad\qquad\qquad\qquad\qquad\quad 3\,CO_2(g) \end{array}$$

or

$$3\,HCO_3^-(aq) + H_3C_6H_5O_7(aq) \rightleftharpoons 3\,CO_2(g) + C_6H_5O_7^{3-}(aq) + 3\,H_2O(l)$$

21-70. The number of moles in 500 milligrams of $Al(OH)_3$ is

$$\text{moles of } Al(OH)_3 = (0.500\text{ g})\left(\frac{1\text{ mol } Al(OH)_3}{78.00\text{ g } Al(OH)_3}\right) = 0.00641\text{ mol}$$

At neutralization (the equivalence point)

$$\text{moles of } OH^-(aq) = \text{moles of } H_3O^+(aq)$$

There are three moles of OH^- per mole of $Al(OH)_3$, and so we have

$$\text{moles of } OH^- = (0.00641 \text{ mol } Al(OH)_3)\left(\frac{3 \text{ mol } OH^-}{1 \text{ mol } Al(OH)_3}\right) = 0.0192 \text{ mol}$$

Thus we have

$$\text{moles of } H_3O^+ = 0.0192 \text{ mol}$$

The volume of $HCl(aq)$ neutralized is given by

$$V = \frac{0.0192 \text{ mol}}{0.10 \text{ mol}\cdot\text{L}^{-1}} = 0.19 \text{ L} = 190 \text{ mL}$$

21-72. Because butyric acid is a monoprotic acid, at the equivalence point

$$\text{moles of acid} = \text{moles of base} = (0.100 \text{ mol}\cdot\text{L}^{-1})(0.0624 \text{ L}) = 0.00624 \text{ mol}$$

We have the correspondence

$$0.550 \text{ g butyric acid} \mathrel{\hat{=}} 0.00624 \text{ mol butyric acid}$$

Dividing by 0.00624, we have

$$88.1 \text{ g butyric acid} \mathrel{\hat{=}} 1.00 \text{ mol butyric acid}$$

Thus, the molecular mass of butyric acid is 88.1.

21-74. The equation for the reaction is

$$H_2C_2O_4(aq) + 2\,NaOH(aq) \longrightarrow Na_2C_2O_4(aq) + 2\,H_2O(l)$$

The number of millimoles of $H_2C_2O_4(aq)$ initially is given by

$$n = MV = (0.10 \text{ M})(25.0 \text{ mL}) = 2.50 \text{ mmol}$$

It requires twice as many millimoles of $NaOH(aq)$ to neutralize the $H_2C_2O_4(aq)$, or 5.00 mmol. The volume of 0.10 M $NaOH(aq)$ required is given by

$$V = \frac{n}{M} = \frac{5.00 \text{ mmol}}{0.10 \text{ M}} = 50.0 \text{ mL}$$

21-76. The number of moles of base used is

$$\text{moles of base} = MV = (0.135 \text{ mol}\cdot\text{L}^{-1})(0.1472 \text{ L}) = 0.0199 \text{ mol}$$

The number of moles of citric acid available is

$$\text{moles of } C_6H_8O_7 = (1.270 \text{ g})\left(\frac{1 \text{ mol } C_6H_8O_7}{192.13 \text{ g } C_6H_8O_7}\right) = 0.006610 \text{ mol}$$

and so we find

$$\left(\frac{0.0199 \text{ mol } OH^-}{0.006610 \text{ mol } C_6H_8O_7}\right)\left(\frac{1 \text{ } H_3O^+}{1 \text{ mol } OH^-}\right) = \frac{3.01 \text{ mol } H_3O^+}{1 \text{ mol } C_6H_8O_7}$$

Thus, there are three acidic protons per molecule of citric acid. We normally write the formula of citric acid as $H_3C_6H_5O_7$.

21-78. Because $CH_3COOH(aq)$ is a monoprotic acid, at the equivalence point

$$\text{millimoles of acid} = \text{millimoles of base} = (0.400\text{ M})(38.5\text{ mL}) = 15.4\text{ mmol}$$

The concentration of acetic acid in the vinegar solution is given by

$$M_a = \frac{15.4\text{ mmol}}{21.0\text{ mL}} = 0.733\text{ M}$$

We now calculate the mass percentage of the acetic acid. For convenience, consider a 100-mL sample of vinegar. The number of moles of acetic acid in 100 mL of vinegar is

$$n = MV = (0.733\text{ mol}\cdot\text{L}^{-1})(0.100\text{ L}) = 0.0733\text{ mol}$$

The mass of acetic acid in 100 mL of vinegar is

$$\text{mass of CH}_3\text{COOH} = (0.0733\text{ mol})\left(\frac{60.05\text{ g CH}_3\text{COOH}}{1\text{ mol CH}_3\text{COOH}}\right) = 4.40\text{ g}$$

The mass of 100 mL of vinegar solution is

$$\text{mass of vinegar} = d\,V = (1.060\text{ g}\cdot\text{mL}^{-1})(100\text{ mL}) = 106.0\text{ g}$$

The mass percentage of acetic acid is

$$\text{mass }\% = \frac{\text{mass of acetic acid}}{\text{mass of vinegar solution}} \times 100 = \frac{4.40\text{ g}}{106.0\text{ g}} \times 100 = 4.15\%$$

21-80. We shall estimate the pH of the solution by using the Henderson–Hasselbalch equation

$$\text{pH} = \text{p}K_a + \log\frac{[\text{base}]_0}{[\text{acid}]_0}$$

We see that the conjugate acid is $H_2PO_4^-(aq)$ and the conjugate base is $HPO_4^{2-}(aq)$. The value of $\text{p}K_a$ for $H_2PO_4^-(aq)$ is $-\log(6.2\times10^{-8}) = 7.21$. We can now do each part of the problem in turn.

(a) $[\text{acid}]_0 = 0.050\text{ M}$ $\quad [\text{base}]_0 = 0.050\text{ M}$

$$\text{pH} = 7.21 + \log\frac{0.050\text{ M}}{0.050\text{ M}} = 7.21 + 0.0 = 7.21$$

(b) $[\text{acid}]_0 = 0.050\text{ M}$ $\quad [\text{base}]_0 = 0.10\text{ M}$

$$\text{pH} = 7.21 + \log\frac{0.10\text{ M}}{0.050\text{ M}} = 7.21 + 0.30 = 7.51$$

(c) $[\text{acid}]_0 = 0.10\text{ M}$ $[\text{base}]_0 = 0.050\text{ M}$

$$\text{pH} = 7.21 + \log\frac{0.050\text{ M}}{0.10\text{ M}} = 7.21 - 0.30 = 6.91$$

21-82. Because the acid is a monoprotic acid, the number of millimoles of the unknown acid is given by

$$\text{millimoles of acid} = \text{millimoles of base} = (37.5\text{ mL})(0.200\text{ M}) = 7.50\text{ mmol}$$

and its concentration is given by

$$M_{\text{acid}} = \frac{7.50\text{ mmol}}{50.0\text{ mL}} = 0.150\text{ M}$$

The equation for the neutralization reaction is

$$\text{HA}(aq) + \text{NaOH}(aq) \longrightarrow \text{NaA}(aq) + \text{H}_2\text{O}(l)$$

where HA is the formula of the acid. After the addition of 15.0 mL of 0.200 M NaOH(aq), we have (15.0 mL)(0.200 M) = 3.00 mmol of $\text{A}^-(aq)$ and (50.0 mL)(0.150 M) − 3.00 mmol = 4.50 mmol of HA(aq) remaining. The pH is given by the Henderson–Hasselbalch equation

$$\text{pH} = \text{p}K_\text{a} + \log\frac{[\text{base}]_0}{[\text{acid}]_0}$$

or

$$4.67 = \text{p}K_\text{a} + \log\frac{3.00\text{ mmol}}{4.50\text{ mmol}} = \text{p}K_\text{a} - 0.176$$

or $\text{p}K_\text{a} = 4.85$.

21-84. The equation for the titration reaction is

$$\text{C}_5\text{H}_5\text{N}(aq) + \text{H}_3\text{O}^+(aq) \longrightarrow \text{C}_5\text{H}_5\text{NH}^+(aq) + \text{H}_2\text{O}(l)$$

Because one mole of acid reacts with one mole of base, the volume of acid required to reach the equivalence point is given by

$$V_\text{a} = \frac{M_\text{b}V_\text{b}}{M_\text{a}} = \frac{(0.098\text{ M})(17.5\text{ mL})}{0.117\text{ M}} = 14.7\text{ mL}$$

The total volume at the equivalence point is 17.5 mL + 14.7 mL = 32.2 mL. The stoichiometric concentration of $\text{C}_5\text{H}_5\text{NH}^+(aq)$ at the equivalence point is

$$[\text{C}_5\text{H}_5\text{NH}^+]_0 = \frac{(0.098\text{ M})(17.5\text{ mL})}{32.2\text{ mL}} = 0.0533\text{ M}$$

The equation for the reaction equilibrium is

$$\text{C}_5\text{H}_5\text{NH}^+(aq) + \text{H}_2\text{O}(l) \rightleftharpoons \text{C}_5\text{H}_5\text{N}(aq) + \text{H}_3\text{O}^+(aq)$$

Setting up a concentration table:

Concentration	$C_5H_5NH^+(aq)$	+	$H_2O(l)$	$\rightleftharpoons$	$H_3O^+(aq)$	+	$C_5H_5N(aq)$
Initial	0.0533 M		–		≈ 0 M		0 M
Change	$-x$		–		$+x$		$+x$
Equilibrium	0.0533 M $-x$		–		x		x

Substitution of these values into the K_a expression (Table 20.6) yields

$$K_a = \frac{[C_5H_5N]\,[H_3O^+]}{[C_5H_5NH^+]} = \frac{x^2}{0.0533\text{ M} - x} = 5.9 \times 10^{-6}\text{ M}$$

Using the method of successive approximations yields

$$x = [H_3O^+] = 5.6 \times 10^{-4}\text{ M}$$

The pH of the solution is

$$\text{pH} = -\log([H_3O^+]/\text{M}) = \log(5.6 \times 10^{-4}) = 3.25$$

21-86. For $CH_3COOH(aq)$, the value of pK_a is 4.74 (Table 20.6). Using the Henderson–Hasselbalch equation

$$\text{pH} = \text{p}K_a + \log\frac{[\text{base}]_0}{[\text{acid}]_0} = \text{p}K_a + \log\frac{[CH_3COO^-]_0}{[CH_3COOH]_0}$$

or

$$4.52 = 4.74 + \log\frac{[CH_3COO^-]_0}{[CH_3COOH]_0}$$

Thus,

$$\log\frac{[CH_3COO^-]}{[CH_3COOH]} = 4.52 - 4.74 = -0.22$$

Taking the antilogarithm of both sides yields

$$\frac{[CH_3COO^-]}{[CH_3COOH]} = 10^{-0.22} = 0.60$$

or

$$\frac{\text{moles } CH_3COO^-}{\text{moles } CH_3COOH} = 0.60$$

$NaOH(aq)$ reacts with $CH_3COOH(aq)$ according to the net ionic equation

$$CH_3COOH(aq) + OH^-(aq) \longrightarrow CH_3COO^-(aq) + H_2O(l)$$

and so we see that

$$\text{moles } OH^-(aq) = \text{moles } CH_3COO^-(aq)$$

and

$$\text{moles CH}_3\text{COOH}(aq) = \text{initial moles CH}_3\text{COOH}(aq) - \text{moles OH}^-(aq)$$

Substituting these into the above expression yields

$$\frac{\text{moles OH}^-(aq)}{\text{initial moles CH}_3\text{COOH}(aq) - \text{moles OH}^-(aq)} = \frac{\text{moles OH}^-(aq)}{(0.500\text{ L})(0.120\text{ M}) - \text{moles OH}^-(aq)} = 0.60$$

Solving for moles $OH^-(aq)$, we have moles $OH^-(aq) = 0.0225$ mol. The corresponding mass of $NaOH(s)$ is 0.900 grams.

21-88. We use the Henderson–Hasselbalch equation

$$\text{pH} = \text{p}K_a + \log\frac{[\text{base}]_0}{[\text{acid}]_0} = \text{p}K_a + \log\frac{[\text{NaNO}_2]_0}{[\text{HNO}_2]_0}$$

with $\text{p}K_a = 3.25$ (Table 20.6). Substituting in the given values, we have

$$3.70 = 3.25 + \log\frac{[\text{NaNO}_2]_0}{[0.200\text{ M}]_0}$$

Solving for $[NaNO_2]_0$, we have

$$[\text{NaNO}_2]_0 = 0.564\text{ M}$$

The number of moles of $NaNO_2(s)$ required is

$$\text{moles of NaNO}_2 = (0.564\text{ mol}\cdot\text{L}^{-1})(0.300\text{ L}) = 0.169\text{ mol}$$

The mass of $NaNO_2(s)$ in 0.169 moles is

$$\text{mass} = (0.169\text{ mol})\left(\frac{69.00\text{ g NaNO}_2}{1\text{ mol NaNO}_2}\right) = 11.7\text{g}$$

21-90. The relevant equation is

$$\text{HNO}_2(aq) + \text{KOH}(aq) \longrightarrow \text{KNO}_2(aq) + \text{H}_2\text{O}(l)$$

The number of millimoles of $HNO_2(aq)$ and $KOH(aq)$ initially are

$$\text{millimoles of HNO}_2\text{ initially} = (0.150\text{ M})(50.0\text{ L}) = 7.50\text{ mmol}$$
$$\text{millimoles of KOH initially} = (0.200\text{ M})(25.0\text{ L}) = 5.00\text{ mmol}$$

After the addition of $KOH(aq)$ to $HNO_2(aq)$, the number of millimoles of $KNO_2(aq)$ produced is

$$\text{millimoles of KNO}_2 = 5.00\text{ mmol}$$

The value of $\text{p}K_a$ for the conjugate acid-base-pair HNO_2/NO_2^- is 3.25 (Table 20.6). The Henderson–Hasselbalch equation is

$$\text{pH} = \text{p}K_a + \log\frac{[\text{NO}_2^-]_0}{[\text{HNO}_2]_0}$$

The value of the pH of the final solution is

$$\text{pH} = 3.25 + \log \frac{5.00 \text{ mmol}}{7.50 \text{ mmol} - 5.00 \text{ mmol}} = 3.55$$

21-92.* (a) From the given values of $\text{p}K_{a_1}$ and $\text{p}K_{a_2}$, we find that the values of K_{a_1} and K_{a_2} for phosphorous acid are 5×10^{-2} M and 2.0×10^{-7} M, respectively. Because $\text{p}K_{a_2} \gg \text{p}K_{a_1}$, we can ignore the contribution of $\text{p}K_{a_2}$ to the value of $[H_3O^+]$. Setting up a concentration table:

Concentration	$H_3PO_3(aq)$	+	$H_2O(l)$	$\rightleftharpoons$	$H_2PO_3^-(aq)$	+	$H_3O^+(aq)$
Initial	3.0 M		–		0 M		≈ 0 M
Change	$-x$		–		$+x$		$+x$
Equilibrium	$3.0 \text{ M} - x$		–		x		x

Substituting these expressions into the K_a expression yields

$$K_a = \frac{[H_2PO_3^-]\,[H_3O^+]}{[H_3PO_3]} = \frac{x^2}{3.0 \text{ M} - x} = 5 \times 10^{-2} \text{ M}$$

The method of successive approximations yields $x = [H_3O^+] = 0.4$ M, and so the pH of the solution is pH = 0.4.

(b) After the addition of 25.0 mL of 3.0 M $KOH(aq)$, the number of moles of $KOH(aq)$ is half that of the initial number of moles of $H_3PO_3(aq)$, and so we are at the first midpoint. The pH is therefore

$$\text{pH} = \text{p}K_{a_1} = 1.3$$

(c) After adding 50.0 mL of 3.0 M $KOH(aq)$, the number of moles of $KOH(aq)$ added is equal to the initial number of moles of $H_3PO_3(aq)$, and so we are at the first equivalence point, and we have a (50.0 mL) (3.0 M) = 1.5 M solution of $H_2PO_3^-(aq)$. $H_2PO_3^-(aq)$ is amphoteric, and so can act as either an acid or a base according to

$$H_2PO_3^-(aq) + H_2O(l) \rightleftharpoons HPO_3^{2-}(aq) + H_3O(aq) \quad \text{(acid)}$$

$$H_2PO_3^-(aq) + H_2O(l) \rightleftharpoons H_3PO_3(aq) + OH^-(aq) \quad \text{(base)}$$

Using the results from Problem 20-95 for an amphoteric species, we have that

$$\text{pH} = \frac{\text{p}K_{a_1} + \text{p}K_{a_2}}{2} = \frac{1.3 + 6.70}{2} = 4.0$$

(d) After the addition of 75.0 mL of 3.0 M $KOH(aq)$, we have

$$\text{millimoles } OH^- = (75.0 \text{ mL})(3.0 \text{ M}) = 225 \text{ mmol}$$

The initial number of millimoles of $H_3PO_3(aq)$ is 150 millimoles. To convert all the $H_3PO_3(aq)$ to $H_2PO_3^-(aq)$ requires 150 millimoles of $OH^-(aq)$, and so we have an additional 225 mmol – 150 mmol = 75 mmol of $OH^-(aq)$. This $OH^-(aq)$ can further react with the $H_2PO_3^-(aq)$ to produce $HPO_3^{2-}(aq)$ according to

$$H_2PO_3^-(aq) + OH^-(aq) \rightleftharpoons HPO_3^{2-}(aq) + H_2O(l)$$

and so we have

$$\text{millimoles of } H_2PO_3^- = 150 \text{ mmol} - 75 \text{ mmol} = 75 \text{ mmol}$$

and

$$\text{millimoles of } HPO_3^{2-} = 75 \text{ mmol}$$

Thus we are at the second midpoint, and so

$$\text{pH} = \text{p}K_{a_2} = 6.70$$

(e) After the addition of 100.0 mL of 3.0 M KOH(aq), the total number of moles of $OH^-(aq)$ added is twice the initial number of moles of $H_3PO_3(aq)$, and so we are at the second equivalence point. The stoichiometric concentration of $HPO_3^{2-}(aq)$ is given by

$$[HPO_3^{2-}] = \frac{150 \text{ mmol}}{150 \text{ mL}} = 1.0 \text{ M}$$

The reaction that occurs is described by the equation

$$HPO_3^{2-}(aq) + H_2O(l) \rightleftharpoons H_2PO_3^-(aq) + OH^-(aq)$$

where the value of K_{b_2} for this reaction equation is given by

$$K_{b_2} = \frac{K_w}{K_{a_2}} = \frac{1.00 \times 10^{-14} \text{ M}^2}{2.0 \times 10^{-7} \text{ M}} = 5.0 \times 10^{-8} \text{ M}$$

Setting up a concentration table:

Concentration	$HPO_3^{2-}(aq)$	+	$H_2O(l)$	$\rightleftharpoons$	$H_2PO_3^-(aq)$	+	$OH^-(aq)$
Initial	1.0 M		–		0 M		≈ 0 M
Change	$-x$		–		$+x$		$+x$
Equilibrium	1.00 M $-x$		–		x		x

Substituting these values into the K_{b_2} expression yields

$$K_{b_2} = \frac{[H_2PO_3^-][OH^-]}{[HPO_3^{2-}]} = \frac{x^2}{1.0 \text{ M} - x} = 5 \times 10^{-8} \text{ M}$$

Using the method of successive approximations, we have that $x = [OH^-] = 2.2 \times 10^{-4}$ M, and so the pOH is pOH = 3.66 and pH = 10.34.

A sketch of the titration curve is

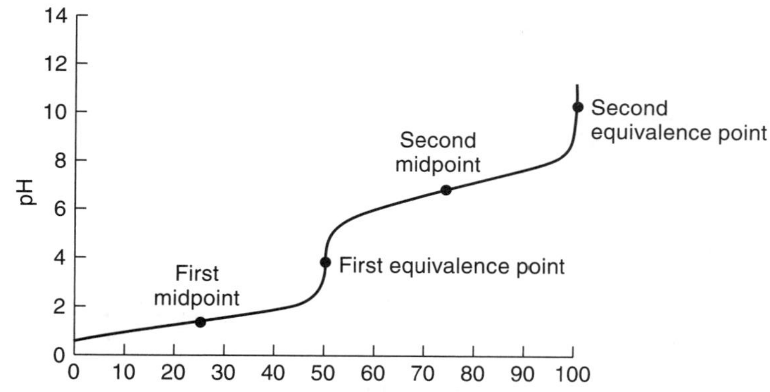

CHAPTER 22. Solubility and Precipitation Reactions

REVIEW OF SOLUBILITY RULES

22-2. (a) insoluble, rule 5 (b) soluble, rule 4 (c) soluble, rule 1 (d) insoluble, rule 3

22-4. (a) insoluble, rule 3 (b) soluble, rule 2 (c) insoluble, rule 3 (d) soluble, rule 2

22-6. (a) $CaSO_4$ is insoluble by rule 6

$$H_2SO_4(aq) + Ca(ClO_4)_2(aq) \longrightarrow CaSO_4(s) + 2\,HClO_4(aq)$$

$$Ca^{2+}(aq) + SO_4^{2-}(aq) \longrightarrow CaSO_4(s)$$

(b) no reaction

(c) $Hg_2(C_6H_5COO)_2$ is insoluble by rule 3.

$$Hg_2(NO_3)_2(aq) + 2\,NaC_6H_5COO(aq) \longrightarrow Hg_2(C_6H_5COO)_2(s) + 2\,NaNO_3(aq)$$

$$Hg_2^{2+}(aq) + 2\,C_6H_5COO^-(aq) \longrightarrow Hg_2(C_6H_5COO)_2(s)$$

(d) Ag_2SO_4 is insoluble by rule 3.

$$Na_2SO_4(aq) + 2\,AgF(aq) \longrightarrow Ag_2SO_4(s) + 2\,NaF(aq)$$

$$2\,Ag^+(aq) + SO_4^{2-}(aq) \longrightarrow Ag_2SO_4(s)$$

K_{sp} CALCULATIONS

22-8. The solubility equilibrium is described by

$$TlCl(s) \rightleftharpoons Tl^+(aq) + Cl^-(aq)$$

The solubility product expression is

$$K_{sp} = [Tl^+][Cl^-] = 1.9 \times 10^{-4}\ M^2$$

If $TlCl(s)$ is equilibrated with pure water, then at equilibrium, we have

$$[Tl^+] = [Cl^-] = s$$

where s is the solubility of $TlCl(s)$ in pure water. Thus,

$$K_{sp} = (s)(s) = s^2 = 1.9 \times 10^{-4}\ M^2$$

$$s = 1.38 \times 10^{-2}\ M$$

The solubility in grams per liter is

$$s = (0.0138\ mol{\cdot}L^{-1})\left(\frac{239.8\ g\ TlCl}{1\ mol\ TlCl}\right) = 3.3\ g{\cdot}L^{-1}$$

22-10. The solubility equilibrium is described by

$$PbBr_2(s) \rightleftharpoons Pb^{2+}(aq) + 2\,Br^-(aq)$$

The solubility product expression is

$$K_{sp} = [Pb^{2+}]\,[Br^-]^2 = 6.6 \times 10^{-6}\ M^3$$

If $PbBr_2$ is equilibrated with pure water, then at equilibrium, we have

$$[Br^-] = 2\,[Pb^{2+}]$$

Setting up a concentration table

Concentration	$PbBr_2(s)$	+	$H_2O(l)$	$\rightleftharpoons$	$Pb^{2+}(aq)$	+	$2\,Br^-(aq)$
Initial	–		–		≈ 0 M		≈ 0 M
Change	–		–		$+s$		$+2s$
Equilibrium	–		–		s		$2s$

Combining these results with the K_{sp} expression, we have

$$K_{sp} = (s)(2s)^2 = 4s^3 = 6.6 \times 10^{-6}\ M^3$$

$$s = (1.65 \times 10^{-6}\ M^3)^{1/3} = 1.18 \times 10^{-2}\ M$$

The solubility in grams per liter is

$$s = (1.18 \times 10^{-2}\ mol{\cdot}L^{-1})\left(\frac{367\ g\ PbBr_2}{1\ mol\ PbBr_2}\right) = 4.3 \times 10^{-3}\ g{\cdot}L^{-1}$$

22-12. The solubility equilibrium is described by

$$LiF(s) \rightleftharpoons Li^+(aq) + F^-(aq)$$

From the reaction stoichiometry, at equilibrium, $[Li^+] = [F^-]$ = solubility of $LiF(s)$. The solubility of $LiF(s)$ is

$$s = \frac{0.13\ g}{0.100\ L} = (1.3\ g{\cdot}L^{-1})\left(\frac{1\ mol\ LiF}{25.94\ g\ LiF}\right) = 0.050\ M$$

Substituting the values of $[Li^+]$ and $[F^-]$ the K_{sp} expression, we have

$$K_{sp} = [Li^+][F^-] = (0.050\text{ M})(0.050\text{ M}) = 2.5 \times 10^{-3}\text{ M}^2$$

22-14. The solubility equilibrium is described by

$$Pb(IO_3)_2(s) \rightleftharpoons Pb^{2+}(aq) + 2\,IO_3^-(aq)$$

The solubility of $Pb(IO_3)_2(s)$ in pure water is

$$s = (0.76\text{ g}\cdot\text{L}^{-1})\left(\frac{1\text{ mol }Pb(IO_3)_2}{557.0\text{ g }Pb(IO_3)_2}\right) = 1.4 \times 10^{-3}\text{ M}$$

From the reaction stoichiometry, we have

$$[IO_3^-] = 2\,[Pb^{2+}]$$

Using the K_{sp} expression, we compute

$$K_{sp} = [Pb^{2+}][IO_3^-]^2 = (1.4 \times 10^{-3}\text{ M})(2.8 \times 10^{-3}\text{ M})^2 = 1.1 \times 10^{-8}\text{ M}^3$$

There is a large discrepancy with the value of K_{sp}, 3.7×10^{-13} M^3, in Table 22.1 due to additional reactions, such as the formation of $Pb(OH)_2$, occurring in the solution. The K_{sp} expression considers only the formation of $Pb(IO_3)_2$. See Example 22-3.

COMMON-ION EFFECT

22-16. The solubility equilibrium that describes the solubility of barium chromate is

$$BaCrO_4(s) \rightleftharpoons Ba^{2+}(aq) + CrO_4^{2-}(aq)$$

and the solubility product expression (Table 22.1) is

$$K_{sp} = [Ba^{2+}][CrO_4^{2-}] = 1.2 \times 10^{-10}\text{ M}^2$$

The initial concentration of $CrO_4^{2-}(aq)$ is from the $(NH)_2CrO_4(aq)$ solution or 0.0553 M. The only source of $Ba^{2+}(aq)$ is from the $BaCrO_4(s)$ that dissolves. Setting up a concentration table

Concentration	$BaCrO_4(s)$	+ $H_2O(l)$	⇌ $Ba^{2+}(aq)$	+ $CrO_4^{2-}(aq)$
Initial	–	–	0 M	0.0553 M
Change	–	–	$+s$	$+s$
Equilibrium	–	–	s	0.0553 M $+s$

Substituting the equilibrium concentrations into the K_{sp} expression, we obtain

$$K_{sp} = (0.553\text{ M} + s)(s) = 1.2 \times 10^{-10}\text{ M}^2$$

Because $BaCrO_4(s)$ is a slightly soluble salt, we expect the value of s to be small. Therefore, we neglect s compared to 0.0553 M, and write

$$(0.0553\text{ M})(s) = 1.2 \times 10^{-10}\text{ M}^2$$

$$s = 2.17 \times 10^{-9}\text{ M}$$

Note that s is small compared to 0.0553 M, and therefore, our assumption that 0.0553 M + $s \approx$ 0.0553 M is acceptable. The solubility of $BaCrO_4(s)$ in grams per liter is

$$s = (2.17 \times 10^{-9}\ \text{mol}\cdot\text{L}^{-1})\left(\frac{253.3\ \text{g BaCrO}_4}{1\ \text{mol BaCrO}_4}\right) = 5.5 \times 10^{-7}\ \text{g}\cdot\text{L}^{-1}$$

22-18. The solubility equilibrium that describes the solubility of $CaSO_4(s)$ is

$$\text{CaSO}_4(s) \rightleftharpoons \text{Ca}^{2+}(aq) + \text{SO}_4^{2-}(aq)$$

and the solubility product expression (Table 22.1) is

$$K_{sp} = [\text{Ca}^{2+}]\,[\text{SO}_4^{-}] = 4.9 \times 10^{-5}\ \text{M}^2$$

The only source of $Ca^{2+}(aq)$ is from the $CaSO_4(s)$ that dissolves. If we let s be the solubility of $CaSO_4(s)$ in the solution, then

$$[\text{Ca}^{2+}] = s$$

The $SO_4^{2-}(aq)$ in solution is due to the 0.25 M $Na_2SO_4(aq)$ plus the $CaSO_4(s)$ that dissolves. At equilibrium, we have

$$[\text{SO}_4^{2-}] = 0.25\ \text{M} + s$$

Substituting these expressions into the K_{sp} expression yields

$$K_{sp} = (0.25\ \text{M} + s)(s) = 4.9 \times 10^{-5}\ \text{M}^2$$

Assuming that s is small compared to 0.25 M yields

$$s \approx \frac{4.9 \times 10^{-5}\ \text{M}^2}{0.25\ \text{M}} = 1.96 \times 10^{-4}\ \text{M}$$

We see that $s \ll 0.25$ M. The solubility of $CaSO_4(s)$ in grams per liter is

$$s = (1.96 \times 10^{-4}\ \text{mol}\cdot\text{L}^{-1})\left(\frac{136.14\ \text{g CaSO}_4}{1\ \text{mol CaSO}_4}\right) = 0.027\ \text{g}\cdot\text{L}^{-1}$$

22-20. The solubility equilibrium that describes the solubility of $Ag_2SO_4(s)$ is

$$\text{Ag}_2\text{SO}_4(s) \rightleftharpoons 2\,\text{Ag}^{+}(aq) + \text{SO}_4^{2-}(aq)$$

and the solubility product expression (Table 22.1) is

$$K_{sp} = [\text{Ag}^{+}]^2[\text{SO}_4^{2-}] = 1.2 \times 10^{-5}\ \text{M}^3$$

The only source of $Ag^{+}(aq)$ is from the $Ag_2SO_4(s)$ that dissolves; thus, at equilibrium,

$$[\text{Ag}^{+}] = 2s$$

The $SO_4^{2-}(aq)$ is due to the 0.15 M $K_2SO_4(aq)$ plus the $Ag_2SO_4(s)$ that dissolves. At equilibrium, we have

$$[\text{SO}_4^{2-}] = 0.15\ \text{M} + s$$

Substitution of the above expressions for $[Ag^+]$ and $[SO_4^{2-}]$ into the K_{sp} expression yields

$$K_{sp} = (0.15\text{ M} + s)(2s)^2 = 1.2 \times 10^{-5}\text{ M}^3$$

Assuming that s is small compared to 0.15 M yields

$$s = 4.5 \times 10^{-3}\text{ M}$$

The solubility of $Ag_2SO_4(s)$ in grams per liter is

$$s = (4.5 \times 10^{-3}\text{ mol}\cdot\text{L}^{-1})\left(\frac{311.8\text{ g Ag}_2\text{SO}_4}{1\text{ mol Ag}_2\text{SO}_4}\right) = 1.4\text{ g}\cdot\text{L}^{-1}$$

FORMATION OF SOLUBLE COMPLEX IONS

22-22. $Cu(NO_3)_2(aq) + 2\,NH_3(aq) + 2\,H_2O(l) \longrightarrow Cu(OH)_2(s) + 2\,NH_4NO_3(aq)$
$Cu(OH)_2(s) + 4\,NH_3(aq) \longrightarrow [Cu(NH_3)_4]^{2+}(aq) + 2\,OH^-(aq)$

22-24. The equation for the solubility of $AgI(s)$ in $NH_3(aq)$,

$$AgI(s) + 2\,NH_3(aq) \rightleftharpoons 2\,[Ag(NH_3)_2]^+(aq) + I^-(aq)$$

is the sum of two equations

$$AgI(s) \rightleftharpoons Ag^+(aq) + I^-(aq) \qquad K_{sp} = 8.5 \times 10^{-17}\text{ M}^2$$

$$Ag^+(aq) + 2\,NH_3(aq) \rightleftharpoons 2\,[Ag(NH_3)_2]^+(aq) \qquad K_f = 2.0 \times 10^7\text{ M}^{-2}$$

The equilibrium constant for the solubility of $AgBr(s)$ in $NH_3(aq)$ is

$$K = K_{sp}K_f = (8.5 \times 10^{-17}\text{ M}^2)(2.0 \times 10^7\text{ M}^{-2}) = 1.7 \times 10^{-9}$$

We set up a concentration table

Concentration	$AgI(s)$	+	$2\,NH_3(aq)$	$\rightleftharpoons$	$[Ag(NH_3)_2]^+(aq)$	+	$I^-(aq)$
Initial	—		0.60 M		0		≈ 0
Change	—		$-2x$		$+x$		$+x$
Equilibrium	—		0.60 M $-2x$		x		x

The equilibrium-constant expression is

$$K = \frac{[Ag(NH_3)_2^+]\,[I^-]}{[NH_3]^2} = \frac{x^2}{(0.60\text{ M} - 2x)^2} = 1.7 \times 10^{-9}$$

Taking the square root of both sides gives

$$\frac{x}{0.60\text{ M} - 2x} = \pm 4.12 \times 10^{-5}$$

We reject the negative root because it leads to a negative concentration. Thus, we have

$$x = 2.47 \times 10^{-5}\text{ M}$$

Because the value of K_{sp} for AgI(s) is smaller than 1.7×10^{-9}, the solubility of AgI(s) in $NH_3(aq)$ is essentially equal to $[Ag(NH_3)_2^+]$. The solubility of AgI(s) in grams per liter is

$$s = (2.47 \times 10^{-5}\ \text{mol}\cdot\text{L}^{-1})\left(\frac{234.8\ \text{g AgI}}{1\ \text{mol AgI}}\right) = 5.8 \times 10^{-3}\ \text{g}\cdot\text{L}^{-1}$$

22-26. The equation for the solubility of AgBr(s) in $NH_3(aq)$,

$$\text{AgBr}(s) + 2\,\text{NH}_3(aq) \rightleftharpoons [\text{Ag(NH}_3)_2]^+(aq) + \text{Br}^-(aq)$$

is the sum of two equations

$$\text{AgBr}(s) \rightleftharpoons \text{Ag}^+(aq) + \text{Br}^-(aq) \qquad K_{sp} = 5.4 \times 10^{-13}\ \text{M}^2$$

$$\text{Ag}^+(aq) + 2\,\text{NH}_3(aq) \rightleftharpoons 2\,[\text{Ag(NH}_3)_2]^+(aq) \qquad K_f = 2.0 \times 10^7\ \text{M}^{-2}$$

The equilibrium constant for the solubility of AgBr(s) in $NH_3(aq)$ is

$$K = K_{sp}K_f = (5.4 \times 10^{-13}\ \text{M}^2)(2.0 \times 10^7\ \text{M}^{-2}) = 1.1 \times 10^{-5}$$

We set up a concentration table

Concentration	AgBr(s)	+	$2\,NH_3(aq)$	$\rightleftharpoons$	$[Ag(NH_3)_2]^+(aq)$	+	$Br^-(aq)$
Initial	–		0.200 M		0		0.200 M
Change	–		$-2x$		$+x$		$+x$
Equilibrium	–		0.200 M $-2x$		x		0.200 M $+x$

The value of the equilibrium constant is given by

$$K = K_{sp}K_f = (5.4 \times 10^{-13}\ \text{M})(2.0 \times 10^7\ \text{M}^{-2}) = 1.08 \times 10^{-5}$$

The equilibrium-constant expression is

$$K = \frac{[\text{Ag(NH}_3)_2^+]\,[\text{Cl}^-]}{[\text{NH}_3]^2} = \frac{(x)(0.200\ \text{M} + x)}{(0.200\ \text{M} - 2x)^2} = 1.08 \times 10^{-5}\ \text{M}$$

Neglecting x with respect to 0.200 M, we obtain

$$s = 2.2 \times 10^{-6}\ \text{M}$$

The solubility of AgBr(s) in the solution in grams per liter is

$$s = (2.2 \times 10^{-6}\ \text{mol}\cdot\text{L}^{-1})\left(\frac{187.8\ \text{g AgBr}}{1\ \text{mol AgBr}}\right) = 4.1 \times 10^{-4}\ \text{g}\cdot\text{L}^{-1}$$

SOLUBILITY AND pH

22-28. The equilibrium expression is

$$\text{MgC}_2\text{O}_4(s) \rightleftharpoons \text{Mg}^{2+}(aq) + \text{C}_2\text{O}_4^{2-}(aq)$$

Recall that $C_2O_4^{2-}(aq)$ is the conjugate base of the weak acid $HC_2O_4^-(aq)$.

(a) The solubility is increased. The added $H_3O^+(aq)$ reacts with $C_2O_4^{2-}(aq)$, thereby reducing the concentration of $C_2O_4^{2-}(aq)$ and causing a shift in the equilibrium from left to right.

(b) The solubility is slightly decreased owing to a shift to the left in the equilibrium

$$C_2O_4^{2-}(aq) + H_2O(l) \rightleftharpoons HC_2O_4^-(aq) + OH^-(aq)$$

which leads to an increase in $[C_2O_4^{2-}]$.

(c) The solubility is decreased; an increase in $[Mg^{2+}]$ shifts the equilibrium from right to left (common-ion effect).

22-30. The following compounds are more soluble at lower pH for the reasons given:

(a) $PbCrO_4$; $CrO_4^{2-}(aq)$ is the conjugate base of the weak acid $HCrO_4^-(aq)$, also because of the equilibrium

$$2\,CrO_4^{2-}(aq) + 2\,H_3O^+(aq) \rightleftharpoons Cr_2O_7^{2-}(aq) + 2\,H_2O(l)$$

(b) $Ag_2C_2O_4$; $C_2O_4^{2-}(aq)$ is the conjugate base of the weak acid $HC_2O_4^-(aq)$.

(c) Ag_2O; because of the reaction

$$Ag_2O(aq) + 2\,H_3O^+(aq) \rightleftharpoons 2\,Ag^+(aq) + 2\,H_2O(l)$$

22-32. The relevant reaction equations are

$$AgC_6H_5COO(s) \rightleftharpoons Ag^+(aq) + C_6H_5COO^-(aq) \qquad K_{sp} = 2.5 \times 10^{-5}\ M^2$$

$$H_3O^+(aq) + C_6H_5COO^-(aq) \rightleftharpoons C_6H_5COOH(aq) + H_2O(l) \qquad K_a = 6.3 \times 10^{-5}\ M$$

Adding these two equations, we obtain

$$AgC_6H_5COO(s) + H_3O^+(aq) \rightleftharpoons Ag^+(aq) + C_6H_5COOH(aq) + H_2O(l)$$

The equilibrium constant for this equation is

$$K_c = \frac{K_{sp}}{K_a} = \frac{2.5 \times 10^{-5}\ M^2}{6.3 \times 10^{-5}\ M} = 0.40\ M$$

At pH = 4.00, $[H_3O^+] = 1.0 \times 10^{-4}$ M, and so we can set up a concentration table

Concentration	$AgC_6H_5COO(s)$ +	$H_3O^+(aq)$	$\rightleftharpoons$ $Ag^+(aq)$ +	$C_6H_5COOH(aq)$	$H_2O(l)$
Initial	–	1.0×10^{-4} M	0	0	–
Change	–	buffered none	$+s$	$+s$	–
Equilibrium	–	1.0×10^{-4} M	s	s	–

Substituting these values into the equilibrium expression, we get

$$K_c = \frac{[Ag^+][C_6H_5COOH]}{[H_3O^+]} = \frac{s^2}{1.0 \times 10^{-4}\ M} = 0.40\ M$$

Taking the positive root, we find that $s = 6.3 \times 10^{-4}$ M, and so

$$s = (6.3 \times 10^{-4}\ M)(229.0\ g{\cdot}mol^{-1}) = 0.14\ g{\cdot}L^{-1}$$

Q_{sp} CALCULATIONS

22-34. The concentration of $Ag^+(aq)$ immediately after mixing is

$$[Ag^+]_0 = \frac{(0.20\ M)(50.0\ mL)}{200.0\ mL} = 0.050\ M$$

and the concentration of $SO_4^{2-}(aq)$ immediately after mixing is

$$[SO_4^{2-}]_0 = \frac{(0.10\ M)(150.0\ mL)}{200.0\ mL} = 0.075\ M$$

The relevant reaction equation is

$$Ag_2SO_4(s) \rightleftharpoons 2\,Ag^+(aq) + SO_4^{2-}(aq)$$

The value of Q_{sp} for this equation is

$$Q_{sp} = [Ag^+]_0^2[SO_4^{2-}]_0 = (0.050\ M)^2(0.075\ M) = 1.9 \times 10^{-4}\ M^3$$

The value of K_{sp} for $Ag_2SO_4(s)$ is $1.2 \times 10^{-5}\ M^3$ (Table 22.1); thus

$$\frac{Q_{sp}}{K_{sp}} = \frac{1.9 \times 10^{-4}\ M^3}{1.2 \times 10^{-5}\ M^3} = 16 > 1$$

Thus, $Ag_2SO_4(s)$ will precipitate from the solution.

22-36. The concentration of $Ag^+(aq)$ immediately after mixing is

$$[Ag^+]_0 = \frac{(0.50\ M)(50.0\ mL)}{100.0\ mL} = 0.25\ M$$

and the concentration of $Br^-(aq)$ immediately after mixing is

$$[Br^-]_0 = \frac{(1.00 \times 10^{-4}\ M)(50.0\ mL)}{100.0\ mL} = 5.0 \times 10^{-5}\ M$$

The relevant reaction equation is

$$AgBr(s) \rightleftharpoons Ag^+(aq) + Br^-(aq)$$

The value of Q_{sp} for this equation is

$$Q_{sp} = [Ag^+]_0[Br^-]_0 = (0.25\ M)(5.0 \times 10^{-5}\ M) = 1.3 \times 10^{-5}\ M^2$$

The value of K_{sp} for $AgBr(s)$ is $5.4 \times 10^{-13}\ M^3$ (Table 22.1); thus

$$\frac{Q_{sp}}{K_{sp}} = \frac{1.3 \times 10^{-5}\ M^2}{5.4 \times 10^{-13}\ M^2} = 2.4 \times 10^7 > 1$$

Thus, $AgBr(s)$ will precipitate from the solution. Because $[Ag^+]_0 \gg [Br^-]_0$, essentially all the $Br^-(aq)$ is precipitated as $AgBr(s)$, and the final equilibrium value of $[Ag^+]$ will still be essentially 0.25 M. Thus, we have at equilibrium following the precipitation of $AgBr(s)$

$$[Ag^+]_0[Br^-] = K_{sp} = 5.4 \times 10^{-13}\ M^2$$

Therefore,

$$[\mathrm{Br}^-] = \frac{5.4\times10^{-13}\ \mathrm{M}^2}{[\mathrm{Ag}^+]_0} = \frac{5.4\times10^{-13}\ \mathrm{M}^2}{0.25\ \mathrm{M}} = 2.2\times10^{-12}\ \mathrm{M}$$

The number of moles of AgBr(s) that precipitates is equal to the number of moles of Br^-(aq). Thus,

$$\begin{aligned}\text{moles of AgBr}(s)\text{ precipitated} &= (\text{initial moles of Br}^-) - (\text{final moles of Br}^-)\\ &= (5.4\times10^{-5}\ \mathrm{M})(0.100\ \mathrm{L}) - (2.2\times10^{-12}\ \mathrm{M})(0.100\ \mathrm{L})\\ &= 5.4\times10^{-6}\ \mathrm{mol}\end{aligned}$$

The equilibrium concentrations following the precipitation of AgBr(s) are

$$[\mathrm{Ag}^+] = [\mathrm{Ag}^+]_0 = 0.25\ \mathrm{M} \qquad [\mathrm{NO}_3^-] = [\mathrm{Ag}^+]_0 = 0.25\ \mathrm{M}$$

$$[\mathrm{Br}^-] = 2.2\times10^{-12}\ \mathrm{M} \qquad [\mathrm{Na}^+] = [\mathrm{Br}^-]_0 = 5.0\times10^{-5}\ \mathrm{M}$$

22-38. The concentration of Zn^{2+}(aq) immediately after mixing is

$$[\mathrm{Zn}^{2+}]_0 = \frac{(0.30\ \mathrm{M})(10.0\ \mathrm{mL})}{20.0\ \mathrm{mL}} = 0.15\ \mathrm{M}$$

Neglecting the reaction of S^{2-}(aq) with H_2O(l), the concentration of S^{2-}(aq) immediately after mixing is

$$[\mathrm{S}^{2-}]_0 = \frac{(2.00\times10^{-4}\ \mathrm{M})(10.0\ \mathrm{mL})}{20.0\ \mathrm{mL}} = 1.0\times10^{-4}\ \mathrm{M}$$

The relevant reaction equation is

$$\mathrm{ZnS}(s) \rightleftharpoons \mathrm{Zn}^{2+}(aq) + \mathrm{S}^{2-}(aq)$$

The corresponding value of Q_{sp} is

$$Q_{sp} = [\mathrm{Zn}^{2+}]_0[\mathrm{S}^{2-}]_0 = (0.15\ \mathrm{M})(1.0\times10^{-4}\ \mathrm{M}) = 1.5\times10^{-5}\ \mathrm{M}^2$$

The value of K_{sp} for ZnS(s) is $1.6\times10^{-24}\ \mathrm{M}^2$ (Table 22.1); thus

$$\frac{Q_{sp}}{K_{sp}} = \frac{1.5\times10^{-5}\ \mathrm{M}^2}{1.6\times10^{-24}\ \mathrm{M}^2} = 9.4\times10^{18} > 1$$

Thus, ZnS(s) will precipitate from the solution. Because $[\mathrm{Zn}^{2+}]_0 \gg [\mathrm{S}^{2-}]_0$, essentially all the S^{2-}(aq) is precipitated as ZnS(s). The number of moles of ZnS(s) that precipitates is essentially equal to the number of moles of S^{2-}(aq) initially present.

(a) Thus, we have

$$\begin{aligned}\text{moles of ZnS}(s)\text{ precipitated} &= \text{initial moles of S}^{2-}(aq)\\ &= (1.0\times10^{-4}\ \mathrm{M})(0.020\ \mathrm{L}) = 2.0\times10^{-6}\ \mathrm{mol}\end{aligned}$$

The number of milligrams of ZnS(s) in 2.0×10^{-6} mol is

$$\text{mass of ZnS precipitated} = (2.0\times10^{-6}\ \mathrm{mol})\left(\frac{97.44\ \mathrm{g\ ZnS}}{1\ \mathrm{mol\ ZnS}}\right)\left(\frac{1000\ \mathrm{mg}}{1\ \mathrm{g}}\right) = 0.19\ \mathrm{mg}$$

(b) The concentration of $Zn^{2-}(aq)$ at equilibrium is essentially $[Zn^{2+}]_0 = 0.15$ M because only 2.0×10^{-6} mol were used to precipitate $ZnS(s)$. The concentration of $S^{2-}(aq)$ at equilibrium can be found from the K_{sp} expression.

$$K_{sp} = [Zn^{2+}][S^{2-}] = (0.15\text{ M})[S^{2-}] = 1.6 \times 10^{-24}\text{ M}^2$$

$$[S^{2-}] = \frac{1.6 \times 10^{-24}\text{ M}^2}{0.15\text{ M}} = 1.1 \times 10^{-23}\text{ M}$$

This result confirms the statement that essentially all the $S^{2-}(aq)$ is precipitated as $ZnS(s)$.

SELECTIVE PRECIPITATION

22-40. The equations for the two solubility equilibria are (Table 22.1)

$$CaSO_4(s) \rightleftharpoons Ca^{2+}(aq) + SO_4^{2-}(aq) \qquad K_{sp} = 4.9 \times 10^{-5}\text{ M}^2$$

$$BaSO_4(s) \rightleftharpoons Ba^{2+}(aq) + SO_4^{2-}(aq) \qquad K_{sp} = 1.1 \times 10^{-10}\text{ M}^2$$

The solubility-constant expressions are

$$K_{sp} = [Ca^{2+}][SO_4^{2-}] \qquad \text{and} \qquad K_{sp} = [Ba^{2+}][SO_4^{2-}]$$

We wish to have $[Ba^{2+}] = (0.00010)(0.100\text{ M}) = 1.0 \times 10^{-5}$ M. The minimum concentration of $SO_4^{2-}(aq)$ required to achieve this is

$$[SO_4^{2-}] = \frac{1.1 \times 10^{-10}\text{ M}^2}{1.1 \times 10^{-5}\text{ M}} = 1.1 \times 10^{-5}\text{ M}$$

We now must see whether this concentration of $SO_4^{2-}(aq)$ will also precipitate the $Ca^{2+}(aq)$. The value of Q_{sp} for $CaSO_4(s)$ is given by

$$Q_{sp} = [Ca^{2+}][SO_4^{2-}] = (0.10\text{ M})(1.1 \times 10^{-5}\text{ M}) = 1.1 \times 10^{-6}\text{ M}^2$$

Because $Q_{sp} < K_{sp}$ for $CaSO_4(s)$, no $CaSO_4(s)$ will precipitate, and so we can achieve the desired separation.

22-42. The equations for the two solubility equilibria are

$$SrSO_4(s) \rightleftharpoons Sr^{2+}(aq) + SO_4^{2-}(aq) \qquad K_{sp} = 3.4 \times 10^{-7}\text{ M}^2$$

$$PbSO_4(s) \rightleftharpoons Pb^{2+}(aq) + SO_4^{2-}(aq) \qquad K_{sp} = 2.5 \times 10^{-8}\text{ M}^2$$

The expressions for the solubility quotients are

$$Q_{sp} = [Sr^{2+}]_0[SO_4^{2-}]_0 \qquad \text{and} \qquad Q_{sp} = [Pb^{2+}]_0[SO_4^{2-}]_0$$

When $Q_{sp} \le K_{sp}$, no precipitation occurs. We now calculate the concentration of $SO_4^{2-}(aq)$ at which $Q_{sp} = K_{sp}$ for the two equilibria.

$$Q_{sp} = (0.0100\text{ M})[SO_4^{2-}] = 3.4 \times 10^{-7}\text{ M}^2$$

$$[SO_4^{2-}] = 3.4 \times 10^{-5}\text{ M}$$

$$Q_{sp} = (0.0100\ \text{M})\left[\text{SO}_4^{2-}\right] = 2.5 \times 10^{-8}\ \text{M}^2$$

$$\left[\text{SO}_4^{2-}\right] = 2.5 \times 10^{-6}\ \text{M}$$

Thus, when the concentration of $SO_4^{2-}(aq)$ is 3.4×10^{-5} M, $Sr^{2+}(aq)$ does not precipitate. The concentration of $Pb^{2+}(aq)$ at this concentration of $SO_4^{2-}(aq)$ is given by

$$[\text{Pb}^{2+}](3.4 \times 10^{-5}\ \text{M}) = 2.5 \times 10^{-8}\ \text{M}^2$$

Solving for $[Pb^{2+}]$, we have

$$[\text{Pb}^{2+}] = 7.4 \times 10^{-4}\ \text{M}$$

The percentage of $Pb^{2+}(aq)$ that precipitated is

$$\%\ \text{precipitated} = \frac{0.0100\ \text{M} - 7.4 \times 10^{-4}\ \text{M}}{0.0100\ \text{M}} \times 100 = 93\%$$

Thus, 99% of the $Sr^{2+}(aq)$ cannot be separated from $Pb^{2+}(aq)$ by selectively precipitating $Sr^{2+}(aq)$ with $SO_4^{2-}(aq)$.

SEPARATION OF CATIONS AS HYDROXIDES AND SULFIDES

22-44. The K_{sp} expression for $SnS(s)$ is (Table 22.1)

$$K_{sp} = [\text{Sn}^{2+}]\,[\text{S}^{2-}] = 1.0 \times 10^{-25}\ \text{M}^2$$

The solubility of $SnS(s)$ is given by

$$s = [\text{Sn}^{2+}] = \frac{1.0 \times 10^{-25}\ \text{M}^2}{[\text{S}^{2-}]}$$

From Equation 22.25

$$[\text{S}^{2-}] = \frac{1.1 \times 10^{-21}\ \text{M}^3}{[\text{H}_3\text{O}^+]^2}$$

At pH = 2.0

$$[\text{H}_3\text{O}^+] = 10^{-2.0} = 1.0 \times 10^{-2}\ \text{M}$$

and thus,

$$[\text{S}^{2-}] = \frac{1.1 \times 10^{-21}\ \text{M}^3}{(1.0 \times 10^{-2}\ \text{M})^2} = 1.1 \times 10^{-17}\ \text{M}$$

Therefore, the solubility of $SnS(s)$ is

$$s = \frac{1.0 \times 10^{-25}\ \text{M}^2}{1.1 \times 10^{-17}\ \text{M}} = 9.1 \times 10^{-9}\ \text{M}$$

22-46. The K_{sp} expression for $Cu(OH)_2(s)$ is (Table 22.1)

$$K_{sp} = [\text{Cu}^{2+}]\,[\text{OH}^-]^2 = 2.2 \times 10^{-20}\ \text{M}^3$$

The solubility of $Cu(OH)_3(s)$ is given by

$$s = [Cu^{2+}] = \frac{2.2 \times 10^{-20}\ M^3}{[OH^-]^2} = \frac{(2.2 \times 10^{-20}\ M^3)[H_3O^+]^2}{K_w^2}$$

At pH = 4.0

$$[H_3O^+] = 10^{-4.0} = 1.0 \times 10^{-4}\ M$$

Thus, the solubility of $Cu(OH)_2(s)$ at pH = 4.0 is given by

$$s = \frac{(2.2 \times 10^{-20}\ M^3)(1.0 \times 10^{-4}\ M)^2}{(1.00 \times 10^{-14}\ M^2)^2} = 2.2\ M$$

The K_{sp} expression for $Zn(OH)_2(s)$ is (Table 22.1)

$$K_{sp} = [Zn^{2+}][OH^-]^2 = 1.0 \times 10^{-15}\ M^3$$

The solubility of $Zn(OH)_2(s)$ at pH = 4.0 is given by

$$s = [Zn^{2+}] = \frac{1.0 \times 10^{-15}\ M^3}{[OH^-]^2} = \frac{(1.0 \times 10^{-15}\ M^3)[H_3O^+]^2}{K_w^2}$$

$$= \frac{(1.0 \times 10^{-15}\ M^3)(1.0 \times 10^{-4}\ M)^2}{(1.00 \times 10^{-14}\ M^2)^2} = 1.0 \times 10^5\ M$$

Of course, a solubility of 1.0×10^5 M is physically unrealistic, but this result means that $Zn(OH)_2(s)$ is very soluble. Thus, at pH = 4.0, both $Cu(OH)_2(s)$ and $Zn(OH)_2(s)$ are very soluble and so a separation cannot be achieved at pH = 4.0.

22-48. The solubilities of $FeS(s)$ and $PbS(s)$ as a function of $[H_3O^+]$ at $[H_2S] = 0.10$ M are obtained using Equation 22.25 and the value of K_{sp} expression. Thus, for $FeS(s)$, we have

$$s = \frac{K_{sp}[H_3O^+]^2}{1.1 \times 10^{-21}\ M^3} = \frac{(6.3 \times 10^{-18}\ M^2)[H_3O^+]^2}{1.1 \times 10^{-21}\ M^3} = (5.7 \times 10^3\ M^{-1})[H_3O^+]^2$$

whereas for $PbS(s)$, we have

$$s = \frac{K_{sp}[H_3O^+]^2}{1.1 \times 10^{-21}\ M^3} = \frac{(8.0 \times 10^{-28}\ M^2)[H_3O^+]^2}{1.1 \times 10^{-21}\ M^3} = (7.3 \times 10^{-7}\ M^{-1})[H_3O^+]^2$$

Note that at any value of $[H_3O^+]$, $PbS(s)$ is much less soluble than $FeS(s)$. For example, the value of $[H_3O^+]$ at which $s = 1 \times 10^{-6}$ M for $PbS(s)$ is

$$1 \times 10^{-6}\ M = (7.3 \times 10^{-7}\ M^{-1})[H_3O^+]^2$$

or

$$[H_3O^+] = 1.17\ M$$

and the pH is $-\log(1.17) = -0.07$. The solubility of $FeS(s)$ at $[H_3O^+] = 1.17$ M is

$$s = (5.7 \times 10^3\ M^{-1})(1.17\ M)^2 = 7.8 \times 10^3\ M$$

Thus, if the pH of the solution is adjusted to about −1, an effective separation can be achieved. At this pH, the solubility of $PbS(s)$ is 1×10^{-6} M, whereas the solubility of $FeS(s)$ is very large.

22-50. The chemical equations for the solubility equilibria are

$$\text{Cd(OH)}_2(s) \rightleftharpoons \text{Cd}^{2+}(aq) + 2\,\text{OH}^-(aq)$$

$$\text{Fe(OH)}_3(s) \rightleftharpoons \text{Fe}^{3+}(aq) + 3\,\text{OH}^-(aq)$$

For $\text{Cd(OH)}_2(s)$, we have

$$s = [\text{Cd}^{2+}] = \frac{K_{\text{sp}}}{[\text{OH}^-]^2} = \frac{K_{\text{sp}}[\text{H}_3\text{O}^+]^2}{(1.00 \times 10^{-14}\ \text{M}^2)^2}$$
$$= \frac{(7.2 \times 10^{-15}\ \text{M}^3)[\text{H}_3\text{O}^+]^2}{(1.00 \times 10^{-14}\ \text{M}^2)^2} = (7.2 \times 10^{13}\ \text{M}^{-1})[\text{H}_3\text{O}^+]^2$$

For $\text{Fe(OH)}_3(s)$, we have

$$s = [\text{Fe}^{3+}] = \frac{K_{\text{sp}}}{[\text{OH}^-]^3} = \frac{K_{\text{sp}}[\text{H}_3\text{O}^+]^3}{(1.00 \times 10^{-14}\ \text{M}^2)^3}$$
$$= \frac{(2.8 \times 10^{-39}\ \text{M}^4)[\text{H}_3\text{O}^+]^3}{(1.00 \times 10^{-14}\ \text{M}^2)^3} = (2.8 \times 10^{3}\ \text{M}^{-2})[\text{H}_3\text{O}^+]^3$$

$\text{Fe(OH)}_3(s)$ is insoluble for pH > 2, but $\text{Cd(OH)}_2(s)$ is soluble up to pH ≈ 8.

AMPHOTERIC METAL HYDROXIDES

22-52. The two relevant chemical equations are (Tables 22.1 and 22.3)

$$\text{Pb(OH)}_2(s) \rightleftharpoons \text{Pb}^{2+}(aq) + 2\,\text{OH}^-(aq) \qquad K_{\text{sp}} = 1.4 \times 10^{-20}\ \text{M}^3$$

$$\text{Pb(OH)}_2(s) + \text{OH}^-(aq) \rightleftharpoons [\text{Pb(OH)}_3]^-(aq) \quad K_{\text{f}} = 0.08$$

At pH $= 13.00$, $[\text{OH}^-] = 0.10$ M, and so the solubility due to the first equation is

$$s = [\text{Pb}^{2+}] = \frac{K_{\text{sp}}}{[\text{OH}^-]^2} = \frac{1.4 \times 10^{-20}\ \text{M}^3}{(0.10\ \text{M})^2} = 1.4 \times 10^{-18}\ \text{M}$$

The solubility due to the second equation is

$$s = [\text{Pb(OH)}_3^-] = K_{\text{f}}[\text{OH}^-] = (0.08)(0.10\ \text{M}) = 0.008\ \text{M}$$

Thus, the total solubility is 0.008 M, and the contribution from the first equation is negligible.

22-54. The two relevant chemical equations are (Tables 22.1 and 22.3)

$$\text{Pb(OH)}_2(s) \rightleftharpoons \text{Pb}^{2+}(aq) + 2\,\text{OH}^-(aq) \qquad K_{\text{sp}} = 1.4 \times 10^{-20}\ \text{M}^3$$

$$\text{Pb(OH)}_2(s) + \text{OH}^-(aq) \rightleftharpoons [\text{Pb(OH)}_3]^-(aq) \qquad K_{\text{f}} = 0.08$$

The solubility due to the first equation is

$$s = [\text{Pb}^{2+}] = \frac{K_{\text{sp}}}{[\text{OH}^-]^2} = \frac{(1.4 \times 10^{-20}\ \text{M}^3)[\text{H}_3\text{O}^+]^2}{(1.0 \times 10^{-14}\ \text{M}^2)^2} = (1.4 \times 10^{8}\ \text{M}^{-1})[\text{H}_3\text{O}^+]^2$$

The solubility due to the second equation is

$$s = [\mathrm{Pb(OH)_3^-}] = K_f[\mathrm{OH^-}] = \frac{(0.08)(1.0 \times 10^{-14}\ \mathrm{M^2})}{[\mathrm{H_3O^+}]} = \frac{8 \times 10^{-16}\ \mathrm{M^2}}{[\mathrm{H_3O^+}]}$$

The total solubility is given by

$$s_{\text{total}} = [\mathrm{Pb^{2+}}] + [\mathrm{Pb(OH)_3^-}] = (1.4 \times 10^8\ \mathrm{M^{-1}})[\mathrm{H_3O^+}]^2 + \frac{8 \times 10^{-16}\ \mathrm{M^2}}{[\mathrm{H_3O^+}]}$$

We have the following table:

pH	4.00	6.00	8.00	10.00	12.00	14.00	16.00
$s_{\text{total}}/\mathrm{M}$	1.4	1.4×10^{-4}	9.4×10^{-8}	8×10^{-6}	8×10^{-4}	0.08	8

ADDITIONAL PROBLEMS

22-56. A common ion is an ion that is present in two different salts within the same solution. For example, the addition of $\mathrm{NaCl}(s)$ to an $\mathrm{NH_4Cl}(aq)$ solution introduces additional chloride, which is a common ion to both substances.

22-58. One such experiment would be to saturate a solution with the solid, filter off the undissolved solid, measure the volume of the solution, evaporate the solution to dryness, and then weigh the remaining solid residue.

22-60. No. There will always be some small fraction of the original solid remaining in solution. However, in many cases, this fraction is negligible.

22-62. The two relevant reaction equations are

$$\mathrm{AgCl}(s) \rightleftharpoons \mathrm{Ag^+}(aq) + \mathrm{Cl^-}(aq)$$

$$\mathrm{Ag_2CrO_4}(s) \rightleftharpoons 2\,\mathrm{Ag^+}(aq) + \mathrm{CrO_4^{2-}}(aq)$$

From Table 22.1, we have that

$$K_{\text{sp}} = [\mathrm{Ag^+}]\,[\mathrm{Cl^-}] = (s)(s) = s^2 = 1.8 \times 10^{-10}\ \mathrm{M^2}$$

$$s = 1.3 \times 10^{-5}\ \mathrm{M}$$

and

$$K_{\text{sp}} = [\mathrm{Ag^+}]^2\,[\mathrm{CrO_4^{2-}}] = (2s)^2(s) = 4s^3 = 1.1 \times 10^{-12}\ \mathrm{M^3}$$

$$s = 6.5 \times 10^{-5}\ \mathrm{M}$$

Thus, $\mathrm{Ag_2CrO_4}(s)$ is more soluble than AgCl(s) in pure water.

22-64. We have the flowchart

1. Add $\mathrm{NaCl}(aq)$. If $\mathrm{Ag^+}(aq)$ is present, it will precipitate as $\mathrm{AgCl}(s)$.
2. Remove any solid and add $\mathrm{H_2SO_4}(aq)$. If $\mathrm{Ba^{2+}}(aq)$ is present, it will precipitate as $\mathrm{BaSO_4}(s)$.
3. To the remaining solution, add $\mathrm{NaOH}(aq)$ until the solution is just basic. If $\mathrm{Cd^{2=}}(aq)$ is present, it will precipitate as $\mathrm{Cd(OH)_2}(s)$.

Other schemes are also possible.

22-66. The concentration of $S^{2-}(aq)$ is given by Equation 22.25:

$$[S^{2-}] = \frac{1.1 \times 10^{-21}\ M^3}{[H_3O^+]^2}$$

At pH = 2.00, $[H_3O^+] = 1.0 \times 10^{-2}$ M and

$$[S^{2-}] = \frac{1.1 \times 10^{-21}\ M^3}{(1.0 \times 10^{-2} M)^2} = 1.1 \times 10^{-17}\ M$$

At pH = 4.00, $[H_3O^+] = 1.0 \times 10^{-4}$ M and

$$[S^{2-}] = \frac{1.1 \times 10^{-21}\ M^3}{(1.0 \times 10^{-4}\ M)^2} = 1.1 \times 10^{-13}\ M$$

The solubility-product expression of FeS(s) is (Table 22.1)

$$K_{sp} = [Fe^{2+}]\,[S^{2-}] = 6.3 \times 10^{-18}\ M^2$$

Thus, the solubility of FeS(s) is given by

$$s = [Fe^{2+}] = \frac{6.3 \times 10^{-18}\ M^2}{[S^{2-}]}$$

At pH = 2.00, the solubility of FeS(s) is given by

$$s = \frac{6.3 \times 10^{-18}\ M^2}{1.1 \times 10^{-17}\ M} = 0.57\ M$$

At pH = 4.00, the solubility of FeS(s) is given by

$$s = \frac{6.3 \times 10^{-18}\ M^2}{1.1 \times 10^{-13}\ M} = 5.7 \times 10^{-5}\ M$$

The solubility-product expression of CdS(s) is (Table 22.1)

$$K_{sp} = [Cd^{2+}]\,[S^{2-}] = 8.0 \times 10^{-27}\ M^2$$

Thus, the solubility of CdS(s) is given by

$$s = [Cd^{2+}] = \frac{8.0 \times 10^{-27}\ M^2}{[S^{2-}]}$$

At pH = 2.00, the solubility of CdS(s) is given by

$$s = \frac{8.0 \times 10^{-27}\ M^2}{1.1 \times 10^{-17}\ M} = 7.3 \times 10^{-10}\ M$$

At pH = 4.00, the solubility of CdS(s) is given by

$$s = \frac{8.0 \times 10^{-27}\ M^2}{1.1 \times 10^{-13}\ M} = 7.3 \times 10^{-14}\ M$$

Thus, we could separate these compounds at pH = 2.00, but not at pH = 4.00.

22-68. The solubility-product expression for $Pb(OH)_2(s)$ is (Table 22.1)

$$K_{sp} = [Pb^{2+}]\,[OH^-]^2 = 1.4 \times 10^{-20}\ M^3$$

Thus, we have for the solubility as a function of $[H_3O^+]$ or of pH

$$s = [Pb^{2+}] = \frac{1.4 \times 10^{-20}\ M^3}{[OH^-]^2} = \frac{(1.4 \times 10^{-20}\ M^3)[H_3O^+]^2}{K_w^2}$$

$$= \frac{(1.4 \times 10^{-20}\ M^3)[H_3O^+]^2}{(1.00 \times 10^{-14}\ M^2)^2} = 1.4 \times 10^8\ M^{-1}[H_3O^+]^2$$

The solubility-product expression for $Sn(OH)_2(s)$ is (Table 22.1)

$$K_{sp} = [Sn^{2+}]\,[OH^-]^2 = 5.5 \times 10^{-27}\ M^3$$

Thus, we have for the solubility as a function of $[H_3O^+]$ or pH

$$s = [Sn^{2+}] = \frac{5.5 \times 10^{-27}\ M^3}{[OH^-]^2} = \frac{(5.5 \times 10^{-27}\ M^3)[H_3O^+]^2}{K_w^2}$$

$$= \frac{(5.5 \times 10^{-27}\ M^3)[H_3O^+]^2}{(1.00 \times 10^{-14}\ M^2)^2} = 55\ M^{-1}[H_3O^+]^2$$

From the above expressions for s, we see that $Pb(OH)_2(s)$ is much more soluble at a given pH than is $Sn(OH)_2(s)$. Let's find the pH at which s for $Sn(OH)_2(s)$ (the less soluble hydroxide) is 1×10^{-6} M.

$$1 \times 10^{-6}\ M = 55\ M^{-1}[H_3O^+]^2$$

$$[H_3O^+]^2 = \frac{1 \times 10^{-6}\ M}{55\ M^{-1}} = 1.8 \times 10^{-8}\ M^2$$

Thus, $[H_3O^+] = 1.3 \times 10^{-4}$ M and pH = 3.87. When the pH is greater than 3.87, the solubility of $Sn(OH)_2(s)$ is less than 10^{-6} M.
The solubility of $Pb(OH)_2(s)$ at pH = 3.87 is

$$s = 1.4 \times 10^8\ M^{-1}[H_3O^+]^2 = (1.4 \times 10^8\ M^{-1})(1.3 \times 10^{-4}\ M)^2 = 2.5\ M$$

Thus at pH = 3.9, an effective separation can be achieved. If the pH is too high, then we would have to consider the formation of hydroxy complexes.

22-70. The equation for the precipitation reactions are

$$Zn(ClO_4)_2(aq) + 2\,KOH(aq) \longrightarrow Zn(OH)_2(s) + 2\,KClO_4(aq)$$

$$Mg(ClO_4)_2(aq) + 2\,KOH(aq) \longrightarrow Mg(OH)_2(s) + 2\,KClO_4(aq)$$

The $Zn(OH)_2(s)$ precipitate dissolves on further addition of $KOH(aq)$ due to the formation of the $[Zn(CH)_4]^{2-}(aq)$ complex ion described by

$$Zn(OH)_2(s) + 2\,OH^-(aq) \longrightarrow [Zn(OH)_4]^{2-}(aq)$$

The $Mg(OH)_2(s)$ precipitate does not dissolve on further addition of $KOH(aq)$ because it does not undergo a complexation reaction.

22-72. We have for the value of K_{sp} (Table 22.1) for the equilibrium

(1) $AgCl(s) \rightleftharpoons Ag^+(aq) + Cl^-(aq) \qquad K_{sp} = 1.8 \times 10^{-10}\ M^2$
Addition of the solubility equilibrium equation to the complexation equation

(2) $Ag^+(aq) + 2\,NH_3(aq) \rightleftharpoons [Ag(NH_3)_2]^+(aq) \qquad K_f = 2.0 \times 10^7\ M^{-2}$
yields

(3) $AgCl(s) + 2\,NH_3(aq) \rightleftharpoons [Ag(NH_3)_2]^+(aq) + Cl^-(aq)$
for which

$$K_3 = K_{sp}K_f = (1.8 \times 10^{-10}\ M^2)(2.0 \times 10^7\ M^{-2}) = 3.6 \times 10^{-3}$$

Because $K_3 \gg K_{sp}$, we have $[Ag(NH_3)_2^+] \gg [Ag^+]$ and thus the solubility s is equal to

$$s = [Ag(NH_3)_2^+] = [Cl^-]$$

A concentration of 250 milligrams of $AgCl(s)$ in 100 milliliters of solution corresponds to a solubility of

$$s = \frac{(0.250\ g)\left(\dfrac{1\ mol\ AgCl}{143.3\ g\ AgCl}\right)}{0.100\ L} = 0.0174\ M$$

Substituting this value of s into the K_3 expression yields

$$K_3 = \frac{[Ag(NH_3)_2^+]\,[Cl^-]}{[NH_3]^2} = \frac{s^2}{[NH_3]^2} = \frac{(0.0174\ M)^2}{[NH_3]^2} = 3.6 \times 10^{-3}$$

Solving for $[NH_3]$,

$$[NH_3]^2 = \frac{(0.0174\ M)^2}{3.6 \times 10^{-3}}$$

Thus

$$[NH_3] = 0.29\ M$$

22-74. (a) We have from Table 22.1,

(1) $Ag_2CrO_4(s) \rightleftharpoons 2\,Ag^+(aq) + CrO_4^{2-}(aq) \qquad K_{sp1} = 1.1 \times 10^{-12}\ M^3$

(2) $AgBr(s) \rightleftharpoons Ag^+(aq) + Br^-(aq) \qquad K_{sp2} = 5.4 \times 10^{-13}\ M^2$
If we reverse Equation 2, multiply through by 2, and then add the result to Equation 1, we obtain Equation 3:

(3) $Ag_2CrO_4(s) + 2\,Br^-(aq) \rightleftharpoons 2\,AgBr(s) + CrO_4^{2-}(aq)$
Thus,

$$K_3 = \frac{K_{sp1}}{K_{sp2}^2} = \frac{1.1 \times 10^{-12}\ M^3}{(5.4 \times 10^{-13}\ M^2)^2} = 3.8 \times 10^{12}\ M^{-1}$$

(b) We have from Table 22.1,

(1) $PbCO_3(s) \rightleftharpoons Pb^{2+}(aq) + CO_3^{2-}(aq) \qquad K_{sp1} = 7.4 \times 10^{-14}\ M^2$

(2) $CaCO_3(s) \rightleftharpoons Ca^{2+}(aq) + CO_3^{2-}(aq) \qquad K_{sp2} = 3.4 \times 10^{-9}\ M^2$
If we reverse Equation 2 and add the result to Equation 1, then we obtain Equation 3:

(3) $PbCO_3(s) + Ca^{2+}(aq) \rightleftharpoons CaCO_3(s) + Pb^{2+}(aq)$

Thus,

$$K_3 = \frac{K_{sp1}}{K_{sp2}} = \frac{7.4 \times 10^{-14}\ \text{M}^2}{3.4 \times 10^{-9}\ \text{M}^2} = 2.2 \times 10^{-5}$$

22-76. The value of K_f is so much larger than K_{sp} that we can neglect the K_{sp} expression compared to K_f, that is

$$s = [\text{HgI}_4^{2-}] + [\text{Hg}^{2-}] \approx [\text{HgI}_4^{2-}]$$

Thus we have for the complexion reaction

$$\text{HgI}_2(s) + 2\,\text{I}^-(aq) \rightleftharpoons [\text{HgI}_4]^{2-}(aq)$$

We set up the following concentration table:

Concentration	$\text{HgI}_2(s)$	+	$2\,\text{I}^-(aq)$	$\rightleftharpoons$	$[\text{HgI}_4]^{2-}(aq)$
Initial	—		0.10 M		0 M
Change	—		$-2x$		$+x$
Equilibrium	—		0.10 M $-2x$		x

The equilibrium-constant expression is

$$K_f = \frac{[\text{HgI}_4^{2-}]}{[\text{I}^-]^2} = \frac{x}{(0.10\ \text{M} - 2x)^2} = 0.79\ \text{M}^{-1}$$

Rearranging to the standard quadratic equation form yields

$$3.16\ \text{M}^{-1}x^2 - 1.316x + 0.0079\ \text{M} = 0$$

The quadratic formula gives $x = 0.0061$ M and 0.41 M. We reject the $x = 0.41$ M solution as physically impossible because $[\text{I}^-]_0 = 0.10$ M. Thus, the solubility of $\text{HgI}_2(s)$ is 0.0061 M.

22-78. From Table 22.1, we obtain

(1) $\text{Ag}_2\text{SO}_4(s) \rightleftharpoons 2\,\text{Ag}^+(aq) + \text{SO}_4^{2-}(aq) \qquad K_{sp1} = 1.2 \times 10^{-5}\ \text{M}^3$

(2) $\text{CaSO}_4(s) \rightleftharpoons \text{Ca}^{2+}(aq) + \text{SO}_4^{2-}(aq) \qquad K_{sp2} = 4.9 \times 10^{-5}\ \text{M}^2$

Addition of equation 1 to the reverse of equation 2 yields

(3) $\text{Ag}_2\text{SO}_4(s) + \text{Ca}^{2+}(aq) \rightleftharpoons \text{CaSO}_4(s) + 2\,\text{Ag}^+(aq)$

Thus,

$$K_3 = \frac{K_{sp1}}{K_{sp2}} = \frac{1.2 \times 10^{-5}\ \text{M}^3}{4.9 \times 10^{-5}\ \text{M}^2} = 0.24\ \text{M}$$

and

$$K_3 = \frac{[\text{Ag}^+]^2}{[\text{Ca}^{2+}]} = 0.24\,\text{M}$$

Set up the following concentration table:

Concentration	$Ag_2SO_4(s)$	+	$Ca^{2+}(aq)$	$\rightleftharpoons$	$CaSO_4(s)$	+	$2\,Ag^+(aq)$
Initial	–		0 M		–		0.100 M
Change	–		$+x$		–		$-2x$
Equilibrium	–		x		–		0.100 M $-2x$

Thus, at equilibrium,

$$\frac{(0.100\text{ M} - 2x)^2}{x} = 0.24\,\text{M}$$

Expanding and rearranging to the standard quadratic equation form yields

$$4x^2 - (0.64\text{ M})\,x + 0.01\text{ M}^2 = 0$$

Using the quadratic formula yields

$$x = 0.14\text{ M} \qquad \text{and} \qquad 0.018\text{ M}$$

We reject the 0.14 M root because it is greater than the original concentration. Thus, we have

$$[Ca^{2+}] = 0.018\text{ M} \qquad \text{and} \qquad [Ag^+] = 0.100\text{ M} - (2)(0.018\text{ M}) = 0.064\text{ M}$$

22-80. The equilibrium expression that describes the solubility of $CuBr(s)$ in water is (Table 22.1)

(1) $CuBr(s) \rightleftharpoons Cu^+(aq) + Br^-(aq) \qquad K_{sp} = 6.3 \times 10^{-9}\text{ M}^2$
Furthermore, $Cu^+(aq)$ reacts with $NH_3(aq)$ according to

(2) $Cu^+(aq) + 2\,NH_3(aq) \rightleftharpoons [Cu(NH_3)_2]^+(aq) \qquad K_f = 6.3 \times 10^{10}\text{ M}^{-2}$
The sum of these two equations is

(3) $CuBr(s) + 2\,NH_3(aq) \rightleftharpoons [Cu(NH_3)_2]^+(aq) + Br^-(aq)$

This reaction represents the dissolution of $CuBr(s)$ in an ammonia solution. The equilibrium constant for equation 3 is equal to the product of the equilibrium constants for equations 1 and 2:

$$K_3 = K_{sp}K_f = (6.3 \times 10^{-9}\text{ M}^2)(6.3 \times 10^{10}\text{ M}^{-2}) = 397$$

Thus

$$K_3 = \frac{[Cu(NH_3)_2^+]\,[Br^-]}{[NH_3]^2} = 397$$

From the reaction 3 stoichiometry, we have

$$[Cu(NH_3)_2^+] = [Br^-]$$

Also we assume that $[Cu^+] \ll [Cu(NH_3)_2^+]$ because $K_{sp} \ll K_3$, and thus essentially all the $Cu^+(aq)$ in solution is in the form $[Cu(NH_3)_2]^+(aq)$. If we let s be the solubility of $CuBr(s)$ in $NH_3(aq)$, then

$$s = [Br^-]$$

and we have

$$K_3 = \frac{(s)(s)}{[NH_3]^2} = \frac{s^2}{(0.15\ M)^2} = 397$$

Thus, $s = 3.0$ M. We see that CuBr(s) is quite soluble in an ammonia solution.

22-82. The equation for the complexation reaction is

$$Zn(OH)_2(s) + 2\,OH^-(aq) \rightleftharpoons [Zn(OH)_4]^{2-}(aq) \qquad K_f = 0.050\ M^{-1}$$

The equilibrium-constant expression is

$$K_f = \frac{[Zn(OH)_4^{2-}]}{[OH^-]^2} = 0.050\ M^{-1}$$

The solubility of $Zn(OH)_2(s)$ in a basic solution is given by

$$s = [Zn(OH)_4^{2-}] = K_f[OH^-]^2 = 0.050\ M^{-1}[OH^-]^2$$

The stoichiometric concentration of $OH^-(aq)$ in a 0.10 M NaOH(aq) solution is

$$[OH^-]_0 = 0.10\ M$$

At equilibrium, we have from the reaction stoichiometry

$$[OH^-] = 0.10\ M - 2s$$

Thus,

$$s = (0.050\ M^{-1})(0.10\ M - 2s)^2$$

Neglecting $2s$ with respect to 0.10 M, we have

$$s = (0.050\ M^{-1})(0.10\ M)^2 = 5.0 \times 10^{-4}\ M$$

22-84. The solubility equilibrium is described by

$$Ca(OH)_2(s) \rightleftharpoons Ca^{2+}(aq) + 2\,OH^-(aq)$$

From the reaction stoichiometry, at equilibrium we have

$$[Ca^{2+}] = \frac{1}{2}[OH^-]$$

The solubility of $Ca(OH)_2(s)$ is equal to $[Ca^{2+}]$ because each mole of $Ca(OH)_2(s)$ that dissolves yields one mole of $Ca^{2+}(aq)$. We can find the value of $[OH^-]$ from the pH of the solution

$$[H_3O^+] = 10^{-12.45} = 3.55 \times 10^{-13}\ M$$

$$[OH^-] = \frac{K_w}{[H_3O^+]} = \frac{1.00 \times 10^{-14}\ M^2}{3.55 \times 10^{-13}\ M} = 2.82 \times 10^{-2}\ M$$

The solubility of $Ca(OH)_2(s)$ is

$$s = [Ca^{2+}] = \frac{1}{2}[OH^-] = \frac{1}{2}(2.82 \times 10^{-2}\ M) = 1.4 \times 10^{-2}\ M$$

The solubility in grams per liter is

$$s = (1.4 \times 10^{-2}\ \text{mol}\cdot\text{L}^{-1})\left(\frac{74.10\ \text{g Ca(OH)}_2}{1\ \text{mol Ca(OH)}_2}\right) = 1.0\ \text{g}\cdot\text{L}^{-1}$$

22-86. (a) $ZnS(s) \rightleftharpoons Zn^{2+}(aq) + S^{2-}(aq)$
The solubility increases. The $H_3O^+(aq)$ from the added $HNO_3(aq)$ reacts with $S^{2-}(aq)$ to form $HS^-(aq)$, thereby decreasing the concentration of $S^{2-}(aq)$. A decrease in $[S^{2-}]$ shifts the equilibrium from left to right.

(b) $AgI(s) \rightleftharpoons Ag^+(aq) + I^-(aq)$
The solubility increases. Ammonia, $NH_3(aq)$ reacts with $Ag^+(aq)$ to form the soluble ion $[Ag(NH_3)_2]^+(aq)$, thereby reducing the amount of $Ag^+(aq)$. A decrease in $[Ag^+]$ shifts the equilibrium from left to right.

22-88. The net ionic equation for the precipitation reaction is

$$Hg_2^{2+}(aq) + 2\,Cl^-(aq) \rightleftharpoons Hg_2Cl_2(s)$$

The concentration of $Cl^-(aq)$ immediately after mixing is

$$[Cl^-]_0 = \frac{(1.00\ \text{M})(50.0\ \text{mL})}{100.0\ \text{mL}} = 0.500\ \text{M}$$

and the concentration of $Hg_2^{2+}(aq)$ is

$$[Hg_2^{2+}]_0 = \frac{(1.00\ \text{M})(50.0\ \text{mL})}{100.0\ \text{mL}} = 0.500\ \text{M}$$

The value of Q_{sp} is

$$Q_{sp} = [Hg_2^{2+}]_0[Cl^-]_0^2 = (0.500\ \text{M})(0.500\ \text{M})^2 = 0.125\ \text{M}^2$$

Because $Q_{sp} > K_{sp}$ (Table 22.1), precipitation will occur. Because of the reaction stoichiometry, there is an excess of $Hg_2^{2+}(aq)$; thus, essentially all the $Cl^-(aq)$ precipitates, and so $Cl^-(aq)$ is the limiting reactant. From the reaction stoichiometry, one mole of $Hg_2Cl_2(s)$ will precipitate with two moles of $Cl^-(aq)$, and so the equilibrium concentration of $Hg_2^{2+}(aq)$ is essentially

$$\begin{aligned}[Hg_2^{2+}] &= [Hg_2^{2+}]_0 - \frac{1}{2}[Cl^-]_0 \\ &= 0.500\ \text{M} - \frac{1}{2}(0.500\ \text{M}) = 0.250\ \text{M}\end{aligned}$$

The solubility product expression is

$$K_{sp} = [Hg_2^{2+}]\,[Cl^-]^2 = (0.250\ \text{M})[Cl^-]^2 = 1.4 \times 10^{-18}\ \text{M}^3$$

$$[Cl^-] = 2.4 \times 10^{-9}\ \text{M}$$

22-90. The equation for the reaction is

$$3\,Hg_2(NO_3)_2(aq) + 2\,AlCl_3(aq) \longrightarrow 3\,Hg_2Cl_2(s) + 2\,Al(NO_3)_3(aq)$$

After mixing, we have

$$[\text{Cl}^-]_0 = \frac{(3)(0.100\text{ M})(150.0\text{ mL})}{250.0\text{ mL}} = 0.180\text{ M}$$

$$[\text{Hg}_2^{2+}]_0 = \frac{(0.200\text{ M})(100.0\text{ mL})}{250.0\text{ mL}} = 0.0800\text{ M}$$

There is an excess of $Cl^-(aq)$; thus, essentially all the $Hg_2^{2+}(aq)$ precipitates. The equilibrium concentration of $Cl^-(aq)$ is essentially $0.180\text{ M} - (2)(0.0800\text{ M}) = 0.020\text{ M}$. The solubility product expression is

$$K_{\text{sp}} = [\text{Hg}_2^{2+}][\text{Cl}^-]^2 = [\text{Hg}_2^{2+}](0.020\text{ M})^2 = 1.4 \times 10^{-18}\text{ M}^3$$

$$[\text{Hg}_2^{2+}] = 3.5 \times 10^{-15}\text{ M}$$

The fraction of $Hg_2^{2+}(aq)$ that is not precipitated is

$$\text{fraction} = \frac{3.5 \times 10^{-15}\text{ M}}{0.080\text{ M}} = 4.4 \times 10^{-14}$$

22-92. We set up a concentration table

Concentration	$AgI(s)$	+	$2\,NH_3(aq)$	⇌	$[Ag(NH_3)_2]^+(aq)$	+	$I^-(aq)$
Initial	–		14.0 M		0		0
Change	–		$-2s$		$+s$		$+s$
Equilibrium	–		$14.0\text{ M} - 2s$		s		s

The value of the equilibrium constant for the equation for the reaction is

$$K = K_{\text{sp}} K_{\text{f}} = (8.5 \times 10^{-17}\text{ M}^2)(2.0 \times 10^7\text{ M}^{-2}) = 1.7 \times 10^{-9}$$

The equilibrium-constant expression is

$$K = \frac{[\text{Ag(NH}_3)_2^+][\text{I}^-]}{[\text{NH}_3]^2} = 1.7 \times 10^{-9}$$

Thus,

$$K = \frac{s^2}{(14.0\text{ M} - 2s)^2} = 1.7 \times 10^{-9}$$

Taking the square root of both sides and solving for s yields

$$s = 5.8 \times 10^{-4}\text{ M}$$

Thus, silver iodide is not soluble in the aqueous ammonia solution.

22-94.* For $Cr(OH)_3(s)$, the two relevant chemical equations are (Tables 22.1 and 22.3)

$$\text{Cr(OH)}_3(s) \rightleftharpoons \text{Cr}^{3+}(aq) + 3\,\text{OH}^-(aq) \qquad K_{\text{sp}} = 6.3 \times 10^{-31}\text{ M}^4$$

$$\text{Cr(OH)}_3(s) + \text{OH}^-(aq) \rightleftharpoons [\text{Cr(OH)}_4]^-(aq) \qquad K_{\text{f}} = 0.04$$

The solubility due to the first equation is

$$s = [Cr^{3+}] = \frac{K_{sp}}{[OH^-]^3} = \frac{(6.3 \times 10^{-31}\ M^4)[H_3O^+]^3}{(1.0 \times 10^{-14}\ M^2)^3} = (6.3 \times 10^{11}\ M^{-2})[H_3O^+]^3$$

The solubility due to the second equation is

$$s = [Cr(OH)_4^-] = K[OH^-] = \frac{(0.04)(1.0 \times 10^{-14}\ M^2)}{[H_3O^+]} = \frac{4.0 \times 10^{-16}\ M^2}{[H_3O^+]}$$

The total solubility is given by

$$s = [Cr^{3+}] + [Cr(OH)_4^-] = (6.3 \times 10^{11}\ M^{-2})[H_3O^+]^3 + \frac{4.0 \times 10^{-16}\ M^2}{[H_3O^+]}$$

For $Sn(OH)_2(s)$, the two relevant chemical equations are

$$Sn(OH)_2(s) \rightleftharpoons Sn^{2+}(aq) + 2\,OH^-(aq) \qquad K_{sp} = 5.5 \times 10^{-27}\ M^3$$

$$Sn(OH)_2(s) + 2\,OH^-(aq) \rightleftharpoons [Sn(OH)_4]^{2-}(aq) \qquad K_f = 0.01\ M^{-1}$$

The total solubility is given by

$$\begin{aligned} s &= [Sn^{2+}] + [Sn(OH)_4^{2-}] = \frac{K_{sp}}{[OH^-]^2} + K_f[OH^-]^2 \\ &= \frac{(5.5 \times 10^{-27}\ M^3)[H_3O^+]^2}{(1.0 \times 10^{-14}\ M^2)^2} + \frac{(0.01\ M^{-1})(1.0 \times 10^{-14}\ M^2)^2}{[H_3O^+]} \\ &= (55\ M^{-1})[H_3O^+]^2 + \frac{1 \times 10^{-30}\ M^3}{[H_3O^+]^2} \end{aligned}$$

The solubilities are plotted against pH in the following figure:

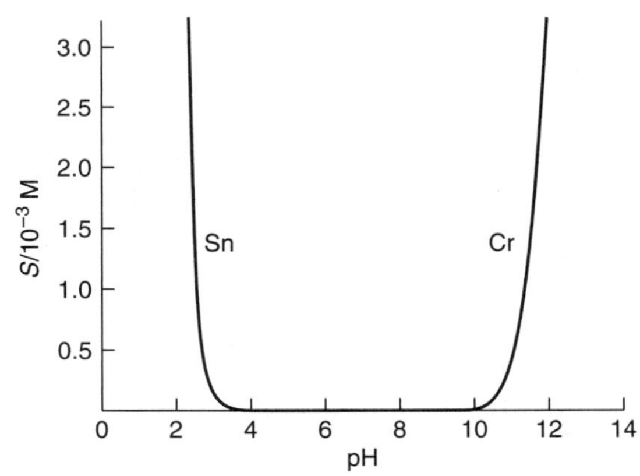

Thus, we see that the best separation is achieved between pH = 4 and pH = 7.

22-96.* Using the method and the results of the previous problem, we have

pH	1.0	2.0	3.0	4.0	5.0	6.0	7.0	8.0	9.0	10.0
s/M	3.3	1.0	0.33	0.11	0.055	0.045	0.044	0.044	0.044	0.044

The plot of pH versus solubility is shown below.

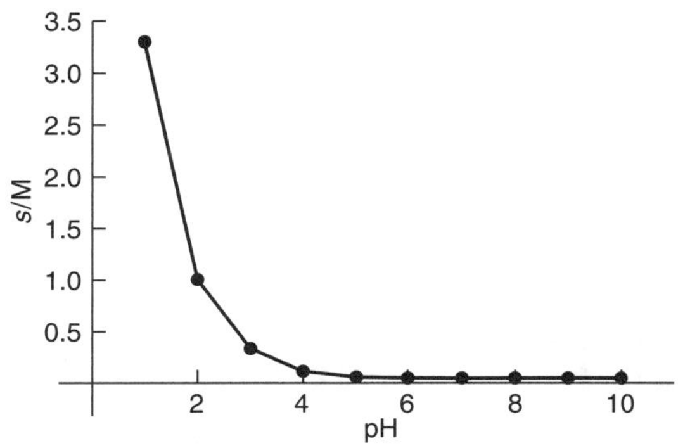

22-98.* (1) For equation 1, we have

$$K_1 = [I_2]$$

The solubility of $I_2(s)$ in pure water is 0.0132 M, thus

$$K_1 = 0.013 \text{ M}$$

(2) For equation 2, we have

$$K_2 = \frac{[I_3^-]}{[I^-]}$$

The solubility of $I_2(s)$ in 0.100 M KI is 0.051 M. Iodine dissolves in KI(aq) as $I_2(aq)$ and $I_3^-(aq)$. Thus, at equilibrium, we have

$$[I_2] + [I_3^-] = 0.051 \text{ M}$$

But $[I_2] = 0.013$ M because the K_1 equilibrium must also be satisfied. Thus, at equilibrium

$$[I_3^-] = 0.051 \text{ M} - 0.013 \text{ M} = 0.038 \text{ M}$$

The concentration of $I^-(aq)$ at equilibrium is given by

$$[I^-] = 0.100 \text{ M} - 0.038 \text{ M} = 0.062 \text{ M}$$

because the formation of $I_3^-(aq)$ decreases $[I^-]$ by an equal amount. The value of K_2 is

$$K_2 = \frac{[I_3^-]}{[I^-]} = \frac{0.038 \text{ M}}{0.062 \text{ M}} = 0.61$$

(3) Equation 3 is obtained by subtracting equation 1 from equation 2, thus

$$K_3 = \frac{K_2}{K_1} = \frac{0.61}{0.013 \text{ M}} = 47 \text{ M}^{-1}$$

CHAPTER 23. Chemical Thermodynamics

ENTROPIES OF FUSION AND VAPORIZATION

23-2. The values of ΔS_{fus} and ΔS_{vap} are given by

$$\Delta S_{\text{fus}} = \frac{\Delta H_{\text{fus}}}{T_{\text{m}}} \qquad \text{and} \qquad \Delta S_{\text{vap}} = \frac{\Delta H_{\text{vap}}}{T_{\text{b}}}$$

The T_{m} and T_{b} are the melting and boiling points, respectively, on the Kelvin temperature scale. Thus, we have for HF

$$\Delta S_{\text{fus}} = \frac{4.577 \times 10^3\,\text{J}\cdot\text{mol}^{-1}}{190.04\,\text{K}} = 24.08\,\text{J}\cdot\text{K}^{-1}\cdot\text{mol}^{-1}$$

$$\Delta S_{\text{vap}} = \frac{25.18 \times 10^3\,\text{J}\cdot\text{mol}^{-1}}{292.69\,\text{K}} = 86.03\,\text{J}\cdot\text{K}^{-1}\cdot\text{mol}^{-1}$$

For HCl

$$\Delta S_{\text{fus}} = \frac{1.991 \times 10^3\,\text{J}\cdot\text{mol}^{-1}}{158.9\,\text{K}} = 12.53\,\text{J}\cdot\text{K}^{-1}\cdot\text{mol}^{-1}$$

$$\Delta S_{\text{vap}} = \frac{17.53 \times 10^3\,\text{J}\cdot\text{mol}^{-1}}{188.3\,\text{K}} = 93.10\,\text{J}\cdot\text{K}^{-1}\cdot\text{mol}^{-1}$$

For HBr

$$\Delta S_{\text{fus}} = \frac{2.406 \times 10^3\,\text{J}\cdot\text{mol}^{-1}}{186.19\,\text{K}} = 12.92\,\text{J}\cdot\text{K}^{-1}\cdot\text{mol}^{-1}$$

$$\Delta S_{\text{vap}} = \frac{19.27 \times 10^3\,\text{J}\cdot\text{mol}^{-1}}{206.2\,\text{K}} = 93.45\,\text{J}\cdot\text{K}^{-1}\cdot\text{mol}^{-1}$$

For HI

$$\Delta S_{\text{fus}} = \frac{2.871 \times 10^3\,\text{J}\cdot\text{mol}^{-1}}{222.24\,\text{K}} = 12.92\,\text{J}\cdot\text{K}^{-1}\cdot\text{mol}^{-1}$$

$$\Delta S_{\text{vap}} = \frac{21.16 \times 10^3\,\text{J}\cdot\text{mol}^{-1}}{237.77\,\text{K}} = 88.99\,\text{J}\cdot\text{K}^{-1}\cdot\text{mol}^{-1}$$

23-4. Hydrogen bonding is not present in $CH_4(l)$, whereas the extent of hydrogen bonding in $H_2O(l)$ is greater than in $NH_3(l)$ because hydrogen bonds to an oxygen atom are stronger than hydrogen bonds to a nitrogen atom. In addition, because the oxygen atom in a H_2O molecule has two lone electron pairs, whereas the nitrogen atom in an NH_3 molecule has one lone electron pair, there are more hydrogen bonds in $H_2O(l)$ than in $NH_3(l)$. Thus, we have

$$CH_4(l) < NH_3(l) < H_2O(l)$$

ENTROPY: MOLECULAR MASS, STRUCTURE, AND PHYSICAL STATE

23-6. (a) The mass and the number of atoms for CO is less than the mass and the number of atoms for CO_2. Thus, we predict for the standard molar entropies of the gases

$$S^\circ(\mathrm{CO}) < S^\circ(\mathrm{CO_2})$$

(b) Because of its ring structure, cyclopropane has less freedom of movement than propane. Thus, we predict for the standard molar entropies of the gases

$$S^\circ(\text{cyclopropane}) < S^\circ(\text{propane})$$

(c) Pentane is a more flexible molecule than neopentane. Thus, we predict for the standard molar entropies of the gases

$$S^\circ(\text{neopentane}) < S^\circ(\text{pentane})$$

23-8. The molecular masses are, approximately, for CH_4, 16.0; for H_2O, 18.0; for NH_3, 17.0, for CH_3OH, 32.0, and for CH_3OD, 33.0. CH_4, H_2O, and NH_3 have about the same mass, but differ in the number of atoms. The larger number of atoms gives rise to a higher entropy. Both CH_3OH and CH_3OD have a greater mass than CH_4, H_2O, and NH_3. Thus, we predict

$$S^\circ(\mathrm{H_2O}) < S^\circ(\mathrm{NH_3}) < S^\circ(\mathrm{CH_4}) < S^\circ(\mathrm{CH_3OH}) < S^\circ(\mathrm{CH_3OD})$$

23-10. Gaseous molecules have a much greater freedom of movement than liquid molecules (higher translational and rotational disorder), and thus $H_2O(g)$ at one bar and 100°C has greater entropy than $H_2O(l)$ at one bar and 100°C.

23-12. (a) The volume of water vapor decreases when the pressure increases at constant temperature. The water molecules have a greater freedom of movement in a larger volume. The entropy will decrease.

(b) The bromine molecules have a greater freedom of movement in the gaseous state. The entropy will increase.

(c) Iodine molecules have a greater thermal disorder at a higher temperature. The entropy will increase.

(d) Iron atoms have a greater thermal disorder at a higher temperature. The entropy will decrease.

VALUES OF ΔS°_{rxn}

23-14. (a) We have three moles of gaseous reactants and no moles of gaseous product ($\Delta n_{gas} = -3$).

(b) We have two moles of gaseous reactants and no moles of gaseous product ($\Delta n_{gas} = -2$).

(c) We have one mole of gaseous reactant and no moles of gaseous products ($\Delta n_{gas} = -1$).

(d) We have four moles of gaseous reactant and two moles of gaseous products ($\Delta n_{gas} = -2$). The value of ΔS°_{rxn} increases as the net change in the number of moles of gas increases; thus,

$$\Delta S^\circ_{rxn}(a) < \Delta S^\circ_{rxn}(b) \approx \Delta S^\circ_{rxn}(d) < \Delta S^\circ_{rxn}(c)$$

23-16. The value of ΔS°_{rxn} is given by

$$\Delta S^\circ_{rxn} = S^\circ[\text{products}] - S^\circ[\text{reactants}]$$

(a) For the equation given

$$\begin{aligned}\Delta S^\circ_{rxn} &= S^\circ[N_2(g)] + 4\,S^\circ[H_2O(g)] - 2\,S^\circ[H_2O_2(l)] - S^\circ[N_2H_4(l)]\\ &= (1)(191.6\,J\cdot K^{-1}\cdot mol^{-1}) + (4)(188.8\,J\cdot K^{-1}\cdot mol^{-1})\\ &\quad - (2)(109.6\,J\cdot K^{-1}\cdot mol^{-1}) - (1)(121.2\,J\cdot K^{-1}\cdot mol^{-1})\\ &= 606.4\,J\cdot K^{-1}\cdot mol^{-1}\end{aligned}$$

(b) For the equation given

$$\begin{aligned}\Delta S^\circ_{rxn} &= 2\,S^\circ[NO(g)] - S^\circ[N_2(g)] - S^\circ[O_2(g)]\\ &= (2)(210.8\,J\cdot K^{-1}\cdot mol^{-1}) - (1)(191.6\,J\cdot K^{-1}\cdot mol^{-1})\\ &\quad - (1)(205.2\,J\cdot K^{-1}\cdot mol^{-1})\\ &= 24.8\,J\cdot K^{-1}\cdot mol^{-1}\end{aligned}$$

(c) For the equation given

$$\begin{aligned}\Delta S^\circ_{rxn} &= 2\,S^\circ[CH_3OH(l)] - 2\,S^\circ[CH_4(g)] - S^\circ[O_2(g)]\\ &= (2)(126.8\,J\cdot K^{-1}\cdot mol^{-1}) - (2)(186.3\,J\cdot K^{-1}\cdot mol^{-1})\\ &\quad - (1)(205.2\,J\cdot K^{-1}\cdot mol^{-1})\\ &= -324.2\,J\cdot K^{-1}\cdot mol^{-1}\end{aligned}$$

(d) For the equation given

$$\begin{aligned}\Delta S^\circ_{rxn} &= S^\circ[C_2H_6(g)] - S^\circ[C_2H_4(g)] - S^\circ[H_2(g)]\\ &= (1)(229.2\,J\cdot K^{-1}\cdot mol^{-1}) - (1)(219.3\,J\cdot K^{-1}\cdot mol^{-1})\\ &\quad - (1)(130.7\,J\cdot K^{-1}\cdot mol^{-1})\\ &= -120.8\,J\cdot K^{-1}\cdot mol^{-1}\end{aligned}$$

SPONTANEITY AND ΔG_{rxn}

23-18. The reaction is spontaneous because it occurs naturally as described by the equation at room temperature. Because it is spontaneous, the sign of ΔG_{rxn} is negative. We learned in Chapter 15 that energy is required for sublimation. Thus, the sign of ΔH_{rxn}, which is equal to ΔH_{sub}, is positive. The value of ΔS_{sub} (solid $\rightarrow$ gas) is also positive; thus, $T\Delta S_{rxn}$ is positive. The reaction is entropy driven because

$$\Delta G_{rxn} - T\Delta S_{rxn} < 0$$

23-20. The value of ΔS°_{rxn} is given by

$$\begin{aligned}\Delta S^\circ_{rxn} &= S^\circ[\text{products}] - S^\circ[\text{reactants}]\\ &= S^\circ[C_2H_5OH(l)] - S^\circ[C_2H_4(g)] - S^\circ[H_2O(l)]\\ &= (1)(160.7\ J\cdot K^{-1}\cdot mol^{-1}) - (1)(219.3\ J\cdot K^{-1}\cdot mol^{-1}) - (1)(70.0\ J\cdot K^{-1}\cdot mol^{-1})\\ &= -128.6\ J\cdot K^{-1}\cdot mol^{-1}\end{aligned}$$

We can calculate ΔG°_{rxn} by using the relationship

$$\begin{aligned}\Delta G^\circ_{rxn} &= \Delta H^\circ_{rxn} - T\Delta S^\circ_{rxn}\\ &= -44.2\ kJ\cdot mol^{-1} - (298.2\ K)(-128.6\ J\cdot K^{-1}\cdot mol^{-1})\left(\frac{1\ kJ}{1000\ J}\right)\\ &= -5.9\ kJ\cdot mol^{-1}\end{aligned}$$

The reaction is spontaneous in the direction

$$C_2H_5OH(l) \longrightarrow C_2H_4(g) + H_2O(l)$$

when $C_2H_5OH(l)$, $C_2H_4(g)$, and $H_2O(l)$ are at 1 bar and 25.0°C.

23-22. The value of ΔS°_{rxn} is given by

$$\begin{aligned}\Delta S^\circ_{rxn} &= 2\,S^\circ[CO(g)] - S^\circ[C(s,\text{ graphite})] - S^\circ[CO_2(g)]\\ &= (2)(197.7\ J\cdot K^{-1}\cdot mol^{-1}) - (1)(5.7\ J\cdot K^{-1}\cdot mol^{-1}) - (1)(213.8\ J\cdot K^{-1}\cdot mol^{-1})\\ &= 175.9\ J\cdot K^{-1}\cdot mol^{-1}\end{aligned}$$

We can calculate ΔG°_{rxn} by using the relationship

$$\begin{aligned}\Delta G^\circ_{rxn} &= \Delta H^\circ_{rxn} - T\Delta S^\circ_{rxn}\\ &= 172.5\ kJ\cdot mol^{-1} - (298.2\ K)(175.9\ J\cdot K^{-1}\cdot mol^{-1})\left(\frac{1\ kJ}{1000\ J}\right) = 120.0\ kJ\cdot mol^{-1}\end{aligned}$$

Thus, when all the reactants and products are at standard conditions, the reaction is spontaneous in the direction

$$2\,CO(g) \longrightarrow C(s) + CO_2(g)$$

The value of ΔG_{rxn} is given by

$$\Delta G_{rxn} = \Delta G^\circ_{rxn} + RT\ln Q$$

where Q is the thermodynamic reaction quotient (no units).

$$Q = \frac{P_{CO}^2}{P_{CO_2}}$$

Substituting in the values for the pressures yields

$$Q = \frac{(5.0\times10^{-4})^2}{20.0} = 1.25\times10^{-8}$$

Thus,

$$\Delta G_{\text{rxn}} = 120.0\ \text{kJ}\cdot\text{mol}^{-1} + \frac{(8.3145\ \text{J}\cdot\text{K}^{-1}\cdot\text{mol}^{-1})(298.2\ \text{K})}{1000\ \text{J}\cdot\text{kJ}^{-1}} \ln(1.25 \times 10^{-8})$$
$$= 120.0\ \text{kJ}\cdot\text{mol}^{-1} - 45.0\ \text{kJ}\cdot\text{mol}^{-1} = 75.0\ \text{kJ}\cdot\text{mol}^{-1}$$

Because $\Delta G_{\text{rxn}} > 0$, the reaction is spontaneous from right to left as written.

23-24. The relation between ΔG_{rxn} and K is

$$\Delta G_{\text{rxn}} = RT \ln \frac{Q}{K}$$

where Q is the thermodynamic reaction quotient (no units). The value of Q is

$$Q = \frac{P^2_{\text{SO}_3}}{P^2_{\text{SO}_2} P_{\text{O}_2}} = \frac{(1.0 \times 10^{-4})^2}{(1.0 \times 10^{-3})^2(0.20)} = 0.050$$

Thus,

$$\Delta G_{\text{rxn}} = \frac{(8.3145\ \text{J}\cdot\text{K}^{-1}\cdot\text{mol}^{-1})(900\ \text{K})}{1000\ \text{J}\cdot\text{kJ}^{-1}} \ln \frac{0.050}{14} = -42.2\ \text{kJ}\cdot\text{mol}^{-1}$$

Because $\Delta G_{\text{rxn}} < 0$, the reaction is spontaneous from left to right at the stated conditions.

EQUILIBRIUM CONSTANTS AND $\Delta G^\circ_{\text{rxn}}$

23-26. We have

$$\Delta G^\circ_{\text{rxn}} = -RT \ln K$$

where K is the (unitless) thermodynamic equilibrium constant. At 527°C, we have

$$\Delta G^\circ_{\text{rxn}} = -(8.3145\ \text{J}\cdot\text{K}^{-1}\cdot\text{mol}^{-1})(800\ \text{K}) \ln(4.6 \times 10^{-3})$$
$$= +3.58 \times 10^4\ \text{J}\cdot\text{mol}^{-1} = 35.8\ \text{kJ}\cdot\text{mol}^{-1}$$

The reaction is spontaneous in the direction

$$\text{CO}(g) + \text{Cl}_2(g) \longrightarrow \text{COCl}_2(g)$$

when $\text{COCl}_2(g)$, $\text{CO}(g)$, and $\text{Cl}_2(g)$ are at standard conditions. The value of ΔG_{rxn} at the other conditions is given by

$$\Delta G_{\text{rxn}} = \Delta G^\circ_{\text{rxn}} + RT \ln Q$$

where Q is the thermodynamic reaction quotient (no units).

$$Q = \frac{[\text{CO}][\text{Cl}_2]}{[\text{COCl}_2]} = \frac{(0.010)(0.010)}{1.00} = 1.0 \times 10^{-4}$$

Thus,

$$\Delta G_{\text{rxn}} = 35.8\ \text{kJ}\cdot\text{mol}^{-1} + \frac{(8.3145\ \text{J}\cdot\text{K}^{-1}\cdot\text{mol}^{-1})(800\ \text{K})}{1000\ \text{J}\cdot\text{kJ}^{-1}} \ln(1.0 \times 10^{-4})$$
$$= 35.8\ \text{kJ}\cdot\text{mol}^{-1} - 61.3\ \text{kJ}\cdot\text{mol}^{-1} = -25.5\ \text{kJ}\cdot\text{mol}^{-1}$$

Because $\Delta G_{\text{rxn}} < 0$, the reaction is spontaneous from left to right as written.

23-28. The value of ΔG°_{rxn} is given by

$$\Delta G^\circ_{rxn} = -RT \ln K$$

where K is the (unitless) thermodynamic equilibrium constant.

$$\begin{aligned}\Delta G^\circ_{rxn} &= -(8.3145\ \text{J}\cdot\text{K}^{-1}\cdot\text{mol}^{-1})(298.2\ \text{K}) \ln(4.0 \times 10^{-8}) \\ &= 4.22 \times 10^4\ \text{J}\cdot\text{mol}^{-1} = 42.2\ \text{kJ}\cdot\text{mol}^{-1}\end{aligned}$$

Because $\Delta G^\circ_{rxn} > 0$, hypochlorous acid will not dissociate spontaneously when $[ClO^-] = [H^+] = [HClO] = 1.00$ M (standard conditions). The value of ΔG_{rxn} at the other conditions is given by

$$\Delta G_{rxn} = RT \ln \frac{Q}{K}$$

where

$$Q = \frac{[ClO^-][H^+]}{[HClO]} = \frac{(1.0 \times 10^{-6})(1.0 \times 10^{-6})}{0.10} = 1.0 \times 10^{-11}$$

Thus,

$$\begin{aligned}\Delta G_{rxn} &= (8.3145\ \text{J}\cdot\text{K}^{-1}\cdot\text{mol}^{-1})(298.2\ \text{K}) \ln \frac{1.0 \times 10^{-11}\ \text{M}}{4.0 \times 10^{-8}\ \text{M}} \\ &= -2.06 \times 10^4\ \text{J}\cdot\text{mol}^{-1} = -20.6\ \text{kJ}\cdot\text{mol}^{-1}\end{aligned}$$

Because $\Delta G_{rxn} < 0$, hypochlorous acid will dissociate spontaneously under these conditions.

23-30. The value of ΔG°_{rxn} is given by

$$\Delta G^\circ_{rxn} = -RT \ln K$$

where K is the (unitless) thermodynamic equilibrium constant.

$$\begin{aligned}\Delta G^\circ_{rxn} &= -(8.3145\ \text{J}\cdot\text{K}^{-1}\cdot\text{mol}^{-1})(298.2\ \text{K}) \ln(1.8 \times 10^{-5}) \\ &= 2.71 \times 10^4\ \text{J}\cdot\text{mol}^{-1} = 27.1\ \text{kJ}\cdot\text{mol}^{-1}\end{aligned}$$

The reaction is spontaneous from right to left when $NH_3(aq)$, $NH_4^+(aq)$, and $OH^-(aq)$ are at standard conditions. The value of ΔG_{rxn} at the other conditions is given by

$$\Delta G_{rxn} = RT \ln \frac{Q}{K}$$

where

$$Q = \frac{[OH^-][NH_4^+]}{[NH_3]} = \frac{(1.0 \times 10^{-6})(1.0 \times 10^{-6})}{0.050} = 2.0 \times 10^{-11}$$

Thus,

$$\begin{aligned}\Delta G_{rxn} &= (8.3145\ \text{J}\cdot\text{K}^{-1}\cdot\text{mol}^{-1})(298.2\ \text{K}) \ln \frac{2.0 \times 10^{-11}}{1.8 \times 10^{-5}} \\ &= -3.40 \times 10^4\ \text{J}\cdot\text{mol}^{-1} = -34.0\ \text{kJ}\cdot\text{mol}^{-1}\end{aligned}$$

Ammonia will react spontaneously with water under these conditions.

23-32. The value of ΔG°_{rxn} is given by

$$\Delta G^\circ_{rxn} = -RT \ln K$$

where K is the (unitless) thermodynamic equilibrium constant.

$$\begin{aligned}\Delta G^\circ_{rxn} &= -(8.3145\ \text{J}\cdot\text{K}^{-1}\cdot\text{mol}^{-1})(298.2\ \text{K}) \ln(3.4 \times 10^{-9}) \\ &= 4.83 \times 10^4\ \text{J}\cdot\text{mol}^{-1} = 48.3\ \text{kJ}\cdot\text{mol}^{-1}\end{aligned}$$

when $[Ca^{2+}] = [CO_3^{2-}] = 1.00$ M (standard conditions). When a solution is prepared in which $[Ca^{2+}] = [CO_3^{2-}] = 1.00$ M, insoluble $CaCO_3(s)$ will precipitate out of the solution because the reaction is spontaneous from right to left as written under these conditions.

23-34. The value of ΔG°_{rxn} is given by

$$\Delta G^\circ_{rxn} = -RT \ln K$$

where K is the (unitless) thermodynamic equilibrium constant.

$$\begin{aligned}\Delta G^\circ_{rxn} &= -(8.3145\ \text{J}\cdot\text{K}^{-1}\cdot\text{mol}^{-1})(298.2\ \text{K}) \ln(2.0 \times 10^{7}) \\ &= -4.17 \times 10^4\ \text{J}\cdot\text{mol}^{-1} = -41.7\ \text{kJ}\cdot\text{mol}^{-1}\end{aligned}$$

The reaction is spontaneous from left to right as written when $Co^{3+}(aq)$, $NH_3(aq)$, and $Co(NH_3)_6^{3+}(aq)$ are at standard conditions. The value of ΔG_{rxn} at other conditions is given by

$$\Delta G_{rxn} = RT \ln \frac{Q}{K}$$

where

$$Q = \frac{[Co(NH_3)_6^{3+}]}{[Co^{3+}]\,[NH_3]^6} = \frac{1.00}{(5.0 \times 10^{-3})(0.10)^6} = 2.0 \times 10^8$$

Thus,

$$\begin{aligned}\Delta G_{rxn} &= (8.3145\ \text{J}\cdot\text{K}^{-1}\cdot\text{mol}^{-1})(298.2\ \text{K}) \ln \frac{2.0 \times 10^8}{2.0 \times 10^7} \\ &= +5.71 \times 10^3\ \text{J}\cdot\text{mol}^{-1} = 5.71\ \text{kJ}\cdot\text{mol}^{-1}\end{aligned}$$

The reaction is spontaneous from right to left as written under these conditions.

CALCULATION OF ΔG°_{rxn} FROM TABULATED DATA

23-36. The value of ΔG°_{rxn} is calculated by using the data in Appendix D or Table 23.1 and the relationship

$$\Delta G^\circ_{rxn} = \Delta G^\circ_f[\text{products}] - \Delta G^\circ_f[\text{reactants}]$$

(a) For the equation given

$$\begin{aligned}\Delta G^\circ_{rxn} &= \Delta G^\circ_f[N_2(g)] + 4\,\Delta G^\circ_f[H_2O(g)] - 2\,\Delta G^\circ_f[H_2O_2(l)] - \Delta G^\circ_f[N_2H_4(l)] \\ &= (1)(0\ \text{kJ}\cdot\text{mol}^{-1}) + (4)(-228.6\ \text{kJ}\cdot\text{mol}^{-1}) \\ &\quad -(2)(-120.4\ \text{kJ}\cdot\text{mol}^{-1}) - (1)(149.3\ \text{kJ}\cdot\text{mol}^{-1}) \\ &= -822.9\ \text{kJ}\cdot\text{mol}^{-1}\end{aligned}$$

Using the relationship of $\Delta G^\circ_{\rm rxn} = -RT\ln K$, we have

$$\ln K = -\frac{\Delta G^\circ_{\rm rxn}}{RT} = -\frac{-822.9\times 10^3\ \rm J\cdot mol^{-1}}{(8.3145\ \rm J\cdot K^{-1}\cdot mol^{-1})(298.2\ K)} = 331.9$$

Thus,

$$K = e^{331.9} = 1\times 10^{144}$$

To get this result on your calculator, you probably have to use the result

$$\log e = 0.43429 \qquad \text{or} \qquad e = 10^{0.43429}$$

Therefore,

$$e^{131.8} = 10^{(331.8)(0.43429)} = 10^{144.14} = 10^{0.14}10^{144} = 1.4\times 10^{144} \approx 1\times 10^{144}$$

(b) For the equation given

$$\begin{aligned}\Delta G^\circ_{\rm rxn} &= 2\,\Delta G^\circ_{\rm f}[\mathrm{NO}(g)] - \Delta G^\circ_{\rm f}[\mathrm{N_2}(g)] - \Delta G^\circ_{\rm f}[\mathrm{O_2}(g)]\\ &= (2)(87.6\ \rm kJ\cdot mol^{-1}) - (1)(0\ kJ\cdot mol^{-1}) - (1)(0\ kJ\cdot mol^{-1})\\ &= 175\ \rm kJ\cdot mol^{-1}\end{aligned}$$

Using the relationship of $\Delta G^\circ_{\rm rxn} = -RT\ln K$, we have

$$\ln K = -\frac{\Delta G^\circ_{\rm rxn}}{RT} = -\frac{175\times 10^3\ \rm J\cdot mol^{-1}}{(8.3145\ \rm J\cdot K^{-1}\cdot mol^{-1})(298.2\ K)} = -70.6$$

Thus,

$$K = e^{-70.6} = 2\times 10^{-31}$$

(c) For the equation given

$$\begin{aligned}\Delta G^\circ_{\rm rxn} &= 2\,\Delta G^\circ_{\rm f}[\mathrm{CH_3OH}(l)] - 2\,\Delta G^\circ_{\rm f}[\mathrm{CH_4}(g)] - \Delta G^\circ_{\rm f}[\mathrm{O_2}(g)]\\ &= (2)(-166.6\ \rm kJ\cdot mol^{-1}) - (2)(-50.5\ kJ\cdot mol^{-1}) - (1)(0\ kJ\cdot mol^{-1})\\ &= -232.2\ \rm kJ\cdot mol^{-1}\end{aligned}$$

Using the relationship of $\Delta G^\circ_{\rm rxn} = -RT\ln K$, we have

$$\ln K = -\frac{\Delta G^\circ_{\rm rxn}}{RT} = -\frac{-232.2\times 10^3\ \rm J\cdot mol^{-1}}{(8.3145\ \rm J\cdot K^{-1}\cdot mol^{-1})(298.2\ K)} = 93.65$$

Thus,

$$K = e^{93.65} = 4.7\times 10^{40}$$

23-38. The value of $\Delta G^\circ_{\rm rxn}$ is given by

$$\Delta G^\circ_{\rm rxn} = \Delta G^\circ_{\rm f}[\text{products}] - \Delta G^\circ_{\rm f}[\text{reactants}]$$

Thus,

$$\begin{aligned}\Delta G^\circ_{\rm rxn} &= 3\,\Delta G^\circ_{\rm f}[\mathrm{Fe}(s)] + 2\,\Delta G^\circ_{\rm f}[\mathrm{CO_2}(g)] - \Delta G^\circ_{\rm f}[\mathrm{Fe_3O_4}(s)] - 2\,\Delta G^\circ_{\rm f}[\mathrm{C}(s,\text{graphite})]\\ &= (3)(0\ \rm kJ\cdot mol^{-1}) + (2)(-394.4\ J\cdot mol^{-1}) - (1)(-1015.4\ kJ\cdot mol^{-1}) - (2)(0\ kJ\cdot mol^{-1})\\ &= 226.6\ \rm kJ\cdot mol^{-1}\end{aligned}$$

The value of $\Delta H^\circ_{\text{rxn}}$ is given by

$$\Delta H^\circ_{\text{rxn}} = \Delta H^\circ_f[\text{products}] - \Delta H^\circ_f[\text{reactants}]$$

Thus,

$$\begin{aligned}\Delta H^\circ_{\text{rxn}} &= 3\,\Delta H^\circ_f[\text{Fe}(s)] + 2\,\Delta H^\circ_f[\text{CO}_2(g)] - \Delta H^\circ_f[\text{Fe}_3\text{O}_4(s)] - 2\,\Delta H^\circ_f[\text{C}(s,\text{ graphite})]\\ &= (3)(0\text{ kJ}\cdot\text{mol}^{-1}) + (2)(-393.5\text{ J}\cdot\text{mol}^{-1}) - (1)(-1118.4\text{ kJ}\cdot\text{mol}^{-1}) - (2)(0\text{ J}\cdot\text{mol}^{-1})\\ &= 331.4\text{ kJ}\cdot\text{mol}^{-1}\end{aligned}$$

Using the relationship of $\Delta G^\circ_{\text{rxn}} = -RT\ln K$, we have

$$\ln K = -\frac{\Delta G^\circ_{\text{rxn}}}{RT} = -\frac{226.6\times10^3\text{ J}\cdot\text{mol}^{-1}}{(8.3145\text{ J}\cdot\text{K}^{-1}\cdot\text{mol}^{-1})(298.2\text{ K})} = -91.39$$

Thus,

$$K = e^{-91.39} = 2.0\times10^{-40}$$

23-40. The value of $\Delta G^\circ_{\text{rxn}}$ is given by

$$\Delta G^\circ_{\text{rxn}} = \Delta G^\circ_f[\text{CCl}_4(l)] + 4\,\Delta G^\circ_f[\text{HCl}(g)] - \Delta G^\circ_f[\text{CH}_4(g)] - 4\,\Delta G^\circ_f[\text{Cl}_2(g)]$$

Thus,

$$\begin{aligned}-396.0\text{ kJ}\cdot\text{mol}^{-1} &= (1)(\Delta G^\circ_f[\text{CCl}_4(l)]) + (4)(-95.3\text{ kJ}\cdot\text{mol}^{-1})\\ &\quad - (1)(-50.5\text{ kJ}\cdot\text{mol}^{-1}) - (4)(0\text{ kJ}\cdot\text{mol}^{-1})\end{aligned}$$

Therefore,

$$-396.0\text{ kJ}\cdot\text{mol}^{-1} = (1)(\Delta G^\circ_f[\text{CCl}_4(l)]) - 330.7\text{ kJ}\cdot\text{mol}^{-1}$$

$$\Delta G^\circ_f[\text{CCl}_4(l)] = -65.3\text{ kJ}\cdot\text{mol}^{-1}$$

which agrees with the value of $\Delta G^\circ_f[\text{CCl}_4(l)]$ given in Appendix D.
The value of $\Delta H^\circ_{\text{rxn}}$ is given by

$$\Delta H^\circ_{\text{rxn}} = \Delta H^\circ_f[\text{CCl}_4(l)] + 4\,\Delta H^\circ_f[\text{HCl}(g)] - \Delta H^\circ_f[\text{CH}_4(g)] - 4\,\Delta H^\circ_f[\text{Cl}_2(g)]$$

Thus,

$$\begin{aligned}-422.8\text{ kJ}\cdot\text{mol}^{-1} &= (1)(\Delta H^\circ_f[\text{CCl}_4(l)]) + (4)(-92.3\text{ kJ}\cdot\text{mol}^{-1})\\ &\quad - (1)(-74.6\text{ kJ}\cdot\text{mol}^{-1}) - (4)(0\text{ kJ}\cdot\text{mol}^{-1})\end{aligned}$$

Therefore,

$$-422.8\text{ kJ}\cdot\text{mol}^{-1} = (1)(\Delta H^\circ_f[\text{CCl}_4(l)]) - 294.6\text{ kJ}\cdot\text{mol}^{-1}$$

$$\Delta H^\circ_f[\text{CCl}_4(l)] = -128.2\text{ kJ}\cdot\text{mol}^{-1}$$

which agrees with the value of $\Delta H^\circ_f[\text{CCl}_4(l)]$ given in Appendix D.

23-42. The equation for the combustion of methane is

$$\text{CH}_4(g) + 2\,\text{O}_2(g) \longrightarrow \text{CO}_2(g) + 2\,\text{H}_2\text{O}(l)$$

The value of ΔG°_{rxn} is given by

$$\begin{aligned}\Delta G^\circ_{rxn} &= \Delta G^\circ_f[\text{products}] - \Delta G^\circ_f[\text{reactants}]\\ &= \Delta G^\circ_f[CO_2(g)] + 2\,\Delta G^\circ_f[H_2O(l)] - \Delta G^\circ_f[CH_4(g)] - 2\,\Delta G^\circ_f[O_2(g)]\\ &= (1)(-394.4\text{ kJ}\cdot\text{mol}^{-1}) + (2)(-237.1\text{ kJ}\cdot\text{mol}^{-1})\\ &\quad - (1)(-50.5\text{ kJ}\cdot\text{mol}^{-1}) - (2)(0\text{ kJ}\cdot\text{mol}^{-1})\\ &= -818.1\text{ kJ}\cdot\text{mol}^{-1}\end{aligned}$$

Thus, the maximum amount of work that can be obtained from the combustion of one mole of methane when $C_2H_6(g)$, $O_2(g)$, $CO_2(g)$, and $H_2O(l)$ are at standard conditions at 25°C is 818.1 kilojoules.

TEMPERATURE DEPENDENCE OF EQUILIBRIUM CONSTANTS

23-44. The van't Hoff equation is

$$\ln\frac{K_2}{K_1} = \frac{\Delta H^\circ_{rxn}}{R}\left(\frac{T_2 - T_1}{T_1 T_2}\right)$$

Substituting the values of the second and fourth sets of data (arbitrarily), we have

$$\ln\frac{6.97\times10^{-3}\text{ bar}}{1.47\times10^{-3}\text{ bar}} = \frac{\Delta H^\circ_{rxn}(1273\text{ K} - 1173\text{ K})}{(8.3145\text{ J}\cdot\text{K}^{-1}\cdot\text{mol}^{-1})(1273\text{ K})(1173\text{ K})}$$

$$1.556 = (8.054\times10^{-6}\text{ J}^{-1}\cdot\text{mol})\,\Delta H^\circ_{rxn}$$

$$\Delta H^\circ_{rxn} = 1.93\times10^{5}\text{ J}\cdot\text{mol}^{-1} = 193\text{ kJ}\cdot\text{mol}^{-1}$$

The value calculated from the data in Appendix D is 192.9 kJ·mol^{-1}. The agreement with the data from Appendix D is very good. The discrepancy is due to the (slight) temperature dependence of ΔH°_{rxn}. A more sophisticated treatment of the data using all the data sets simultaneously (called the method of least squares) gives $\Delta H^\circ_{rxn} = 193$ kJ·mol^{-1}.

23-46. The value of ΔH°_{rxn} is given by

$$\begin{aligned}\Delta H^\circ_{rxn} &= 2\,\Delta H^\circ_f[HI(g)] - \Delta H^\circ_f[H_2(g)] - \Delta H^\circ_f[I_2(g)]\\ &= (2)(26.5\text{ kJ}\cdot\text{mol}^{-1}) - (1)(0\text{ kJ}\cdot\text{mol}^{-1}) - (1)(62.4\text{ J}\cdot\text{mol}^{-1})\\ &= -9.4\text{ kJ}\cdot\text{mol}^{-1}\end{aligned}$$

Using the van't Hoff equation, we have

$$\ln\frac{K_p}{58.0} = \frac{(-9.4\times10^{3}\text{ J}\cdot\text{mol}^{-1})(773.2\text{ K} - 673.2\text{ K})}{(8.3145\text{ J}\cdot\text{K}^{-1}\cdot\text{mol}^{-1})(773.2\text{ K})(673.2\text{ K})} = -0.22$$

Taking antilogarithms of both sides yields

$$\frac{K_p}{58.0} = 0.80$$

Thus,

$$K_p = 46 \qquad \text{at } 400.0^\circ\text{C}$$

CLAPEYRON–CLAUSIUS EQUATION

23-48. The Clapeyron–Clausius equation is

$$\ln \frac{P_2}{P_1} = \frac{\Delta H^\circ_{vap}}{R}\left(\frac{T_2 - T_1}{T_1 T_2}\right)$$

Let $P_2 = 455$ Torr, $T_2 = 293.2$ K, and $T_1 = 277.2$ K to write

$$\ln \frac{455 \text{ Torr}}{P_1} = \frac{(26.52 \times 10^3 \text{ J}\cdot\text{mol}^{-1})(16.0 \text{ K})}{(8.3145 \text{ J}\cdot\text{K}^{-1}\cdot\text{mol}^{-1})(277.2 \text{ K})(293.2 \text{ K})} = 0.628$$

Taking antilogarithms of both sides yields

$$\frac{455 \text{ Torr}}{P_1} = e^{0.628} = 1.87$$

Thus,

$$P_1 = \frac{455 \text{ Torr}}{1.87} = 243 \text{ Torr}$$

23-50. The Clapeyron–Clausius equation is

$$\ln \frac{P_2}{P_1} = \frac{\Delta H^\circ_{vap}}{R}\left(\frac{T_2 - T_1}{T_1 T_2}\right)$$

At the normal boiling point, $P_2 = 760.0$ Torr, $T_2 = 629.88$ K, $T_1 = 298.2$ K, and $\Delta H_{vap} = 59.11 \text{ kJ}\cdot\text{mol}^{-1}$, thus,

$$\ln \frac{760.0 \text{ Torr}}{P_1} = \frac{(59.11 \times 10^3 \text{ J}\cdot\text{mol}^{-1})(629.88 \text{ K} - 298.2 \text{ K})}{(8.3145 \text{ J}\cdot\text{K}^{-1}\cdot\text{mol}^{-1})(629.88 \text{ K})(298.2 \text{ K})} = 12.55$$

Taking antilogarithms of both sides yields

$$\frac{760.0 \text{ Torr}}{P_1} = e^{12.55} = 2.8 \times 10^5$$

Thus,

$$P_1 = \frac{760 \text{ Torr}}{2.8 \times 10^5} = 2.7 \times 10^{-3} \text{ Torr} \quad \text{at } 25^\circ\text{C}$$

If we now let $T_1 = 373.2$ K, then the corresponding values are

$$\ln \frac{760.0 \text{ Torr}}{P_1} = \frac{(59.11 \times 10^3 \text{ J}\cdot\text{mol}^{-1})(629.88 \text{ K} - 373.2 \text{ K})}{(8.3145 \text{ J}\cdot\text{K}^{-1}\cdot\text{mol}^{-1})(629.88 \text{ K})(373.2 \text{ K})} = 7.763$$

Thus

$$\frac{760.0 \text{ Torr}}{P_1} = e^{7.763} = 2.35 \times 10^3$$

or

$$P_1 = \frac{760.0 \text{ Torr}}{2.35 \times 10^3} = 0.323 \text{ Torr} \quad \text{at } 100^\circ\text{C}$$

Thus, we see that P_{vap} of $Hg(l)$ is negligible at both temperatures.

23-52. Let $P_1 = 48.1$ Torr, $P_2 = 133.0$ Torr, $T_1 = 273.2$ K, and $T_2 = 293.2$ K and write the Clapeyron–Clausius equation as

$$\ln \frac{133.0 \text{ Torr}}{48.1 \text{ Torr}} = \frac{(\Delta H^\circ_{\text{vap}})(293.2 \text{ K} - 273.2 \text{ K})}{(8.3145 \text{ J}\cdot\text{K}^{-1}\cdot\text{mol}^{-1})(273.2 \text{ K})(293.2 \text{ K})}$$

$$1.017 = (3.003 \times 10^{-5} \text{ J}^{-1}\cdot\text{mol})\Delta H_{\text{vap}}$$

Solving for $\Delta H^\circ_{\text{vap}}$ gives

$$\Delta H^\circ_{\text{vap}} = 3.387 \times 10^4 \text{ J}\cdot\text{mol}^{-1} = 33.87 \text{ kJ}\cdot\text{mol}^{-1}$$

23-54. The value of $\Delta H^\circ_{\text{vap}} = 33.87 \text{ kJ}\cdot\text{mol}^{-1}$ from Problem 23–52. Now use the Clapeyron–Clausius equation with $P_1 = 133.0$ Torr, $P_2 = 760.0$ Torr, and $T_1 = 293.2$ K and write

$$\ln \frac{760.0 \text{ Torr}}{133.0 \text{ Torr}} = \frac{(33.87 \times 10^3 \text{ J}\cdot\text{mol}^{-1})(T_2 - 293.2 \text{ K})}{(8.3145 \text{ J}\cdot\text{K}^{-1}\cdot\text{mol}^{-1})(T_2)(293.2 \text{ K})}$$

or

$$0.1254\, T_2 = T_2 - 293.2 \text{ K}$$

or

$$T_2 = \frac{293.2 \text{ K}}{0.8746} = 335.2 \text{ K} = 62.0^\circ\text{C}$$

23-56. Let P_2 be the vapor pressure of NaCl(l) at 1100°C and P_1 be that at 900°C. Then, the Clapeyron–Clausius equation is

$$\ln \frac{P_2}{P_1} = \frac{(180 \times 10^3 \text{ J}\cdot\text{mol}^{-1})(200 \text{ K})}{(8.3145 \text{ J}\cdot\text{K}^{-1}\cdot\text{mol}^{-1})(1173 \text{ K})(1373 \text{ K})} = 2.69$$

Taking antilogarithms of both sides yields

$$\frac{P_2}{P_1} = e^{2.69} = 15$$

ADDITIONAL PROBLEMS

23-58. Even when a reaction is spontaneous, it may not occur at a detectable rate; spontaneous is not synonymous with immediate. In the case of most combustible fuels the activation energy (see Chapter 18) is greater than the thermal energy available at room temperature, and so they may be stored (in the absence of a spark or flame) indefinitely.

23-60. A room does not spontaneously become messy; making the room messy requires an input of energy from an external source.

23-62. (a) The value of $\Delta G^\circ_{\text{rxn}}$ is given by

$$\begin{aligned}\Delta G^\circ_{\text{rxn}}\left(\text{HI}(g) \rightleftharpoons \frac{1}{2}\text{H}_2(g) + \frac{1}{2}\text{I}_2(g)\right) &= \frac{1}{2}\Delta G^\circ_{\text{f}}[\text{H}_2(g)] + \frac{1}{2}\Delta H^\circ_{\text{f}}[\text{I}_2(g)] - \Delta H^\circ_{\text{f}}[\text{HI}(g)] \\ &= \frac{1}{2}\{\Delta G^\circ_{\text{f}}[\text{H}_2(g)] + \Delta H^\circ_{\text{f}}[\text{I}_2(g)] - 2\Delta H^\circ_{\text{f}}[\text{HI}(g)]\} \\ &= \frac{1}{2}\Delta G^\circ_{\text{rxn}}(2\text{HI}(g) \rightleftharpoons \text{H}_2(g) + \text{I}_2(g))\end{aligned}$$

Notice that the value of the thermodynamic equilibrium constant likewise is $(1.08 \times 10^{-2})^{1/2} = 0.104$.

(b) How we choose to write the reaction equation has no effect on the physical reaction and how we describe it.

(c) If we reverse the equation for the reaction, then the sign of ΔG_{rxn} will change.

23-64. The value of $\Delta S^\circ_{\text{rxn}}$ is given by

$$\Delta S^\circ_{\text{rxn}} = S^\circ[\text{products}] - S^\circ[\text{reactants}]$$

(a) For the equation given

$$\begin{aligned}\Delta S^\circ_{\text{rxn}} &= S^\circ[\text{I}_2(g)] - S^\circ[\text{I}_2(s)]\\ &= (1)(260.7\,\text{J}\cdot\text{K}^{-1}\cdot\text{mol}^{-1}) - (1)(116.1\,\text{J}\cdot\text{K}^{-1}\cdot\text{mol}^{-1})\\ &= 144.6\,\text{J}\cdot\text{K}^{-1}\cdot\text{mol}^{-1}\end{aligned}$$

(b) For the equation given

$$\begin{aligned}\Delta S^\circ_{\text{rxn}} &= S^\circ[\text{BaO}(s)] + S^\circ[\text{CO}_2(g)] - S^\circ[\text{BaCO}_3(s)]\\ &= (1)(72.1\,\text{J}\cdot\text{K}^{-1}\cdot\text{mol}^{-1}) + (1)(213.8\,\text{J}\cdot\text{K}^{-1}\cdot\text{mol}^{-1})\\ &\quad - (1)(112.1\,\text{J}\cdot\text{K}^{-1}\cdot\text{mol}^{-1})\\ &= 173.8\,\text{J}\cdot\text{K}^{-1}\cdot\text{mol}^{-1}\end{aligned}$$

(c) For the equation given

$$\begin{aligned}\Delta S^\circ_{\text{rxn}} &= S^\circ[\text{CH}_3\text{Cl}(g)] + S^\circ[\text{HCl}(g)] - S^\circ[\text{CH}_4(g)] - S^\circ[\text{Cl}_2(g)]\\ &= (1)(234.6\,\text{J}\cdot\text{K}^{-1}\cdot\text{mol}^{-1}) + (1)(186.9\,\text{J}\cdot\text{K}^{-1}\cdot\text{mol}^{-1})\\ &\quad - (1)(186.3\,\text{J}\cdot\text{K}^{-1}\cdot\text{mol}^{-1}) - (1)(223.1\,\text{J}\cdot\text{K}^{-1}\cdot\text{mol}^{-1})\\ &= 12.1\,\text{J}\cdot\text{K}^{-1}\cdot\text{mol}^{-1}\end{aligned}$$

(d) For the equation given

$$\begin{aligned}\Delta S^\circ_{\text{rxn}} &= 2\,S^\circ[\text{NaCl}(s)] + S^\circ[\text{Br}_2(l)] - 2\,S^\circ[\text{NaBr}(s)] - S^\circ[\text{Cl}_2(g)]\\ &= (2)(72.1\,\text{J}\cdot\text{K}^{-1}\cdot\text{mol}^{-1}) + (1)(152.2\,\text{J}\cdot\text{K}^{-1}\cdot\text{mol}^{-1})\\ &\quad - (2)(86.8\,\text{J}\cdot\text{K}^{-1}\cdot\text{mol}^{-1}) - (1)(223.1\,\text{J}\cdot\text{K}^{-1}\cdot\text{mol}^{-1})\\ &= -100.3\,\text{J}\cdot\text{K}^{-1}\cdot\text{mol}^{-1}\end{aligned}$$

23-66. The value of ΔH_{vap} in Table 15.3 is the value at 25°C, not at the normal boiling point of $\text{CH}_2\text{Cl}(l)$ (40°C).

23-68. We have the relationship $\Delta G^\circ_{\text{rxn}} = -RT\ln K$. Thus,
(a) $\Delta G^\circ_{\text{rxn}} > 0$ requires that $K < 1$
(b) $\Delta G^\circ_{\text{rxn}} = 0$ requires that $K = 1$
(c) $\Delta G^\circ_{\text{rxn}} < 0$ requires that $K > 1$

23-70. The van't Hoff equation is

$$\ln\frac{K_2}{K_1} = \frac{\Delta H^\circ_{\text{rxn}}}{R}\left(\frac{T_2 - T_1}{T_1 T_2}\right)$$

Substituting the values of the last two sets of data, for example, we have

$$\ln\frac{1.77}{1.34} = \frac{\Delta H^\circ_{\text{rxn}}(1273\text{ K} - 1173\text{ K})}{(8.3145\text{ J}\cdot\text{K}^{-1}\cdot\text{mol}^{-1})(1273\text{ K})(1173\text{ K})}$$

$$0.278 = (2.05\times10^{-6}\text{ J}^{-1}\cdot\text{mol})\,\Delta H^\circ_{\text{rxn}}$$

Thus,

$$\Delta H^\circ_{\text{rxn}} = 3.95\times10^{4}\text{ J}\cdot\text{mol}^{-1} = 39.5\text{ kJ}\cdot\text{mol}^{-1}$$

The value calculated from the data in Appendix D is 41.2 $\text{kJ}\cdot\text{mol}^{-1}$. The discrepancy is due to the large temperature difference between 25°C and the data in the problem. A more sophisticated treatment of the data using all the data sets simultaneously (called the method of least squares) gives $\Delta H^\circ_{\text{rxn}} = 34.9\text{ kJ}\cdot\text{mol}^{-1}$.

23-72. (a) The equation for which we wish to calculate the equilibrium constant is

$$\text{AgCl}(s) \rightleftharpoons \text{Ag}^+(aq) + \text{Cl}^-(aq)$$

The value of $\Delta G^\circ_{\text{rxn}}$ is given by

$$\begin{aligned}\Delta G^\circ_{\text{rxn}} &= \Delta G^\circ_f[\text{Ag}^+] + \Delta G^\circ_f[\text{Cl}^-] - \Delta G^\circ_f[\text{AgCl}(s)]\\ &= (1)(77.1\text{ kJ}\cdot\text{mol}^{-1}) + (1)(-131.2\text{ kJ}\cdot\text{mol}^{-1}) - (1)(-109.8\text{ J}\cdot\text{mol}^{-1})\\ &= 55.7\text{ kJ}\cdot\text{mol}^{-1}\end{aligned}$$

The equilibrium constant, K, can be calculated using the relation

$$\ln K = -\frac{\Delta G^\circ_{\text{rxn}}}{RT} = -\frac{55.7\times10^{3}\text{ J}\cdot\text{mol}^{-1}}{(8.3145\text{ J}\cdot\text{K}^{-1}\cdot\text{mol}^{-1})(298.2\text{ K})} = -22.5$$

Thus, $K = 2\times10^{-10}$.

(b) The equation is

$$\text{AgBr}(s) \rightleftharpoons \text{Ag}^+(aq) + \text{Br}^-(aq)$$

The value of $\Delta G^\circ_{\text{rxn}}$ is given by

$$\begin{aligned}\Delta G^\circ_{\text{rxn}} &= \Delta G^\circ_f[\text{Ag}^+] + \Delta G^\circ_f[\text{Br}^-] - \Delta G^\circ_f[\text{AgBr}(s)]\\ &= (1)(77.1\text{ kJ}\cdot\text{mol}^{-1}) + (1)(-104.0\text{ kJ}\cdot\text{mol}^{-1}) - (1)(-96.9\text{ J}\cdot\text{mol}^{-1})\\ &= 70.0\text{ kJ}\cdot\text{mol}^{-1}\end{aligned}$$

The equilibrium constant, K, can be calculated using the relation

$$\ln K = -\frac{\Delta G^\circ_{\text{rxn}}}{RT} = -\frac{70.0\times10^{3}\text{ J}\cdot\text{mol}^{-1}}{(8.3145\text{ J}\cdot\text{K}^{-1}\cdot\text{mol}^{-1})(298.2\text{ K})} = -28.2$$

Thus, $K = 6\times10^{-13}$.

23-74. We use the van't Hoff equation

$$\ln\frac{K_p(T_2)}{K_p(T_1)} = \frac{\Delta H^\circ_{\text{rxn}}}{R}\left(\frac{T_2 - T_1}{T_1 T_2}\right)$$

Substituting the values given in the statement of the problem, we have

$$\ln\frac{K_p(1200\text{ K})}{0.236}=\left(\frac{34.78\times10^3\text{ J·mol}^{-1}}{8.3145\text{ J·mol}^{-1}\text{·K}^{-1}}\right)\left(\frac{1200\text{ K}-800\text{ K}}{(800\text{ K})(1200\text{ K})}\right)=1.743$$

or

$$K_p(1200\text{ K})=(0.236)e^{1.743}=1.35$$

23-76. The value of ΔG°_{rxn} is given by

$$\begin{aligned}\Delta G^\circ_{rxn}&=6\,\Delta G^\circ_f[CO_2(g)]+6\,\Delta G^\circ_f[H_2O(l)]-\Delta G^\circ_f[C_6H_{12}O_6(s)]-6\,\Delta G^\circ_f[O_2(g)]\\&=(6)(-394.4\text{ kJ·mol}^{-1})+(6)(-237.1\text{ kJ·mol}^{-1})\\&\quad-(1)(-916\text{ kJ·mol}^{-1})-(6)(0\text{ kJ·mol}^{-1})\\&=-2873\text{ kJ·mol}^{-1}\end{aligned}$$

The maximum work that can be obtained is equal to ΔG°_{rxn}. Thus, for the complete combustion of one mole of glucose under standard conditions, the maximum work that can be obtained is 2873 kJ.

23-78. The equation is

$$AgCl(s)\rightleftharpoons Ag^+(aq)+Cl^-(aq)$$

We can calculate ΔH°_{rxn} using the data given and the van't Hoff equation,

$$\ln\frac{K_2}{K_1}=\frac{\Delta H^\circ_{rxn}}{R}\left(\frac{T_2-T_1}{T_1T_2}\right)$$

Substituting these values, we have

$$\ln\frac{2.15\times10^{-8}\text{ M}^2}{13.2\times10^{-10}\text{ M}^2}=\frac{\Delta H^\circ_{rxn}(50.0\text{ K})}{(8.3145\text{ J·K}^{-1}\text{·mol}^{-1})(373.2\text{ K})(323.2\text{ K})}$$

$$2.790=(4.99\times10^{-5}\text{ J}^{-1}\text{·mol})\,\Delta H^\circ_{rxn}$$

Thus,

$$\Delta H^\circ_{rxn}=55.9\times10^4\text{ J·mol}^{-1}=55.9\text{ kJ·mol}^{-1}$$

23-80. The equation is

$$CuCl_2\cdot2H_2O(s)\rightleftharpoons CuCl_2\cdot H_2O(s)+H_2O(g)$$

Thus, $K=P_{H_2O}$. We can calculate ΔH°_{rxn} using the data given and the van't Hoff equation,

$$\ln\frac{K_2}{K_1}=\frac{\Delta H^\circ_{rxn}}{R}\left(\frac{T_2-T_1}{T_1T_2}\right)$$

Thus,

$$\ln\frac{91.2\text{ Torr}}{3.72\text{ Torr}}=\frac{\Delta H^\circ_{rxn}(42.0\text{ K})}{(8.3145\text{ J·K}^{-1}\text{·mol}^{-1})(291.2\text{ K})(333.2\text{ K})}$$

$$3.199=(5.21\times10^{-5}\text{ J}^{-1}\text{·mol})\,\Delta H^\circ_{rxn}$$

Thus,

$$\Delta H^\circ_{\text{rxn}} = 61.4 \times 10^4\ \text{J}\cdot\text{mol}^{-1} = 61.4\ \text{kJ}\cdot\text{mol}^{-1}$$

23-82. The vapor pressure equilibrium for bromine is given by

$$\text{Br}_2(l) \rightleftharpoons \text{Bg}_2(g) \qquad K_\text{p} = P_{\text{Br}_2}$$

We can calculate the value of K from the value of $\Delta G^\circ_{\text{rxn}}$

$$\begin{aligned}\Delta G^\circ_{\text{rxn}} &= \Delta G^\circ_\text{f}[\text{Br}_2(g)] - \Delta G^\circ_\text{f}[\text{Br}_2(l)] \\ &= (1)(3.142\ \text{kJ}\cdot\text{mol}^{-1}) - (1)(0\ \text{kJ}\cdot\text{mol}^{-1}) \\ &= 3.142\ \text{kJ}\cdot\text{mol}^{-1} = 3.142 \times 10^3\ \text{J}\cdot\text{mol}^{-1}\end{aligned}$$

The equilibrium constant is given by

$$\ln K = -\frac{\Delta G^\circ_{\text{rxn}}}{RT} = -\frac{3.142 \times 10^3\ \text{J}\cdot\text{mol}^{-1}}{(8.3145\ \text{J}\cdot\text{K}^{-1}\cdot\text{mol}^{-1})(298.2\ \text{K})} = -1.267$$

Therefore, $K = 0.282$ and $K_\text{p} = 0.282$ bar.

23-84. Start with

$$\ln \frac{P_2}{P_1} = \frac{\Delta H^\circ_{\text{sub}}}{R}\left(\frac{T_2 - T_1}{T_1 T_2}\right)$$

Let $P_1 = 1.0 \times 10^2$ kPa, $T_1 = -78.6^\circ\text{C} = 194.6$ K, and $T_2 = -100.0^\circ\text{C} = 173.2$ K and write

$$\ln \frac{P_2}{1.0 \times 10^2\ \text{kPa}} = \frac{(25.2 \times 10^3\ \text{J}\cdot\text{mol}^{-1})(173.2\ \text{K} - 194.6\ \text{K})}{(8.3145\ \text{J}\cdot\text{K}^{-1}\cdot\text{mol}^{-1})(173.2\ \text{K})(194.6\ \text{K})} = -1.92$$

$$\frac{P_2}{1.0 \times 10^2\ \text{kPa}} = e^{-1.92} = 0.15$$

$$P_2 = 15\ \text{kPa}$$

The result, 15 kPa, is 7% in error of the experimental result of 14 kPa, suggesting that applying the Clapeyron–Clausius equation to the sublimation of a solid gives only an adequate result.

23-86. The equation for the reaction is

$$\frac{1}{2}\text{H}_2(g) + \frac{1}{2}\text{Cl}_2(g) \longrightarrow \text{HCl}(g)$$

(a) The value of the thermodynamic equilibrium constant is given by

$$\ln K = -\frac{\Delta G^\circ_{\text{rxn}}}{RT} = -\frac{-95.3 \times 10^3\ \text{J}\cdot\text{mol}^{-1}}{(8.3145\ \text{J}\cdot\text{K}^{-1}\cdot\text{mol}^{-1})(298.2\ \text{K})} = 38.4$$

where $\Delta G^\circ_{\text{rxn}} = \Delta G^\circ_\text{f}$. Thus,

$$K = 5 \times 10^{16}$$

The value of the concentration quotient is

$$Q = \frac{P_{HCl}}{P_{H_2}^{1/2} P_{Cl_2}^{1/2}} = \frac{0.31}{(3.5)^{1/2}(1.5)^{1/2}} = 0.14$$

The value of ΔG_{rxn} for the reaction is

$$\Delta G_{rxn} = RT \ln \frac{Q}{K} = (8.3145\,\text{J}\cdot\text{K}^{-1}\cdot\text{mol}^{-1})(298.2\text{ K}) \ln \frac{0.14}{5 \times 10^{16}}$$
$$= -1.00 \times 10^5\,\text{J}\cdot\text{mol}^{-1} = -100\,\text{kJ}\cdot\text{mol}^{-1}$$

(b) The reaction is more favorable under the conditions given because the magnitude of ΔG_{rxn} is greater than that of ΔG°_{rxn}.

23-88.* Assuming the process to be reversible, from Equations 23.1 and 23.3a we have

$$\Delta S_{engine} = \Delta S_h + \Delta S_c = \frac{q_{rev,h}}{T_h} - \frac{q_{rev,c}}{T_c} \qquad (1)$$

Because $\Delta U_{engine} = q_{rev,h} - q_{rev,c} - w = 0$, the work done by the engine is given by

$$w = q_{rev,h} - q_{rev,c} \qquad (2)$$

From equation (1) we have

$$\frac{q_{rev,c}}{T_h} = \frac{q_{rev,h}}{T_c}$$

or

$$q_{rev,c} = q_{rev,h}\left(\frac{T_c}{T_h}\right)$$

Substituting this expression for $q_{rev,c}$ into equation (2) yields

$$w = q_{rev,h} - q_{rev,h}\left(\frac{T_c}{T_h}\right) = q_{rev,h}\left(1 - \frac{T_c}{T_h}\right)$$

Thus,

$$\text{maximum efficiency} = \frac{w}{q_{rev,h}} = 1 - \frac{T_c}{T_h} = \frac{T_h - T_c}{T_h} \qquad (3)$$

Because the efficiency of a reversible process is the maximum efficiency obtainable, for an irreversible process we have

$$\text{efficiency} = \frac{w}{q_{rev,h}} < \frac{T_h - T_c}{T_h} \qquad (4)$$

The maximum efficiency of a steam engine that extracts energy in the form of heat from boiling water at 100°C and releases it into its surroundings at 20°C is given by equation (3) as

$$\text{maximum efficiency} = \frac{T_h - T_c}{T_h} = \frac{373\ \text{K} - 293\ \text{K}}{373\ \text{K}} = 0.21$$

or 21%. In practice, the efficiency would be less due to factors such as friction. Equation (3) indicates that a greater efficiency is obtained by engines working with a higher value of T_h or a lower value of T_c. Notice that if the two temperatures are the same ($T_h = T_c$), then no mechanical energy is produced and $w/q = 0$. In order for the engine to achieve 100% efficiency ($w/q = 1$), the value of T_c must equal zero kelvin, which is an unattainable temperature. Thus, in practice, it is never possible for an engine to achieve 100% efficiency.

CHAPTER 24. Oxidation-Reduction Reactions

OXIDATION STATES

24-2. We use the rules given in Section 24-1.

(a) The oxidation state of the hydrogen atom is +1 by rule 6. The oxidation state of each oxygen atom is $+1 + 2x = 0$ or $x = -1/2$ by rule 2.

(b) The Lewis formula for $S_2O_8^{2-}$ is

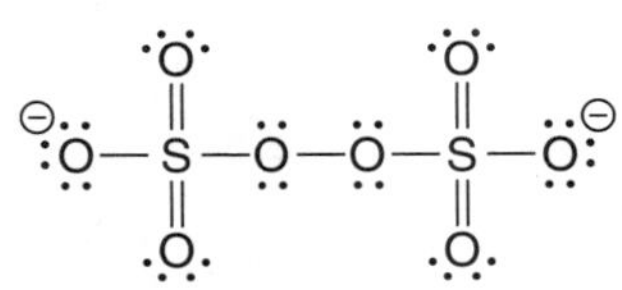

Because the electronegativity of an oxygen atom is greater than that of a sulfur atom, we have

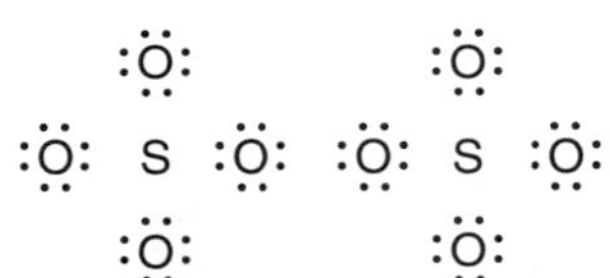

The six terminal oxygen atoms have an oxidation state of $6 - 8 = -2$. The oxidation state of the two oxygen atoms linking the sulfur atoms is $6 - 7 = -1$. The oxidation state of each sulfur atom is $6 - 0 = +6$.

(c) The oxidation state of each oxygen atom is $3x = -1$ or $x = -1/3$ by rule 2.

(d) The Lewis formula of H_3COOCH_3 is

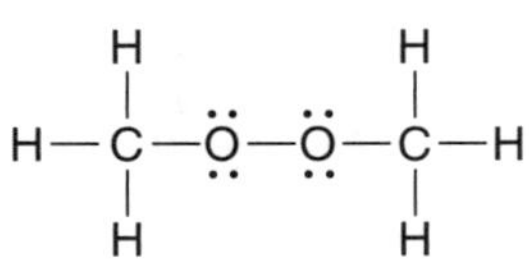

Because an oxygen atom is more electronegative than a carbon atom, we apportion the oxygen valence electrons as follows

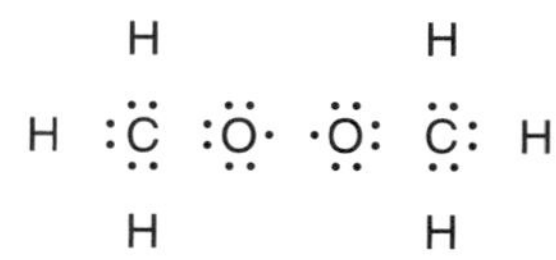

and the oxidation state of each oxygen atom is $6 - 7 = -1$.

24-4. We use the rules given in Section 24-1. The oxidation state of each oxygen atom is -2 by rule 7.

(a) The oxidation state of each chlorine atom is $2x + (-2) = 0$ or $x = +1$ by rule 2.

(b) The oxidation state of the chlorine atom is $x + (2)(-2) = 0$ or $x = +4$ by rule 2.

(c) The oxidation state of each chlorine atom is $2x + (7)(-2) = 0$ or $x = +7$ by rule 2.

(d) The oxidation state of each chlorine atom is $2x + (5)(-2) = 0$ or $x = +5$ by rule 2.

24-6. We use the rules given in Section 24-1.

(a) The oxidation state of a H atom is $+1$ by rule 3; the oxidation state of an O atom is -2 by rule 7; and the oxidation state of an Fe atom is given by $(3)(-2) + 2x = 0$ or $x = +3$ by rule 2.

(b) The oxidation state of an O atom is -2 by rule 7; and the oxidation state of an Fe atom is given by $3x + (4)(-2) = 0$ or $x = +8/3$ by rule 2 or equivalently, two iron atoms at $+3$ and one iron atom at $+2$.

(c) The oxidation state of a K atom is $+1$ by rule 3. The Lewis formula of a CN^- ion is

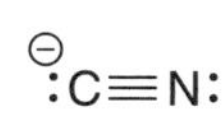

The electronegativity of a nitrogen atom is greater than that of a carbon atom, and so we apportion the valence electrons as

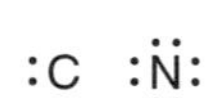

The oxidation states are: C, $6 - 2 = +4$ and N, $5 - 8 = -3$.

(d) The oxidation state of a K atom is $+1$ by rule 3. The Lewis formula of CNO^- is

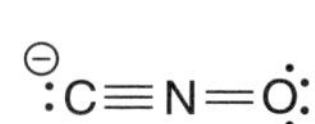

The electronegativity order is $O > N > C$. Thus, we apportion the valence electrons as

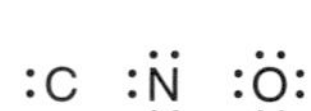

Therefore, the oxidation states are: C, $4 - 4 = 0$; N, $5 - 4 = +1$; and O, $6 - 8 = -2$.

24-8. (a) The Lewis formula for CS_2 is

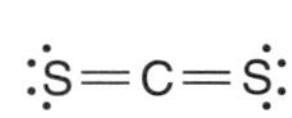

The electronegativity of a sulfur atom is greater than that of a carbon atom; thus, we apportion the electrons as follows

The oxidation states are: S, $6 - 8 = -2$ and C, $4 - 0 = +4$.

(b) The Lewis formula for CH_3SSCH_3 is

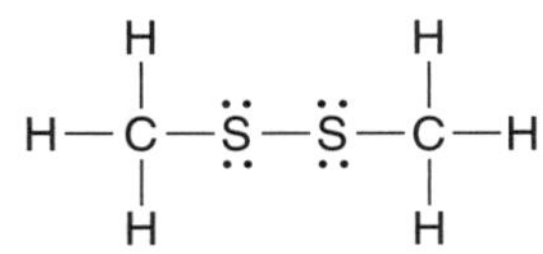

Because the electronegativity of a sulfur atom is greater than that of a carbon atom and that of a carbon atom is greater than that of a hydrogen atom, we apportion the valence electrons as follows:

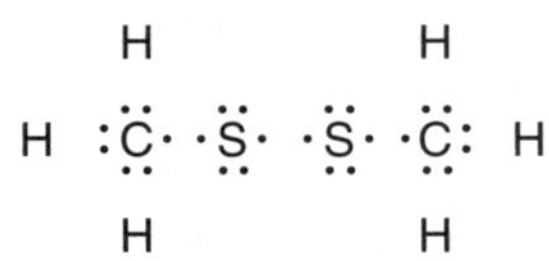

The oxidation states are: H, $1 - 0 = +1$; S, $6 - 7 = -1$; and C, $4 - 6 = -2$.

(c) The Lewis formula for $HCONH_2$ is

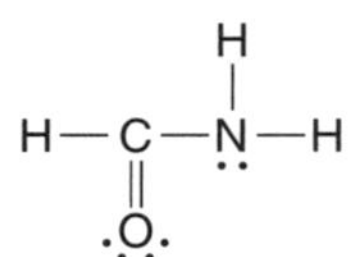

The electronegativity order is O > N > C > H. Thus, we have

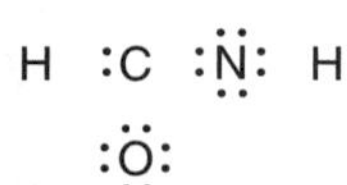

Therefore, the oxidation states are: H, $1 - 0 = +1$; C, $4 - 2 = +2$; O, $6 - 8 = -2$; and N, $5 - 8 = -3$.

(d) The Lewis formula is given in the problem. The electronegativity order is N > S > C > H. Thus, we apportion the valence electrons as

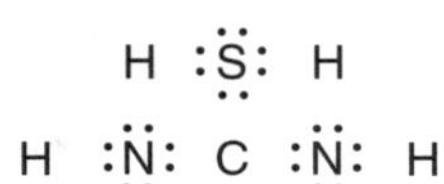

Therefore, the oxidation states are: H, $1 - 0 = +1$; C, $4 - 0 = +4$; N, $5 - 8 = -3$; and S, $6 - 8 = -2$.

TRANSITION METAL COMPOUNDS

24-10. (a) The oxidation state of the titanium atom in TiF_6^{2-} is +4 [$x + 4(-1) = -2$ or $x = +4$]; thus, the titanium atom has four valence electrons giving $4 + (6 \times 7) + 2 = 48$ valence electrons. Placing the titanium atom as the central atom, we have

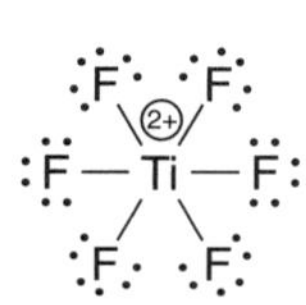

(b) The oxidation state of the titanium atom in $TiBr_5^-$ is +4 $[x + 5(-1) = -1$ or $x = +4]$; thus, the titanium atom has four valence electrons giving $4 + (5 \times 7) + 2 = 40$ valence electrons. Placing the titanium atom as the central atom, we have

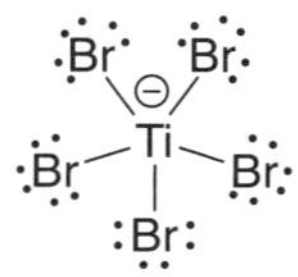

24-12. The shapes of the species are

(a) :Cl—Hg—Cl: AX_2 linear

(b) 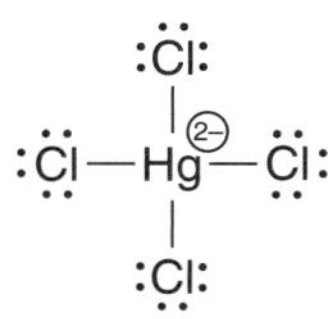 AX_4 tetrahedral

(c) 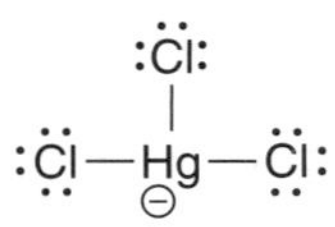 AX_3 trigonal planar

OXIDIZING AGENTS AND REDUCING AGENTS

24-14. The oxidation state of S decreases from +6 in $Na_2SO_4(s)$ to −2 in $Na_2S(s)$ Thus, S is reduced and $Na_2SO_4(s)$ acts as the oxidizing agent. The oxidation state of C increases from 0 in $C(s)$ to +2 in $CO(g)$. Thus, C is oxidized and $C(s)$ acts as the reducing agent.

24-16. The oxidation state of Cl decreases from +4 in $ClO_2(g)$ to +3 in $NaClO_2(s)$. Thus, Cl is reduced and $ClO_2(g)$ acts as the oxidizing agent. The oxidation state of C increases from 0 in $C(s)$ to +4 in $CaCO_3(s)$. Thus, C is oxidized and $C(s)$ acts as the reducing agent.

24-18. (a) The oxidation state of In increases from +1 in $In^+(aq)$ to +3 in $In^{3+}(aq)$. Thus, In is oxidized and $In^+(aq)$ acts as the reducing agent. The oxidation state of Fe decreases from +3 in $Fe^{3+}(aq)$ to +2 in $Fe^{2+}(aq)$. Thus, Fe is reduced and $Fe^{3+}(aq)$ acts as the oxidizing agent. The half reaction equations are

$$In^+(aq) \longrightarrow In^{3+}(aq) + 2\,e^- \quad \text{(oxidation half reaction)}$$

$$2\,Fe^{3+}(aq) + 2\,e^- \longrightarrow 2\,Fe^{2+}(aq) \quad \text{(reduction half reaction)}$$

(b) The oxidation state of S increases from −2 in $H_2S(g)$ to 0 in $S(s)$. Thus, S is oxidized and $H_2S(g)$ acts as the reducing agent. The oxidation state of Cl decreases from +1 in $ClO^-(aq)$ to −1 in $Cl^-(aq)$. Thus, Cl is reduced and $ClO^-(aq)$ acts as the oxidizing agent. The half reaction equations are

$$H_2S(aq) \longrightarrow S(s) + 2\,H^+(aq) + 2\,e^- \quad \text{(oxidation half reaction)}$$

$$ClO^-(aq) + 2\,H^+(aq) + 2\,e^- \longrightarrow Cl^-(aq) + H_2O(l) \quad \text{(reduction half reaction)}$$

24-20. (a) Fe is oxidized from +2 to +3 and Cr is reduced from +6 to +3.
(b) No species is oxidized or reduced (just a precipitation reaction).
(c) O is reduced from 0 to −2 and C is oxidized from −4 to +4.

(d) No species is oxidized or reduced (just an acid-base reaction).
(e) Br is oxidized from -1 to 0 and Cl is reduced from 0 to -1.

24-22. For the reactions in Problem 23-20, we have that

(a) Fe is oxidized $(+2 \rightarrow +3)$
Cr is reduced $(+6 \rightarrow +3)$
$Cr_2O_7^{2-}(aq)$ is the oxidizing agent
$Fe^{2+}(aq)$ is the reducing agent

(b) not an oxidation-reduction reaction

(c) C is oxidized $(-4 \rightarrow +4)$
O is reduced $(0 \rightarrow -2)$
$O_2(g)$ is the oxidizing agent
$CH_4(g)$ is the reducing agent

(d) not an oxidation-reduction reaction

(e) Br is oxidized $(-1 \rightarrow 0)$
Cl is reduced $(0 \rightarrow -1)$
$Cl_2(g)$ is the oxidizing agent
$NaBr(aq)$ is the reducing agent

24-24. The hydrogen atoms in $LiAlH_4(s)$ are in an unusual oxidation state (-1). The common oxidation state for a hydrogen atom is $+1$, and so the hydrogen atoms in $LiAlH_4$ are easily oxidized from -1 to $+1$, making $LiAlH_4(s)$ a strong reducing agent.

BALANCING HALF REACTIONS

24-26. The steps are

(a) $H_2BO_3^- \longrightarrow BH_4^-$

$H_2BO_3^- + 8\,H^+ \longrightarrow BH_4^- + 3\,H_2O$

$H_2BO_3^-(aq) + 8\,H^+(aq) + 8e^- \longrightarrow BH_4^-(aq) + 3\,H_2O(l)$

(b) $2\,ClO_3^- \longrightarrow Cl_2$

$2\,ClO_3^- + 12\,H^+ \longrightarrow Cl_2 + 6\,H_2O$

$2\,ClO_3^-(aq) + 12\,H^+(aq) + 10e^- \longrightarrow Cl_2(g) + 6\,H_2O(l)$

(c) $Cl_2 \longrightarrow 2\,HClO$

$Cl_2 + 2\,H_2O \longrightarrow 2\,HClO + 2\,H^+$

$Cl_2(g) + 2\,H_2O(l) \longrightarrow 2\,HClO(aq) + 2\,H^+(aq) + 2\,e^-$

24-28. The steps are

(a) $OsO_4 + 8\,H^+ \longrightarrow Os + 4\,H_2O$

$OsO_4(s) + 8\,H^+(aq) + 8\,e^- \longrightarrow Os(s) + 4\,H_2O(l)$

(b) $S + 6\,OH^- \longrightarrow SO_3^{2-} + 3\,H_2O$

$S(s) + 6\,OH^-(aq) \longrightarrow SO_3^{2-}(aq) + 3\,H_2O(l) + 4\,e^-$

(c) $Sn + 2\,H_2O \longrightarrow HSnO_2^- + 3\,H^+$

$Sn(s) + 2\,H_2O(l) \longrightarrow HSnO_2^-(aq) + 3\,H^+(aq) + 2\,e^-$

24-30. The steps are

(a) $[Au(CN)_2]^- \longrightarrow Au + 2\,CN^-$

$[Au(CN)_2]^-(aq) + e^- \longrightarrow Au(s) + 2\,CN^-(aq)$

(b) $MnO_4^- + 4\,H^+ \longrightarrow MnO_2 + 2\,H_2O$

$MnO_4^-(aq) + 4\,H^+(aq) + 3\,e^- \longrightarrow MnO_2(s) + 2\,H_2O(l)$

(c) $Cr(OH)_3 \longrightarrow CrO_4^{2-} + 3\,OH^-$

$Cr(OH)_3 + 8\,OH^- \longrightarrow CrO_4^{2-} + 4\,H_2O + 3\,OH^-$

$Cr(OH)_3 + 5\,OH^- \longrightarrow CrO_4^{2-} + 4\,H_2O$

$Cr(OH)_3(s) + 5\,OH^-(aq) \longrightarrow CrO_4^{2-}(aq) + 4\,H_2O(l) + 3\,e^-$

BALANCING OXIDATION-REDUCTION EQUATIONS

24-32. (a) We begin by identifying the species that is oxidized and the species that is reduced.

$$ZnS \longrightarrow S + Zn^{2+} \quad \text{(oxidation)}$$
$$NO_3^- \longrightarrow NO \quad \text{(reduction)}$$

The various steps are

$$ZnS \longrightarrow S + Zn^{2+} \quad \text{(oxidation)}$$
$$NO_3^- \longrightarrow NO + 2\,H_2O \quad \text{(reduction)}$$

$$ZnS \longrightarrow S + Zn^{2+} \quad \text{(oxidation)}$$
$$NO_3^- + 4\,H^+ \longrightarrow NO + 2\,H_2O \quad \text{(reduction)}$$

$$ZnS \longrightarrow S + Zn^{2+} + 2\,e^- \quad \text{(oxidation)}$$
$$NO_3^- + 4\,H^+ + 3\,e^- \longrightarrow NO + 2\,H_2O \quad \text{(reduction)}$$

$$3\,ZnS \longrightarrow 3\,S + 3\,Zn^{2+} + 6\,e^- \quad \text{(oxidation)}$$
$$2\,NO_3^- + 8\,H^+ + 6\,e^- \longrightarrow 2\,NO + 4\,H_2O \quad \text{(reduction)}$$

Thus, the complete balanced equation is

$$3\,ZnS(s) + 2\,NO_3^-(aq) + 8\,H^+(aq) \longrightarrow 3\,S(s) + 3\,Zn^{2+}(aq) + 2\,NO(g) + 4\,H_2O(l)$$

oxidizing agent	$NO_3^-(aq)$	reducing agent	$ZnS(s)$
species oxidized	S	species reduced	N

(b) The two half reaction equations balanced with respect to the elements other than oxygen and hydrogen are

$$HNO_2 \longrightarrow NO_3^- \quad \text{(oxidation)}$$
$$MnO_4^- \longrightarrow Mn^{2+} \quad \text{(reduction)}$$

The various steps are

$$\begin{array}{ll} HNO_2 \longrightarrow NO_3^- & \text{(oxidation)} \\ MnO_4^- \longrightarrow Mn^{2+} + 4\,H_2O & \text{(reduction)} \end{array}$$

$$\begin{array}{ll} HNO_2 + H_2O \longrightarrow NO_3^- + 3\,H^+ & \text{(oxidation)} \\ MnO_4^- + 8\,H^+ \longrightarrow Mn^{2+} + 4\,H_2O & \text{(reduction)} \end{array}$$

$$\begin{array}{ll} HNO_2 + H_2O \longrightarrow NO_3^- + 3\,H^+ + 2\,e^- & \text{(oxidation)} \\ MnO_4^- + 8\,H^+ + 5\,e^- \longrightarrow Mn^{2+} + 4\,H_2O & \text{(reduction)} \end{array}$$

$$\begin{array}{ll} 5\,HNO_2 + 5\,H_2O \longrightarrow 5\,NO_3^- + 15\,H^+ + 10\,e^- & \text{(oxidation)} \\ 2\,MnO_4^- + 16\,H^+ + 10\,e^- \longrightarrow 2\,Mn^{2+} + 8\,H_2O & \text{(reduction)} \end{array}$$

Thus, the complete balanced equation is

$$2\,MnO_4^-(aq) + 5\,HNO_2(aq) + H^+(aq) \longrightarrow 5\,NO_3^-(aq) + 2\,Mn^{2+}(aq) + 3\,H_2O(l)$$

oxidizing agent $MnO_4^-(aq)$ reducing agent $HNO_3(aq)$

species oxidized N species reduced Mn

24-34. (a) The two half reaction equations balanced with respect to the elements other than oxygen and hydrogen are

$$\begin{array}{ll} CoCl_2 \longrightarrow Co(OH)_3 + 2\,Cl^- & \text{(oxidation)} \\ Na_2O_2 \longrightarrow 2\,Na^+ & \text{(reduction)} \end{array}$$

The various steps are

$$\begin{array}{ll} CoCl_2 + 3\,OH^- \longrightarrow Co(OH)_3 + 2\,Cl^- & \text{(oxidation)} \\ Na_2O_2 + 2\,H_2O \longrightarrow 2\,Na^+ & \text{(reduction)} \end{array}$$

$$\begin{array}{ll} CoCl_2 + 3\,OH^- \longrightarrow Co(OH)_3 + 2\,Cl^- & \text{(oxidation)} \\ Na_2O_2 + 2\,H_2O \longrightarrow 4\,OH^- + 2\,Na^+ & \text{(reduction)} \end{array}$$

$$\begin{array}{ll} CoCl_2 + 3\,OH^- \longrightarrow Co(OH)_3 + 2\,Cl^- + e^- & \text{(oxidation)} \\ Na_2O_2 + 2\,H_2O + 2\,e^- \longrightarrow 4\,OH^- + 2\,Na^+ & \text{(reduction)} \end{array}$$

$$\begin{array}{ll} 2\,CoCl_2 + 6\,OH^- \longrightarrow 2\,Co(OH)_3 + 4\,Cl^- + 2\,e^- & \text{(oxidation)} \\ Na_2O_2 + 2\,H_2O + 2\,e^- \longrightarrow 4\,OH^- + 2\,Na^+ & \text{(reduction)} \end{array}$$

Thus, the complete balanced equation is

$$2\,CoCl_2(s) + Na_2O_2(aq) + 2\,H_2O(l) + 2\,OH^-(aq) \longrightarrow 2\,Co(OH)_3(s) + 4\,Cl^-(aq) + 2\,Na^+(aq)$$

(b) The two half reaction equations balanced with respect to the elements other than oxygen and hydrogen are

$$\begin{array}{ll} C_2O_4^{2-} \longrightarrow 2\,CO_2 & \text{(oxidation)} \\ MnO_2 \longrightarrow Mn^{2+} & \text{(reduction)} \end{array}$$

The various steps are

$$\begin{array}{ll} C_2O_4^{2-} \longrightarrow 2\,CO_2 & \text{(oxidation)} \\ MnO_2 + 4\,H^+ \longrightarrow Mn^{2+} + 2\,H_2O & \text{(reduction)} \end{array}$$

$$\begin{array}{ll} C_2O_4^{2-} \longrightarrow 2\,CO_2 + 2\,e^- & \text{(oxidation)} \\ MnO_2 + 4\,H^+ + 2\,e^- \longrightarrow Mn^{2+} + 2\,H_2O & \text{(reduction)} \end{array}$$

Thus, the complete balanced equation is

$$C_2O_4^{2-}(aq) + MnO_2(s) + 4\,H^+(aq) \longrightarrow Mn^{2+}(aq) + 2\,CO_2(g) + 2\,H_2O(l)$$

24-36. (a) The two half reaction equations balanced with respect to the elements other than oxygen and hydrogen are

$$\begin{array}{ll} 2\,I^- \longrightarrow I_2 & \text{(oxidation)} \\ Cr_2O_7^{2-} \longrightarrow 2\,Cr^{3+} & \text{(reduction)} \end{array}$$

The various steps are

$$\begin{array}{ll} 2\,I^- \longrightarrow I_2 & \text{(oxidation)} \\ Cr_2O_7^{2-} + 14\,H^+ \longrightarrow 2\,Cr^{3+} + 7\,H_2O & \text{(reduction)} \end{array}$$

$$\begin{array}{ll} 2\,I^- \longrightarrow I_2 + 2\,e^- & \text{(oxidation)} \\ Cr_2O_7^{2-} + 14\,H^+ + 6\,e^- \longrightarrow 2\,Cr^{3+} + 7\,H_2O & \text{(reduction)} \end{array}$$

$$\begin{array}{ll} 6\,I^- \longrightarrow 3\,I_2 + 6\,e^- & \text{(oxidation)} \\ Cr_2O_7^{2-} + 14\,H^+ + 6\,e^- \longrightarrow 2\,Cr^{3+} + 7\,H_2O & \text{(reduction)} \end{array}$$

Thus, the complete balanced equation is

$$Cr_2O_7^{2-}(aq) + 6\,I^-(aq) + 14\,H^+(aq) \longrightarrow 2\,Cr^{3+}(aq) + 3\,I_2(s) + 7\,H_2O(l)$$

(b) The two half reaction equations balanced with respect to the elements other than oxygen and hydrogen are

$$\begin{array}{ll} CuS \longrightarrow S + Cu^{2+} & \text{(oxidation)} \\ NO_3^- \longrightarrow NO & \text{(reduction)} \end{array}$$

The various steps are

$$\begin{array}{ll} CuS \longrightarrow S + Cu^{2+} & \text{(oxidation)} \\ NO_3^- + 4\,H^+ \longrightarrow NO + 2\,H_2O & \text{(reduction)} \end{array}$$

$$\begin{array}{ll} CuS \longrightarrow S + Cu^{2+} + 2\,e^- & \text{(oxidation)} \\ NO_3^- + 4\,H^+ + 3\,e^- \longrightarrow NO + 2\,H_2O & \text{(reduction)} \end{array}$$

$$\begin{array}{ll} 3\,CuS \longrightarrow 3\,S + 3\,Cu^{2+} + 6\,e^- & \text{(oxidation)} \\ 2\,NO_3^- + 8\,H^+ + 6\,e^- \longrightarrow 2\,NO + 4\,H_2O & \text{(reduction)} \end{array}$$

Thus, the complete balanced equation is

$$3\,CuS(s) + 2\,NO_3^-(aq) + 8\,H^+(aq) \longrightarrow 3\,S(s) + 3\,Cu^{2+}(aq) + 2\,NO(g) + 4\,H_2O(l)$$

24-38. (a) The two half reaction equations balanced with respect to the elements other than oxygen and hydrogen are

$$\begin{array}{ll} N_2H_4 \longrightarrow N_2 & \text{(oxidation)} \\ Cu(OH)_2 \longrightarrow Cu & \text{(reduction)} \end{array}$$

The various steps are

$$\begin{array}{ll} N_2H_4 + 4\,OH^- \longrightarrow N_2 + 4\,H_2O & \text{(oxidation)} \\ Cu(OH)_2 \longrightarrow Cu + 2\,OH^- & \text{(reduction)} \end{array}$$

$$\begin{array}{ll} N_2H_4 + 4\,OH^- \longrightarrow N_2 + 4\,H_2O + 4\,e^- & \text{(oxidation)} \\ Cu(OH)_2 + 2\,e^- \longrightarrow Cu + 2\,OH^- & \text{(reduction)} \end{array}$$

$$\begin{array}{ll} N_2H_4 + 4\,OH^- \longrightarrow N_2 + 4\,H_2O + 4\,e^- & \text{(oxidation)} \\ 2\,Cu(OH)_2 + 4\,e^- \longrightarrow 2\,Cu + 4\,OH^- & \text{(reduction)} \end{array}$$

Thus, the complete balanced equation is

$$N_2H_4(aq) + 2\,Cu(OH)_2(s) \longrightarrow N_2(g) + 2\,Cu(s) + 4\,H_2O(l)$$

(b) The two half reaction equations balanced with respect to the elements other than oxygen and hydrogen are

$$\begin{array}{ll} H_3AsO_3 \longrightarrow H_3AsO_4 & \text{(oxidation)} \\ I_2 \longrightarrow 2\,I^- & \text{(reduction)} \end{array}$$

The various steps are

$$\begin{array}{ll} H_3AsO_3 + H_2O \longrightarrow H_3AsO_4 + 2\,H^+ & \text{(oxidation)} \\ I_2 \longrightarrow 2\,I^- & \text{(reduction)} \end{array}$$

$$\begin{array}{ll} H_3AsO_3 + H_2O \longrightarrow H_3AsO_4 + 2\,H^+ + 2\,e^- & \text{(oxidation)} \\ I_2 + 2\,e^- \longrightarrow 2\,I^- & \text{(reduction)} \end{array}$$

Thus, the complete balanced equation is

$$H_3AsO_3(aq) + I_2(aq) + H_2O(l) \longrightarrow H_3AsO_4(aq) + 2\,I^-(aq) + 2\,H^+(aq)$$

24-40. (a) The two half reaction equations balanced with respect to the elements other than oxygen and hydrogen are

$$\begin{array}{ll} SO_3^{2-} \longrightarrow SO_4^{2-} & \text{(oxidation)} \\ Co(OH)_2 \longrightarrow Co & \text{(reduction)} \end{array}$$

The various steps are

$$\begin{array}{ll} SO_3^{2-} + 2\,OH^- \longrightarrow SO_4^{2-} + H_2O & \text{(oxidation)} \\ Co(OH)_2 \longrightarrow Co + 2\,OH^- & \text{(reduction)} \end{array}$$

$$\begin{array}{ll} SO_3^{2-} + 2\,OH^- \longrightarrow SO_4^{2-} + H_2O + 2\,e^- & \text{(oxidation)} \\ Co(OH)_2 + 2\,e^- \longrightarrow Co + 2\,OH^- & \text{(reduction)} \end{array}$$

Thus, the complete balanced equation is

$$\mathrm{Co(OH)_2}(s) + \mathrm{SO_3^{2-}}(aq) \longrightarrow \mathrm{SO_4^{2-}}(aq) + \mathrm{Co}(s) + \mathrm{H_2O}(l)$$

(b) The two half reaction equations balanced with respect to the elements other than oxygen and hydrogen are

$$\begin{aligned} &3\,\mathrm{I^-} \longrightarrow \mathrm{I_3^-} &&\text{(oxidation)}\\ &\mathrm{IO_3^-} \longrightarrow \mathrm{I_3^-} &&\text{(reduction)} \end{aligned}$$

The various steps are

$$\begin{aligned} &3\,\mathrm{I^-} \longrightarrow \mathrm{I_3^-} &&\text{(oxidation)}\\ &3\,\mathrm{IO_3^-} + 18\,\mathrm{H^+} \longrightarrow \mathrm{I_3^-} + 9\,\mathrm{H_2O} &&\text{(reduction)}\\[6pt] &3\,\mathrm{I^-} \longrightarrow \mathrm{I_3^-} + 2\,\mathrm{e^-} &&\text{(oxidation)}\\ &3\,\mathrm{IO_3^-} + 18\,\mathrm{H^+} + 16\,\mathrm{e^-} \longrightarrow \mathrm{I_3^-} + 9\,\mathrm{H_2O} &&\text{(reduction)}\\[6pt] &24\,\mathrm{I^-} \longrightarrow 8\,\mathrm{I_3^-} + 16\,\mathrm{e^-} &&\text{(oxidation)}\\ &3\,\mathrm{IO_3^-} + 18\,\mathrm{H^+} + 16\,\mathrm{e^-} \longrightarrow \mathrm{I_3^-} + 9\,\mathrm{H_2O} &&\text{(reduction)} \end{aligned}$$

Thus, the complete balanced equation is

$$3\,\mathrm{IO_3^-}(aq) + 24\,\mathrm{I^-}(aq) + 18\,\mathrm{H^+}(aq) \longrightarrow 9\,\mathrm{I_3^-}(aq) + 9\,\mathrm{H_2O}(l)$$

CALCULATIONS INVOLVING OXIDATION-REDUCTION EQUATIONS

24-42. The number of millimoles of $\mathrm{Ce^{4+}}(aq)$ is given by

$$\text{millimoles of } \mathrm{Ce^{+}} = MV = (0.0965\ \mathrm{M})(37.5\ \mathrm{mL}) = 3.62\ \mathrm{mmol}$$

which is equal to the number of millimoles of $\mathrm{Fe^{2+}}$ according to the equation for the reaction. The concentration of $\mathrm{Fe^{2+}}(aq)$ in the sample is given by

$$\mathrm{M} = \frac{3.62\ \mathrm{mmol}}{35.0\ \mathrm{mL}} = 0.103\ \mathrm{M}$$

The mass of iron in the sample is

$$\text{mass of iron} = (3.62 \times 10^{-3}\ \mathrm{mol})\left(\frac{55.845\ \mathrm{g\ Fe}}{1\ \mathrm{mol\ Fe}}\right) = 0.202\ \mathrm{g} = 202\ \mathrm{mg}$$

24-44. We must first balance the equation for the reaction. The solution is acidic, and so we have for the ions that undergo oxidation and reduction

$$\mathrm{Fe^{2+} \longrightarrow Fe^{3+}} \qquad \text{(oxidation)}$$
$$\mathrm{MnO_4^- \longrightarrow Mn^{2+}} \qquad \text{(reduction)}$$

$$\mathrm{Fe^{2+} \longrightarrow Fe^{3+}} \qquad \text{(oxidation)}$$
$$\mathrm{MnO_4^- + 8\,H^+ \longrightarrow Mn^{2+} + 4\,H_2O} \qquad \text{(reduction)}$$

$$\mathrm{Fe^{2+} \longrightarrow Fe^{3+} + e^-} \qquad \text{(oxidation)}$$
$$\mathrm{MnO_4^- + 8\,H^+ + 5\,e^- \longrightarrow Mn^{2+} + 4\,H_2O} \qquad \text{(reduction)}$$

$$\mathrm{5\,Fe^{2+} \longrightarrow 5\,Fe^{3+} + 5\,e^-} \qquad \text{(oxidation)}$$
$$\mathrm{MnO_4^- + 8\,H^+ + 5\,e^- \longrightarrow Mn^{2+} + 4\,H_2O} \qquad \text{(reduction)}$$

Thus, the complete balanced equation is

$$\mathrm{MnO_4^-}(aq) + 5\,\mathrm{Fe^{2+}}(aq) + 8\,\mathrm{H^+}(aq) \longrightarrow \mathrm{Mn^{2+}}(aq) + 5\,\mathrm{Fe^{3+}}(aq) + 4\,\mathrm{H_2O}(l)$$

or, including the spectator ions,

$$\mathrm{KMnO_4}(aq) + 5\,\mathrm{FeCl_2}(aq) + 8\,\mathrm{HCl}(aq) \longrightarrow \mathrm{MnCl_2}(aq) + 5\,\mathrm{FeCl_3}(aq) + \mathrm{KCl}(aq) + 4\,\mathrm{H_2O}(l)$$

The number of moles of $\mathrm{KMnO_4}(aq)$ that oxidizes the $\mathrm{FeCl_2}(aq)$ is

$$n = MV = (0.0512\ \mathrm{M})\,(0.0316\ \mathrm{L}) = 1.618 \times 10^{-3}\ \mathrm{mol}$$

The number of moles of $\mathrm{FeCl_2}$ that reacts with 1.613×10^{-3} moles of $\mathrm{KMnO_4}(aq)$ is

$$\text{moles of FeCl}_2 = (1.618 \times 10^{-3}\ \text{mol of KMnO}_4)\left(\frac{5\ \text{mol FeCl}_2}{1\ \text{mol KMnO}_4}\right) = 8.09 \times 10^{-3}\ \text{mol}$$

The number of moles of iron in 8.09×10^{-3} moles of $\mathrm{FeCl_2}(aq)$ is

$$\text{moles of Fe} = (8.09 \times 10^{-3}\ \text{mol FeCl}_2)\left(\frac{1\ \text{mol Fe}}{1\ \text{mol FeCl}_2}\right) = 8.09 \times 10^{-3}\ \text{mol}$$

The mass of iron is given by

$$\text{mass of Fe} = (8.09 \times 10^{-3}\ \text{mol Fe})\left(\frac{55.845\ \text{g Fe}}{1\ \text{mol Fe}}\right) = 0.452\ \text{g}$$

The percentage of iron in the ore is

$$\%\ \text{iron} = \frac{0.452\ \text{g Fe}}{4.24\ \text{g ore}} \times 100 = 10.7\%$$

24-46. We must first balance the equation for the reaction. The solution is basic, and so we have for the ions that undergo oxidation and reduction

$$\begin{aligned} &C \longrightarrow CO_3^{2-} && \text{(oxidation)} \\ &ClO_2 \longrightarrow ClO_2^- && \text{(reduction)} \end{aligned}$$

$$\begin{aligned} &C + 6\,OH^- \longrightarrow CO_3^{2-} + 3\,H_2O && \text{(oxidation)} \\ &ClO_2 \longrightarrow ClO_2^- && \text{(reduction)} \end{aligned}$$

$$\begin{aligned} &C + 6\,OH^- \longrightarrow CO_3^{2-} + 3\,H_2O + 4\,e^- && \text{(oxidation)} \\ &ClO_2 + e^- \longrightarrow ClO_2^- && \text{(reduction)} \end{aligned}$$

$$\begin{aligned} &C + 6\,OH^- \longrightarrow CO_3^{2-} + 3\,H_2O + 4\,e^- && \text{(oxidation)} \\ &4\,ClO_2 + 4\,e^- \longrightarrow 4\,ClO_2^- && \text{(reduction)} \end{aligned}$$

Thus, the complete balanced equation is

$$6\,OH^-(aq) + C(s) + 4\,ClO_2(g) \longrightarrow 4\,ClO_2^-(aq) + CO_3^{2-}(aq) + 3\,H_2O(l)$$

or, including the spectator ions,

$$4\,NaOH(aq) + Ca(OH)_2(aq) + C(s) + 4\,ClO_2(g) \longrightarrow 4\,NaClO_2(aq) + CaCO_3(aq) + 3\,H_2O(l)$$

The number of moles of $NaClO_2(aq)$ in 1.00 metric tons is

$$\begin{aligned} \text{moles of NaClO}_2 &= (1.00 \text{ metric ton})\left(\frac{10^3 \text{ kg}}{1 \text{ metric ton}}\right)\left(\frac{10^3 \text{ g}}{1 \text{ kg}}\right)\left(\frac{1 \text{ mol NaClO}_2}{90.44 \text{ g NaClO}_2}\right) \\ &= 1.106 \times 10^4 \text{ mol} \end{aligned}$$

The number of moles of $ClO_2(g)$ required is

$$\text{moles of ClO}_2 = (1.106 \times 10^4 \text{ mol of NaClO}_2)\left(\frac{4 \text{ mol ClO}_2}{4 \text{ mol NaClO}_2}\right) = 1.106 \times 10^4 \text{ mol}$$

The mass of $ClO_2(g)$ required is given by

$$\begin{aligned} \text{mass of ClO}_2 &= (1.106 \times 10^4 \text{ mol ClO}_2)\left(\frac{67.452 \text{ g ClO}_2}{1 \text{ mol ClO}_2}\right) \\ &= 7.46 \times 10^5 \text{ g} = 7.46 \times 10^2 \text{ kg} = 0.746 \text{ metric ton} \end{aligned}$$

24-48. The two half reaction equations balanced with respect to the elements other than oxygen and hydrogen are

$$\begin{aligned} &As_4O_6 \longrightarrow As_4O_{10} \\ &I_3^- \longrightarrow 3\,I^- \end{aligned}$$

The steps in balancing the first half reaction are

$$\begin{aligned} &As_4O_6 + 4\,H_2O \longrightarrow As_4O_{10} + 8\,H^+ \\ &As_4O_6 + 4\,H_2O \longrightarrow As_4O_{10} + 8\,H^+ + 8\,e^- \end{aligned}$$

The steps in balancing the second half reaction are

$$I_3^- + 2e^- \longrightarrow 3I^-$$

Multiply the second half reaction by 4 and add the two half reactions. Add the phases to obtain

$$4I_3^-(aq) + As_4O_6(s) + 4H_2O(l) \longrightarrow 12I^-(aq) + As_4O_{10}(s) + 8H^+(aq)$$

A 0.1021 gram sample of $As_4O_6(s)$ corresponds to

$$\text{millimoles of } As_4O_6 = (0.1021\text{ g})\left(\frac{1\text{ mol }As_4O_6}{395.68\text{ g }As_4O_6}\right)\left(\frac{10^3\text{ mmol}}{1\text{ mol}}\right) = 0.2580\text{ mmol}$$

According to the balanced equation for the reaction, it requires

$$\text{millimoles of } I_3^- = (0.2580\text{ mmol }As_4O_6)\left(\frac{4\text{ mmol }I_3^-}{1\text{ mmol }As_4O_6}\right) = 1.032\text{ mmol}$$

The concentration of I_3^- is

$$M = \frac{1.032\text{ mmol}}{36.55\text{ mL}} = 0.02824\text{ M}$$

ADDITIONAL PROBLEMS

24-50. The species that loses electrons in an oxidation-reduction reaction is oxidized. The reactant that contains the species that gains electrons (is reduced) is the oxidizing agent.

24-52. No. Although sodium is a more reactive metal than iron, it is so reactive that it reacts violently with water.

24-54. The oxidation states of the nitrogen atoms are as follows:

(a) NH_3: H(+1); N(−3) (b) N_2: N(0) (c) NH_2: H(+1); N(−2)

(d) N_2O:N(+1); O(−2) (e) NO: N(+2); O(−2)

24-56. We use the rules given in Section 24-1.

(a) By analogy with oxygen, the oxidation state of a Se atom is −2 by rule 7, and the oxidation state of a Mo atom is given by $x + (2)(-2) = 0$ or $x = +4$ by rule 2.

(b) By analogy with oxygen, the oxidation state of a S atom is −2 by rule 7, and the oxidation state of a Si atom is given by $x + (2)(-2) = 0$ or $x = +4$ by rule 2. We cannot use carbon as an analogy for silicon because carbon has various oxidation states.

(c) By analogy with aluminum, the oxidation state of a Ga atom is +3 and the oxidation state of an As atom is $x + (+3) = 0$ or $x = -3$ by rule 2. We cannot use nitrogen or phosphorus as an analogy for arsenic because both nitrogen and phosphorus have various oxidation states.

(d) The oxidation state of a K atom is +1 by rule 3; the oxidation state of an O atom is −2 by rule 7; and the oxidation state of a S atom is given by $(2)(+1) + 2x + (3)(-2) = 0$ or $x = +2$ by rule 2.

24-58. The two half reactions are

$$Tl^{+} \longrightarrow Tl^{3+}$$
$$IO_3^- + Cl^- \longrightarrow ICl_2^-$$

The steps in balancing the oxidation half reaction equation are

$$Tl^{+} \longrightarrow Tl^{3+} + 2\,e^-$$

The steps in balancing the reduction half reaction equation are

$$IO_3^- + 2\,Cl^- \longrightarrow ICl_2^-$$
$$IO_3^- + 2\,Cl^- \longrightarrow ICl_2^- + 3\,H_2O$$
$$IO_3^- + 2\,Cl^- + 6\,H^+ \longrightarrow ICl_2^- + 3\,H_2O$$
$$IO_3^- + 2\,Cl^- + 6\,H^+ + 4\,e^- \longrightarrow ICl_2^- + 3\,H_2O$$

Multiply the oxidation half reaction equation by 2, add the two half reaction equations, and add the phases to obtain.

$$2\,Tl^{+}(aq) + IO_3^-(aq) + 2\,Cl^-(aq) + 6\,H^+(aq) \longrightarrow 2\,Tl^{3+}(aq) + ICl_2^-(aq) + 3\,H_2O(l)$$

24-60. The two half reaction equations balanced with respect to the elements other than oxygen and hydrogen are

$$S_2O_8^{2-} \longrightarrow 2\,SO_4^{2-}$$
$$\longrightarrow O_2$$

The steps in balancing the reduction half reaction equation are

$$S_2O_8^{2-} \longrightarrow 2\,SO_4^{2-}$$
$$S_2O_8^{2-} + 2\,e^- \longrightarrow 2\,SO_4^{2-}$$

The steps in balancing the oxidation half reaction equation are

$$2\,H_2O \longrightarrow O_2$$
$$2\,H_2O \longrightarrow O_2 + 4\,H^+$$
$$2\,H_2O \longrightarrow O_2 + 4\,H^+ + 4\,e^-$$

Multiply the reduction half-reaction equation by 2, add the two half reaction equations, and add the phases to obtain

$$2\,S_2O_8^{2-}(aq) + 2\,H_2O(l) \longrightarrow 4\,SO_4^{2-}(aq) + O_2(g) + 4\,H^+(aq)$$

24-62. The two half reaction equations balanced with respect to the elements other than oxygen and hydrogen are

$$KMnO_4 \longrightarrow Mn^{2+} + K^+$$
$$Na_2C_2O_4 \longrightarrow CO_2 + 2\,Na^+$$

The steps in balancing the reduction half reaction equation are

$$\mathrm{KMnO_4 \longrightarrow Mn^{2+} + K^+ + 4\,H_2O}$$
$$\mathrm{KMnO_4 + 8\,H^+ \longrightarrow Mn^{2+} + K^+ + 4\,H_2O}$$
$$\mathrm{KMnO_4 + 8\,H^+ + 5\,e^- \longrightarrow Mn^{2+} + K^+ + 4\,H_2O}$$

The steps in balancing the oxidation half reaction equation are

$$\mathrm{Na_2C_2O_4 \longrightarrow 2\,CO_2 + 2\,Na^+ + 2\,e^-}$$

Multiply the reduction half reaction equation by 2 and the oxidation half reaction equation by 5 to obtain

$$\mathrm{2\,KMnO_4 + 16\,H^+ + 10\,e^- \longrightarrow 2\,Mn^{2+} + 2\,K^+ + 8\,H_2O}$$
$$\mathrm{5\,Na_2C_2O_4 \longrightarrow 10\,CO_2 + 10\,Na^+ + 10\,e^-}$$

Add the two half reaction equations to obtain

$$\mathrm{2\,KMnO_4 + 16\,H^+ + 5\,Na_2C_2O_4 \longrightarrow 2\,Mn^{2+} + 2\,K^+ + 8\,H_2O + 10\,CO_2 + 10\,Na^+}$$

Add the phases to obtain

$$\mathrm{2\,KMnO_4}(aq) + \mathrm{16\,H^+}(aq) + \mathrm{5\,Na_2C_2O_4}(aq) \longrightarrow \mathrm{2\,Mn^{2+}}(aq) + \mathrm{2\,K^+}(aq) + \mathrm{8\,H_2O}(l) + \mathrm{10\,CO_2}(g) + \mathrm{10\,Na^+}(aq)$$

24-64. The two half reaction equations balanced with respect to the elements other than oxygen and hydrogen are

$$\mathrm{MnO_4^- \longrightarrow Mn^{2+}}$$
$$\mathrm{Mo^{3+} \longrightarrow MoO_2^{2+}}$$

The steps in balancing the reduction half reaction equation are

$$\mathrm{MnO_4^- + 8\,H^+ \longrightarrow Mn^{2+} + 4\,H_2O}$$
$$\mathrm{MnO_4^- + 8\,H^+ + 5\,e^- \longrightarrow Mn^{2+} + 4\,H_2O}$$

The steps in balancing the oxidation half reaction equation are

$$\mathrm{Mo^{3+} + 2\,H_2O \longrightarrow MoO_2^{2+} + 4\,H^+}$$
$$\mathrm{Mo^{3+} + 2\,H_2O \longrightarrow MoO_2^{2+} + 4\,H^+ + 3\,e^-}$$

Multiply the reduction half reaction equation by 3 and the oxidation half reaction equation by 5 to obtain

$$\mathrm{3\,MnO_4^- + 24\,H^+ + 15\,e^- \longrightarrow 3\,Mn^{2+} + 12\,H_2O}$$
$$\mathrm{5\,Mo^{3+} + 10\,H_2O \longrightarrow 5\,MoO_2^{2+} + 20\,H^+ + 15\,e^-}$$

Add the two half reaction equations and add the phases to obtain

$$\mathrm{3\,MnO_4^-}(aq) + \mathrm{4\,H^+}(aq) + \mathrm{5\,Mo^{3+}}(aq) \longrightarrow \mathrm{3\,Mn^{2+}}(aq) + \mathrm{5\,MoO_2^{2+}}(aq) + \mathrm{2\,H_2O}(l)$$

The number of millimoles of $MoO_4^{2-}(aq)$ in the original sample is given by

$$\text{millimoles of } MoO_4^{2-} = (20.85\text{ mL})\left(\frac{0.0955\text{ mmol }MnO_4^-}{1\text{ mL}}\right)\left(\frac{5\text{ mmol }Mo^{3+}}{3\text{ mmol }MnO_4^-}\right)$$
$$\times\left(\frac{1\text{ mmol }MoO_4^{2-}}{1\text{ mmol }Mo^{3+}}\right)$$
$$= 3.32\text{ mmol}$$

and so the concentration of $MoO_4^{2-}(aq)$ is

$$M = \frac{3.32\text{ mmol}}{32.15\text{ mL}} = 0.103\text{ M}$$

24-66. The quantity of $I_2(s)$ produced is given by

$$\text{millimoles of } I_2 = (0.06500\text{ M})(12.65\text{ mL})\left(\frac{1\text{ mmol }I_2}{2\text{ mmol }Na_2S_2O_3}\right) = 0.4111\text{ mmol}$$

The quantity of $IO_3^-(aq)$ that produced 0.4111 millimoles of $I_2(s)$ is given by

$$\text{millimoles of } IO_3^- = (0.4111\text{ mmol }I_2)\left(\frac{1\text{ mmol }IO_3^-}{3\text{ mmol }I_2}\right) = 0.1370\text{ mmol}$$

This quantity represents an amount of $IO_3^-(aq)$ that was added in excess. The rest of it is tied up in the $La(IO_3)_3(s)$ precipitate. The total amount of $IO_3^-(aq)$ added was

$$\text{total millimoles of } IO_3^- = (0.1105\text{ M})(40.00\text{ mL}) = 4.420\text{ mmol}$$

The quantity of $IO_3^-(aq)$ that precipitated $La^{3+}(aq)$ was

$$\text{millimoles of } IO_3^- \text{ precipitated} = 4.420\text{ mmol} - 0.1370\text{ mmol} = 4.283\text{ mmol}$$

The equation for the precipitation reaction is

$$La^{3+}(aq) + 3\,IO_3^-(aq) \longrightarrow La(IO_3)_3(s)$$

and so

$$\text{millimoles of } La^{3+} = (4.283\text{ mmol }IO_3^-)\left(\frac{1\text{ mmol }La^{3+}}{3\text{ mmol }IO_3^-}\right) = 1.428\text{ mmol}$$

The quantity of $La_2(SO_4)_3(s)$ in the sample is given by

$$\text{mass of } La_2(SO_4)_3 = (1.428\text{ mmol }La^{3+})\left(\frac{1\text{ mmol }La_2(SO_4)_3}{2\text{ mmol }La^{3+}}\right)\left(\frac{10^{-3}\text{ mol}}{1\text{ mmol}}\right)$$
$$\times\left(\frac{566.0\text{ g }La_2(SO_4)_3}{1\text{ mol }La_2(SO_4)_3}\right)$$
$$= 0.4041\text{ g}$$

and the mass percentage of $La_2(SO_4)_3$ in the original sample is

$$\text{mass \% } La_2(SO_4)_3 = \frac{0.4041\text{ g }La_2(SO_4)_3}{3.651\text{ g sample}} \times 100 = 11.07\%$$

24-68. The quantity of $H_2SO_4(aq)$ neutralized is given by

$$\text{moles of } H_2SO_4 = (0.00250\text{ M})(0.01850\text{ L})\left(\frac{1\text{ mol } H_2SO_4}{2\text{ mol NaOH}}\right)$$
$$= 2.313 \times 10^{-5}\text{ mol}$$

According to the equation for the reaction between $H_2O_2(aq)$ and $SO_2(g)$, this is also equal to the number of moles of $SO_2(g)$. The mass of $SO_2(g)$ is given by

$$\text{mass of } SO_2 = (2.313 \times 10^{-5}\text{ mol})\left(\frac{64.06\text{ g } SO_2}{1\text{ mol } SO_2}\right) = 1.482 \times 10^{-3}\text{ g}$$

and the mass percentage of $SO_2(g)$ in the air sample is

$$\text{mass \% } SO_2 = \frac{1.482 \times 10^{-3}\text{ g}}{812.1\text{ g}} \times 100 = 1.825 \times 10^{-4}\%$$

24-70. The equations for the reactions that take place are

$$H_2S(g) + Cd^{2+}(aq) \longrightarrow CdS(s) + 2\,H^+(aq)$$
$$CdS(s) + I_2(aq) \longrightarrow CdI_2(aq) + S(s)$$
$$I_2(aq) + 2\,Na_2S_2O_3(aq) \longrightarrow Na_2S_4O_6(aq) + 2\,NaI(aq)$$

The excess of $I_2(aq)$ is given by

$$\text{excess millimoles of } I_2 = (0.0750\text{ M})(7.65\text{ mL})\left(\frac{1\text{ mmol } I_2}{2\text{ mmol } Na_2S_2O_3}\right)$$
$$= 0.287\text{ mmol}$$

The total quantity of $I_2(s)$ used is

$$\text{total millimoles of } I_2 = (0.0115\text{ M})(30.00\text{ mL}) = 0.345\text{ mmol}$$

and the quantity of $I_2(s)$ that reacts with the CdS(s) is

$$\text{millimoles of } I_2 \text{ reacted} = 0.345\text{ mmol} - 0.287\text{ mmol} = 0.058\text{ mmol}$$

According to the above equations, this is equal to the number of millimoles of $H_2S(s)$, and so the mass of $H_2S(g)$ in the air is

$$\text{mass of } H_2S = (0.058\text{ mmol})\left(\frac{10^{-3}\text{ mol}}{1\text{ mmol}}\right)\left(\frac{34.08\text{ g } H_2S}{1\text{ mol } H_2S}\right)$$
$$= 1.98 \times 10^{-3}\text{ g}$$

and the mass percentage of $H_2S(g)$ in the air is

$$\text{mass \% } H_2S = \left(\frac{1.98 \times 10^{-3}\text{ g}}{10.75\text{ g}}\right) \times 100 = 0.018\%$$

24-72.* Titrate with $KMnO_4(aq)$ to determine the amount of $Fe^{2+}(aq)$. Then use a reducing agent such as powdered zinc to change all the $Fe^{3+}(aq)$ back to $Fe^{2+}(aq)$ and titrate again.

CHAPTER 25. # Electrochemistry

CELL SETUPS

25-2. The cell diagram is sketched below. Electrons flow from the negative electrode to the positive electrode in the external circuit. The reaction at the negative electrode is

$$Mn(s) \longrightarrow Mn^{2+}(aq) + 2e^-$$

Positive ions, $Mn^{2+}(aq)$, are produced at the negative electrode. The reaction at the positive electrode is

$$Cr^{2+}(aq) + 2e^- \longrightarrow 2\,Cr(s)$$

Positive ions, $Cr^{2+}(aq)$, are consumed at the positive electrode. We write the cell diagram with the oxidation half reaction occurring at the left electrode and the reduction half reaction occurring at the right electrode. Thus, the cell diagram is

$$Mn(s)|MnSO_4(aq)||CrSO_4(aq)|Cr(s)$$

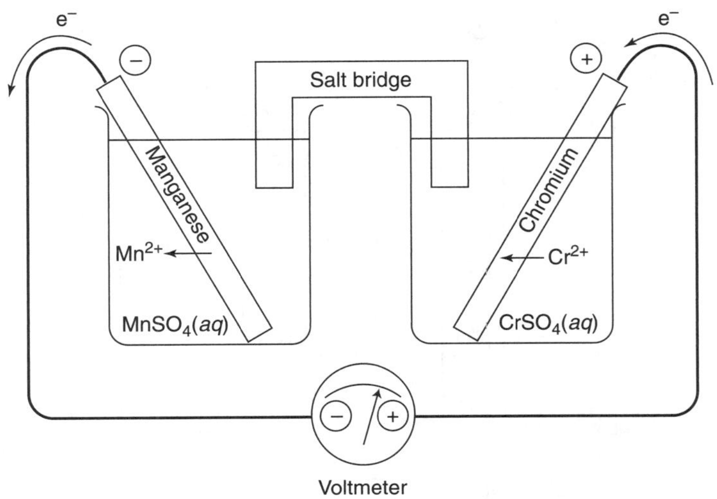

25-4. The cell diagram is sketched below. The reaction at the negative electrode is

$$Co(s) \longrightarrow Co^{2+}(aq) + 2e^-$$

The reaction at the positive electrode is

$$Pb^{2+}(aq) + 2e^- \longrightarrow Pb(s)$$

The cell diagram is

$$Co(s)|Co(NO_3)_2(aq)||Pb(NO_3)_2(aq)|Pb(s)$$

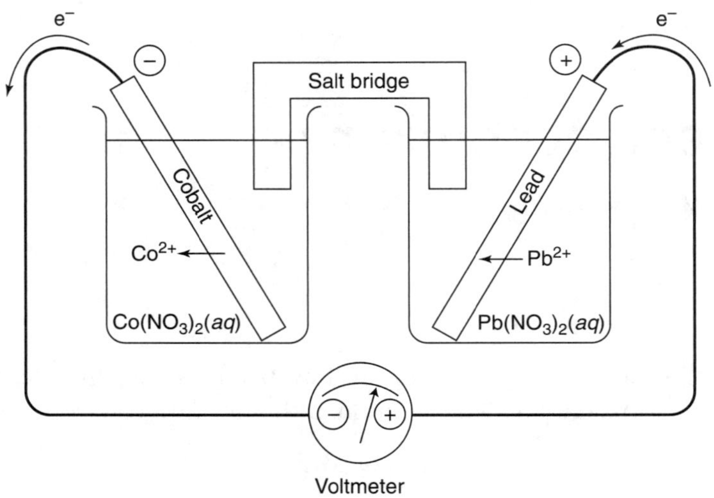

25-6. The cell diagram is sketched below. The reaction at the negative electrode is

$$Zn(s) \longrightarrow Zn^{2+}(aq) + 2e^-$$

The reaction at the positive electrode is

$$Hg_2Cl_2(s) + 2e^- \longrightarrow 2\,Hg(l) + 2\,Cl^-(aq)$$

The cell diagram is

$$Zn(s)|ZnCl_2(aq)|Hg_2Cl_2(aq)|Hg(l)|Pt(s)$$

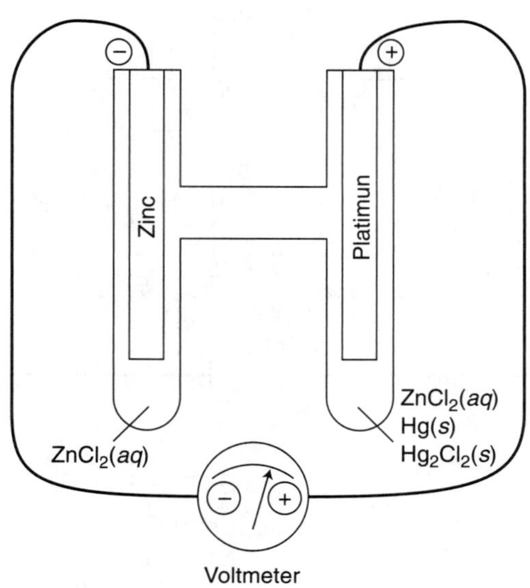

CELL DIAGRAMS

25-8. Oxidation takes place at the left electrode; thus,

$$Sn(s) \longrightarrow Sn^{2+}(aq) + 2\,e^-$$

The equation for the reduction half reaction at the right electrode is

$$Ag^+(aq) + e^- \longrightarrow Ag(s)$$

The net cell reaction equation is the sum of the two electrode half reaction equations. Multiplying the reduction half reaction equation by two and adding the oxidation half reaction equation yields the net cell reaction equation.

$$Sn(s) + 2\,Ag^+(aq) \longrightarrow Sn^{2+}(aq) + 2\,Ag(s)$$

25-10. Oxidation takes place at the left electrode. Thus, the equation for the half reaction at the left electrode is

$$Cu(s) \longrightarrow Cu^{2+}(aq) + 2\,e^-$$

The equation for the reduction half reaction at the right electrode is

$$Ag^+(aq) + e^- \longrightarrow Ag(s)$$

The net cell reaction equation is the sum of the two electrode half reaction equations. Multiplying the reduction half reaction equation by two and adding the oxidation half reaction equation yields the net cell reaction equation.

$$Cu(s) + 2\,Ag^+(aq) \longrightarrow Cu^{2+}(aq) + 2\,Ag(s)$$

25-12. Oxidation takes place at the left electrode; thus,

$$Pb(s) + SO_4^{2-}(aq) \longrightarrow PbSO_4(s) + 2\,e^-$$

The $SO_4^{2-}(aq)$ comes from the $K_2SO_4(aq)$ solution in contact with $Pb(s)$ and $PbSO_4(s)$. The equation for the reduction half reaction at the right electrode is

$$Hg_2SO_4(s) + 2\,e^- \longrightarrow 2\,Hg(l) + SO_4^{2-}(aq)$$

The net cell reaction is

$$Pb(s) + Hg_2SO_4(s) \longrightarrow 2\,Hg(l) + PbSO_4(s)$$

ELECTROCHEMICAL CELLS AND LE CHÂTELIER'S PRINCIPLE

25-14. (a) An increase in the concentration of $Ag^+(aq)$ drives the reaction from left to right; thus, the cell voltage increases.

(b) An increase in $PbSO_4(s)$ has no effect on the cell voltage.

(c) An increase in the concentration of $SO_4^{2-}(aq)$ increases the driving force of the cell reaction from left to right; thus, the cell voltage increases.

(d) No change. This increases the cell current, not the cell voltage.

25-16. (a) An increase in the amount of Pb(*s*) has no effect on the cell voltage.

(b) A decrease in the pH corresponds to an increase in $[H^+]$, which drives the reaction from right to left. The cell voltage decreases.

(c) A dilution of the cell electrolyte decreases $[H^+]$ and $[SO_4^{2-}]$, which drives the reaction from left to right. The cell voltage increases.

(d) Added NaOH(*s*) reacts with $H^+(aq)$ thereby causing a decrease in $[H^+]$, which drives the reaction from left to right. The cell voltage increases.

25-18. (a) An increase in $[Ag^+]$ drives the reaction from left to right. The cell voltage increases.

(b) An increase in $[Fe^{3+}]$drives the reaction from right to left. The cell voltage decreases.

(c) A two-fold decrease in both $[Fe^{3+}]$ and $[Fe^{2+}]$ does not change the value of Q and thus has no effect on the cell voltage.

(d) A decrease in Ag(*s*) has no effect on the cell voltage.

(e) A decrease in $[Fe^{2+}]$ drives the reaction from right to left. The cell voltage decreases.

(f) The added NaCl(*s*) precipitates the $Ag^+(aq)$ as AgCl(*s*) thereby decreasing $[Ag^+]$, which drives the reaction from right to left. The cell voltage decreases.

THE NERNST EQUATION

25-20. (a) The two half reaction equations are

$$\mathrm{Cu}(s) \longrightarrow \mathrm{Cu}^{2+}(aq) + 2\,\mathrm{e}^-$$

$$\mathrm{Mg}^{2+}(aq) + 2\,\mathrm{e}^- \longrightarrow \mathrm{Mg}(s)$$

Thus, $\nu_e = 2$.

(b) The two half reaction equations are

$$2\,\mathrm{Na}(s) \longrightarrow 2\,\mathrm{Na}^+(aq) + 2\,\mathrm{e}^-$$

$$2\,\mathrm{H_2O}(l) + 2\,\mathrm{e}^- \longrightarrow \mathrm{H_2}(g) + 2\,\mathrm{OH}^-(aq)$$

Thus, $\nu_e = 2$.

25-22. We shall use the Nernst equation in the form of Equation 25.10

$$\ln K = \frac{\nu_e E^\circ_{\mathrm{cell}}}{0.02570\,\mathrm{V}}$$

where K is the (unitless) thermodynamic equilibrium constant. The value of ν_e for the reaction is 2, and thus,

$$\ln K = \frac{(2)\,(0.22\ \mathrm{V})}{0.02570\,\mathrm{V}} = 17$$

Solving for K yields

$$K = e^{17} = 2.7 \times 10^7$$

25-24. We first calculate E°_{cell} by using the Nernst equation, Equation 25.13

$$E_{\text{cell}} = E^\circ_{\text{cell}} - \left(\frac{0.02570\,\text{V}}{\nu_e}\right) \ln Q$$

where Q is the (dimensionless) thermodynamic reaction quotient. Thus, using the given equation for the reaction,

$$E_{\text{cell}} = E^\circ_{\text{cell}} - \left(\frac{0.02570\,\text{V}}{\nu_e}\right) \ln \frac{[\text{Co}^{2+}]/\text{M}}{[\text{Sn}^{2+}]/\text{M}}$$

For this reaction $\nu_e = 2$. Thus, using the given concentrations, we have

$$0.168\ \text{V} = E^\circ_{\text{cell}} - \left(\frac{0.02570\,\text{V}}{2}\right) \ln \frac{0.020}{0.18}$$

$$0.168\ \text{V} = E^\circ_{\text{cell}} + 0.028\ \text{V}$$

Solving for E°_{cell},

$$E^\circ_{\text{cell}} = 0.140\ \text{V}$$

Using Equation 25.10,

$$\ln K = \frac{\nu_e E^\circ_{\text{cell}}}{0.02570\,\text{V}}$$

and the above value of E°_{cell}, we compute

$$\ln K = \frac{(2)(0.140\ \text{V})}{0.0592\ \text{V}} = 10.9$$

Thus,

$$K = e^{10.9} = 5.4 \times 10^4$$

25-26. Application of the Nernst equation, Equation 25.13, to the cell reaction equation yields

$$E_{\text{cell}} = E^\circ_{\text{cell}} - \left(\frac{0.02570\,\text{V}}{\nu_e}\right) \ln \frac{[\text{Zn}^{2+}]/\text{M}}{[\text{Hg}_2^{2+}]/\text{M}}$$

Substituting in the given concentrations with $\nu_e = 2$ yields

$$1.553\ \text{V} = E^\circ_{\text{cell}} - \left(\frac{0.02570\,\text{V}}{2}\right) \ln \frac{0.50}{0.30}$$

$$1.553\ \text{V} = E^\circ_{\text{cell}} - 0.0066\ \text{V}$$

Solving for E°_{cell},

$$E^\circ_{\text{cell}} = 1.560\ \text{V}$$

Using Equation 25.10,

$$\ln K = \frac{\nu_e E^\circ_{\text{cell}}}{0.02570\,\text{V}}$$

and the above value of E°_{cell}, we compute

$$\ln K = \frac{(2)(1.560\text{ V})}{0.0592\text{ V}} = 52.70$$

Thus,

$$K = 10^{52.70} = 5.0 \times 10^{52}$$

25-28. The oxidation half reaction equation is

$$H_2(g) \longrightarrow 2\,H^+(aq) + 2\,e^-$$

The reduction half reaction equation is

$$Cd^{2+}(aq) + 2\,e^- \longrightarrow Cd(s)$$

The cell reaction equation is

$$H_2(g) + Cd^{2+}(aq) \longrightarrow 2\,H^+(aq) + Cd(s)$$

Application of the Nernst equation, Equation 25.13, to the cell reaction equation yields

$$E_{\text{cell}} = E^\circ_{\text{cell}} - \left(\frac{0.02570\,\text{V}}{\nu_e}\right) \ln \frac{[H^+]^2/\text{M}^2}{(P_{H_2}/\text{bar})([Cd^{2+}]/\text{M})}$$

$$\begin{aligned} E_{\text{cell}} &= -0.403\text{ V} - \left(\frac{0.02570\,\text{V}}{2}\right) \ln \frac{(0.10)^2}{(0.10)(2.5 \times 10^{-3})} \\ &= -0.403\text{ V} - 0.0474\text{ V} = -0.450\text{ V} \end{aligned}$$

25-30. The equation for the half reaction at the left electrode (oxidation) is

$$H_2(g) \longrightarrow 2\,H^+(\text{aq, }0.030\text{ M}) + 2\,e^-$$

The equation for half reaction at the right electrode (reduction) is

$$2\,H^+(\text{aq, }0.250\text{ M}) + 2\,e^- \longrightarrow 2\,H_2(g)$$

The equation for the overall reaction is

$$H^+(\text{aq, }0.250\text{ M}) \longrightarrow H^+(\text{aq, }0.030\text{ M}) \quad \text{(a concentration cell)}$$

Application of the Nernst equation, Equation 25.13, to the cell reaction yields

$$\begin{aligned} E_{\text{cell}} &= E^\circ_{\text{cell}} - \left(\frac{0.02570\,\text{V}}{2}\right) \ln \frac{0.030}{0.250} \\ &= 0\text{ V} + 0.0272\text{ V} = 0.0272\text{ V} \end{aligned}$$

CONCENTRATIONS FROM CELL MEASUREMENTS

25-32. The equation for the reaction that takes place at the left electrode (oxidation) is

$$\mathrm{Zn}(s) \longrightarrow \mathrm{Zn}^{2+}(aq, 0.100\ \mathrm{M}) + 2\,\mathrm{e}^-$$

and that at the right electrode (reduction) is

$$\mathrm{Zn}^{2+}(aq, c) + 2\,\mathrm{e}^- \longrightarrow \mathrm{Zn}(s)$$

The cell reaction equation is

$$\mathrm{Zn}^{2+}(aq, c) \longrightarrow \mathrm{Zn}^{2+}(aq, 0.100\ \mathrm{M})$$

Application of the Nernst equation, Equation 25.13, to the cell reaction ($\nu_e = 2$) yields

$$E_{\mathrm{cell}} = E^{\circ}_{\mathrm{cell}} - \left(\frac{0.02570\ \mathrm{V}}{2}\right) \ln \frac{0.100}{c/\mathrm{M}}$$

or

$$20.0\ \mathrm{mV} = -(12.9\ \mathrm{mV}) \ln \frac{0.100}{c/\mathrm{M}}$$

or

$$\ln \frac{c/\mathrm{M}}{0.100} = \frac{20.0}{12.9} = 1.55$$

or

$$c = 0.47\ \mathrm{M}.$$

25-34. Oxidation occurs at the left electrode

$$\mathrm{Zn}(s) \longrightarrow \mathrm{Zn}^{2+}(aq) + 2\,\mathrm{e}^-$$

Reduction occurs at the right electrode

$$2\,\mathrm{Ag}^+(aq) + 2\,\mathrm{e}^- \longrightarrow 2\,\mathrm{Ag}(s)$$

The cell reaction equation is

$$\mathrm{Zn}(s) + 2\,\mathrm{Ag}^+(aq) \longrightarrow \mathrm{Zn}^{2+}(aq) + 2\,\mathrm{Ag}(s)$$

Application of the Nernst equation, Equation 25.13, to the cell reaction with $\nu_e = 2$ yields

$$E_{\mathrm{cell}} = E^{\circ}_{\mathrm{cell}} - \left(\frac{0.02570\ \mathrm{V}}{2}\right) \ln \frac{[\mathrm{Zn}^{2+}]/\mathrm{M}}{[\mathrm{Ag}^+]^2/\mathrm{M}^2}$$

Thus, using the data given

$$1.502\ \mathrm{V} = 1.561\ \mathrm{V} - \left(\frac{0.02570\ \mathrm{V}}{2}\right) \ln \frac{[\mathrm{Zn}^{2+}]/\mathrm{M}}{(0.10)^2}$$

Thus

$$\ln \frac{[\mathrm{Zn}^{2+}]/\mathrm{M}}{(0.10)^2} = \frac{(2)(1.561\ \mathrm{V} - 1.502\ \mathrm{V})}{0.02570\ \mathrm{V}} = 4.6$$

and

$$[\mathrm{Zn}^{2+}]/\mathrm{M} = (0.10)^2 e^{4.6} = (0.10)^2(99) = 0.99\ \mathrm{M}$$

USE OF TABULATED E°_{red} VALUES

25-36. (a) The reduction half reaction equation is

$$2\,H_2O(l) + 2\,e^- \longrightarrow 2\,OH^-(aq) + H_2(g) \qquad E^\circ_{red} = -0.828\text{ V}$$

The oxidation half reaction equation is

$$2\,Na(s) \longrightarrow 2\,Na^+(aq) + 2\,e^- \qquad E^\circ_{ox} = -E^\circ_{red} = -(-2.71)\text{ V}$$

where the E°_{red} values for the half reaction equations are obtained from Table 25.3 and Appendix G. The value of E°_{cell} for the equation is

$$\begin{aligned} E^\circ_{cell} &= E^\circ_{red}[\,H_2O\,|\,OH^- + H_2] + E^\circ_{ox}[Na\,|\,Na^+] \\ &= -0.828\text{ V} + 2.71\text{ V} = 1.88\text{ V} \end{aligned}$$

(b) The reduction half reaction equation is

$$2\,H^+(aq) + 2\,e^- \longrightarrow H_2(g) \qquad E^\circ_{red} = 0\text{ V}$$

The oxidation half reaction equation is

$$Pd(s) \longrightarrow Pd^{2+}(aq) + 2\,e^- \qquad E^\circ_{ox} = -E^\circ_{red} = -(+0.951\text{ V})$$

where the E°_{red} values for the half reaction equations are obtained from Table 25.3 and Appendix G. The value of E°_{cell} for the equation is

$$\begin{aligned} E^\circ_{cell} &= E^\circ_{red}[\,H^+\,|\,H_2] + E^\circ_{ox}[Pd\,|\,Pd^{2+}] \\ &= 0\text{ V} - 0.951\text{ V} = -0.951\text{ V} \end{aligned}$$

(c) The reduction half reaction equation is

$$O_2(g) + 2\,H_2O(l) + 4\,e^- \longrightarrow 4\,OH^-(aq) \qquad E^\circ_{red} = +0.401\text{ V}$$

The oxidation half reaction equation is

$$4\,Fe(OH)_2(s) + 4\,OH^-(aq) \longrightarrow 4\,Fe(OH)_3(s)(aq) + 4\,e^- \qquad E^\circ_{ox} = -E^\circ_{red} = -(-0.56)\text{ V}$$

where the E°_{red} values for the half reaction equations are obtained from Table 25.3 and Appendix G. The value of E°_{cell} for the equation is

$$\begin{aligned} E^\circ_{cell} &= E^\circ_{red}[\,O_2\,|\,OH^-] + E^\circ_{ox}[Fe(OH)_2\,|\,Fe(OH)_3] \\ &= 0.401\text{ V} + 0.56\text{ V} = 0.96\text{ V} \end{aligned}$$

25-38. The oxidation half reaction equation at the left electrode is

$$Cu^+(aq) \longrightarrow Cu^{2+}(aq) + e^-$$

The reduction half reaction equation at the right electrode is

$$NO_3^-(aq) + 2\,H^+(aq) + e^- \longrightarrow NO_2(g) + H_2O(l)$$

The value of E°_{cell} for the complete cell is given by

$$E^\circ_{cell} = E^\circ_{red}[\,NO_3^-\,|\,NO_2] + E^\circ_{ox}[Cu^{2+}\,|\,Cu^+]$$

From Appendix G we obtain

$$E^\circ_{\text{ox}}[\text{Cu}^{2+} \,|\, \text{Cu}^+] = -E^\circ_{\text{red}}[\text{Cu}^{2+} \,|\, \text{Cu}^+] = -(0.153 \text{ V})$$

Thus,

$$E^\circ_{\text{cell}} = 0.65 \text{ V} = E^\circ_{\text{red}}[\text{NO}_3^- \,|\, \text{NO}_2] - 0.153 \text{ V}$$

$$E^\circ_{\text{red}}[\text{NO}_3^- \,|\, \text{NO}_2] = +0.80 \text{ V}$$

25-40. The oxidation half reaction equation is

$$2\,\text{H}_2\text{O}(l) \longrightarrow \text{H}_2\text{O}_2(aq) + 2\,\text{H}^+(aq) + 2\,\text{e}^-$$

and the reduction half reaction equation is

$$\text{S}_2\text{O}_8^{2-}(aq) + 2\,\text{e}^- \longrightarrow 2\,\text{SO}_4^{2-}(aq)$$

The value of E°_{cell} for the complete cell is given by

$$E^\circ_{\text{cell}} = E^\circ_{\text{red}}[\text{S}_2\text{O}_8^{2-} \,|\, \text{SO}_4^{2-}] + E^\circ_{\text{ox}}[\text{H}_2\text{O} \,|\, \text{H}_2\text{O}_2]$$

From Appendix G, we obtain

$$E^\circ_{\text{cell}} = 2.010 \text{ V} - 1.776 \text{ V} = +0.234 \text{ V}$$

Thus, the net reaction equation

$$\text{S}_2\text{O}_8^{2-}(aq) + \text{H}_2\text{O}(l) \longrightarrow \text{H}_2\text{O}_2(aq) + 2\,\text{SO}_4^{2-}(aq) + 2\,\text{H}^+(aq)$$

has a positive value of E°_{cell} and thus is spontaneous when $Q = 1$. An aqueous solution of potassium peroxodisulfate is not stable over a long period of time because $\text{S}_2\text{O}_8^{2-}(aq)$ is capable of oxidizing water.

25-42. The oxidation half reaction equation is

$$\text{Cu}(s) \longrightarrow \text{Cu}^{2+}(aq) + 2\,\text{e}^-$$

and the reduction half reaction equation is

$$2\,\text{Ag}^+(aq) + 2\,\text{e}^- \longrightarrow 2\,\text{Ag}(s)$$

The value of E°_{cell} for the complete cell is given by

$$E^\circ_{\text{cell}} = E^\circ_{\text{red}}[\text{Ag}^+ \,|\, \text{Ag}] + E^\circ_{\text{ox}}[\text{Cu} \,|\, \text{Cu}^{2+}]$$

From Appendix G, we have

$$E^\circ_{\text{cell}} = 0.7996 \text{ V} - 0.342 \text{ V} = +0.458 \text{ V}$$

From Equation 25.10, we have

$$\ln K = \frac{\nu_e E^\circ_{\text{cell}}}{0.02570 \text{ V}} = \frac{(2)(0.458 \text{ V})}{0.02570 \text{ V}} = 35.6$$

$$K = e^{35.6} = 3.0 \times 10^{15}$$

Application of the Nernst equation to the cell equation with $\nu_e = 2$ yields

$$E_{cell} = 0.458\text{ V} - \left(\frac{0.02570\text{ V}}{2}\right)\ln\frac{[\text{Cu}^{2+}]/\text{M}}{[\text{Ag}^+]^2/\text{M}^2}$$

Thus, using the concentrations given, we obtain

$$E_{cell} = 0.458\text{ V} - \left(\frac{0.02570\text{ V}}{2}\right)\ln\frac{(1.00\times 10^{-4})}{(0.10)^2}$$
$$= 0.458\text{ V} + 0.0592\text{ V} = 0.517\text{ V} > 0$$

Hence, the reaction is spontaneous under the given conditions.

25-44. The oxidation half reaction equation is

$$\text{Cd}(s) \rightleftharpoons \text{Cd}^{2+}(aq) + 2\,\text{e}^-$$

and the reduction half reaction equation is

$$\text{Pb}^{2+}(aq) + 2\,\text{e}^- \rightleftharpoons \text{Pb}(s)$$

The value of E°_{cell} is given by

$$E^\circ_{cell} = E^\circ_{red}[\,\text{Pb}^{2+}\,|\,\text{Pb}] + E^\circ_{ox}[\text{Cd}\,|\,\text{Cd}^{2+}]$$

From Appendix G, we have

$$E^\circ_{cell} = -0.126\text{ V} - (-0.403)\text{ V} = 0.277\text{ V}$$

Application of the Nernst equation to the cell equation with $\nu_e = 2$ yields

$$E_{cell} = E^\circ_{cell} - \left(\frac{0.02570\text{ V}}{2}\right)\ln\frac{[\text{Cd}^{2+}]/\text{M}}{[\text{Pb}^{2+}]/\text{M}}$$

Thus,

$$E_{cell} = 0.277\text{ V} - \left(\frac{0.02570\text{ V}}{2}\right)\ln\frac{0.010}{0.10}$$
$$= 0.277\text{ V} + 0.0296\text{ V} = 0.307\text{ V}$$

25-46. The oxidation half reaction equation in acidic soultion is

$$\text{BH}_4^-(aq) + 3\,\text{H}_2\text{O}(l) \longrightarrow \text{H}_2\text{BO}_3^-(aq) + 8\,\text{H}^+(aq) + 8\,\text{e}^-$$

Because the reaction is carried out in basic solution, we convert the above half reaction equation to basic solution by adding 8 $\text{OH}^-(aq)$ to both sides of the equation to get

$$\text{BH}_4^-(aq) + 8\,\text{OH}^-(aq) \longrightarrow \text{H}_2\text{BO}_3^-(aq) + 5\,\text{H}_2\text{O}(l) + 8\,\text{e}^-$$

The reduction half reaction equation is

$$8\,O_2(g) + 8\,e^- \longrightarrow 8\,O_2^-(aq) \qquad E^\circ_{red}[O_2\,|\,O_2^-] = -0.56\ V$$

We can obtain the value of $E^\circ_{red}[\,H_2BO_3^-\,|\,BH_4^-]$ from the value of E°_{cell}

$$E^\circ_{cell} = E^\circ_{red}[O_2\,|\,O_2^-] + E^\circ_{ox}[\,H_2BO_3^-\,|\,BH_4^-]$$
$$= E^\circ_{red}[O_2\,|\,O_2^-] - E^\circ_{red}[\,H_2BO_3^-\,|\,BH_4^-]$$

Thus,

$$0.68\ V = -0.56\ V - E^\circ_{red}[\,H_2BO_3^-\,|\,BH_4^-]$$

and

$$E^\circ_{red}[\,H_2BO_3^-\,|\,BH_4^-] = -1.24\ V$$

E_{cell} AND SPONTANEITY

25-48. The oxidation half reaction equation is

$$4\,Fe^{2+}(aq) \longrightarrow 4\,Fe^{3+}(aq) + 4\,e^- \qquad E^\circ_{ox} = -(+0.771\ V) = -0.771\ V$$

and the reduction half reaction equation is

$$O_2(g) + 4\,H^+(aq) + 4\,e^- \longrightarrow 2\,H_2O(l) \qquad E^\circ_{red} = +1.229\ V$$

The equation for the complete reaction is

$$4\,Fe^{2+}(aq) + O_2(g) + 4\,H^+(aq) \longrightarrow 4\,Fe^{3+}(aq) + 2\,H_2O(l)$$

and

$$E^\circ_{cell} = 1.229\ V - 0.771\ V = 0.458\ V$$

Application of the Nernst equation to the cell equation with $\nu_e = 4$ yields

$$E_{cell} = 0.458\ V - \left(\frac{0.02570\ V}{4}\right)\ln\frac{[Fe^{3+}]^4/M^4}{([Fe^{2+}]^4/M^4)(P_{O_2}/bar)([H^+]^4/M^4)}$$

Substitution of the given values yields

$$E_{cell} = 0.458\ V - \left(\frac{0.02570\ V}{4}\right)\ln\frac{(0.10)^4}{(0.10)^4(0.20)(1.0\times 10^{-2})^4}$$
$$= 0.458\ V - 0.1287\ V = 0.329\ V > 0$$

The positive value of E_{cell} means that the air oxidation of $Fe^{2+}(aq)$ is spontaneous under the stated conditions.

25-50. The reduction half reaction equation is

$$O_3(g) + H_2O(l) + 2\,e^- \longrightarrow O_2(g) + 2\,OH^-(aq) \qquad E^\circ_{red} = +1.24\text{ V}$$

The oxidation half reaction equation is

$$Cu(s) + 2\,OH^-(aq) \longrightarrow Cu(OH)_2(s) + 2\,e^- \qquad E^\circ_{ox} = -(-0.222\text{ V}) = 0.222\text{ V}$$

The equation for the complete reaction is

$$O_3(g) + H_2O(l) + Cu(s) \longrightarrow Cu(OH)_2(s) + O_2(g)$$

and

$$E^\circ_{cell} = 1.24\text{ V} + 0.222\text{ V} = 1.46\text{ V}$$

Application of the Nernst equation to the cell equation with $\nu_e = 2$ yields

$$E_{cell} = 1.46\text{ V} - \left(\frac{0.02570\text{ V}}{2}\right)\ln\frac{P_{O_2}/\text{bar}}{P_{O_3}/\text{bar}}$$

Thus, we have at $P_{O_2} = 0.20$ bar and $P_{O_3} = 1.00 \times 10^{-4}$ bar,

$$E_{cell} = 1.46\text{ V} - \left(\frac{0.02570\text{ V}}{2}\right)\ln\frac{0.20}{1.00\times10^{-4}}$$
$$= 1.46\text{ V} - 0.09772\text{ V} = 1.36\text{ V}$$

Because $E_{cell} > 0$, the oxidation of $Cu(s)$ by $O_3(g)$ is a spontaneous process under the stated conditions.

CELLS AND ΔG_{rxn}

25-52. The value of ΔG_{rxn} is given by Equation 25.5.

$$\Delta G_{rxn} = -\nu_e F E_{cell}$$

Two moles of electrons are transferred in this reaction; thus, $\nu_e = 2$.

$$\Delta G_{rxn} = -(2)(96\,500\text{ C}\cdot\text{mol}^{-1})(2.03\text{V}) = -392\,000\text{ J}\cdot\text{mol}^{-1} = -392\text{ kJ}\cdot\text{mol}^{-1}$$

25-54. The value of ΔG°_{rxn} is given by

$$\Delta G^\circ_{rxn} = -\nu_e F E^\circ_{cell}$$

In this reaction six moles of iron are oxidized from an oxidation state of +2 to an oxidation state of +3. Thus, the reaction requires six moles of electrons and so $\nu_e = 6$.

$$\Delta G^\circ_{rxn} = -(6)(96\,500\text{ C}\cdot\text{mol}^{-1})(0.461\text{V}) = -267\,000\text{ J}\cdot\text{mol}^{-1} = -267\text{ kJ}\cdot\text{mol}^{-1}$$

25-56. (a) The oxidation half reaction equation is

$$Zn(s) \longrightarrow Zn^{2+}(aq) + 2\,e^-$$

and the reduction half reaction equation is

$$Cu^{2+}(aq) + 2\,e^- \longrightarrow Cu(s)$$

The standard cell voltage E°_{cell} is given by

$$\begin{aligned} E^\circ_{cell} &= E^\circ_{red}[\,Cu^{2+}\,|\,Cu] + E^\circ_{ox}[Zn\,|\,Zn^{2+}] \\ &= 0.342\text{ V} + (0.762\text{ V}) = 1.104\text{ V} \end{aligned}$$

The value of ΔG°_{rxn} is given by

$$\Delta G^\circ_{rxn} = -\nu_e F E^\circ_{cell}$$

Two moles of electrons are transferred in this reaction; thus, $\nu_e = 2$.

$$\Delta G^\circ_{rxn} = -(2)(96\,485\text{ C}\cdot\text{mol}^{-1})(1.104\text{V}) = -213\,000\text{ J}\cdot\text{mol}^{-1} = -213.0\text{ kJ}\cdot\text{mol}^{-1}$$

(b) The oxidation half reaction equation is

$$Ag(s) \longrightarrow Ag^+(aq) + e^-$$

and the reduction half reaction equation is

$$Fe^{3+}(aq) + e^- \longrightarrow Fe^{2+}(aq)$$

The standard cell voltage E°_{cell} is given by

$$\begin{aligned} E^\circ_{cell} &= E^\circ_{red}[\,Fe^{3+}\,|\,Fe^{2+}] + E^\circ_{ox}[Ag\,|\,Ag^+] \\ &= 0.771\text{ V} + (-0.7996\text{ V}) = -0.029\text{ V} \end{aligned}$$

The value of ΔG°_{rxn} is given by

$$\Delta G^\circ_{rxn} = -\nu_e F E^\circ_{cell}$$

One mole of electrons is transferred in this reaction; thus, $\nu_e = 1$.

$$\Delta G^\circ_{rxn} = -(1)(96\,485\text{ C}\cdot\text{mol}^{-1})(-0.029\text{ V}) = 2800\text{ J}\cdot\text{mol}^{-1} = 2.8\text{ kJ}\cdot\text{mol}^{-1}$$

25-58. The equation for the cell equation is

$$Co(s) + 2\,Ag^+(aq) \longrightarrow Co^{2+}(aq) + 2\,Ag(s)$$

The standard cell voltage E°_{cell} is given by

$$\begin{aligned} E^\circ_{cell} &= E^\circ_{red}[\,Ag^+\,|\,Ag] + E^\circ_{ox}[Co\,|\,Co^{2+}] \\ &= 0.7996\text{ V} + 0.28\text{ V} = 1.08\text{ V} \end{aligned}$$

The value of ΔG°_{rxn} is given by

$$\Delta G^\circ_{rxn} = -\nu_e F E^\circ_{cell}$$

Two moles of electrons are transferred in this reaction; thus, $\nu_e = 2$.

$$\Delta G^\circ_{rxn} = -(2)(96\,500\ \text{C}\cdot\text{mol}^{-1})(1.08\ \text{V}) = -208\,000\ \text{J}\cdot\text{mol}^{-1} = -208\ \text{kJ}\cdot\text{mol}^{-1}$$

The value of ΔG_{rxn} is given by

$$\begin{aligned}\Delta G_{rxn} &= \Delta G^\circ_{rxn} + RT\ln Q = \Delta G^\circ_{rxn} + RT\ln\frac{[\text{Co}^{2+}]/\text{M}}{[\text{Ag}^+]^2/\text{M}^2}\\ &= -208\ \text{kJ}\cdot\text{mol}^{-1} + (8.3145\ \text{J}\cdot\text{K}^{-1}\cdot\text{mol}^{-1})(298.2\ \text{K})\ln\frac{0.0155}{(1.50)^2}\\ &= -208\ \text{kJ}\cdot\text{mol}^{-1} - 12.34\ \text{kJ}\cdot\text{mol}^{-1} = -220\ \text{kJ}\cdot\text{mol}^{-1}\end{aligned}$$

The value of E_{cell} for the cell under the conditions given is

$$\begin{aligned}E_{cell} &= \frac{-\Delta G_{rxn}}{\nu_e F}\\ &= -\frac{-220\times10^3\ \text{J}\cdot\text{mol}^{-1}}{(2)(96\,500\ \text{C}\cdot\text{mol}^{-1})}\\ &= 1.14\ \text{V}\end{aligned}$$

FARADAY'S LAWS

25-60. We shall use Equation 25.31:

$$m = \left(\frac{It}{F}\right)\left(\frac{M}{\nu_e}\right)$$

The half reaction equation for the deposition of Be(s) is

$$\text{Be}^{2+}(aq) + 2\,\text{e}^- \longrightarrow \text{Be}(s)$$

and therefore, $\nu_e = 2$. Substitution of the data given into Equation 25.31 yields

$$\begin{aligned}m &= \frac{(5.0\ \text{A})(1.0\ \text{h})\left(\frac{3600\ \text{s}}{1\ \text{h}}\right)\left(\frac{9.012\ \text{g Be}}{1\ \text{mol Be}}\right)}{(9.65\times10^4\ \text{C}\cdot\text{mol}^{-1})(2)}\\ &= 0.84\ \text{g}\end{aligned}$$

In cancelling the units in the expression for *It*, recall that $1\ \text{A} = 1\ \text{C}\cdot\text{s}^{-1}$.

25-62. The mass deposited is calculated using Equation 25.31:

$$m = \left(\frac{It}{F}\right)\left(\frac{M}{\nu_e}\right)$$

The electrode reaction for the deposition of Au(s) is

$$\text{Au(CN)}_2^-(aq) + \text{e}^- \longrightarrow \text{Au}(s) + 2\,\text{CN}^-(aq)$$

and thus, $\nu_e = 1$. The time required to deposit 200.0 milligrams of Au(s) using a current of 30.0 mA is

$$\begin{aligned} t &= \frac{mF\nu_e}{IM} \\ &= \frac{(0.2000\text{ g})(96\,500\text{ C}\cdot\text{mol}^{-1})(1)}{(0.0300\text{ A})(197.0\text{ g}\cdot\text{mol}^{-1})} \\ &= 3266\text{ s} = 54\text{ min} \end{aligned}$$

25-64. The mass of gas evolved is calculated using Equation 25.31:

$$m = \left(\frac{It}{F}\right)\left(\frac{M}{\nu_e}\right)$$

The electrode reaction for the evolution of $O_2(g)$ is

$$2\,H_2O(l) \longrightarrow O_2(g) + 4\,H^+(aq) + 4\,e^-$$

and thus, $\nu_e = 4$. Substitution of the data given into Equation 25.31 yields

$$\begin{aligned} m &= \frac{(30.35\text{ A})(2.00\text{ h})\left(\dfrac{3600\text{ s}}{1\text{ h}}\right)\left(\dfrac{32.00\text{ g O}_2}{1\text{ mol O}_2}\right)}{(9.65\times 10^4\text{ C}\cdot\text{mol}^{-1})(4)} \\ &= 18.1\text{ g} \end{aligned}$$

The number of moles of $O_2(g)$ produced is

$$n = (18.1\text{ g})\left(\frac{1\text{ mol O}_2}{32.00\text{ g O}_2}\right) = 0.566\text{ mol}$$

The volume of $O_2(g)$ produced is calculated by using the ideal-gas equation

$$\begin{aligned} V &= \frac{nRT}{P} \\ &= \frac{(0.566\text{ mol})(0.0821\text{ L}\cdot\text{atm}\cdot\text{K}^{-1}\cdot\text{mol}^{-1})(298.2\text{ K})}{1.00\text{ atm}} \\ &= 13.9\text{ L} \end{aligned}$$

ADDITIONAL PROBLEMS

25-66. The salt bridge provides an electrical connection between the two half reactions in the cell. It permits the internal passage of the electrical current carried by the ions to occur and at the same time prevents the cell from internally short-circuiting. If the salt bridge is removed, the cell will not produce a voltage.

25-68. The reaction carried out in an electrochemical cell spontaneously produces a current. The reaction carried out in an electrolytic cell is not spontaneous and requires an input of an electrical current from an external source to occur.

25-70. We have for Table 10.8,

Metal	reactivity	$E^\circ_{\text{red}}/\text{V}$
Li		-3.041
K	react directly with cold	-2.931
Ba	water and vigorously	-2.912
Ca	with dilute acids to produce H_2	-2.868
Na		-2.71
Mg		-2.372
Al		-1.662
Mn	react with hot water and	-1.185
Zn	acids to produce H_2	-0.7618
Cr		$-0.744\ (+3 \rightarrow 0)$ or $-0.913\ (+2 \rightarrow 0)$
Fe		$-0.447\ (+2 \rightarrow 0)$ or $-0.037\ (+3 \rightarrow 0)$
Co		-0.28
Ni	react with acids to	-0.257
Sn	produce H_2	-0.1375
Pb		-0.1262
Cu		$+0.3419$
Hg	do not react with water	$+0.851\ (+2 \rightarrow 0)$ or $+0.7973\ (+1 \rightarrow 0)$
Ag	or acids to produce H_2	$+0.7996$

The standard reduction potential for cerium is -2.336 V, which lies in between that of magnesium and aluminum.

25-72. The number of coulombs involved is given by

$$\text{charge} = It = (0.10 \times 10^{-3}\ \text{A})(1000\ \text{h})(3600\ \text{s}\cdot\text{h}^{-1}) = 360\ \text{C}$$

The half reaction equation (from Appendix G) is

$$\text{Ag}_2\text{O}(s) + \text{H}_2\text{O}(l) + 2\,\text{e}^- \rightleftharpoons 2\,\text{Ag}(s) + 2\,\text{OH}^-(aq)$$

The mass of silver deposited is given by

$$\text{mass} = \left(\frac{360\ \text{C}}{96\ 500\ \text{C}\cdot\text{mol}^{-1}}\right)\left(\frac{2\ \text{mol Ag}}{2\ \text{mol e}^-}\right)(107.9\ \text{g}\cdot\text{mol}^{-1}) = 0.402\ \text{g}$$

We could also have used Equation 25.31 with $\nu_e = 1$ because ν_e is the coefficient of the electrons in the half reaction required to produce *one* formula unit of substance (in this case, silver metal), and so we must divide the stated equation through by two.

25-74. The equation for the electrochemical production of aluminum is

$$2\,\text{Al}_2\text{O}_3(soln) + 3\,\text{C}(s) \longrightarrow 4\,\text{Al}(l) + 3\,\text{CO}_2(g)$$

(a) The fraction of Al(s) by mass in $\text{Al}_2\text{O}_3(s)$ is

$$\text{fraction of Al} = \frac{(2)(26.981)}{(2)(26.981) + (3)(15.9994)} = \frac{53.96}{101.96} = 0.529$$

The mass of $Al_2O_3(s)$ required to produce 2.0×10^6 metric tons of $Al(s)$ is

$$\text{mass Al}_2\text{O}_3 = \frac{(2.0 \times 10^6 \text{ metric ton})\left(\dfrac{1000 \text{ kg}}{1 \text{ metric ton}}\right)}{0.529}$$

$$= 3.8 \times 10^9 \text{ kg}$$

The mass of bauxite is

$$(0.55)\,(\text{mass of bauxite}) = \text{mass of Al}_2\text{O}_3 = 3.8 \times 10^9 \text{ kg}$$

$$\text{mass of bauxite} = \left(\frac{3.8 \times 10^9 \text{ kg}}{0.55}\right)\left(\frac{1 \text{ metric ton}}{1000 \text{ kg}}\right) = 6.9 \times 10^6 \text{ metric ton}$$

(b) The number of moles of Al(s) in two million metric tons of Al(s) is

$$\text{moles of Al} = (2.0 \times 10^6 \text{ metric ton})\left(\frac{1000 \text{ kg}}{1 \text{ metric ton}}\right)\left(\frac{1000 \text{ g}}{1 \text{ kg}}\right)\left(\frac{1 \text{ mol Al}}{26.98 \text{ g Al}}\right)$$

$$= 7.4 \times 10^{10} \text{ mol}$$

Each atom of aluminum requires three electrons, so the number of coulombs required to deposit 7.4×10^{10} mol of $Al(s)$ is given by

$$\text{number of coulombs required} = (3)\,(7.4 \times 10^{10} \text{ mol})\,(96\,500 \text{ C·mol}^{-1}) = 2.1 \times 10^{16} \text{ C}$$

For a cell voltage of 5.0 V, this amounts to $(1 \text{ J} = 1 \text{ C·V})$

$$\text{number of joules required} = (5.0 \text{ V})\,(2.1 \times 10^{16} \text{ C}) = 1.1 \times 10^{17} \text{ J} = 1.1 \times 10^{14} \text{ kJ}$$

(c) The fraction of total energy generated this represents is given by

$$\text{fraction} = \frac{\text{energy to produce Al}}{\text{total energy produced}} = \frac{1.1 \times 10^{14} \text{ kJ}}{13 \times 10^{15} \text{ kJ}} = 0.0085 = 0.85\%$$

25-76. Copper does not react with acids to produce $H_2(g)$ because its standard reduction potential is positive (+0.342 V). (See also Table 10.8.) It does react with nitric acid, however, because nitric acid acts as an oxidizing agent according to

$$3\,\text{Cu}(s) + 8\,\text{HNO}_3(aq) \longrightarrow 3\,\text{Cu(NO}_3)_2(aq) + 2\,\text{NO}(g) + 4\,\text{H}_2\text{O}(l) \quad E^\circ_{\text{cell}} = 0.615 \text{ V}$$

Zinc, however, does react with acids to produce $H_2(g)$, and so we have

$$\text{Zn}(s) + 2\,\text{HCl}(aq) \longrightarrow \text{ZnCl}_2(aq) + \text{H}_2(g)$$

Thus, the core of a modern penny that is scratched would be removed leaving behind a thin copper metal shell.

25-78. The oxidation half reaction equation is

$$\text{Ag}(s) + \text{Br}^-(aq) \longrightarrow \text{AgBr}(s) + \text{e}^-$$

and the reduction half reaction equation is

$$Ag^+(aq) + e^- \longrightarrow Ag(s)$$

The equation for the net cell reaction is

$$Ag^+(aq) + Br^-(aq) \longrightarrow AgBr(s)$$

and thus $\nu_e = 1$. We shall use Equation 25.10

$$\ln K = \frac{\nu_e E^\circ_{cell}}{0.02570\ \text{V}}$$

where

$$K = \frac{1}{[Ag^+]\,[Br^-]/M^2}$$

Substituting in the value of E°_{cell} yields

$$\ln K = \frac{(1)\,(0.728\ \text{V})}{0.02570\ \text{V}} = 28.3$$

$$K = e^{28.3} = 2.0 \times 10^{12}$$

$$K_{sp} = \frac{M^2}{K} = \frac{M^2}{2.0 \times 10^{12}} = 5.0 \times 10^{-13}\ M^2$$

25-80. The value of ΔG°_{rxn} is given by

$$\begin{aligned}\Delta G^\circ_{rxn} &= 3\,\Delta G^\circ_f[H_2O(l)] + 2\,\Delta G^\circ_f[CO_2(g)] \\ &\quad - \Delta G^\circ_f[CH_3CH_2OH(l)] - 3\,\Delta G^\circ_f[O_2(g)] \\ &= 3(-237.1\ \text{kJ}\cdot\text{mol}^{-1}) + 2(-394.4\ \text{kJ}\cdot\text{mol}^{-1}) - (-174.8\ \text{kJ}\cdot\text{mol}^{-1}) - 0\ \text{kJ}\cdot\text{mol}^{-1} \\ &= -1325.3\ \text{kJ}\cdot\text{mol}^{-1}\end{aligned}$$

The unbalanced equations for the two half reactions are

$$CH_3CH_2OH(l) \longrightarrow 2\,CO_2(g) \quad \text{(oxidation)}$$
$$O_2(g) \longrightarrow H_2O(l) \quad \text{(reduction)}$$

The balanced equations are

$$CH_3CH_2OH(l) + 3\,H_2O(l) \longrightarrow 2\,CO_2(g) + 12\,H^+(aq) + 12\,e^-$$

$$3\ O_2(g) + 12\,H^+(aq) + 12\,e^- \longrightarrow 6\,H_2O(l)$$

Therefore, a total of 12 electrons are transferred for the reaction as described by

$$CH_3CH_2OH(l) + 3\,O_2(g) \longrightarrow 2\,CO_2(g) + 3\,H_2O(l)$$

and so

$$E^{\circ}_{\text{cell}} = -\frac{\Delta G^{\circ}_{\text{rxn}}}{\nu_e F} = \frac{1325.3 \times 10^3\ \text{J}\cdot\text{mol}^{-1}}{(12)(96\,485\ \text{C}\cdot\text{mol}^{-1})} = 1.1447\ \text{V}$$

25-82. The voltage measured between the Zn(s) and Cu(s) rods will be zero because the cell is short-circuited. The reducing agent Zn(s) must be physically separated from the oxidizing agent $CuSO_4(aq)$. In this case, the Zn(s) rod is in the $Cu^{2+}(aq)$ solution.

25-84. We shall use the Nernst equation in the form given by Equation 25.6,

$$E_{\text{cell}} = -\left(\frac{RT}{\nu_e F}\right)\ln\frac{Q}{K}$$

The cell reaction will take place until $Q = K$, which occurs when equilibrium has been reached:

$$E_{\text{cell}} = -\left(\frac{RT}{\nu_e F}\right)\ln 1 = 0$$

Thus, at equilibrium, the cell voltage is zero and the cell is dead.

25-86.* The equation for the half reaction in which $O_2(g)$ is evolved is (see Appendix G)

$$2\,H_2O(l) \longrightarrow O_2(g) + 4\,H^+(aq) + 4\,e^- \qquad \text{(oxidation)}$$

and the equation for half reaction of the deposition of Ni(s) is

$$Ni^{+x}(aq) + (x)\,e^- \longrightarrow Ni(s) \qquad \text{(reduction)}$$

Now

$$39.12\ \text{g Ni}(s) \approx 0.6665\ \text{mol Ni}(s)$$

and

$$16.00\ \text{g O}_2(g) \approx 0.5000\ \text{mol O}_2(g)$$

Multiplying the oxidation half reaction equation by x and the reduction half reaction equation by 4, we obtain the balanced equation for the overall reaction

$$4\,Ni^{+x}(aq) + 2x\,H_2O(l) \rightleftharpoons x\,O_2(g) + 4\,Ni(s) + 4x\,H^+(aq)$$

Thus, we see that

$$\frac{x\ \text{mol O}_2(g)}{4\ \text{mol Ni}(s)} = \frac{0.5000\ \text{mol O}_2(g)}{0.6665\ \text{mol Ni}(s)}$$

or $x = 3$.

25-88.* The oxidation half reaction is

$$Pb(s) + SO_4^{2-}(aq) \longrightarrow PbSO_4(s) + 2\,e^-$$

and the reduction half reaction is

$$SO_4^{2-}(aq) + PbO_2(s) + 4\,H^+(aq) + 2\,e^- \longrightarrow PbSO_4(s) + 2\,H_2O(l)$$

(a) The equation for the net cell reaction is

$$Pb(s) + PbO_2(s) + 4\,H^+(aq) + 2\,SO_4^{2-}(aq) \rightleftharpoons 2\,PbSO_4(s) + 2\,H_2O(l)$$

(b) The standard cell voltage E°_{cell} is given by

$$\begin{aligned} E^\circ_{cell} &= E^\circ_{red}[\,PbO_2\,|\,PbSO_4] + E^\circ_{ox}[Pb\,|\,PbSO_4] \\ &= 1.691\text{ V} - (-0.359\text{ V}) = 2.050\text{ V} \end{aligned}$$

(c) The value of ΔG°_{rxn} is given by

$$\Delta G^\circ_{rxn} = -\nu_e F E^\circ_{cell}$$

Two moles of electrons are transferred in this reaction; thus, $\nu_e = 2$.

$$\Delta G^\circ_{rxn} = -(2)(96\,485\text{ C}\cdot\text{mol}^{-1})(2.050\text{ V}) = -396\,600\text{ J}\cdot\text{mol}^{-1} = -396.6\text{ kJ}\cdot\text{mol}^{-1}$$

(d) The value of E_{cell} is given by the Nernst equation

$$E_{cell} = E^\circ_{cell} - \left(\frac{0.02570\text{ V}}{\nu_e}\right)\ln Q$$

where Q is the thermodynamic reaction quotient

$$Q = \frac{1}{([H^+]/M)^4([SO_4^{2-}]/M)^2}$$

We must now determine the values of $[H^+]$ and $[SO_4^{2-}]$ in a 10.0 M $H_2SO_4(aq)$ solution. This is similar to the calculation that we did in Example 20-17 and the previous problem. We have

$$\begin{array}{ccc} H_2SO_4(aq) + H_2O(l) \rightleftharpoons & H_3O^+(aq) & + \; HSO_4^-(aq) \\ & 10.0\text{ M} & 10.0\text{ M} \end{array}$$

$$\begin{array}{ccc} HSO_4^-(aq) + H_2O(l) \rightleftharpoons & H_3O^+(aq) & + \; SO_4^{2-}(aq) \\ 10.0\text{ M} - x & 10.0\text{ M} + x & x \end{array}$$

which gives

$$K_{a_2} = \frac{(10.0\text{ M} + x)(x)}{10.0\text{ M} - x} = 0.010\text{ M}$$

Solving for x yields $x = 0.00998$ M, so that we have $[H^+] = 10.0$ M and ($[SO_4^{2-}] = 0.00998$ M).

$$Q = \frac{1}{([H^+]/M)^4([SO_4^{2-}]/M)^2} = \frac{1}{(10.0)^4(0.00998)^2} = 1.00$$

Substituting the values into the Nernst equation yields

$$E_{cell} = 2.05\text{ V} - \left(\frac{0.02570\text{ V}}{2}\right)\ln 1.00 = 2.05\text{ V}$$

(e) The number of cells in a 12 V battery is given by

$$\frac{12\ \text{V}}{2.05\ \text{V}\cdot\text{cell}^{-1}} = 6\ \text{cells}$$

25-90.* The values of E°_{red} are not additive, but the values of $\Delta G^\circ_{\text{red}} = -\nu_e F E^\circ_{\text{red}}$ are additive because the Gibbs energy is a thermodynamic state function. Thus,

$$\Delta G^\circ_{\text{red}}[\text{Cr}^{3+} \mid \text{Cr}^{2+}] = -F E^\circ_{\text{red}}[\text{Cr}^{3+} \mid \text{Cr}^{2+}] = \Delta G^\circ_{\text{red}}[\text{Cr}^{2+} \mid \text{Cr}] = -2F E^\circ_{\text{red}}[\text{Cr}^{2+} \mid \text{Cr}]$$

Then use

$$\begin{aligned} E^\circ_{\text{red}}[\text{Cr}^{3+} \mid \text{Cr}] &= -\frac{\Delta G^\circ_{\text{red}}[\text{Cr}^{3+} \mid \text{Cr}]}{3F} \\ &= \frac{E^\circ_{\text{red}}[\text{Cr}^{3+} \mid \text{Cr}^{2+}] + 2\,E^\circ_{\text{red}}[\text{Cr}^{2+} \mid \text{Cr}]}{3} \\ &= \frac{-0.407\ \text{V} + (2)(-0.913\ \text{V})}{3} = -0.744\ \text{V} \end{aligned}$$

CHAPTER 26. The Chemistry of the Transition Metals

TRANSITION METAL CHEMISTRY

26-2. The equations for the production of titanium metal from its ore are

$$TiO_2(s) + C(s) + 2\,Cl_2(g) \xrightarrow{\text{high T}} TiCl_4(l) + CO_2(g)$$

$$TiCl_4(l) + 2\,Mg(l) \longrightarrow Ti(s) + 2\,MgCl_2(l)$$

26-4. Vanadium(V) oxide, $V_2O_5(s)$, is a catalyst in the production of sulfuric acid by the contact process.

26-6. The balanced equations for the reactions of a chromous bubbler are

$$2\,Cr^{3+}(aq) + Zn(s) \longrightarrow 2\,Cr^{2+}(aq) + Zn^{2+}(aq)$$

and

$$4\,Cr^{2+}(aq) + 4\,H^{+}(aq) + O_2(g) \longrightarrow 4\,Cr^{3+}(aq) + 2\,H_2O(l)$$

26-8. The equation for the reaction is

$$TiCl_4(l) + 2\,H_2O(l) \longrightarrow \underset{\text{(a smoke)}}{TiO_2(s)} + 4\,HCl(g)$$

The $TiO_2(s)$ is highly dispersed and appears as a smoke.

26-10. See Section 26-3 of the text.

26-12. Selenium and tellurium are by-products of copper production.

26-14. Twenty-four-karat gold is 100 percent gold. We have the following relationship between 14-karat gold and 24-karat gold:

$$\frac{14 \text{ karat gold}}{24 \text{ karat gold}} = \frac{x}{100\%}$$

Solving for x, we have that $x = 58\%$. Thus 14-karat gold is 58 percent gold.

26-16. The number of grams of $FeCr_2O_4(s)$ in 100.0 grams of the ore is given by

$$\text{mass of } FeCr_2O_4 = (100 \text{ g ore})\left(\frac{65.0 \text{ g } FeCr_2O_4}{100.0 \text{ g ore}}\right) = 65.0 \text{ g}$$

The number of moles of $FeCr_2O_4(s)$ in 100.0 grams of the ore is given by

$$\text{moles of } FeCr_2O_4 = (65.0 \text{ g } FeCr_2O_4)\left(\frac{1 \text{ mol } FeCr_2O_4}{223.84 \text{ g } FeCr_2O_4}\right) = 0.290 \text{ mol}$$

The number of grams of chromium that can be obtained from 0.290 moles of $FeCr_2O_4(s)$ is given by

$$\text{mass of Cr} = (0.290 \text{ mol } FeCr_2O_4)\left(\frac{2 \text{ mol Cr}}{1 \text{ mol } FeCr_2O_4}\right)\left(\frac{52.00 \text{ g Cr}}{1 \text{ mol Cr}}\right) = 30.2 \text{ g}$$

26-18. 10 000 metric tons of ore corresponds to 5000 metric tons of $Fe_2O_3(s)$ and 5000 metric tons of $SiO_2(s)$. Thus,

$$\text{mass of iron} = (5.0 \times 10^9 \text{ g } Fe_2O_3)\left(\frac{1 \text{ mol } Fe_2O_3}{159.7 \text{ g } Fe_2O_3}\right)\left(\frac{2 \text{ mol Fe}}{1 \text{ mol } Fe_2O_3}\right)\left(\frac{55.85 \text{ g Fe}}{1 \text{ mol Fe}}\right)$$
$$= 3.50 \times 10^9 \text{ g} = 3.50 \times 10^6 \text{ kg} = 3500 \text{ metric ton}$$

$$\text{mass of slag} = (5.0 \times 10^9 \text{ g } SiO_2)\left(\frac{1 \text{ mol } SiO_2}{60.08 \text{ g } SiO_2}\right)\left(\frac{1 \text{ mol } CaSiO_3}{1 \text{ mol } SiO_2}\right)\left(\frac{116.2 \text{ g } CaSiO_3}{1 \text{ mol } CaSiO_3}\right)$$
$$= 9.67 \times 10^9 \text{ g} = 9.67 \times 10^6 \text{ kg} = 9670 \text{ metric ton}$$

26-20. The mass of ore required is given by

$$\% \text{ Cu} = \frac{\text{mass Cu}}{\text{mass ore}} \times 100$$

or

$$\text{mass of ore} = \frac{\text{mass Cu} \times 100}{\%\text{Cu}} = \frac{91 \text{ metric ton} \times 100}{0.25\%} = 36\,400 \text{ metric ton}$$

ELECTRON CONFIGURATIONS AND OXIDATION STATES

26-22. (a) $1s^22s^22p^63s^23p^63d^6$ or $[Ar]3d^6$

(b) $1s^22s^22p^63s^23p^6$ or [Ar]

(c) $1s^22s^22p^63s^23p^63d^{10}4s^24p^64d^{10}4f^{14}5s^25p^65d^8$ or $[Xe]4f^{14}5d^8$

(d) $1s^22s^22p^63s^23p^63d^{10}$ or $[Ar]3d^{10}$

26-24. (a) Re(III) has one fewer electron than Re(II), and so it has four $5d$ electrons.

(b) Sc(III) has one fewer electron than Sc(II), and so it has no $3d$ electrons.

(c) Ru(IV) has two fewer electrons than Ru(II), and so it has four $4d$ electrons.

(d) Hg(II) has ten $5d$ electrons.

26-26. (a) The d^3 ions with a +2 oxidation state are V(II), Nb(II), and Ta(II).

(b) The d^8 ions with a +1 oxidation state are those ions which are d^7 ions in a +2 oxidation state. Thus, the answer is Co(I), Rh(I), and Ir(I).

(c) The d^0 ions with a +4 oxidation state are those ions which are d^2 ions in a +2 oxidation state. Thus, the answer is Ti(IV), Zr(IV), and Hf(IV).

26-28. (a) $[Ir(H_2O)_6]^{3+}$. The charge on the H_2O ligand is zero. The overall charge of the complex ion is +3, and so if x is the charge on Ir, then

$$x + 6(0) = +3$$

$$x = +3$$

(b) $[Co(NH_3)_3(CO)_3]^{3+}$. The charge on the NH_3 ligand is 0 and the charge on the CO ligand is 0. The overall charge of the complex ion is +3, and so if x is the charge on Co, then

$$x + 3(0) + 3(0) = +3$$

$$x = +3$$

(c) $[CuCl_4]^{2-}$. The charge on the Cl^- ligand is −1. The overall charge of the complex ion is −2, and so if x is the charge on Cu, then

$$x + 4(-1) = -2$$

$$x = +2$$

(d) $[Ni(CN)_4]^{2-}$. The charge on the CN^- ligand is −1. The overall charge of the complex ion is −2, and so if x is the charge on Ni, then

$$x + 4(-1) = -2$$

$$x = +2$$

26-30. (a) $[Mo(CO)_4Cl_2]^+$. The charge on the CO ligand is 0 and the charge on the Cl^- ligand is −1. The overall charge of the complex ion is +1, and so if x is the charge on Mo, then

$$x + 4(0) + 2(-1) = +1$$

$$x = +3$$

(b) $[Ta(NO_2)_3Cl_3]^{3-}$. The charge on the NO_2^- ligand is −1 and the charge on the Cl^- ligand is −1. The overall charge of the complex ion is −3, and so if x is the charge on Ta, then

$$x + 3(-1) + 3(-1) = -3$$

$$x = +3$$

(c) $[Co(CN)_6]^{3-}$. The charge on the CN^- ligand is −1. The overall charge of the complex ion is −3, and so if x is the charge on Co, then

$$x + 6(-1) = -3$$

$$x = +3$$

(d) $[Ni(CO)_4]$. The charge on the CO ligand is 0. The overall charge of the complex molecule is 0, and so if x is the charge on Ni, then

$$x + 4(0) = 0$$

$$x = 0$$

IONS FROM COMPLEX SALTS

26-32. (a) one mole of $[Cr(NH_3)_6]^{3+}(aq)$ and three moles of $Br^-(aq)$
The name of the complex ion is hexaamminechromium(III).

(b) one mole of $[Pt(NH_3)_3Cl_3]^+(aq)$ and one mole of $Cl^-(aq)$
The name of the complex ion is triamminetrichloroplatinum(IV).

(c) one mole of $[Mo(H_2O)_6]^{3+}(aq)$ and three moles of $Br^-(aq)$
The name of the complex ion is hexaaquamolybdenum(III).

(d) four moles of $K^+(aq)$ and one mole of $[Cr(CN)_6]^{4-}(aq)$
The name is the complex ion is hexacyanochromate(II).

26-34. The key point is that only the chloride ions that exist in solution as $Cl^-(aq)$, and not the chloride ions that are complexed with the platinum ions, are precipitated by $Ag^+(aq)$ as AgCl(s). Let's look at each case in turn.

$PtCl_2 \cdot 4\,NH_3$: Because both chloride ions per formula unit are precipitated by $Ag^+(aq)$, both chloride ions must exist in solution as $Cl^-(aq)$. The chemical formula of the complex salt must be $[Pt(NH_3)_4]Cl_2$.

$PtCl_2 \cdot 3\,NH_3$: One of the chloride ions must be complexed to the platinum ion because it is not precipitated by $Ag^+(aq)$. The chemical formula of the complex salt must be $[Pt(NH_3)_3Cl]Cl$.

$PtCl_2 \cdot 2\,NH_3$: Both chloride ions must be complexed to the platinum ion because they are not precipitated by $Ag^+(aq)$. The chemical formula of the complex salt must be $[Pt(NH_3)_2Cl_2]$.

CHEMICAL FORMULAS AND NAMES

26-36. (a) The complex ion is $[Fe(CN)_6]^{3-}$. If the oxidation state of Fe is denoted by x, then $x + 6(-1) = -3$, or $x = +3$. The name of the compound is potassium hexacyanoferrate(III).

(b) The complex molecule is $[Ni(CO)_4]$. If the oxidation state of Ni is denoted by x, then $x + 4(0) = 0$, or $x = 0$. The name of the compound is tetracarbonylnickel(0).

(c) The complex ion is $[Ru(H_2O)_6]^{3+}$. If the oxidation state of Ru is denoted by x, then $x + 6(0) = +3$, or $x = +3$. The name of the compound is hexaaquaruthenium(III) chloride.

(d) The complex ion is $[Al(OH)_4]^-$. If the oxidation state of Al is denoted by x, then $x + 4(-1) = -1$, or $x = +3$. The name of the compound is sodium tetrahydroxoaluminate(III).

26-38. (a) The complex ion is $[Au(CN)_4]^-$. Denoting the oxidation state of gold by x, we have $x + 4(-1) = -1$, or $x = +3$. The compound is called sodium tetracyanoaurate(III).

(b) The complex ion is $[Cr(H_2O)_6]^{3+}$. Denoting the oxidation state of chromium by x, we have $x + 6(0) = +3$, or $x = +3$. The compound is called hexaaquachromium(III) chloride.

(c) The complex ion is $[V(en)_3]^{3+}$. Denoting the oxidation state of vanadium by x, we have $x + 3(0) = +3$, or $x = +3$. The compound is called tris(ethylenediamine)vanadium(III) chloride.

(d) The complex ion is $[Cu(NH_3)_6]^{2+}$. Denoting the oxidation state of copper by x, we have $x + 6(0) = +2$, or $x = +2$. The molecule is called hexaamminecopper(II) chloride.

26-40. (a) The complex consists of a central nickel atom with a bromide ion, a chloride ion, and two cyanide ions, CN^-, as ligands. The oxidation state of the nickel atom is +2, and so the charge on the complex ion is $+2 + (-1) + (-1) + 2(-1) = -2$. The formula of the compound is $Na_2[Ni(CN)_2BrCl]$.

(b) The complex ion consists of a central cobalt atom with four ONO^- ions as ligands. The oxidation state of the cobalt atom is +2, and so the charge on the complex ion is $+2 + 4(-1) = -2$. The formula of the compound is $Rb_2[Co(NO_2)_4]$.

(c) The complex ion consists of a central vanadium atom with six chloride ions as ligands. The oxidation state of the vanadium atom is +3, and so the charge on the complex ion is $+3 + 6(-1) = -3$. The formula of the compound is $K_3[VCl_6]$.

(d) The complex ion consists of a central chromium atom with five ammonia molecules and one chloride ion as ligands. The oxidation state of the chromium atom is +3, and so the charge on the complex ion is $+3 + 5(0) + (-1) = +2$. The formula of the compound is $[Cr(NH_3)_5Cl](CH_3COO)_2$.

26-42. (a) The complex ion consists of a central iron atom with six cyanide ions as ligands. The oxidation state of the iron atom is +2, and so the charge on the complex ion is $+2 + 6(-1) = -4$. The formula of the compound is $Ba_2[Fe(CN)_6]$.

(b) The complex ion consists of a central cobalt atom with a chloride ion, a hydroxide ion, and two ethylenediamine molecules as ligands. The oxidation state of the cobalt is +3, and so the charge on the complex ion is $+3 + (-1) + (-1) + 2(0) = +1$. The formula of the compound is $[CoCl(OH)(en)_2]NO_3$.

(c) The complex ion consists of a central platinum atom with two nitrite ions and two oxalate ions as ligands. The oxidation state of the platinum atom is +4, and so the charge on the complex ion is $+4 + 2(-1) + 2(-2) = -2$. The formula of the compound is $Li_2[Pt(NO_2)_2(ox)_2]$.

(d) The complex ion consists of a central vanadium atom with two ethylenediamine molecules and one oxalate ion as ligands. The oxidation state of the vanadium atom is +3, and so the charge on the complex ion is $+3 + 2(0) + (-2) = +1$. The formula of the compound is $[V(en)_2ox]CH_3COO$.

ISOMERS

26-44. (a) The complex is octahedral. There are two isomers, *cis* and *trans*.

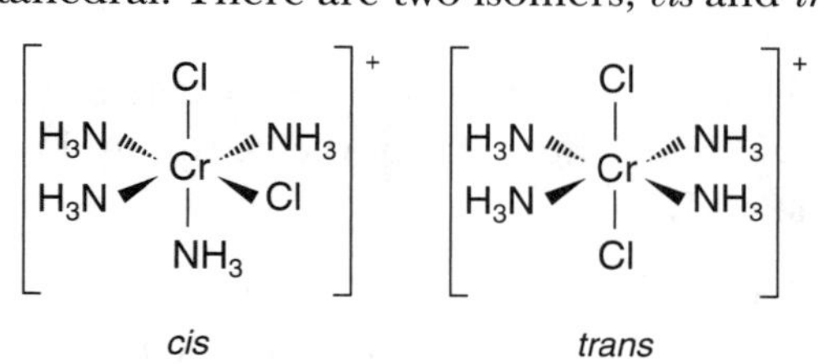

(b) The complex is octahedral. There is only one form because all the vertices on an octahedron are equivalent.

(c) The complex is tetrahedral and so there is only one form.

(d) There are two isomers, *cis* and *trans*.

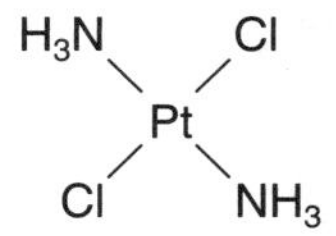

trans

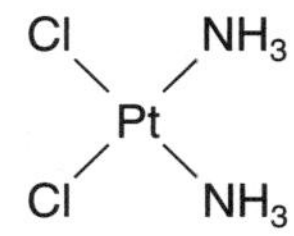

cis

26-46. (a) The structure of the complex ion is octahedral. The possible arrangements around the central palladium ion are

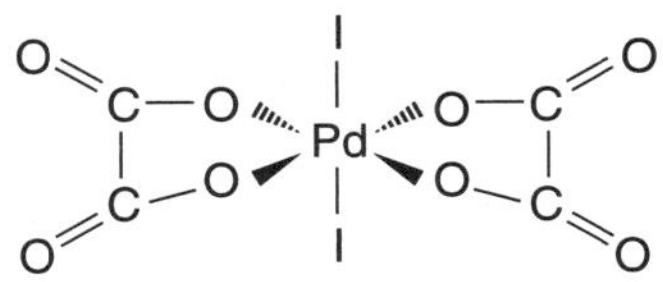

trans,trans

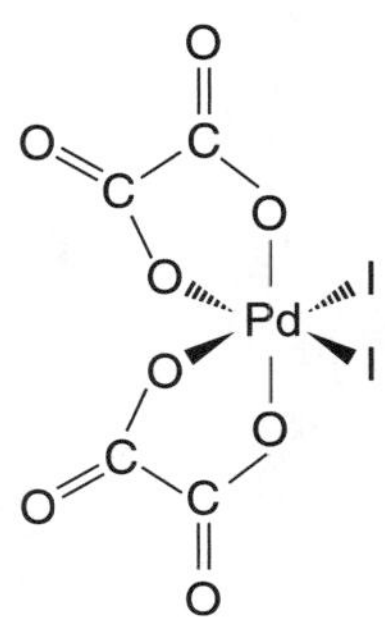

cis,cis

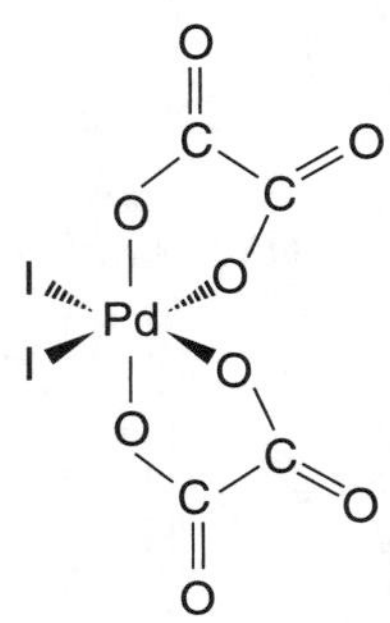

cis,cis

(b) The structure of the complex ion is octahedral. The possible arrangements around the central platinum ion are

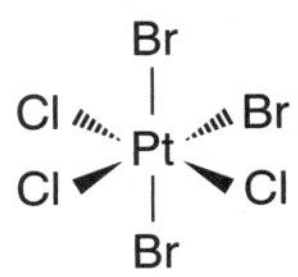

cis,trans

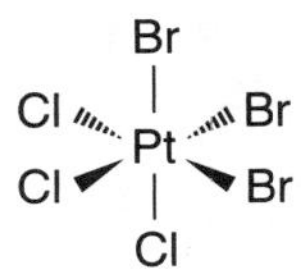

cis,cis

HIGH-SPIN AND LOW-SPIN COMPLEXES

26-48. (a) Nickel(II) is a d^8 ion. The d-orbital electron configuration of a high-spin Ni^{2+} complex is

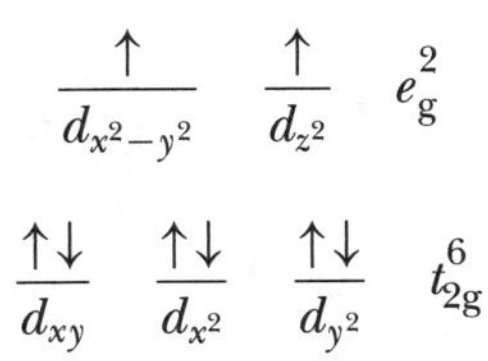

or $t_{2g}^6 e_g^2$.

(b) Manganese(II) is a d^5 ion. The d-orbital electron configuration of a high-spin Mn^{2+} complex is

$$\begin{array}{cccc} & \underline{\uparrow} & \underline{\uparrow} & e_g^2 \\ & d_{x^2-y^2} & d_{z^2} & \\ \underline{\uparrow} & \underline{\uparrow} & \underline{\uparrow} & t_{2g}^3 \\ d_{xy} & d_{x^2} & d_{y^2} & \end{array}$$

or $t_{2g}^3 e_g^2$.

(c) Iron(III) is a d^5 ion. The d-orbital electron configuration of a low-spin Fe^{3+} complex is

$$\begin{array}{cccc} & \underline{\quad} & \underline{\quad} & e_g^0 \\ & d_{x^2-y^2} & d_{z^2} & \\ \underline{\uparrow\downarrow} & \underline{\uparrow\downarrow} & \underline{\uparrow} & t_{2g}^5 \\ d_{xy} & d_{x^2} & d_{y^2} & \end{array}$$

or $t_{2g}^5 e_g^0$.

(d) Titanium(IV) is a d^0 ion. The d-orbital electron configuration of Ti^{4+} is

$$\begin{array}{cccc} & \underline{\quad} & \underline{\quad} & e_g^0 \\ & d_{x^2-y^2} & d_{z^2} & \\ \underline{\quad} & \underline{\quad} & \underline{\quad} & t_{2g}^0 \\ d_{xy} & d_{x^2} & d_{y^2} & \end{array}$$

or simply $t_{2g}^0 e_g^0$.

(e) Nickel(II) is a d^8 ion. The d-orbital electron configuration of a Ni^{2+} complex is

$$\begin{array}{cccc} & \underline{\uparrow} & \underline{\uparrow} & e_g^2 \\ & d_{x^2-y^2} & d_{z^2} & \\ \underline{\uparrow\downarrow} & \underline{\uparrow\downarrow} & \underline{\uparrow\downarrow} & t_{2g}^6 \\ d_{xy} & d_{x^2} & d_{y^2} & \end{array}$$

or $t_{2g}^6 e_g^2$.

26-50. We set up the following table:

	Ion	Oxidation state of the metal	x in d^x	Low-spin d-orbital electron configuration (Number of upaired electrons)	High-spin d-orbital electron configuration (Number of) unpaired electrons)
(a)	$[Mn(NH_3)_6]^{3+}$	Mn(III)	4	$t_{2g}^4 e_g^0$ (2)	$t_{2g}^3 e_g^1$ (4)
(b)	$[Rh(CN)_6]^{3-}$	Rh(III)	6	$t_{2g}^6 e_g^0$ (0)	$t_{2g}^4 e_g^2$ (4)
(c)	$[Co(C_2O_4)_3]^{4-}$	Co(II)	7	$t_{2g}^6 e_g^1$ (1)	$t_{2g}^5 e_g^2$ (3)
(d)	$[IrBr_6]^{4-}$	Ir(II)	7	$t_{2g}^6 e_g^1$ (1)	$t_{2g}^5 e_g^2$ (3)
(e)	$[Ru(NH_3)_6]^{3+}$	Ru(III)	5	$t_{2g}^5 e_g^0$ (1)	$t_{2g}^3 e_g^2$ (5)

Thus we see that $[Mn(NH_3)_6]^{3+}$ is low spin; $[Rh(CN)_6]^{3-}$ is low spin; $[Co(C_2O_4)_3]^{4-}$ is high spin; $[IrBr_6]^{4-}$ is high spin; and $[Ru(NH_3)_6]^{3+}$ is low spin.

PARAMAGNETISM IN COMPLEX IONS

26-52. (a) The oxidation state of the rhodium atom in $[Rh(NH_3)_6]^{3+}$ is +3. Rhodium(III) is a d^6 ion and the complex is a low-spin complex. Thus the d electron configuration is $t_{2g}^6 e_g^0$, and so there are no unpaired electrons in $[Ru(NH_3)_6]^{3+}$.

(b) The oxidation state of the iron atom in $[FeF_6]^{3-}$ is +3. Iron(III) is a d^5 ion and F^- gives rise to high-spin complexes. Thus the d electron configuration is $t_{2g}^3 e_g^2$, and so there are five unpaired electrons in $[FeF_6]^{3-}$.

(c) The oxidation state of the iridium atom in $[Ir(H_2O)_6]^{3+}$ is +3. Iridium(III) is d^6 and the complex is a low-spin complex. Thus the d electron configuration is $t_{2g}^6 e_g^0$, and so there are no unpaired electrons in $[Ir(H_2O)_6]^{3+}$.

26-54. Both complexes contain an iron ion in the +2 oxidation state. Iron(II) is a d^6 ion. Both complexes have an octahedral structure because each involves six ligands. The low-spin and high-spin d-electron configurations are

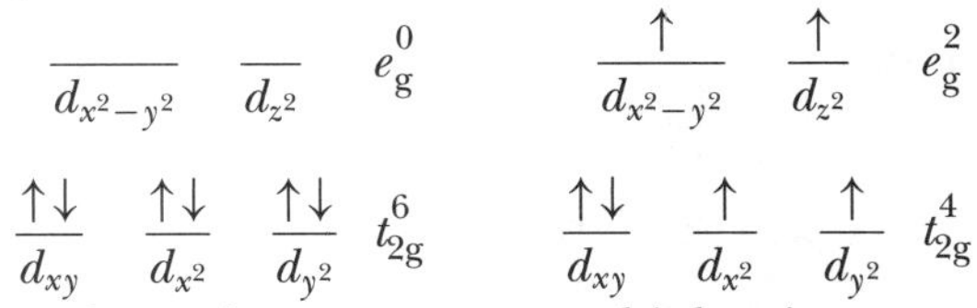

low spin high spin

There are no unpaired electrons in the low-spin configuration; thus, a complex ion in this configuration is diamagnetic. There are four unpaired electrons in the high-spin configuration; thus, a complex in this configuration is paramagnetic. We see that the complex ion, $[Fe(H_2O)_6]^{2+}$ is high-spin, whereas the complex ion $[Fe(CN)_6]^{3-}$ is low-spin.

ADDITIONAL PROBLEMS

26-56. (a) The charge on the complex ion is given by $+3 + 6(-1) = -3$. The formula is $[Co(NO_2)_6]^{3-}$.

(b) The charge on the complex ion is given by $+4 + 2(-1) + 2(0) = +2$. The formula is *trans*-$[PtCl_2(en)_2]^{2+}$.

(c) The charge on the complex ion is given by $+2 + 5(-1) + 0 = -3$. The formula is $[Fe(CN)_5CO]^{3-}$.

(d) The charge on the complex ion is given by $+3 + 2(-1) + 2(-1) = -1$. The formula is *trans*-$[AuCl_2I_2]^-$.

26-58. (a) The geometric isomers are

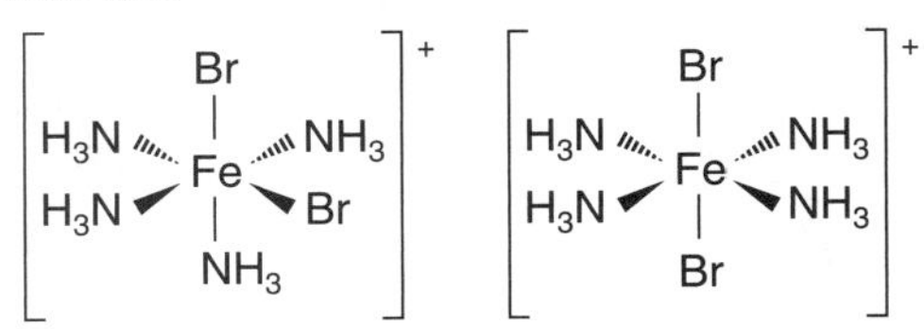

cis *trans*

(b) The geometric isomers are

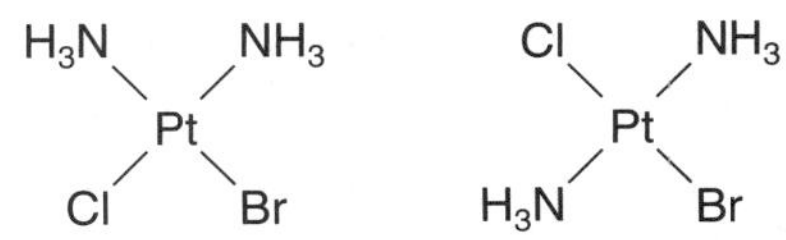

cis *trans*

(c) The geometric isomers are

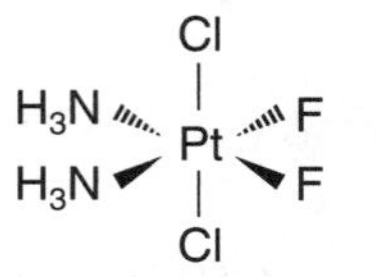

cis,trans,cis

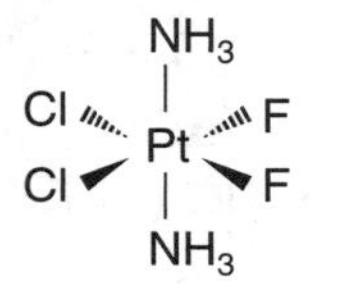

trans,cis,cis

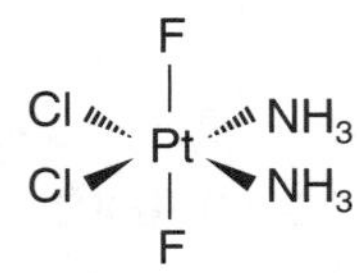

cis,cis,trans

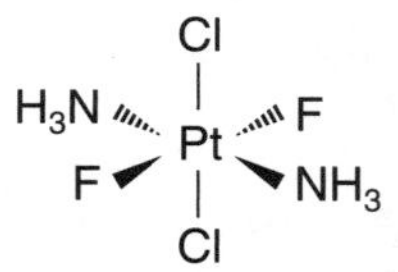

trans,trans,trans

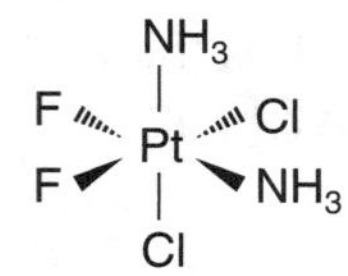

cis,cis,cis

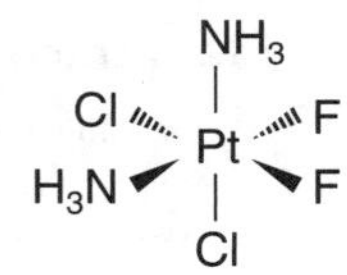

cis,cis,cis

(d) The optical isomers are

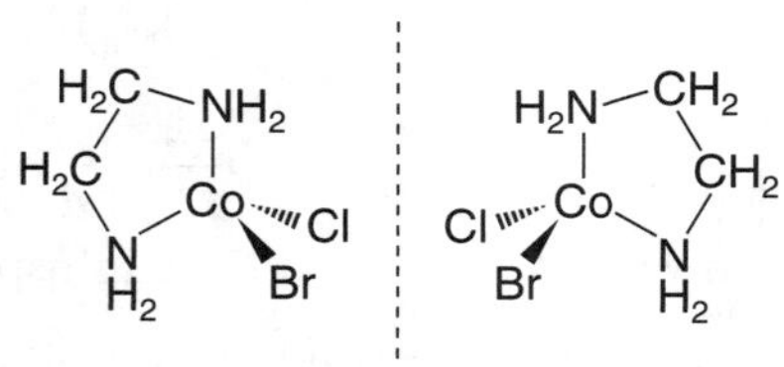

optical isomers

26-60. We set up the following table:

	Lewis formula		Class of molecule	Molecular shape
(a)	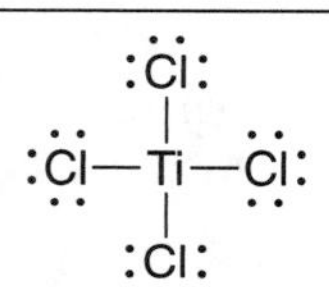		AX_4	Tetrahedral
(b)	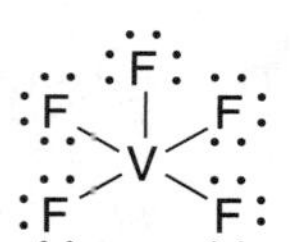		AX_5	Trigonal bipyramidal
(c)	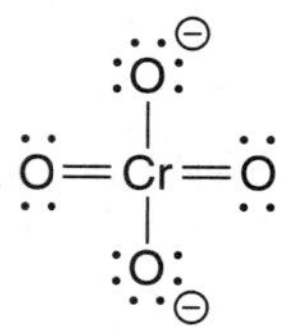	plus other resonance forms	AX_4	Tetrahedral
(d)	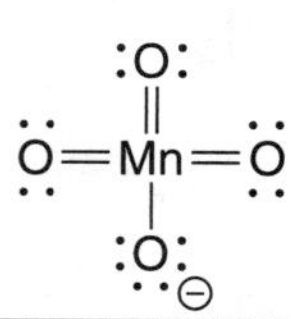	plus other resonance forms	AX_4	Tetrahedral

26-62. Because X is a neutral ligand, the iron atom in the complex ion $[Fe(X)_6]^{2+}$ is in a +2 oxidation state, and so is a d^6 ion. Similarly, the nickel atom in the complex ion $[Ni(X)_6]^{2+}$ is in a +2 oxidation state, and so is a d^8 ion. Because d^6 ions in octahedral complexes can be high spin or low spin, but d^8 ions in octahedral complexes can be only high spin and because low spin d^6

ions are diamagnetic, while high spin d^6 ions are paramagnetic, the iron complex is better suited for this experiment.

26-64. The glycinate ion, $H_2NCH_2COO^-$, has an amine group at one end like the ethylenediamine ligand and a carboxylic acid group at the other end like the oxalate ligand. Thus we predict (correctly) that the glycinate ion is a bidentate ligand. To demonstrate this, we can perform an experiment to determine the stoichiometric ratio of the reaction between the glycinate ligand and a transition metal that forms an octahedral complex to see how many ligands bind to each metal atom.

26-66.* tetraamminediiodoplatinum(IV) tetraiodoplatinate(II)

26-68.* There are two optical isomers and no geometric isomers of the $[Pt(en)_3]^{4+}$ complex ion.

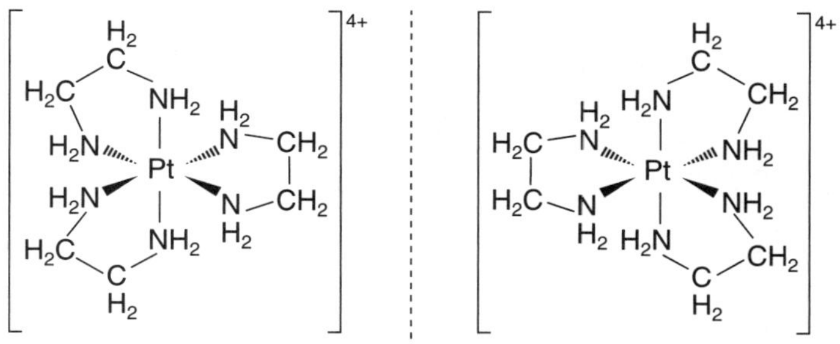

optical isomers

Notice that rotating the right-hand isomer by 180° results in a nonsuperimposable structure, as shown.

26-70.* The equation for the reaction is

$$Pb_2[Fe(CN)_6](s) + 4\,I^-(aq) \rightleftharpoons 2\,PbI_2(s) + [Fe(CN)_6]^{4-}(aq) \qquad (1)$$

and its equilibrium constant is

$$K_1 = \frac{[Fe(CN)_6]^{4-}]}{[I^-]^4} = \frac{0.11\text{ M}}{(0.57\text{ M})^4} = 1.04\text{ M}^{-3}$$

The equations for the dissolution of both $Pb_2[Fe(CN)_6](s)$ and $PbI_2(s)$ are

$$Pb_2[Fe(CN)_6](s) \rightleftharpoons 2\,Pb^{2+}(aq) + [Fe(CN)_6]^{4-}(aq) \qquad (2)$$

$$PbI_2(s) \rightleftharpoons Pb^{2+}(aq) + 2\,I^-(aq) \qquad (3)$$

with

$$K_{sp_2} = [Pb^{2+}]^2[Fe(CN)_6]^{4-}]$$

and

$$K_{sp_3} = [Pb^{2+}][I^-]^2$$

If we multiply equation 3 by 2, then reverse it, and add equation 2, then we obtain equation 1. Thus

$$K_1 = \frac{K_{sp_2}}{K_{sp_3}^2}$$

and so

$$K_{sp_2} = K_1 K_{sp_3}^2 = (1.04\text{ M}^{-3})(9.8 \times 10^{-9}\text{ M}^3)^2 = 1.0 \times 10^{-16}\text{ M}^3$$